Curt Jackson

Elements of Computer Mathematics

Elements of Computer Mathematics

Sandra Talbot
Mattatuck Community College

Harold Baker
Adjunct Faculty
Waterbury State Technical College

Lawrence Gilligan
College of Applied Science
University of Cincinnati

Brooks/Cole Publishing Company
Monterey, California

Brooks/Cole Publishing Company
A Division of Wadsworth, Inc.

Printed in the United States of America

10 9 8 7 6 5 4 3

Library of Congress Cataloging in Publication Data

Talbot, Sandra.
Elements of computer mathematics.

Includes index.
1. Electronic data processing—Mathematics.
I. Baker, Harold, 1941– . II. Gilligan, Lawrence G. III. Title.
QA76.9.M35T35 1985 519.4 84-21388
ISBN 0-534-04392-5

Sponsoring Editor: Craig Barth
Production Service: Bernard Scheier
Manuscript Editor: Caroline Eastman
Interior Design: Peter Martin, Design Office
Cover Design: Victoria A. Van Deventer
Art Coordinator: Bernard Scheier
Interior Illustrations: Anco/Boston
Photo Researcher: Judy Mason
Typesetting: Syntax International
Cover Printing: Phoenix Color Corporation
Printing and Binding: R. R. Donnelley & Sons Company

Chapter photo credits:
1 Digital Equipment Corporation
3 Smithsonian Institution
4 Digital Equipment Corporation
5 The Bettmann Archive Inc
8 General Motors Corporation
9 The Bettmann Archive Inc
10 General Motors Corporation

About the Cover: Phillip A. Harrington is a well-known photographer who specializes in photomicrography. He does annual report work and special projects for such companies as General Electric, Schering Plough, C. R. Bard, Revlon, AT&T and RCA. His work has appeared in articles, or on the covers, of such magazines as Omni, Science Digest, Medical Economics, Time, and Newsweek. His photography is frequently used to enhance exhibitions, ad campaigns, and books of great variety. Harrington is a fellow of the New York Microscopical Society.

Phillip Harrington shot this wafer (containing 256k chips) with a Nikon 35mm model EL camera, using a Vivitar macro lens. He used Kodachrome tungsten 40 ASA film.

Cover photo: Phillip A. Harrington/Fran Heyl Associates, Courtesy of AT&T Technologies.

Preface

Teachers of mathematics courses for computer science and data processing students have long perceived the need for a textbook specifically designed for these students. *Elements of Computer Mathematics* represents our effort to fill this void. It presents the mathematical concepts needed by students in computer-related fields in a context that is relevant to today's technology. Interesting and meaningful computer applications are presented wherever appropriate to motivate the student and aid in understanding the mathematics.

Orientation

This text is designed for a one-semester course at two-year technical or community colleges. It is also suitable for use as an introductory mathematics text for students in computer-related fields at four-year colleges and in high schools which offer specialized courses in data processing or computer science. The text assumes completion of a full year of high school algebra or the equivalent. It can also be easily adapted for the student with a more substantial mathematics background.

Because of the broad range of mathematical abilities and backgrounds found among computer programming and data processing students, the authors present concepts and processes in a manner that is complete and self-contained. All concepts are clearly and completely explained in concrete terms, with reference to real-life examples whenever possible. Emphasis is on the "how to" aspect of problem solving. The mathematical theory behind a particular technique is presented when it is relevant and appropriate; it is avoided when it would simply cause confusion without any real benefit to the student.

Chapter Organization and Interdependence

Organization

The text covers a wide variety of topics suitable for any syllabus. There is also room for flexibility in the selection of topics by the instructor, both with respect to the specific topics presented and the order of presentation. The flowchart indicates both the independence and interdependence of chapters.

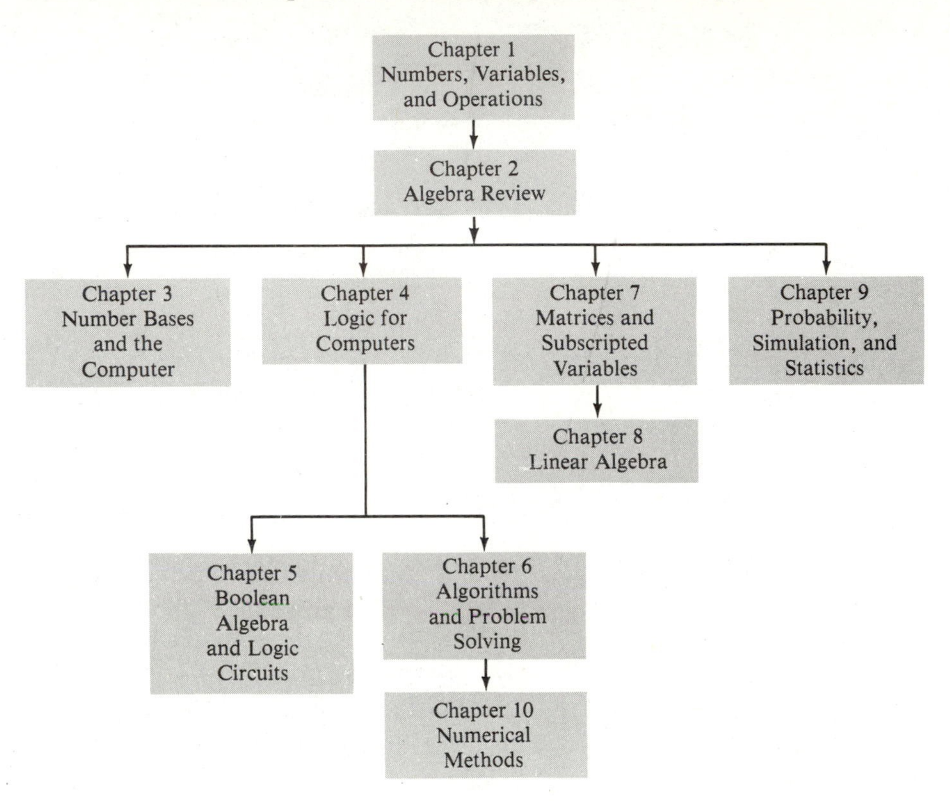

Topic Coverage

Real Numbers: Chapter 1 deals with the real number system, its subsets, and their properties. Concepts are presented as they relate to computer storage and manipulation of data.

Algebra: Chapter 2 encompasses a thorough review of the basic concepts of algebra. Topics covered include exponents and radicals, solving first and second degree equations, factoring, algebraic fractions, graphing, and functions. Particular attention is paid to the use of the function concept in computer programming.

Number Bases: The binary, octal, and hexadecimal systems of numeration are discussed in detail in Chapter 3, using the properties of the decimal system for illustration. The relationship of these systems to calculation and data representation in computers is explored.

Logic and Boolean Algebra: In Chapters 4 and 5, the basic concepts of mathematical logic, Boolean algebra, and logic networks are studied. Discussion of the application of these concepts to computer programming and design is an important part of both chapters.

Pseudocode and Flowcharting: The process of algorithm development using pseudocode and flowcharts, with emphasis on structured programming techniques, is explained through the use of many examples in Chapter 6.

Matrices and Linear Algebra: Using many concrete examples, matrices and their use in solving systems of linear equations are presented in Chapters 7 and 8. An optional section on the use of matrices in BASIC programming is included. Chapter 8 contains a brief introduction to linear programming problems and the simplex algorithm as a practical application of matrices.

Probability and Statistics: Chapter 9 presents a summary of the basic concepts of probability and statistics. Of particular interest to the programming student is the discussion of the use of random numbers in computer simulations.

Numerical Methods: The text concludes with a presentation, in Chapter 10, of several iterative techniques for solving single equations and systems of equations. Flowcharts are presented for most of the methods discussed, as tools for the student with sufficient programming background as well as aids in understanding the techniques.

Relevance to Computers

The authors have endeavored to present a coherent collection of topics that are interesting and helpful to the computer science or data processing student. Particular attention is given to the relationship of mathematical concepts to computer science and computer programming, with many specific examples and applications presented. Differences between what may be expected on the basis of theoretical mathematics and what one may encounter in doing mathematics on a computer which is handicapped by finite precision are pointed out and explained. The text offers many examples of arithmetic operations on a computer, and their expected and unexpected results.

Pedagogical Features

Examples

A strength of the text is its numerous completely worked out and carefully explained examples. Color highlighting is used to emphasize important points in the explanations of examples and to make the solutions easy to read and understand.

Exercises

At the end of each section of each chapter is a set of practice exercises. The problems in each exercise set are presented in order of increasing difficulty and are closely patterned after the examples in the body of the chapter. Students are referred to specific examples for help in solving the practice exercises, thus giving them algorithms to follow for their solutions. Answers to the odd-numbered exercises are contained in an Appendix.

Highlighting of Important Ideas

The text contains many devices to help the student identify key terms, concepts, properties, and algorithms. At their first appearance in the text, important terms are printed in **boldface** type. Boxes are placed around important definitions and properties, so that the student can easily locate them. Summary lists of properties and steps for solving specific types of problems are also boxed for clear identification.

Chapter Reviews

Each chapter concludes with a Chapter Review, as a further study aid. A summary of the chapter, in one or two paragraphs, is presented first. This is followed by a vocabulary list containing all the important new terms encountered in the chapter. A Chapter Test affords students an opportunity to test their mastery of the material covered in the chapter. The Appendix contains the answers to all questions in the Chapter Tests.

Supplemental Material

An Instructor's Manual is available from the publisher. It contains sample tests for each chapter.

A Solutions Manual which contains complete solutions for all the even-numbered exercises in the text is also available. At the instructor's discretion, it may be purchased by students as a supplement to the text.

Acknowledgements

The authors wish to express their appreciation to the following educators whose reviews of their work provided encouragement and constructive criticism. They are: Albert Bordonaro, Waterbury State Technical College; Kenneth Chapman, Michael J. Owens Technical College; Richard Dixon, Mohave Community College; Greg Maybury, Parkland College; Robert Seaver, Lorain Community College; Peter Wursthorn, Greater Hartford Community College. We also thank the editorial staff of Brooks/Cole and their associates whose efforts, advice, and patience made the publication of this book possible, especially Craig Barth, Bernard Scheier, Caroline Eastman, Judy Blamer, Judy Macdonald.

Our families deserve a special thanks for their encouragement, loyalty, and support, as do our typists Roseann La Paglia and Susan Gilligan. We thank also those who have helped us, by their influence on our lives and careers, to prepare for this undertaking, especially

Gene, Jennifer, and Amanda Talbot
Henry Pronovost, Mattatuck Community College
Dr. Salvatore Anastasio, S.U.N.Y. at New Paltz
Dr. Theron Rockhill, S.U.N.Y. at Brockport
John Bobko, Waterbury State Technical College
Professor Taylor Booth, University of Connecticut
Stephen M. Colwell, Waterbury State Technical College
Kenneth DeRego, President, Hartford State Technical College
Francis Fisher, Litchfield High School
George Miller, Central Connecticut State University
Russell Walter, author of *The Secret Guide to Computers*

Finally, Sandra Talbot and Harold Baker thank LG^2 for his role as instigator of and inspiration for this project.

Sandra Talbot
Harold Baker
Lawrence Gilligan

Contents

CHAPTER

1

The computer is a powerful tool. It can "crunch" numbers and manipulate huge quantities of information with far greater speed than any person. But, like the human beings who conceive of and build them, computers have limitations. No computer can "think" of the value of pi to as many places as you or I can. This chapter examines some of the basic types of numbers and their properties. It will make you aware of, and thus prepare you to deal with, some of the computer's limitations.

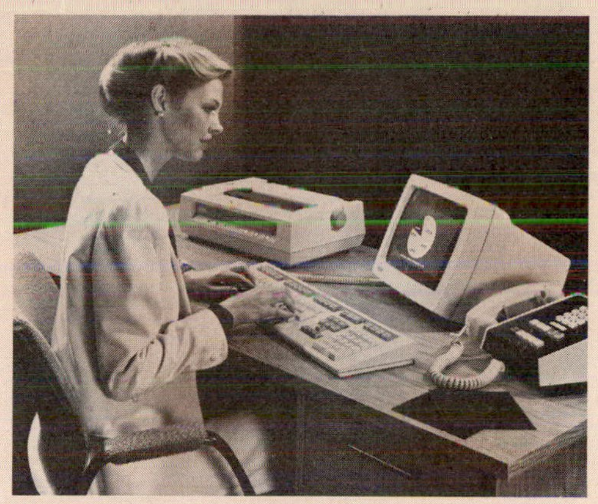

Numbers, Variables, and Operations

The typical human brain weighs only a few pounds, yet it can conceive of difficult concepts such as sizes of infinity, negative numbers, and numbers with never-ending decimal expansions. Although no one can explain how the brain deals with these concepts or how its memory "works," the computer's memory *is* explainable and is discussed in this chapter.

The reader is assumed to be familiar with the real number system. Figure 1.1 and the definitions following are meant as a review of these concepts.

FIGURE 1.1

The real number system.

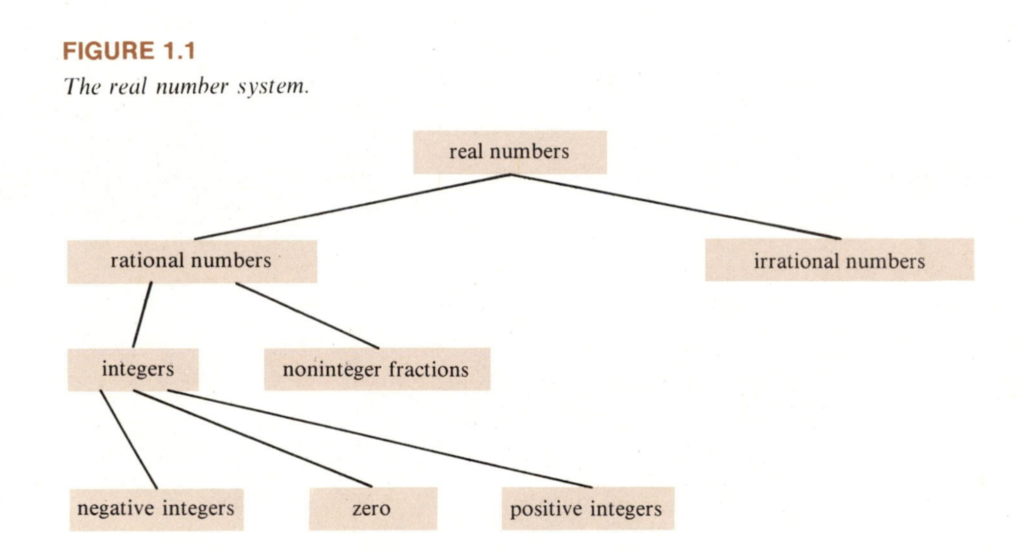

Positive integers: $\{1, 2, 3, \ldots\}$
Zero: $\{0\}$
Negative integers: $\{\ldots, -3, -2, -1\}$
Rational numbers: All numbers that can be written as a fraction $\frac{a}{b}$, where a and $b(\neq 0)$ are integers (or, equivalently, a number with either a terminating or repeating decimal expansion). For example, $\frac{1}{3}$, 0.424242 . . . , and -3.7.
Irrational numbers: Numbers whose decimal expansions are nonterminating and nonrepeating. Examples: 4.12112111211112 . . . , $\pi = 3.415926535\ldots$, $\sqrt{2}$, $-\sqrt{3}$ (see Figure 1.2).

FIGURE 1.2
The real number line.

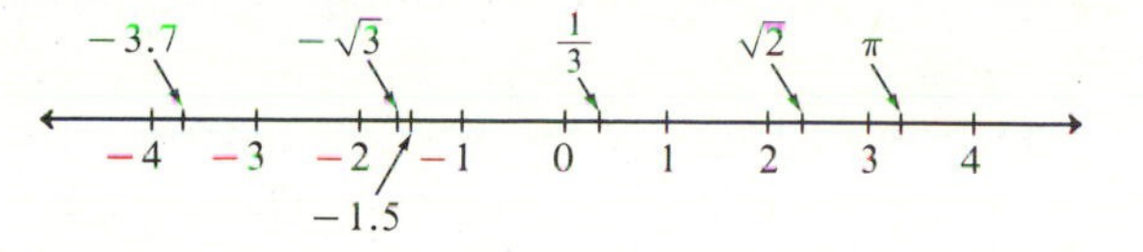

FIGURE 1.3
A 1000-cell memory.

A computer's memory is basically a sequence of cells, called **bytes**. These bytes are numbered (labeled) consecutively. The label is called the cell's **address**. The first address is usually a 0; a 1000-cell memory is represented in Figure 1.3.

Each byte can store a data item and, of course, the number of cells is finite. In this chapter we shall study one type of data item that computers are capable of storing—numbers. The different types of numbers that we use are stored differently in a computer. The remainder of this chapter introduces and describes the various types of numbers, the arithmetic of these numbers, and the rounding and approximating of numbers, all relative to the *finite brain*, the computer.

1.1 Types of Numbers

Numbers as Labels

Numbers permeate our lives. A telephone number (343-6692), a zip code (06751), and a social security number (127-40-5072) really serve only as *labels*; we do not expect to do arithmetic with them. The number 400, as used in "the 400-meter dash," is a label describing a track event. One would not ordinarily multiply it, add it, or perform any other arithmetic operation on it. Another example of a number used as a label is the product or stock number in a plant or store's inventory records. Today books are identified by special numbers called ISBN numbers, like 0-534-04392-5. The ISBN serves as an identifier or stock number; it is not the object of arithmetic operations. It is important in some computer languages, such as Pascal, PL/1, and COBOL, to pay careful attention to data *types*, and especially to the way in which numbers are being used.

Since this is a mathematics book, our chief concern will not be with numbers as labels. We shall concentrate on numbers that are added, subtracted, and operated on using the other operations of arithmetic—numbers that are used in *calculations.*

This chapter introduces the main types of numbers encountered in working with computers. The chart in Figure 1.1 illustrates the relationships among these types of numbers. The set (collection) of real numbers includes several subsets (subcollections)—for example, rational numbers, irrational numbers, and integers. Each subset has its own set of properties. The chart shows that the integers are a subset of the rational numbers and that the major subsets of the reals are the rational and irrational numbers. The hierarchy is as follows:

1. The real numbers are divided into the rationals and the irrationals. These two sets are mutually exclusive.
2. The rational numbers can be further subdivided into integers and nonintegers.
3. The integers consist of the negative integers, zero, and the positive integers.

Note that zero is a number and is a member of the collection of integers, even though it is neither a positive nor a negative integer. Thus, it is an integer, a rational number, and a real number. It is not just a "place-holder."

There is another set of numbers, called the complex numbers, that contains the real numbers. In fact, the complex numbers include both real and imaginary numbers. We shall not stress imaginary numbers in this text, but they are discussed briefly at the end of this section.

We now introduce the major number types that are either the object or the result of arithmetic operations.

The Integer Type

We have already seen what is meant by an integer. The analysis of an integer type used by a computer is that an integer

1. consists of digits (0, 1, 2, 3, 4, 5, 6, 7, 8, 9),
2. has a sign (only required when negative), and
3. has no decimal point.

In computers commas are not used to break up an integer, so the number 15,674 is written as 15674.

An important property of integers is **closure**. If you add (or subtract or multiply) two integers, the result is another integer. We say the integers are closed under the operation of addition (or subtraction or multiplication). Integers are *not* closed under the operation of division since the quotient of two integers is not always an integer. For example, $7 \div 2 = \frac{7}{2} = 3.5$, which is not an integer. In computer programming, the concept of closure is important because in many languages variables are *declared* to be certain data types and the results of operations must be consistent with that predeclared type. This problem is addressed in Section 1.2 and there is further discussion of how integers are stored internally in Chapter 3.

The Real Type

We have already seen examples of real numbers that are rational (repeating or terminating decimal expansions) and that are irrational (nonrepeating, nonterminating decimal expansions). In the computer, irrational real numbers are *approximated* using a predetermined number of decimal places. (Remember, the computer has only a finite number of cells.) So, in the way that we may have approximated the irrational number π as $\frac{22}{7}$ or 3.1416 for certain applications in geometry, the computer approximates an irrational number like $\sqrt{5}$ as 2.23 or 2.23606798 (depending on the computer, of course). Note that the computer also approximates *rational* numbers whose decimal expansions are nonterminating (numbers like $\frac{1}{3} = 0.33333\ldots$ with repeating decimal expansions).

EXAMPLE 1

Determine whether each of the following real numbers is an integer, a rational number, or an irrational number. Would the computer's representation of the number be approximate?

a. -5 **b.** $\frac{5}{6}$ **c.** $\sqrt{7}$

Solution

a. -5 is an *integer* (and also a rational number); it need not be approximated in a computer.

b. $\frac{5}{6}$ is a *rational* number. Because its decimal expansion is infinite ($\frac{5}{6} = 0.8333333\ldots$), it would be approximated by a computer.

c. $\sqrt{7}$ is an *irrational* number, and it would have to be approximated.

The real numbers are closed under the operation of addition (or subtraction or multiplication) because the sum (or difference or product) of any two real numbers is still a real number. The real numbers are *not* closed under division because division by zero is undefined. If A and B are real numbers, the programmer using the calculation

$$C = A/B$$

should be very careful to provide an "escape" (and not try to calculate C) if B is zero. We highlight the division property of zero below; it is extremely important.

DIVISION PROPERTY OF ZERO
If $B = 0$, A/B is *undefined* and *not* a real number. (In other words, division by zero is *not allowed*.)

The Complex Type

The complex numbers include all the real numbers and another number type called imaginary numbers. Thus, every real number is also complex, but not every complex number is real.

Complex numbers are of the form $a + bi$, where a and b are real numbers, and $i = \sqrt{-1}$. If $b = 0$, then the number $a + bi$ is real. If $b \neq 0$, then the number is called an **imaginary number**. Imaginary numbers can occur as roots of some types of algebraic equations, such as quadratics, which will be studied in Chapter 2.

Suppose, for example, that we wanted to describe the numbers $7 + 6\sqrt{-1}$, $2\sqrt{-1}$, and -3 with reference to their number type or types. Since $7 + 6\sqrt{-1}$ is of the form $a + bi$, with $a = 7$ and $b = 6$ (and, by definition, $i = \sqrt{-1}$), this number is complex and imaginary. Now consider the number $2\sqrt{-1}$. This is also of the form $a + bi$, this time with $a = 0$ and $b = 2$; so it is also complex and imaginary. A number like $2\sqrt{-1}$, of the form $a + bi$ with $a = 0$, is called a **pure imaginary number**. Finally, -3 can also be written in the form $a + bi$; this time $a = -3$ and $b = 0$. Since $b = 0$, -3 is complex and real.

EXAMPLE 2

Give the number type(s) of each of the following numbers:

a. 4 **b.** 0.33 **c.** $2i$ **d.** $-3 - 4i$

Solution

a. $4 = a + bi$, with $a = 4$ and $b = 0$. Therefore, 4 is complex, but since $b = 0$, 4 is *not* imaginary. The number 4 is also a real number, a rational number, and an integer.

b. $0.33 = a + bi$, with $a = 0.33$ and $b = 0$. So, 0.33 is also complex but not imaginary. It is a real number and a rational number.

c. $2i = a + bi$, with $a = 0$ and $b = 2$. Therefore, $2i$ is a complex number; in fact, it is a pure imaginary number.

d. $-3 - 4i = a + bi$, with $a = -3$ and $b = -4$; so this number is complex. Since $b \neq 0$, it is *not* a real number.

The importance of considering imaginary numbers in the context of data processing lies in the fact that most computer systems cannot handle imaginary numbers directly. Imaginary numbers result from the extraction of the square roots of negative numbers. For example, $\sqrt{-4}$, $\sqrt{-1}$, and $\sqrt{-3}$ are imaginary numbers. Most computer systems generate an error message when one attempts to take the square root of a negative number.

We summarize the points of this section as follows:

1. There are three basic types of numbers: integer, real, and complex.
2. Not all real numbers can be completely stored in the finite number of "cells" in a computer's memory. Consider Table 1.1.

TABLE 1.1

Number	Type	Stored in a Computer With Four Cells	Eight Cells
$\frac{3}{4}$	real (rational)	.7500	.75000000
$\frac{1}{17}$	real (rational)	.0588	.05882353
$\frac{1}{3}$	real (rational)	.3333	.33333333
$\sqrt{0.5}$	real (irrational)	.7071	.70710678

3. Division by zero is undefined. Caution must be taken as to operations on approximate numbers. For example, the actual value of $(\frac{1}{3} - 0.3333)$ is not zero, but in a (hypothetical) computer storing only four digits, $\frac{1}{3}$ is stored as .3333 and $\frac{1}{3} - 0.3333$ *would be* zero! In our (hypothetical) computer an attempt to divide by $(\frac{1}{3} - 0.3333)$ would cause an error message.

4. Taking an even root of a negative number results in complex numbers. An odd root of a negative number ($\sqrt[3]{-64} = -4$, $\sqrt[5]{-32} = -2$) is a real number.

EXERCISES FOR SECTION 1.1

In Exercises 1 through 25, (a) label each number as integer, real, complex, or undefined. (Note that some numbers are of more than one type.) Also, (b) assuming a four-digit computer, state whether or not the number can be fully represented in the computer (see Example 1).

1. 3 **2.** $\sqrt{5}$ **3.** -4
4. $1/9$ **5.** $\sqrt{9}$ **6.** $\sqrt{-9}$
7. $-\sqrt{9}$ **8.** 0.343456 **9.** $1/2 - 1/8$
10. $9.2 - 1/3$ **11.** 0 **12.** $5/0$
13. $0/0$ **14.** $\sqrt{0}$ **15.** $0/4$
16. $-\sqrt{4}$ **17.** $-\sqrt{16}$ **18.** $\pi - 3$
19. $22/7$ **20.** 8.9 **21.** 3.1415927
22. $\sqrt[3]{-8}$ **23.** $452/100$ **24.** $\sqrt{-8}$
25. $-\frac{6}{5}$

In Exercises 26 through 35, indicate whether the statement is true or false.

26. Every integer is a real number.
27. Every real number is a complex number.
28. Every real number has a repeating decimal expansion.
29. The real numbers are closed under subtraction.
30. The rational numbers are closed under division.
31. The number π is irrational.
32. Every rational number can be expressed as the quotient of two integers.
33. Every irrational number has a nonrepeating, nonterminating decimal expansion.
34. Every real number is a rational number.
35. $\frac{A}{A} = 1$ for every real number, A.

In Exercises 36 through 53, indicate the value of each expression, if possible, and its type(s). (There may be more than one answer.) Also indicate which operations, if any, would result in a computer error (see Example 2).

36. $\sqrt{9}$ **37.** $-\sqrt{9}$ **38.** $\sqrt{16}$

39. $\sqrt{-4}$	**40.** $\sqrt{0}$	**41.** $\sqrt{-16}$
42. $-\sqrt{16}$	**43.** $\sqrt{25}$	**44.** $\sqrt{0.01}$
45. $\sqrt{7}$	**46.** $\sqrt{-0.01}$	**47.** $\sqrt{169}$
48. 6/0	**49.** 0/0	**50.** $-\sqrt{64}$
51. $\sqrt{-64}$	**52.** $\sqrt{-1}$	**53.** $\sqrt{-169}$

1.2 Identifiers and Assignment Statements

Identifiers are sequences of characters used to represent or name things. In computers, an identifier can refer to a storage location or cell or to its contents. Examples of identifiers are X, SUM, PI, RATE, Y2. Identifiers can be classified as being used either (1) for constants or (2) for variables.

The value of a **constant** does not change in a discussion or in a computer program. Numbers, such as 3.1, -5, 4.22, are examples of constants, as are identifiers assigned to a number, such as PI = 3.1416. In this text constants will be numerical, although it *is* possible to have string constants, such as "HAROLD", "HIT RETURN TO CONTINUE", or "THE VALUE IS".

Variables are identifiers whose values may change in the execution of a program. Variables can be given an initial value in an assignment statement. Some examples of assignment statements (from BASIC) are:

LET X = 5	Read "X is assigned the value 5."
LET Y2 = 6 * X	Read "Y2 is assigned 6 times the latest value that was assigned to X."
LET Z = Z + 1	Read "Z is assigned 1 plus the latest value that was assigned to Z."

In some languages, a distinction is made between assignment using an equals sign and the use of that sign as a relational operator. For example, assigning the value 5 to X is accomplished differently in FORTRAN, APL, and Pascal.

X = 5	FORTRAN
X ← 5	APL
X := 5	Pascal

In any of the three cases, the assignment is really read from right to left. The expression or value on the right is first evaluated and the identifier on the left "receives" that value. The statement Z + 1 = Z would, therefore, be an invalid assignment statement.

EXAMPLE 1

What is the value of X at the end of the following BASIC program?

```
10  LET T = 2
20  LET Y = 5
30  LET X = Y + T
40  LET X = X + 1
50  END
```

Solution

In line 30, X is assigned the value 7; in line 40, X is assigned the value 7 + 1 or 8. The final value of X is 8.

In some computer languages (Pascal, for example) a programmer must declare the type of variables being used before using them. In the next example, we see a possible error that can occur when an identifier's outcome is not fully anticipated by a programmer.

EXAMPLE 2

Suppose that in a Pascal program the identifiers RESULT, X, and Y are declared to be of type integer and RESULT = X/Y. What is the value of RESULT if

a. X = 6 and Y = 2 **b.** X = 7 and Y = 2

Solution

a. RESULT = 6/2 = 3, and no problem occurs.

b. The "actual" value of 7/2 is 3.5; but if RESULT has been declared to be of type integer, then one of two things will occur (depending on the computer): (1) an error message (because 3.5 is not an integer) or (2) RESULT *may* be assigned the value 3 (3.5 "converted" to an integer by ignoring the decimal portion).

EXERCISES FOR SECTION 1.2

In Exercises 1 through 15, determine whether or not the statement is a valid assignment statement.

1. SUM = 5

2. SUM = SUM + 1

3. A = X * Y

4. X * Y = A

5. PROD = PROD * N

6. Y + 1 = Y

7. B + C = B

8. T ← X + Y

9. X + Y ← T

10. B = X + Y/Z

11. X = X + X

12. 0 = X

13. Q + 1 = X − 5

14. 7 := Q

15. Q := 7 − X

In Exercises 16 through 21, determine the value of X at the conclusion of each BASIC program fragment (see Example 1).

16.
```
10  LET T = 7
20  LET Y = 6
30  LET X = Y + T
40  LET X = X + 1
50  END
```

17.
```
10  LET Z = 0.5
20  LET Y = 0.6
30  LET X = Z + Y
40  LET X = X + 1.5
50  END
```

18.
```
10  LET T = 7
20  LET Y = 6
30  LET X = Y + T
40  LET X = X + X
50  END
```

19.
```
10  LET Z = 0.5
20  LET Y = 0.6
30  LET X = Z + Y
40  LET X = X + X
50  END
```

20.
```
10  LET T = 7
20  LET Y = 6
30  LET X = 2
40  LET X = X * X
50  LET X = Y + T
60  END
```

21.
```
10  LET T = 7
20  LET Y = 6
30  LET X = 2
40  LET X = Y + T
50  LET X = X + X
60  END
```

In Exercises 22 through 25, assume the following declarations of identifiers have been made in Pascal:

```
SUM1: INTEGER        SUM2: REAL
   X: INTEGER
   Y: INTEGER
```

Find the value of the indicated variable; assume the variables have been initialized as follows:

```
SUM1 := 0        SUM2 := 0.0
   X := 5
   Y := 6
```

(See Example 2.)

22. Find SUM1 if SUM1 := Y/2

23. Find SUM1 if SUM1 := X/2

24. Find SUM2 if SUM2 := Y/X
25. Find SUM2 if SUM2 := X/2.0

26. How does a variable differ from a constant?
27. *Variable Names.* In many forms of BASIC, variables are limited to two characters in length with the first character a letter. In most FORTRAN versions, variables may be six characters long (again, the first is usually a letter). The restrictions on the variable identifiers differ from machine to machine as well as from language to language. Determine the restrictions on variables for a machine and language to which you have access.

1.3 The Order of Operations

In card games, each card is usually assigned a rank. For most games this ranking, or hierarchy, is ace, king, queen, jack, ten, etc. In algebra and arithmetic there is a ranking known as the **order of operations**, the **rules of precedence**, or the **hierarchy of operations**. Some computer languages, such as PL/1, Pascal, BASIC, and FORTRAN, assume these rules; some others, such as APL, have no precedence rules. Here, we shall discuss the order of operations of addition (+), subtraction (−), multiplication (·), division (/), and exponentiation (^). In Chapter 2, we shall reexamine the order of operations and incorporate other operations.

Exponentiation, multiplication, division, addition, and subtraction obey the following hierarchy:

Operation	Priority
Exponentiation (^)	Do first, before any multiplication, division, addition, or subtraction.
Multiplication and division (· and / have equal precedence)	When all exponentiation is complete, then do all multiplication and division together, proceeding from left to right.
Addition and subtraction (+ and − have equal precedence)	Lowest priority. After all exponentiation, multiplication, and division, do all additions and subtractions together, from left to right.

Use of these rules is illustrated in the following examples:

EXAMPLE 1

Evaluate $6 + 2 \cdot 5$.

Solution

First, scan the expression to notice all operations. Since multiplication is done *before* addition, evaluate $2 \cdot 5$ first; then perform the addition:

$$6 + 2 \cdot 5$$
$$6 + 10$$
$$16$$

EXAMPLE 2

Evaluate $6/3 + 2 \cdot 5$

Solution

The operations are division, addition, and multiplication. Perform the division and multiplication first, from left to right:

$6/3 + 2 \cdot 5$	Evaluate first $6/3$, then $2 \cdot 5$.
$2 + 10$	$6/3 = 2$; $2 \cdot 5 = 10$.
12	

EXAMPLE 3

Evaluate $5 + 6/2 \cdot 3$.

Solution

$5 + 6/2 \cdot 3$	Evaluate $6/2$ first.
$5 + 3 \cdot 3$	Next, $3 \cdot 3 = 9$.
$5 + 9$	
14	

EXAMPLE 4

Evaluate 4 − 6 · 3 + 2^3.

Solution

4 − 6 · 3 + 2^3	First, 2^3 = $2^3 = 8$.
$4 - 6 \cdot 3 + 8$	Next, $6 \cdot 3 = 18$.
$4 - 18 + 8$	$4 - 18 = -14$.
$-14 + 8$	
-6	

EXAMPLE 5

Evaluate 10 − 24/2^3.

Solution

10 − 24/2^3	First, 2^3 = 8.
10 − 24/8	Next, 24/8 = 3.
10 − 3	
7	

EXAMPLE 6

Evaluate 10 − 24/2^3 · 3.

Solution

10 − 24/2^3 · 3	First, 2^3 = 8.
10 − 24/8 · 3	Next, 24/8 = 3.
10 − 3 · 3	3 · 3 = 9.
10 − 9	
1	

When parentheses are used, any indicated operations *within* parentheses are performed first. Compare Examples 7, 8, and 9.

EXAMPLE 7

Evaluate 6 · 2/4 · 3.

Solution

6 · 2/4 · 3
12/4 · 3
3 · 3
9

EXAMPLE 8

Evaluate 6 · 2/(4 · 3).

Solution

6 · 2/(4 · 3)	First, evaluate the operation in parentheses, 4 · 3 = 12.
6 · 2/12	
12/12	
1	

EXAMPLE 9

Evaluate 6 · (2/4) · 3.

Solution

6 · (2/4) · 3	First, perform the operation in parentheses, 2/4 = 0.5.
6 · 0.5 · 3	
3 · 3	
9	

EXAMPLE 10

Evaluate 10 − 24/(2^3 · 3).

Solution

10 − 24/(2^3 · 3)	Within parentheses, 2^3 = 8.
10 − 24/(8 · 3)	Then 8 · 3 = 24.
10 − 24/24	
10 − 1	
9	

Notice that the use of parentheses may change the outcome.

EXAMPLE 11

Evaluate X + Z^Y · X when X = 5, Y = 2, and Z = 4.

Solution

X + Z^Y · X	Substitute the values for X, Y, and Z.
5 + 4^2 · 5	
5 + 16 · 5	
5 + 80	
85	

EXAMPLE 12

Evaluate

a. (−2)^4 + 3 **b.** −2^4 + 3

Solution

a. (−2)^4 + 3 Exponentiation first. The parentheses indicate that the exponent "belongs" to the quantity −2.

16 + 3

19

b. $-2\text{^}4 + 3$

The way a computer handles an expression like $-2\text{^}4$ varies from one computer to another. Some computers assume that the negative sign belongs with the 2 (it is "part of the number"). In this case, the result would be the same as for part a:

$$-2\text{^}4 + 3 = 16 + 3$$
$$= 19$$

Some computers see the negative sign as being separate from the number (a sign of operation, in this case subtraction). They will perform the exponentiation on the 2 first.

$-2\text{^}4 + 3$	Raise 2 to the fourth power.
$-16 + 3$	Add (-16) and $(+3)$.
-13	

To avoid confusion and possible error, it is safest to use parentheses in a situation like this one. If a programmer wants the computer to evaluate the negative of (2^4), the expression should be written −(2^4).

EXERCISES FOR SECTION 1.3

In Exercises 1 through 20, evaluate the given expression (see Examples 1–6).

1. $7 + 3 \cdot 2$

2. $7 \cdot 3 + 2$

3. $15/5 \cdot 3$

4. $15/3 \cdot 5$

5. $3 \cdot 5/15$

6. $5 - 2 \cdot 3$

7. $2 \cdot 3 - 5$

8. $12/2 + 3 \cdot 2$

9. $2 + 12 \cdot 2/2$

10. $2 + 12/2 \cdot 2$

11. $5 - 6/2/3$

12. $6/2 - 5/3$

13. $5 - 6 \cdot 2 + 3\text{^}2$

14. $4 - 7 \cdot 10 + 5\text{^}3$

15. $2 - 3\text{^}3 + 4\text{^}3$

16. $72 - 48/2\text{^}3$

17. $72 - 48/2\text{^}3 \cdot 3$

18. $2 \cdot 27/3\text{^}2$

19. $27/3\text{^}2 \cdot 3$

20. $3 \cdot 3\text{^}2/27$

In Exercises 21 through 30, evaluate each expression (see Examples 7–10).

21. $5 \cdot (4/2) \cdot 2$

22. $5 \cdot 4/(2 \cdot 2)$

23. $(5 \cdot 4)/(2 \cdot 2)$

24. $12 - 48/(2\text{^}3 \cdot 2)$

25. $12 - (48/2 \cdot 3)\text{^}2$

26. $(12 - 48)/2 + 14$

27. $12 - 48/(2 + 14)$

28. $12 - (48/2) + 14$

29. $(12 - 48)/(2 + 14)$

30. $5/(2\text{^}3\text{^}2 - 80)$

In Exercises 31 through 40, find the value of X using $A = 2$, $B = 6$, $C = -4$, $D = 4$, $Y = 0$ (see Example 11).

31. $X = (-B + D)/2 \cdot A$

32. $X = (-B + D)/(2 \cdot A)$

33. $X = (-B + D)/2/A$

34. $X = B + Y\text{^}A$

35. $X = (B + Y)\text{^}A$

36. $X = C + B\text{^}A + D$

37. $X = (C + B)\text{^}A + D$

38. $X = (C + B)\text{^}(A + D)$

39. $X = A \cdot B/C \cdot D$

40. $X = (A \cdot B)/C \cdot D$

In algebra, multiplication is usually *implicit*. That is, $5(x + 2)$ represents the multiplication of 5 with $x + 2$ even though no multiplication symbol appears. The BASIC equivalent of $5(x + 2)$ is $5 * (X + 2)$. In Exercises 41 through 45, rewrite each given algebraic expression in an equivalent BASIC form.

41. $(a + b)(c + d)$

42. $3x + 4y$

43. $\dfrac{2x + y}{3}$

44. $\dfrac{x + y}{4z}$

45. $x + \dfrac{3(x + 1)}{2y}$

1.4 Scientific and Exponential Notation

Sometimes the results of calculations are very large or very small, and a computer may output the number as

```
3.45E+11
```

This means $3.45 \cdot 10^{11}$ or 345,000,000,000. In other words, E+11 means "times 10 raised to the eleventh power."

When does a computer use this E-notation? It depends on the computer. One popular microcomputer can store nine digits, so if 1234567890 is entered, 1.23456789E+09 is displayed. This type of E-format is more often called **scientific notation**.

SCIENTIFIC NOTATION

A positive number is said to be expressed in scientific notation if it is written in the form

$$N \cdot 10^a$$

where $1 \le N < 10$ and a is an integer.

Some examples follow.

EXAMPLE 1

Write 1492 in scientific notation.

Solution

Using the definition, we must find N, a number between 1 and 10 derived from 1492 by merely moving the decimal point. Notice that if we move the decimal point in 1492. three places to the left, we have

$$1.492.$$

With $N = 1.492$, the power of 10 we must multiply N by to preserve the magnitude of the original number is 10^3 (the three-place movement of the decimal point).

$$1492 = 1.492 \cdot 10^3$$

EXAMPLE 2

Write 7,694,250,000 (a) in scientific notation and (b) in E-notation.

Solution

In each case below, recall that we do not use commas when writing numbers.

a. To convert 7694250000. to scientific notation, we move the decimal point nine places (why nine?) to the left:

$$7694250000 = 7.694250 \cdot 10^9$$

b. $7.694250 \cdot 10^9$ is written as 7.69425E+09 by most computers.

EXAMPLE 3

Write −765,421,000,000 (a) in scientific notation and (b) in E-notation.

Solution

a. The definition at the beginning of Section 1.4 is only for positive numbers. For a negative number such as -765421000000, we find N such that $-10 < N \leq -1$. We have

$$-765421000000 = -7.65421 \cdot 10^{11}$$

b. $-7.65421 \cdot 10^{11} = -7.65421\text{E}+11$

EXAMPLE 4

Express 0.002576 in E-format.

Solution

By moving the decimal point three places *to the right*, we obtain $N = 2.576$. In scientific notation, we have:

$$0.002576 = 2.576 \cdot 10^{-3}$$

Note that the exponent has a minus sign because 2.576 must be made *smaller* and 10^{-3} in effect returns the decimal point to its original position.

In E-format, $0.002576 = 2.576\text{E}-03$

Thus far, we have seen several different ways of displaying a number. Observe that 1492 can be written as $1492 = 1.492 \cdot 10^3 = 1.492\text{E}+03$.

It is also equal to each of the following:

$$1492 = 14.92 \cdot 10^2 = 0.1492 \cdot 10^4$$

This last form is known as the **normalized exponential form**. A number in normalized exponential form is expressed as $N \cdot 10^a$, where $0.1 \leq N < 1$ if the number is positive and $-1 < N \leq -0.1$ for a negative number; a still represents an integer.

EXAMPLE 5

Write 0.0338 (a) in scientific notation, (b) in E-format, and (c) in normalized exponential form.

Solution

a. $0.0338 = 3.38 \cdot 10^{-2}$
b. $0.0338 = 3.38\text{E} - 02$
c. $0.0338 = 0.338 \cdot 10^{-1}$

A fourth format is the **normalized E-format** and we would write 0.0338 = 0.338E−01.

EXAMPLE 6

Write −56.42 in all four formats.

Solution

scientific notation: $-5.642 \cdot 10^1$
(scientific) E-format: −5.642E+01
normalized exponential form: $-0.5642 \cdot 10^2$
normalized E-format: −0.5642E+02

Whenever numbers are written as the product of some number N and a power of 10, the number N is called the **mantissa**. Its value depends, of course, on whether scientific or normalized form is used. Table 1.2 summarizes the different formats and their mantissas.

TABLE 1.2

Real Number	Normalized E-Format	Scientific E-Format	Mantissa
576.2	0.5762E+03		0.5762
		5.762E+02	5.762
−0.00076	−0.76E−03		−0.76
		−7.6E−04	−7.6
186000	0.186E+06		0.186
		1.86E+05	1.86

Significant Digits

Recall that computers store and display only a finite number of digits. If −113476562800 is entered into a microcomputer that stores nine digits, the resulting display might look like

```
−1.13476563E+11
```

If 0098 is entered, the leading two zeros will be ignored and 98 is displayed. The leading two zeros are not significant. In the case of 0.0054, the display .0054 would appear, even though the zeros are not significant. The rules for determining the significant digits in a number are as follows:

RULES FOR SIGNIFICANT DIGITS

1. Nonzero digits are always significant.
2. Zero is a significant digit if it is between other significant digits.
3. Zero is *not* significant when it comes before all the nonzero digits.
4. Trailing zeros are significant only if the number contains an embedded decimal point.

EXAMPLE 7

How many significant digits does each number have?

a. 5.08 **b.** 5302 **c.** 156000

Solution

a. 5.08 has three significant digits.
b. 5302 has four significant digits.
c. Since there is no decimal point in 156000, the three trailing zeros are not significant, and we say that 156000 has three significant digits.

In part c of Example 7, notice that if we wanted to express 156000 with four significant digits, then we could use scientific notation and write $1.560 \cdot 10^5$. To express 156000 as a number with five significant digits, we could write $1.5600 \cdot 10^5$.

EXAMPLE 8

How many significant digits does each of the following numbers have?

a. 0.00304 **b.** $6.14 \cdot 10^{-8}$

Solution

a. 0.00*304* has three significant digits.
b. *6.14* $\cdot 10^{-8}$ has three significant digits. When a number is expressed in scientific notation, only the mantissa has to be examined to determine which digits are significant.

Truncation

To truncate means to "chop off." Computers may *truncate* numbers that are the result of mathematical operations. Some computers drop off the least significant digits (those significant digits that are rightmost) if they cannot be stored in memory. Suppose a computer can store only four digits. Then 7653.9 would be stored as 7653 if truncation occurs. (Truncation should not be confused with rounding. Rounding 7653.9 to four significant digits, would give 7654.)

The effects of truncation error can be illustrated with a simple example of adding three numbers, in Example 9. First, we define the **relative error**, e_r:

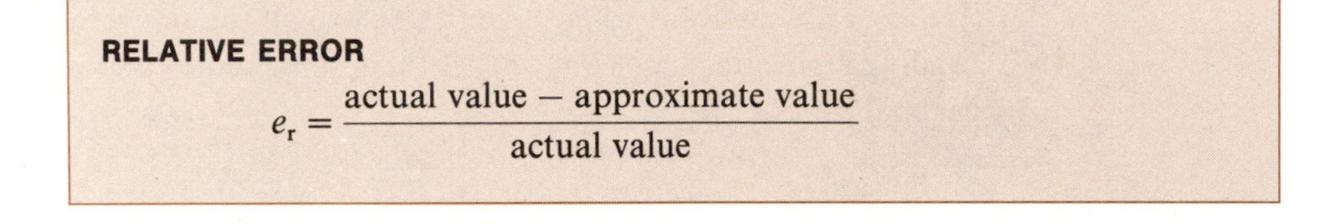

RELATIVE ERROR

$$e_r = \frac{\text{actual value} - \text{approximate value}}{\text{actual value}}$$

EXAMPLE 9

Suppose a computer truncates all numbers to two digits.

a. What would be the result of the following calculation?

$$3.9 + 2.8 + 4.9$$

b. What is the relative error?

Solution

a.
$$\begin{array}{r} 3.9 \\ +\,2.8 \\ \hline 6.7 \\ +\,4.9 \\ \hline 11.6 \end{array}$$

Truncation would occur and 11.6 would be stored as 11.

b. $e_r = \dfrac{11.6 - 11}{11.6} = \dfrac{0.6}{11.6} = 0.051724$

The error obtained in Example 9 can be compounded if calculations are performed repeatedly, as the next example illustrates.

EXAMPLE 10

Suppose a computer truncates all numbers to two digits.

a. What would be the result of the following calculation?

$$3.9 + 2.8 + 4.9 + 3.9 + 2.8 + 4.9$$

b. What is the relative error?

Solution

a. The result of adding the first three numbers has already been found to be 11.

$$\begin{array}{l} 3.9 \\ 2.8 \\ \underline{4.9} \\ 11.6 \rightarrow 11 \quad \text{Truncation occurs.} \\ \quad \underline{+\ 3.9} \\ \quad 14.9 \rightarrow 14 \quad \text{Truncation occurs.} \\ \quad\quad \underline{+\ 2.8} \\ \quad\quad 16.8 \rightarrow 16 \quad \text{Truncation occurs.} \\ \quad\quad\quad \underline{+\ 4.9} \\ \quad\quad\quad 20.9 \rightarrow 20 \quad \text{Truncation occurs.} \end{array}$$

Thus, the (truncated) result is 20.

b. The actual result of the calculation is 23.2.

$$e_r = \frac{23.2 - 20}{23.2} = \frac{3.2}{23.2} = 0.137931$$

(Notice that the error has more than doubled.)

Table 1.3 points out the difference between rounding and truncating.

TABLE 1.3

Real Numbers	Rounded to Four Significant Digits	Truncated to Four Significant Digits
764.29	764.3	764.2
0.025555	0.02556	0.02555
186272	$1.863 \cdot 10^5$	$1.862 \cdot 10^5$
25.985	25.99	25.98

EXERCISES FOR SECTION 1.4

In Exercises 1 through 25, write each number (a) in scientific notation and (b) in E-format (see Examples 1–4).

1. 1562
2. 2485
3. 175
4. 42,000
5. 573,000,000
6. 123,000,000
7. −2,300,000
8. −578,000
9. −1200
10. −67
11. 0.00000145
12. 0.0000573
13. −0.0576
14. −0.00001598
15. −0.005002
16. −35.46
17. 80,000
18. −80,000
19. 0.000000973
20. 93,000,000
21. −0.00000004002
22. 860,000,000,000,000
23. 3005
24. −50,020
25. −0.000006002

In Exercises 26 through 40, write each number (a) in normalized exponential form and (b) in normalized E-format (see Examples 5 and 6).

26. 1562
27. 180
28. 42,000
29. 573,000,000
30. −2,300,000
31. −578,000
32. 0.00561
33. 0.00000350
34. −0.00204
35. −35.46
36. 80,000
37. −80,000
38. −0.000000149
39. −50,020
40. −0.00000603

Computers are being designed to operate at faster and faster speeds. In Exercises 41 through 44 are given units used to express CPU (central processing unit) speeds. Convert each to units of seconds written both in scientific notation and in scientific E-notation.

41. 1 millisecond = 1 thousandth of a second
42. 1 microsecond = 1 millionth of a second
43. 1 nanosecond = 1 billionth of a second
44. 1 picosecond = 1 trillionth of a second

In Exercises 45 through 55 determine how many significant digits each number has (see Examples 7 and 8).

45. 7.02
46. 8.001
47. 8.0010
48. 5301
49. 5300
50. 5300.0
51. $6.15 \cdot 10^{8}$
52. 0.000101
53. 0.00300
54. $1.57 \cdot 10^{-2}$
55. −0.00010304

In Exercises 56 through 60 assume each calculation is done on a computer that truncates all numbers to four digits. (a) Find the result of the calculation done on the computer, and (b) find the relative error (see Examples 9 and 10).

56. $0.6789 + 0.5013 + 1.866$

57. $3.859 + 9.146 + 1.999$

58. $\frac{2}{3} + \frac{2}{3} + \frac{2}{3}$

59. $\frac{2}{3} + \frac{2}{3} + \frac{2}{3} + \frac{2}{3} + \frac{2}{3} + \frac{2}{3}$

60. $0.6789 + 0.5013 + 1.866 + 0.6789 + 0.5013 + 1.866 + 0.6789 + 0.5013 + 1.866$
Compare with Exercise 56.

1.5 The Axioms of Algebra and Computer Arithmetic

From previous mathematics courses, we know that, if A, B, and C are real numbers,

$$(A + B) + C = A + (B + C) = A + B + C$$

This is known as the **associative property of addition**. Although this property applies to mathematics, we want to investigate whether the property holds within the finite "brain" of a computer. First, we introduce some basic concepts that may help to clarify future examples.

The computer's "brain" is known as the **central processing unit (CPU)**. For our purposes, we shall refer to two main components of the CPU, the **arithmetic logic unit (ALU)** and the **primary memory**. We shall assume, in our hypothetical computer, that we can store three digits in primary memory but the ALU can work with four-digit numbers. Usually, the ALU carries out results of arithmetic operations to at least one place more than it stores in its primary memory.

Now, let us investigate the associativity of addition in our hypothetical computer.

EXAMPLE 1

Let $A = 85.9$, $B = 44.4$, and $C = 2.9$. Do the following evaluations on our hypothetical three-digit computer.

a. Find $(A + B) + C$ **b.** Find $A + (B + C)$

Solution

a. $A + B = 85.9 + 44.4 = 130.3$ — When this is done by the ALU, the result is *stored* as 130.

$(A + B) + C = 130 + 2.9 = 132.9$ — But this is stored as 132.

Thus, $(A + B) + C = 132$

b. $B + C = 44.4 + 2.9 = 47.3$

$A + (B + C) = 85.9 + 47.3 = 133.2$ — But 133 is the stored value.

This example points out that, in general, $(A + B) + C \neq A + (B + C)$, in computer arithmetic. The reader should note, however, that whether $(A + B) + C = A + (B + C)$ depends on the numbers and the computer's capabilities. For example, if $A = 12$, $B = 13$, and $C = 15$, then $(A + B) + C = A + (B + C)$. Since most computers hold eight or more significant digits, the type of error illustrated by Example 1 is not too likely, but you should be aware that it *can* happen.

The next example examines the associativity of multiplication in a computer.

EXAMPLE 2

Let $A = 7.9$, $B = 4.6$, and $C = 2.1$ Do the following evaluations on our hypothetical three-digit computer.

a. Find $(AB)C$ **b.** Find $A(BC)$

Solution

a. $(AB) = A \cdot B = 7.9 \cdot 4.6 = 36.34$ — This is truncated to 36.3.

$(AB)C = 36.3 \cdot 2.1 = 76.23$ — This is truncated to 76.2.

Thus, $(AB)C = 76.2$ in the three-digit computer.

b. $(BC) = B \cdot C = 4.6 \cdot 2.1 = 9.66$

$A(BC) = 7.9 \cdot 9.66 = 76.314$ — This is truncated to 76.3.

The point of both Examples 1 and 2 is that the axioms we know from mathematics do not necessarily apply to the truncated arithmetic done in a computer.

Other types of errors occur because of the way computers store numbers internally. This system, called the binary number system, and its potential for numerical error is described in Chapter 3. We conclude this section with some elementary (yet baffling) calculations performed on a popular microcomputer. The discrepancy between the displayed result and the actual result is the topic of part of Chapter 3.

Direction	Displayed Result	Actual Result
PRINT 30.27 − 30	.270000003	0.27
PRINT 30.125 − 30	.125	0.125
PRINT 30.63 − 30	.629999995	0.63
PRINT 40.54 − 40	.539999992	0.54
PRINT 40.54 − 30	10.54	10.54
PRINT 40.875 − 40	.875	0.875
PRINT 40.87 − 40	.870000005	0.87
PRINT 654.321 − 654	.321000099	0.321
PRINT 765.4321 − 765	.432100296	0.4321
PRINT 87.011 − 87	.0109999776	0.011

EXERCISES FOR SECTION 1.5

In Exercises 1 through 6, find (a) $(A + B) + C$ and (b) $A + (B + C)$ assuming arithmetic on our hypothetical three-digit computer (see Example 1).

1. $A = 87.1 \quad B = 9.9 \quad C = 3.9$

2. $A = 31.6 \quad B = 83.4 \quad C = 4.7$

3. $A = 67.9 \quad B = 34.7 \quad C = 39.9$

4. $A = 36.8 \quad B = 41.2 \quad C = 91.9$

5. $A = 37.5 \quad B = 47.7 \quad C = 10.9$

6. $A = 11.1 \quad B = 12.2 \quad C = 13.3$

In Exercises 7 through 12, find (a) $(AB)C$ and (b) $A(BC)$ using the hypothetical computer (see Example 2).

7. $A = 7.8 \quad B = 4.2 \quad C = 2.1$

8. $A = 8.1 \quad B = 3.5 \quad C = 1.2$

9. $A = 1.1 \quad B = 3.7 \quad C = 1.2$

10. $A = 1.2 \quad B = 1.1 \quad C = 3.7$

11. $A = 10.6 \quad B = 3.4 \quad C = 1.1$

12. $A = 11.2 \quad B = 1.2 \quad C = 4.5$

In Exercises 13 through 16, examine the distributive property. That is, find (a) $A(B + C)$ and (b) $AB + AC$ using the arithmetic of our hypothetical computer.

13. $A = 3.4 \quad B = 7.1 \quad C = 2.9$

14. $A = 3.4 \quad B = 8.7 \quad C = 9.9$

15. $A = 11.4 \quad B = 8.7 \quad C = 12.4$

16. $A = 10.9 \quad B = 12.5 \quad C = 1.5$

17. According to the results of Exercises 13 through 16, in computer arithmetic is it always true that $A(B + C) = AB + AC$?

18. In algebra, we can find that an *actual* solution to $x^2 - 30x + 1 = 0$ is $R = (30 - \sqrt{896})/2$. Using the fact that in our hypothetical computer the truncated value of $\sqrt{896}$ would be 29.9, find the value of R.

19. a. Using the hypothetical computer's arithmetic, find $R^2 - 30R + 1$. (See Exercise 18.)
    b. Explain why the answer to part a is *not* zero.

20. a. Follow the procedure of Exercise 18 to find the value of $R = (30 + \sqrt{896})/2$.
    b. Explain why $R^2 - 30R + 1$ is not zero.

1.6 Chapter Review

Because the computer can store in its memory only a finite number of digits, computer arithmetic sometimes gives unexpected results in seemingly simple computations. This chapter addressed the various types of numbers used by the computer and some of the problems that can arise because of the finite nature of the computer's "memory."

The chapter began with definitions and descriptions of the basic number types: real—including integer, rational, and irrational—and complex. Identifiers and assignment statements were studied in relationship to the storage and manipulation of numbers in a computer. The precedence rules for arithmetic operations were discussed in Section 1.3.

Sections 1.4 and 1.5 dealt with some of the problems that arise and adjustments that are made because of the limitations of computer storage. These included the concepts of scientific notation, significant digits, and truncation vs. rounding. Finally, some of the idiosyncrasies of computers were presented in Section 1.5.

VOCABULARY

real number
rational number
irrational number
integer
byte
memory address
closure
complex number
imaginary number
identifier
constant
variable
assignment statement
priority of operations
rules of precedence
hierarchy of operators
E-format
scientific notation

normalized exponential form
normalized E-format
mantissa
significant digits
truncation
rounding
relative error

CHAPTER TEST

1. Identify each of the numbers below as integer, real, complex, rational, or irrational. More than one label may apply to a given number (for example, the number 28 is integer, rational, real, and complex).

   a. $\sqrt{4}$ b. $\sqrt{-4}$ c. 0.125
   d. 6 e. 3.14 f. 22/7
   g. $\sqrt{7}$ h. $3 + 2\sqrt{-1}$

2. Evaluate each of the following expressions following the rules for priority of arithmetic operations.

   a. $2 + 6 \cdot 5$ b. $\sqrt{0}$
   c. $4 + 6/5$ d. 12 − 2^4/2
   e. $6 + 2 \cdot 32/(6 + 2)$

3. Write each number below (a) in scientific notation and (b) in E-format.

   a. 432 b. 43200000
   c. −4320 d. 0.000432
   e. 0.432

4. How many significant digits does each of the following numbers have?

   a. 6.18 b. 4502 c. 120000

5. Suppose a computer truncates all numbers to two digits. What would be the result of the following calculation in such a hypothetical computer: 2.8 + 3.7 + 14.6? What is the relative error?

6. Which of the following numbers can be *completely* stored in a computer's memory?

   a. 4/5 b. 22/7 c. $\sqrt{14}$

7. What is the value of *X* after the following BASIC program is RUN?

```
10  LET X = 5
20  LET X = X + 3
30  LET X = X/2
40  END
```

8. What type of number is a zip code? If the variable *X* is used to store the zip code 06759, and *X* is of type integer, what will be stored in the computer's memory?

9. Given $x = 2$, $y = 4$, $z = 6$, evaluate each of the following.

   a. *y*^*x* + *z*
   b. $x + y + z/[z - (x + y)]$

10. Explain the difference between rounding a number and truncating a number.

CHAPTER

2

Many business, scientific, and engineering problems deal with equations in some context. A graph like the ones at the right gives a picture of an equation and acts as an aid in interpreting the equation. In this chapter, you will review methods of solving and graphing some common types of equations.

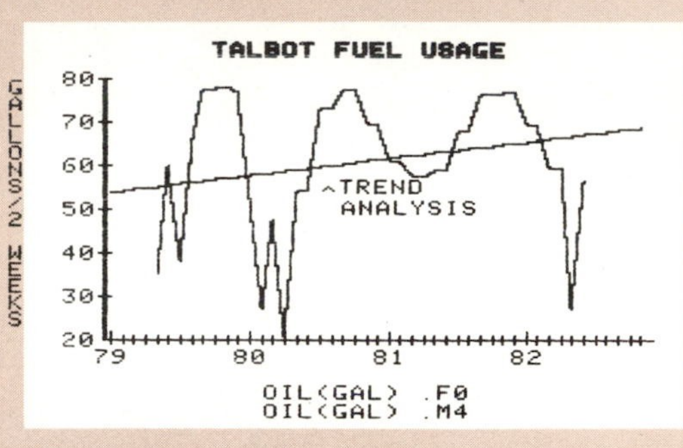

Algebra Review

A computer can do *only* what someone tells it (via programs) to do. Many programming applications, especially in business, do not require "heavy-duty" math. But programming applications often do make use of formulas of varying degrees of complexity. In this chapter we shall review the mathematics necessary to deal with such formulas.

Consider, for example, the formulas involved in calculating a simple payroll. Provision must be made for paying time-and-one-half for overtime; deductions must be calculated as percentages of gross pay; net pay must be determined by subtracting the final deduction amount from the gross pay. Each of these calculations involves the use of one or more formulas.

Since a computer can perform such calculations only by following a programmer's instructions, it is important that programming students understand the basics of algebraic notation and the manipulation of algebraic symbols.

2.1 Exponents and Radicals

Some of the basic concepts of algebra—for example, variables and the order of operations—have already been reviewed in Chapter 1. In this chapter, we shall study additional concepts, beginning with one touched upon in the last chapter, that of an exponent.

Definition of a Positive Integer Exponent

When we say that we want to *raise a number* (say 3) *to a positive integer power* (say 4), we mean that we intend to write a product in which the given number, called the **base**, appears as a factor a number of times equal to the power, or **exponent**. We write

$$3^4 = \underbrace{3 \cdot 3 \cdot 3 \cdot 3}$$

3 appears as a factor four times.

In this expression, 3 is the **base** and 4 is the **exponent**. Note that the exponent is written slightly above and to the right of the base (it is also usually smaller) in order to distinguish exponentiation (the use of exponents) from multiplication.

EXAMPLE 1

Evaluate each of the following:

a. 2^3 **b.** $(-3)^2$ **c.** $(-2)^3$ **d.** 10^5 **e.** $\left(\frac{3}{4}\right)^3$ **f.** 1^9

Solution

a. $2^3 = \underbrace{2 \cdot 2 \cdot 2}_{\text{2 is used as a factor three times.}} = 8$

b. $(-3)^2 = \underbrace{(-3) \cdot (-3)}_{\text{-3 appears as a factor two times.}} = 9$

c. $(-2)^3 = (-2) \cdot (-2) \cdot (-2)$
$= 4 \cdot (-2) = -8$

d. $10^5 = 10 \cdot 10 \cdot 10 \cdot 10 \cdot 10$
$= 100{,}000$

e. $\left(\frac{3}{4}\right)^3 = \left(\frac{3}{4}\right) \cdot \left(\frac{3}{4}\right) \cdot \left(\frac{3}{4}\right)$
$= \left(\frac{9}{16}\right) \cdot \left(\frac{3}{4}\right) = \frac{27}{64}$

f. $1^9 = 1 \cdot 1 \cdot 1 \cdot 1 \cdot 1 \cdot 1 \cdot 1 \cdot 1 \cdot 1$
$= 1$

If we use a variable instead of a constant as the base, exponentiation carries the same meaning:

EXPONENTIATION

$$x^a = \underbrace{x \cdot x \cdot x \cdot \cdots \cdot x}_{a \text{ factors}}$$

Also, if x is any base, x^1 is defined to be x. So, $5^1 = 5$; $(-17)^1 = -17$, and so on.

Properties of Positive Integer Exponents

For now, in discussing the properties of exponents, we shall limit ourselves to exponents that are *positive integers*. Later in this section we shall extend the discussion to exponents in general, using the properties we discover here as a starting point.

Suppose x is any base, and we wish to form the product

$$x^3 \cdot x^5$$

Since $x^3 = x \cdot x \cdot x$ and $x^5 = x \cdot x \cdot x \cdot x \cdot x$, we have

$$x^3 \cdot x^5 = (x \cdot x \cdot x) \cdot (x \cdot x \cdot x \cdot x \cdot x)$$

Since x appears as a factor *eight times* (that is, $3 + 5$ times) in this expression, it follows that

$$x^3 \cdot x^5 = x^8 \qquad \text{or} \qquad x^3 \cdot x^5 = x^{3+5}$$

Now consider the product $x^a \cdot x^b$, where a and b are *any* positive integers. Note that the *base x is the same* for both exponents.

$$x^a \cdot x^b = \underbrace{(x \cdot x \cdot \cdots \cdot x)}_{a \text{ factors}} \cdot \underbrace{(x \cdot x \cdot \cdots \cdot x)}_{b \text{ factors}}$$

$$= \underbrace{x \cdot x \cdot \cdots \cdot x}_{(a+b) \text{ factors}}$$

So, $x^a \cdot x^b = x^{a+b}$.

EXAMPLE 2

Simplify each of the following expressions, if possible:

a. $3^2 \cdot 3^3$ **b.** $x^4 \cdot x$ **c.** $a^b \cdot a^{2b}$ **d.** $(2x^2) \cdot (3x^3)$ **e.** $x^3 \cdot y^5$

Solution

a. $3^2 \cdot 3^3 = 3^{2+3} = 3^5 \; (= 3 \cdot 3 \cdot 3 \cdot 3 \cdot 3 = 243)$

b. $x^4 \cdot x = x^4 \cdot x^1$ Since $x^1 = x$.

$= x^{4+1} = x^5$

c. $a^b \cdot a^{2b} = a^{b+2b} = a^{3b}$

d. $(2x^2) \cdot (3x^3) = (2 \cdot 3) \cdot (x^2 \cdot x^3)$ By the commutative and associative properties of multiplication.

$= 6 \cdot (x^{2+3})$

$= 6x^5$

e. $x^3 \cdot y^5$ cannot be simplified, since the bases are different.

Next, consider the following quotient:

$$x^7 \div x^3, \qquad \text{or} \qquad \frac{x^7}{x^3}$$

Note that *the bases are again the same.* If we write this expression out the long way,

$$\frac{x^7}{x^3} = \frac{x \cdot x \cdot x \cdot x \cdot x \cdot x \cdot x}{x \cdot x \cdot x}$$

which reduces to $x \cdot x \cdot x \cdot x$ (canceling three xs in the numerator with the three in the denominator). So,

$$\frac{x^7}{x^3} = x^4 \qquad \text{or} \qquad \frac{x^7}{x^3} = x^{7-3}$$

In general, if $a > b$ (read *a is greater than b*), then

$$\frac{x^a}{x^b} = \frac{\overbrace{x \cdot x \cdot x \cdot \cdots \cdot x}^{a \text{ factors}}}{\underbrace{x \cdot x \cdot \cdots \cdot x}_{b \text{ factors}}}$$

$= x^{(a-b)}$ Cancel b xs from the numerator with the b xs in the denominator, leaving $a - b$ of the xs in the numerator.

EXAMPLE 3

Simplify any of the following expressions that can be simplified.

a. $\dfrac{10^7}{10^4}$ **b.** $\dfrac{x^5}{x^2}$ **c.** $\dfrac{x^{7a}}{x^{6a}}$ **d.** $\dfrac{15a^3}{5a^2}$ **e.** $\dfrac{a^5}{b^3}$

Solution

a. $\dfrac{10^7}{10^4} = 10^{7-4} = 10^3 = 1000$

b. $\dfrac{x^5}{x^2} = x^{5-2} = x^3$

c. $\dfrac{x^{7a}}{x^{6a}} = x^{7a-6a} = x^{1a} = x^a$

d. $\dfrac{15a^3}{5a^2} = \underset{(15 \div 5)}{3}a^{3-2} = 3a^1 = 3a$ Since $a^1 = a$.

e. $\dfrac{a^5}{b^3}$ cannot be simplified since the bases are different.

We have so far discussed two properties of positive integer exponents.

PROPERTIES OF EXPONENTIATION

If x is any real number, then

1. $x^a \cdot x^b = x^{a+b}$
2. $\dfrac{x^a}{x^b} = x^{a-b}, \qquad a > b, x \neq 0$

The other properties to be considered all deal with raising expressions that already contain exponents to a positive integer power. They are

3. $(x^a)^b = x^{a \cdot b}$
4. $(xy)^a = x^a y^a$
5. $\left(\dfrac{x}{y}\right)^a = \dfrac{x^a}{y^a}, \ y \neq 0$

We briefly justify each of these properties with an example. Readers should satisfy themselves that each of these examples can be generalized to apply to any positive integer exponents.

EXAMPLE 4

Justify properties 3, 4 and 5 by evaluating the following expressions:

a. $(x^3)^2$ (Property 3)

b. $(xy)^4$ (Property 4)

c. $\left(\dfrac{x}{y}\right)^4$ (Property 5)

Solution

a. $(x^3)^2 = x^3 \cdot x^3$ — x^3 is used as a factor two times.

$= (x \cdot x \cdot x) \cdot (x \cdot x \cdot x)$ — Here, x appears as a factor six times.

$= x^6$

But $6 = 3 \cdot 2$, so $(x^3)^2 = x^{3 \cdot 2}$

b. $(xy)^4 = (xy) \cdot (xy) \cdot (xy) \cdot (xy)$ — xy is used as a factor four times.

$= (x \cdot x \cdot x \cdot x) \cdot (y \cdot y \cdot y \cdot y)$ — By the associative and commutative properties of multiplication.

$= x^4 \cdot y^4$

c. $\left(\frac{x}{y}\right)^4 = \frac{x}{y} \cdot \frac{x}{y} \cdot \frac{x}{y} \cdot \frac{x}{y}$ — $\frac{x}{y}$ is used as a factor four times.

$= \frac{x \cdot x \cdot x \cdot x}{y \cdot y \cdot y \cdot y}$ — By the definition of multiplication of fractions.

$= \frac{x^4}{y^4}$

Let us use these properties in one more example before going on to zero and negative integers as exponents.

EXAMPLE 5

Using properties 3, 4 and 5, simplify any of the following expressions that can be simplified.

a. $(2^2)^3$ **b.** $(x^{2y})^4$ **c.** $(4a^2b)^2$ **d.** $\left(\frac{3x}{4y^2}\right)^3$

Solution

a. $(2^2)^3 = 2^{2 \cdot 3} = 2^6 = 64$

b. $(x^{2y})^4 = x^{(2y) \cdot 4} = x^{8y}$

c. $(4a^2b)^2 = 4^2 \cdot (a^2)^2 \cdot b^2 = 16a^{2 \cdot 2}b^2 = 16a^4b^2$

d. $\left(\frac{3x}{4y^2}\right)^3 = \frac{(3x)^3}{(4y^2)^3} = \frac{3^3 \cdot x^3}{4^3 \cdot (y^2)^3} = \frac{27x^3}{64y^{2 \cdot 3}} = \frac{27x^3}{64y^6}$

Zero and Negative Integer Exponents

In all the examples we have discussed so far, we have required that the exponents be positive integers. Let us now consider what meaning, if any, can be given to zero and negative integers as exponents. When we define these exponents, we shall want to do so in a way that is consistent with the five properties possessed by positive integer exponents.

Consider first the problem of using zero as an exponent. If x is any *nonzero* real number, we wish to attach some meaning to (*define*) the symbol x^0.

Since $x \neq 0$, we can look at the expression x^a/x^a, where a is any positive integer. Specifically, let's look at x^4/x^4.

By definition,

$$\frac{x^4}{x^4} = \frac{x \cdot x \cdot x \cdot x}{x \cdot x \cdot x \cdot x} = \frac{1}{1} = 1$$

since all of the xs in the numerator cancel with all of the xs in the denominator.

Now let us apply Property 2 to x^4/x^4. This gives us

$$\frac{x^4}{x^4} = x^{(4-4)} = x^0$$

Since $x^4/x^4 = 1$ and $x^4/x^4 = x^0$, consistency requires that we make the following definition:

ZERO AS AN EXPONENT

If x is any *nonzero* real number, then $x^0 = 1$.

Now we take on the task of defining a *negative integer exponent*. We use the fact that any negative integer (for example, -2) can be written as the difference of two positive integers (for example, $-2 = 3 - 5$). Let x be a nonzero real number and consider the quotient x^a/x^b, where a and b are positive integers and $a < b$ (read *a is less than b*). Specifically, let us consider the expression x^3/x^5. By definition,

$$\frac{x^3}{x^5} = \frac{x \cdot x \cdot x}{x \cdot x \cdot x \cdot x \cdot x} = \frac{1}{x \cdot x},$$

since the three xs in the numerator cancel with three of the xs in the denominator. Thus,

$$\frac{x^3}{x^5} = \frac{1}{x^2}$$

If we apply Property 2 to the expression x^3/x^5, we get

$$\frac{x^3}{x^5} = x^{(3-5)} = x^{-2}$$

If we are to be consistent, then we should *define* x^{-2} by

$$x^{-2} = \frac{1}{x^2} \text{ for } x \neq 0$$

This leads to the following general definition:

NEGATIVE EXPONENTS

If x is any *nonzero* real number and a is a positive integer, then

$$x^{-a} = \frac{1}{x^a}$$

We have defined negative integer exponents in terms of positive integer exponents in a way that preserves the properties of exponents. Therefore, all five properties hold for negative integer exponents as well:

PROPERTIES OF EXPONENTIATION

If x and y are nonzero real numbers and a and b are *any* integers, then

1. $x^a \cdot x^b = x^{a+b}$
2. $\frac{x^a}{x^b} = x^{a-b}$
3. $(x^a)^b = x^{a \cdot b}$
4. $(xy)^a = x^a y^a$
5. $\left(\frac{x}{y}\right)^a = \frac{x^a}{y^a}$

EXAMPLE 6

Evaluate each of the following expressions:

a. 2^{-2} **b.** 3^{-1} **c.** $(-2)^{-3}$ **d.** $\left(\frac{1}{4}\right)^{-2}$ **e.** $\left(\frac{17}{165}\right)^0$

Solution

We apply the definitions of negative exponents and zero as an exponent.

a. $2^{-2} = \frac{1}{2^2} = \frac{1}{2 \cdot 2} = \frac{1}{4}$

b. $3^{-1} = \frac{1}{3^1} = \frac{1}{3}$

Note: raising a nonzero number to the -1 power is the equivalent of taking the **reciprocal**, the multiplicative inverse, of that number.

c. $(-2)^{-3} = \frac{1}{(-2)^3} = \frac{1}{(-2)\cdot(-2)\cdot(-2)} = \frac{1}{-8} = -\frac{1}{8}$

d. $\left(\frac{1}{4}\right)^{-2} = \frac{1}{\left(\frac{1}{4}\right)^2} = \frac{1}{\frac{1^2}{4^2}} = \frac{1}{\frac{1}{16}} = 1 \cdot \frac{16}{1} = 16$

e. $\left(\frac{17}{165}\right)^0 = 1$ By the definition of zero as an exponent.

EXAMPLE 7

Use the five properties of exponentiation to simplify any of the following expressions that can be simplified:

a. $\frac{y^{-4}}{y^{-7}}$ b. $\frac{3^6}{3^8}$ c. $\frac{(5x^2)\cdot(-3x^4)^2}{x^{-3}\cdot(10x)^0}$ d. $\left(\frac{x^{2r}x^{r+3}}{x^{3r}}\right)^{-1}$

Solution

a. $\frac{y^{-4}}{y^{-7}} = y^{[-4-(-7)]}$ Property 2

$= y^{(-4+7)} = y^3$

b. $\frac{3^6}{3^8} = 3^{6-8}$ Property 2

$= 3^{-2} = \frac{1}{3^2}$ Definition of negative exponent

$= \frac{1}{9}$

c. $\frac{(5x^2)\cdot(-3x^4)^2}{x^{-3}\cdot(10x)^0} = \frac{5x^2\cdot[(-3)^2\cdot(x^4)^2]}{x^{-3}\cdot(10x)^0}$ Property 4

$= \frac{5x^2\cdot 9x^8}{x^{-3}\cdot(10x)^0}$ Property 3; $(-3)^2 = 9$, $(x^4)^2 = x^8$

$= \frac{45x^{10}}{x^{-3}(10x)^0}$ Property 1; $5\cdot 9 = 45$; $x^2\cdot x^8 = x^{10}$

$= \frac{45x^{10}}{x^{-3}\cdot 1}$ Definition of a^0; $(10x)^0 = 1$

$= 45x^{[10-(-3)]}$ Property 2

$= 45x^{13}$

d. $\left(\dfrac{x^{2r} \cdot x^{r+3}}{x^{3r}}\right)^{-1} = \left(\dfrac{x^{2r+(r+3)}}{x^{3r}}\right)^{-1}$ Property 1

$= \left(\dfrac{x^{3r+3}}{x^{3r}}\right)^{-1}$

$= (x^{(3r+3-3r)})^{-1}$ Property 2

$= (x^3)^{-1} = x^{-3}$ Property 3

$= \dfrac{1}{x^3}$ Definition of negative exponent

Fractional Exponents and Radicals

The final type of exponents we shall consider is fractions. We begin with fractions whose numerators are 1. Most people are familiar with the concept of a **square root**. The square roots of the number 16, for example, written $\pm\sqrt{16}$, are the numbers whose square are 16. That is $\pm\sqrt{16} = a$ if $a^2 = 16$ There are two candidates for $\pm\sqrt{16}$, $+4$ and -4. We shall be interested in the **principal root**, which is the positive one, since this is the root used in computer applications. Thus $\sqrt{16} = 4$.

In general, if n is a positive integer, we can define the ***n*th root** of a number x, denoted $\sqrt[n]{x}$, to be the number a with the property that $a^n = x$, if such a number exists. That is,

> **THE *n*TH ROOT**
> $\sqrt[n]{x} = a$ if and only if $a^n = x$.

Recall from Chapter 1 that, if n is *even*, $\sqrt[n]{x}$ is a real number only if x is zero or positive.

EXAMPLE 8

Evaluate each of the following:

a. $\sqrt{100}$ **b.** $\sqrt[3]{-8}$ **c.** $\sqrt[5]{x^{10}}$

Solution

a. $\sqrt{100} = 10$ since $10^2 = 100$.

b. $\sqrt[3]{-8} = -2$ since $(-2)^3 = -8$.

c. $\sqrt[5]{x^{10}} = x^2$ since $(x^2)^5 = x^{10}$.

For programming applications, it is useful to have a different notation for roots, since the radical symbol ($\sqrt{\ }$) does not appear on computer keyboards. Such a notation does exist in algebra; we substitute $x^{1/n}$ for $\sqrt[n]{x}$.

In BASIC, this would be written X^(1/N) or, for some computers, as X ** (1/N). The parentheses around (1/N) are important here.

Thus we have the following definition:

> If x is a real number and n is a nonzero integer, then
>
> $$x^{1/n} = a \text{ if and only if } a^n = x$$
>
> That is, a is the number whose nth power is x, if such a number exists.

According to Example 8, then, using this new notation

a. $100^{1/2} = 10$

b. $(-8)^{1/3} = -2$

c. $(x^{10})^{1/5} = x^2$ (Compare this with Property 3.)

Finally, we are ready to define fractional exponents in general.

> **FRACTIONAL EXPONENTS**
>
> If x is a nonzero real number and m and n are integers ($m > 0$ and $n \neq 0$,) then
>
> $$x^{m/n} = (x^{1/n})^m \qquad \text{or} \qquad x^{m/n} = (x^m)^{1/n}$$

In radical notation, this becomes

$$x^{m/n} = (\sqrt[n]{x})^m \qquad (\text{or } x^{m/n} = \sqrt[n]{x^m})$$

Fractional exponents obey all the rules for integer exponents.

EXAMPLE 9

Evaluate each of the following:

a. $(100)^{3/2}$ **b.** $(-8)^{2/3}$ **c.** $(x^{10})^{2/5}$ **d.** $(-1)^{2/3}$

Solution

a. $(100)^{3/2} = (100^{1/2})^3 = (10)^3 = 1{,}000$

Or, $(100)^{3/2} = (100^3)^{1/2} = (1{,}000{,}000)^{1/2} = 1{,}000$

b. $(-8)^{2/3} = ((-8)^{1/3})^2 = (-2)^2 = 4$

c. $(x^{10})^{2/5} = x^{(10 \cdot 2/5)} = x^{20/5} = x^4$ Here, we used Property 3.

d. $(-1)^{2/3} = [(-1)^{1/3}]^2 = (-1)^2 = 1$

EXAMPLE 10

Use properties 1 through 5 to simplify each of the following expressions:

a. $(y^{3/5})^{5/3}$ **b.** $x^{1/2} \cdot x^{3/4}$ **c.** $\dfrac{x^{5/6}}{x^{2/3}}$ **d.** $\left(\dfrac{y^4}{4}\right)^{1/2}$

Solution

a. $(y^{3/5})^{5/3} = y^{(3/5)\cdot(5/3)}$ Property 3

$= y^1 = y$

b. $x^{1/2} \cdot x^{3/4} = x^{(1/2)+(3/4)}$ Property 1

$= x^{5/4}$

c. $\dfrac{x^{5/6}}{x^{2/3}} = x^{(5/6)-(2/3)}$ Property 2

$= x^{1/6}$

d. $\left(\dfrac{y^4}{4}\right)^{1/2} = \dfrac{(y^4)^{1/2}}{4^{1/2}}$ Property 5

$= \dfrac{y^{4\cdot(1/2)}}{4^{1/2}}$ Property 3

$= \dfrac{y^2}{2}$

Note: Computer programmers must exercise care in using fractional exponents. In BASIC, for example, the statement to assign the value $\sqrt{4^3}$ to the variable

X would be: LET X = 4^(3/2). The parentheses here are *essential*; because of the rules for priority of operators the statement LET X = 4^3/2 would assign the value 64/2 = 32 to X instead of the desired value, 8.

EXERCISES FOR SECTION 2.1

In Exercises 1 through 12, find the value of the given expression (see Example 1).

1. 1^4 **2.** 2^5 **3.** $(-1)^7$

4. $(-3)^3$ **5.** 10^4 **6.** $(-100)^2$

7. $\left(\frac{1}{10}\right)^4$ **8.** $\left(\frac{-1}{5}\right)^3$ **9.** $\left(\frac{-3}{4}\right)^2$

10. $\left(\frac{2}{3}\right)^3$ **11.** $(0.5)^2$ **12.** $(-0.3)^3$

In Exercises 13 through 24, simplify any of the given expressions that can be simplified (see Example 2).

13. $4 \cdot 4^2$ **14.** $(-3)^3 \cdot (-3)$ **15.** $10^2 \cdot 10^3$

16. $(-2)^3 \cdot (-2)^5$ **17.** $2a^2 \cdot b^5$ **18.** $x^3 \cdot x^4$

19. $b \cdot b^5$ **20.** $-5x^2 \cdot y^3$ **21.** $(\frac{1}{2}a^2) \cdot (-6a)$

22. $(2b^2) \cdot (5b^2)$ **23.** $2^x \cdot 2^{3x}$ **24.** $3^{2a} \cdot 3^3$

In Exercises 25 through 36, simplify the given expression, if it can be simplified (see Example 3).

25. $\frac{3^7}{3^6}$ **26.** $\frac{(-10)^4}{(-10)^2}$ **27.** $\frac{(5)^{2a}}{5}$

28. $\frac{(-3)^{b+2}}{(-3)^2}$ **29.** $\frac{7x^3}{6y^2}$ **30.** $\frac{15a^5}{-5b^2}$

31. $\frac{54x^3}{12x^2}$ **32.** $\frac{3y^5}{6y^2}$ **33.** $\frac{12a^2b^3}{-4ab^2}$

34. $\frac{x^{2a}y^{3b}}{x^a y^b}$ **35.** $\frac{14x^7y^5}{21x^3y^2}$ **36.** $\frac{15a^2b^3}{5x}$

In Exercises 37 through 48, simplify the given expression if possible, using only positive exponents (see Examples 4 and 5).

37. $(x^2)^4$ **38.** $(a^3)^4$ **39.** $(3y^3)^2$

40. $(-5a^2)^3$ **41.** $(-2a^2b)^2$ **42.** $(-3xy^2)^3$

43. $\left(\dfrac{x^3y}{2}\right)^2$

44. $\left(\dfrac{xy^2}{z^3}\right)^4$

45. $\left(\dfrac{3a^2b}{-4z^3}\right)^3$

46. $\left(\dfrac{a^3b^2}{2c}\right)^2$

47. $\left(\dfrac{a^{2b}}{2c^{4x}}\right)^3$

48. $\left(\dfrac{-2x^a}{3y^{3b}}\right)^2$

In Exercises 49 through 60, find the value of the given expression (see Example 6).

49. 1^{-2}

50. 2^{-1}

51. $(-2)^{-3}$

52. $(-1)^{-2}$

53. $\left(\dfrac{1}{10}\right)^{-1}$

54. $\left(\dfrac{1}{2}\right)^{-1}$

55. $\left(\dfrac{19}{751}\right)^0$

56. $(0.1)^{-3}$

57. $\left(\dfrac{2}{3}\right)^{-2}$

58. $\left(-\dfrac{18}{61}\right)^0$

59. $(0.5)^{-2}$

60. $\left(-\dfrac{3}{7}\right)^{-3}$

In Exercises 61 through 72, simplify the given expression if possible using only positive exponents (see Example 7).

61. $\dfrac{2^4}{2^7}$

62. $\dfrac{3^{-2}}{3^{-4}}$

63. $\dfrac{x^{-4}}{y^{-2}}$

64. $\dfrac{y^4}{y^{-6}}$

65. $\left(\dfrac{a^2}{b^2}\right)^{-2}$

66. $\left(\dfrac{x}{y^3}\right)^{-3}$

67. $\left(\dfrac{2x^{-2}y^5 - 5}{3x^2y^{-6} + 1}\right)^0$

68. $\dfrac{x^{-1}y^{-2}}{x^{-3}y^{-4}}$

69. $\left(\dfrac{x^{-1}y^{-2}}{x^{-3}y^{-4}}\right)^{-1}$

70. $\left(\dfrac{3a^3b^{-5}}{6a^2b^{-2}}\right)^{-2}$

71. $\left[\dfrac{(2a^2)(3a^3)}{4a^{-1}}\right]^{-1}$

72. $\left(\dfrac{6a^2b^{-2}}{3a^3b^{-5}}\right)^0$

In Exercises 73 through 84, find the value of the given expression (see Examples 8 and 9).

73. $\sqrt[3]{1000}$

74. $\sqrt{25}$

75. $25^{1/2}$

76. $1000^{1/3}$

77. $(-1000)^{2/3}$

78. $(-1)^{5/7}$

79. $4^{3/2}$

80. $4^{-3/2}$

81. $(-32)^{3/5}$

82. $9^{-1/2}$

83. $9^{-3/2}$

84. $(4^2)^{3/2}$

In Exercises 85 through 96, simplify all expressions not already in their simplest forms (see Example 10).

85. $\left(\dfrac{x^6}{8}\right)^{1/3}$

86. $\left(\dfrac{4y^4}{9}\right)^{1/2}$

87. $\left(\dfrac{x^6}{8}\right)^{-1/3}$

88. $\left(\frac{4y^4}{9}\right)^{-1/2}$ **89.** $\frac{x^{3/4}}{x^{1/2}}$ **90.** $\frac{a^{-1/2}}{a^{-3/4}}$

91. $[(2x^2)^{1/2}]^2$ **92.** $[(-8a)^{1/3}]^3$ **93.** $[(-32a^{10})^{1/3}]^{3/5}$

94. $(5x^{-2})^{-3/2}$ **95.** $\left(\frac{x^2y^4z^6}{a^4b^{10}}\right)^{1/2}$ **96.** $\left(\frac{x^2y^4z^6}{a^4b^{10}}\right)^{-1/2}$

97. Amanda has been offered a part-time job after school. Her prospective employer has suggested two possible methods of paying her for her first week's work:

Method 1: Amanda is paid one dollar the first day, two dollars the second day, etc., her daily wages *doubling* each day

Method 2: She is paid ten dollars the first day and her daily wages are halfed on each succeeding day—\$10/2 on the second day, (\$10/2)/2 on the third day, etc.

If she works five days per week, which method gives her the higher wages for her first week on the job?

98. Suppose Amanda's boss (see Exercise 97) decides to pay her by the week, instead of by the day. He now proposes two new methods of payment, each starting with a salary of \$10 per week:

Method 1: Amanda's weekly salary will *double* each week.

Method 2: Each week's salary will be the *square* of the preceding week's.

Which method of payment should she choose?

2.2 Equations

First and Second Degree in One Variable

An **equation** is simply a statement that two quantities are equal. For example, the statement "$2 + 5 = 7$" is an equation. It states that the quantity $2 + 5$ and the quantity 7 are equal. In algebra, we are usually concerned with equations that contain one or more **variables**—equations like $2x + 3 = 5$. In this section, we discuss equations in *one variable*. We shall study methods of **solving** such equations; that is, methods of finding a value or values that can be substituted for the variable to make the equation a true statement.

First Degree Equations

Let's look first at **first degree equations** in one variable.

> First degree equations in one variable are of the form $ax + b = c$, where a, b and c are real numbers and $a \neq 0$.

A **solution** of such an equation is a value of the variable that, when substituted in the equation, makes it a true statement.

Consider, for example, the equation $2x + 3 = 5$. If we substitute the value 3 for x, we get:

$$2 \cdot 3 + 3 = 5 \qquad \text{or} \qquad 6 + 3 = 5$$

This is obviously not a true statement, so $x = 3$ is *not* a solution of the equation. However, suppose we substitute 1 for x, giving

$$2 \cdot 1 + 3 = 5 \qquad \text{or} \qquad 2 + 3 = 5$$

Since this *is* a true statement, $x = 1$ *is* a solution of the equation.

In attempting to find solutions for first degree equations, there are three situations that can occur.

1. The equation may have exactly *one solution.* Consider, for example, the equation $2x + 3 = 5$. We have seen that $x = 1$ is a solution of this equation. Furthermore, a little thought should convince us that 1 is the *only* solution.
2. The equation may have *no solution.* That is, no matter what value is substituted for the variable, the equation represents a *false* statement. The equation

   $$x + 1 = x + 2$$

   is an example of such a situation. Solving this equation would require finding a number with the following property: adding one to the number gives exactly the same result as adding two to the number. Common sense tells us that such a number does not exist. Therefore, the equation $x + 1 = x + 2$ has *no solution.*
3. The equation may have *infinitely many solutions. Any* value we substitute for the variable makes the equation a true statement. The equation

   $$x + 1 = (x + 3) - 2$$

   is an example of such an equation since the expressions on both sides of the equal sign represent the quantity $x + 1$. Thus, any value we substitute for x makes the equation a true statement.

Methods of Solving First Degree Equations

The easiest equation of the form

$$ax + b = c$$

to solve is one in which $a = 1$; that is, the equation looks like

$$x + b = c$$

In solving such an equation, our aim will be to keep it "balanced" while trying to get it in the form $x =$ a constant. If we add a quantity to one side of the equation, we must add the same quantity to the other side. (In this case, we'll want to add $-b$ to both sides giving us $x = c - b$.)

EXAMPLE 1

Find the solution of the equation $x + 7 = 10$.

Solution

Our aim is to get this equation into the form: $x =$ a constant. We can do this by adding -7 to both sides.

$x + 7 = 10$	Original equation
$(x + 7) + (-7) = 10 + (-7)$	Add -7 to both sides; equality is maintained.
$x = 3$	$7 + (-7) = 0$ and $10 + (-7) = 3$

The solution above uses a basic property of equality.

PROPERTIES OF EQUALITY

1. If the same quantity is *added* to both sides of an equation, equality is preserved. If $a = b$, then $a + c = b + c$.

Any first degree equation that has a solution can be solved by using a combination of this property, the basic properties of real numbers, and the following:

2. If both sides of an equation are *multiplied* by the same *nonzero* quantity, equality is preserved. If $a = b$ and $c \neq 0$, then $ac = bc$.

EXAMPLE 2

Find the solution of the equation $4y = 18$.

Solution

As in Example 1, we wish to substitute for the original equation, an equation of the form

$$y = \text{a constant}$$

This can be accomplished by multiplying both sides by $\frac{1}{4}$.

$4y = 18$	Original equation
$\frac{1}{4}(4y) = \frac{1}{4}(18)$	Multiply both sides by $\frac{1}{4}$ (Note: $\frac{1}{4}$ is the *reciprocal* of 4, the coefficient of y).
$y = 4.5$	$\frac{1}{4} \cdot 4 = 1$; $\frac{1}{4} \cdot 18 = 4.5$

To *check* the solution (that is, to verify that it is correct), we substitute 4.5 for y in the *original* equation:

$$4 \cdot 4.5 = 18$$
$$18.0 = 18 \quad \checkmark$$

Solving most first degree equations involves using a combination of these two properties. At each step in the solution process, we replace the original equation by one that is *equivalent* to it; that is, by one that has the *same solution.* If the equation has a unique solution, the end is reached when the equivalent equation has the form

$$\text{variable} = \text{a constant}$$

since the solution of such an equation is obvious.

The process of solving more complex first degree equations is best illustrated by some examples.

EXAMPLE 3

Solve all of the following first degree equations that have unique solutions:

a. $2x + 5 = -7$ **b.** $-3x + 7 = 5$ **c.** $2y - 5 = 3y + 2$

Solution

a.

$2x + 5 = -7$	
$2x + 5 - 5 = -7 - 5$	Add -5 to both sides.
$2x = -12$	
$\frac{1}{2}(2x) = \frac{1}{2}(-12)$	Multiply both sides by $\frac{1}{2}$.
$x = -6$	

Check

$$2 \cdot (-6) + 5 = -7$$
$$-12 + 5 = -7$$
$$-7 = -7 \quad \checkmark$$

b.
$$-3x + 7 = 5$$
$$-3x + 7 - 7 = 5 - 7 \qquad \text{Add } -7 \text{ to both sides.}$$
$$-3x = -2$$
$$-\tfrac{1}{3}(-3x) = -\tfrac{1}{3}(-2) \qquad \text{Multiply both sides by } -\tfrac{1}{3}\text{, the reciprocal of } -3.$$
$$x = \tfrac{2}{3}$$

Check

$$-3(\tfrac{2}{3}) + 7 = 5$$
$$-2 + 7 = 5$$
$$5 = 5 \quad \checkmark$$

c. $2y - 5 = 3y + 2$

Here, our aim is to move all occurrences of the variable to one side of the equal sign and all constants to the other side. We begin by adding 5 to both sides:

$$2y - 5 + 5 = 3y + 2 + 5$$
$$2y = 3y + 7$$

Now, add $-3y$ to both sides:

$$2y - 3y = -3y + 3y + 7$$
$$-1y = 7 \qquad \text{Combine like terms.}$$
$$-1(-1y) = -1(7) \qquad \text{Multiply both sides by } -1.$$
$$y = -7$$

Check

$$2(-7) - 5 = 3(-7) + 2$$
$$-14 - 5 = -21 + 2$$
$$-19 = -19 \quad \checkmark$$

EXAMPLE 4

Solve any of the following first degree equations that have unique solutions:

a. $5y - 2 + 4y = y + 12$ **b.** $5 + 2(7x - 4) = 7(2x - 3)$
c. $4(x - 3) + 5 = 2(3x - 1)$ **d.** $-12 + 4(z + 2) = 2(2z - 2)$

Solution

The first step here is to *simplify* both sides of the equation by removing parentheses (if necessary) and combining like terms. We can then use properties 1 and 2 to arrive at a solution, if one exists.

a. $5y - 2 + 4y = y + 12$

$9y - 2 = y + 12$	Combine like terms on the left side.
$9y = y + 14$	Add 2 to both sides.
$8y = 14$	Add $-y$ to both sides.
$y = (14)\frac{1}{8} = 1.75$	Multiply both sides by $\frac{1}{8}$.

b. $5 + 2(7x - 4) = 7(2x - 3)$

$5 + 14x - 8 = 14x - 21$	Remove parentheses on both sides.
$14x - 3 = 14x - 21$	Combine like terms on the left.
$14x = -18 + 14x$	Add 3 to both sides.
$0 = -18$	Add $-14x$ to both sides.

The last statement is obviously *false.* When we arrive at such a contradictory statement in attempting to solve an equation, then the original equation has *no solution.*

c. $4(x - 3) + 5 = 2(3x - 1)$

$4x - 12 + 5 = 6x - 2$	Remove parentheses on both sides.
$4x - 7 = 6x - 2$	Combine like terms on the left.
$4x = 6x + 5$	Add 7 to both sides.
$-2x = 5$	Add $-6x$ to both sides.
$x = 5(-\frac{1}{2}) = -2.5$	Multiply both sides by $-\frac{1}{2}$.

d. $-12 + 4(z + 2) = 2(2z - 2)$

$-12 + 4z + 8 = 4z - 4$	Remove parentheses.
$-4 + 4z = 4z - 4$	Combine like terms.
$4z = 4z$	Add 4 to both sides.

The statement above is *always true.* If we continue the normal method of solution, we add $-4z$ to both sides to obtain:

$$\mathbf{0 = 0}$$

Whenever we arrive at the equation $0 = 0$ (or any other equation—like $4z = 4z$—that is clearly always true), the original equation has *infinitely many solutions. Any* value we substitute for z makes the equation a true statement.

The reader should check each solution above to ensure that it is correct.

Second Degree Equations

Second degree equations in one variable represent the next step up in complexity.

> **SECOND DEGREE EQUATIONS**
> A second degree equation in one variable is an equation of the form
>
> $$ax^2 + bx + c = 0$$
>
> where a, b, and c are real numbers and $a \neq 0$.

There are several methods of solving such an equation. We shall consider first one that is strictly numerical, since such a method of solution is most readily used in computer programming applications. It requires only that we know the values of a, b and c.

If a second degree equation has any real number solution at all, it may have two solutions. If the equation is in the form

$$ax^2 + bx + c = 0$$

the solutions are given by the **quadratic formula**:

$$x_1 = \frac{-b + \sqrt{b^2 - 4ac}}{2a} \qquad (1)$$

$$x_2 = \frac{-b - \sqrt{b^2 - 4ac}}{2a} \qquad (2)$$

These equations are often written in shorthand form:

$$x = \frac{-b \pm \sqrt{b^2 - 4ac}}{2a}$$

This version of the quadratic formula is just a compact version of equations (1) and (2) above.

The quantity $b^2 - 4ac$ is called the **discriminant** of the second degree equation. It is this quantity that determines whether a given quadratic (that is, second degree) equation has one solution, two solutions, or no solution.

EXAMPLE 5

Find the discriminant in each of the following second degree equations:

a. $3x^2 - x - 2 = 0$ **b.** $4x^2 - 12x + 9 = 0$ **c.** $x^2 + x + 4 = 0$

Solution

a. Here, $a = 3$, $b = -1$, and $c = -2$. So,

$$\begin{aligned} b^2 - 4ac &= (-1)^2 - 4(3)(-2) \\ &= 1 - 12(-2) \\ &= 1 + 24 \\ &= 25 \quad \text{The discriminant is positive.} \end{aligned}$$

b. In this equation, $a = 4$, $b = -12$, and $c = 9$.

$$\begin{aligned} b^2 - 4ac &= (-12)^2 - 4(4)(9) \\ &= 144 - 16(9) \\ &= 144 - 144 \\ &= 0 \quad \text{The discriminant is zero.} \end{aligned}$$

c. This time, $a = 1$, $b = 1$, and $c = 4$, so that

$$\begin{aligned} b^2 - 4ac &= (1)^2 - 4(1)(4) \\ &= 1 - 16 \\ &= -15 \quad \text{The discriminant is negative.} \end{aligned}$$

Now let us use the quadratic formula to try to solve these same three equations.

EXAMPLE 6

Solve each of the second degree equations of Example 5.

Solution

We have already calculated $b^2 - 4ac$ in each case, so we can plug those results into the quadratic formula.

a. For $3x^2 - x - 2 = 0$, $a = 3$, $b = -1$, $c = -2$, and $b^2 - 4ac = 25$. Using the quadratic formula, we have

$$x_1 = \frac{-(-1) + \sqrt{25}}{2(3)} = \frac{1 + 5}{6} = \frac{6}{6}, \quad x_1 = 1$$

$$x_2 = \frac{-(-1) - \sqrt{25}}{2(3)} = \frac{1 - 5}{6} = \frac{-4}{6}, \quad x_2 = -\tfrac{2}{3}$$

The equation has *two solutions*: $x_1 = 1$ and $x_2 = -\frac{2}{3}$.

Check

Substitute each solution in the original equation.

For $x_1 = 1$

$$3(1)^2 - 1 - 2 = 0$$
$$3(1) - 1 - 2 = 0$$
$$3 - 3 = 0$$
$$0 = 0 \checkmark$$

$x_2 = -\frac{2}{3}$

$$3\left(-\frac{2}{3}\right)^2 - \left(-\frac{2}{3}\right) - 2 = 0$$
$$3\left(\frac{4}{9}\right) + \frac{2}{3} - 2 = 0$$
$$\frac{4}{3} + \frac{2}{3} - 2 = 0$$
$$\frac{6}{3} - 2 = 0$$
$$2 - 2 = 0$$
$$0 = 0 \checkmark$$

b. For $4x^2 - 12x + 9 = 0$, $a = 4$, $b = -12$, $c = 9$, and $b^2 - 4ac = 0$. Substituting in the quadratic formula gives

$$x_1 = \frac{-(-12) + \sqrt{0}}{2(4)} = \frac{12 + 0}{8} = \frac{12}{8} = \frac{3}{2}$$

So, $x_1 = \frac{3}{2}$

$$x_2 = \frac{-(-12) - \sqrt{0}}{8} = \frac{12 - 0}{8} = \frac{12}{8} = \frac{3}{2}$$

So, $x_2 = \frac{3}{2}$

Here, the two solutions obtained by applying the quadratic formula are identical. That is, the equation has only one solution, $x = \frac{3}{2}$ (or $x = 1.5$).

Check

$$4\left(\frac{3}{2}\right)^2 - 12\left(\frac{3}{2}\right) + 9 = 0$$
$$4\left(\frac{9}{4}\right) - 18 + 9 = 0$$
$$9 - 18 + 9 = 0$$
$$-9 + 9 = 0$$
$$0 = 0$$

c. For $x^2 + x + 4 = 0$, $a = 1$, $b = 1$, $c = 4$, and $b^2 - 4ac = -15$. Again, we substitute in the quadratic formula:

$$x_1 = \frac{-1 + \sqrt{-15}}{2(1)} \qquad x_2 = \frac{-1 - \sqrt{-15}}{2(1)}$$

In each case, we need to evaluate $\sqrt{-15}$. But there is no real number whose square is negative. Since we cannot find a real number value for $\sqrt{-15}$, we cannot solve the original equation. This equation has no solution in the real numbers.

We pause here to summarize Examples 5 and 6. We obtained the following results:

In part a, the discriminant ($b^2 - 4ac$) was *positive*, and the equation had *two* solutions.
In part b, the discriminant was *zero*, and the equation had *one solution.*
In part c, the discriminant was *negative*, and the equation had *no solution* in the real numbers.

These results hold in the general case. That is:

The second degree equation $ax^2 + bx + c = 0$ has

two solutions when $b^2 - 4ac > 0$,
one solution when $b^2 - 4ac = 0$, and
no solution when $b^2 - 4ac < 0$.

We conclude this section with another example.

EXAMPLE 7

Solve any of the following second degree equations that have solutions:

a. $4x^2 = 4x - 1$ **b.** $x^2 + x = -1$ **c.** $(x - 2) \cdot (3x + 1) = 0$

Solution

In each case, we must first apply the techniques of the first part of this section (on first degree equations) to rewrite the given equation in the form $ax^2 + bx + c = 0$.

a. $4x^2 = 4x - 1$ becomes

$$4x^2 - 4x + 1 = 0 \qquad \text{Add } -4x + 1 \text{ to both sides.}$$

Now apply the quadratic formula. We have $a = 4$, $b = -4$, and $c = 1$.

$$b^2 - 4ac = (-4)^2 - 4(4)(1) = 16 - 16 = 0$$

There is one solution:

$$x = \frac{-(-4) + \sqrt{0}}{2(4)}$$

$$x = \frac{4}{8}$$

$$x = \tfrac{1}{2}$$

b. $x^2 + x = -1$ becomes

$$x^2 + x + 1 = 0 \qquad \text{Add 1 to both sides.}$$

Here, $a = 1$, $b = 1$, and $c = 1$.

$$\begin{aligned} b^2 - 4ac &= (1)^2 - 4(1)(1) \\ &= 1 - 4 \\ &= -3 \end{aligned}$$

Since $b^2 - 4ac$ is negative, the equation has *no solution.*

c. $(x - 2) \cdot (3x + 1) = 0$

Simplify the left side of the equation by removing parentheses and combining like terms.

$$\begin{aligned} x(3x + 1) - 2(3x + 1) &= 0 \\ 3x^2 + x - 6x - 2 &= 0 \\ 3x^2 - 5x - 2 &= 0 \end{aligned}$$

In this equation, $a = 3$, $b = -5$, and $c = -2$.

$$\begin{aligned} b^2 - 4ac &= (-5)^2 - 4(3)(-2) \\ &= 25 - 12(-2) \\ &= 25 + 24 \\ &= 49 \end{aligned}$$

There are two solutions:

$$x_1 = \frac{-(-5) + \sqrt{49}}{2(3)} \qquad\qquad x_2 = \frac{-(-5) - \sqrt{49}}{2(3)}$$

$$= \frac{5 + 7}{6} = \frac{12}{6} \qquad\qquad = \frac{5 - 7}{6} = \frac{-2}{6}$$

$$x_1 = 2 \qquad\qquad x_2 = -\tfrac{1}{3}$$

The reader is urged to check the results of parts a and c by substituting the solutions in the *original* equations.

EXERCISES FOR SECTION 2.2

In Exercises 1 through 15, find the solution of the given first degree equation (see Examples 1–3).

1. $x - 3 = 3$
2. $x + 4 = 2$
3. $x + 1 = -2$
4. $x - 1 = -3$
5. $5y = 20$
6. $3y = -12$
7. $-4y = 16$
8. $-5y = -10$
9. $4x + 1 = 9$
10. $3x - 2 = 1$
11. $-5x - 3 = -18$
12. $-6x + 1 = 15$
13. $-2x - 3 = 6$
14. $3x + 4 = x$
15. $7x + 10 = 36 - 6x$

In Exercises 16 through 29, find the solution of the given first degree equation if a unique solution exists. If a unique solution does *not* exist, state whether there are infinitely many solutions or no solutions (see Example 4).

16. $\frac{1}{5}(15x + 10) = 0$
17. $\frac{1}{10}(10x - 1) = 0$
18. $2(3x - 1) = 3x$
19. $-4(2x + 3) = 4x$
20. $-4(2x + 3) = 4$
21. $2(3x - 1) = 6$
22. $2x + 3 = 5(x - 1)$
23. $5x - 3 = 5(x - \frac{3}{5})$
24. $3(x - 4) = 5(2x + 1) - 3$
25. $\frac{1}{2}(2x + 4) = \frac{1}{3}(3x - 9)$
26. $4y = \frac{1}{2}(8y + 9)$
27. $3(2x - 1) + 4 = \frac{1}{6}(36x + 6)$
28. $2(2x - 1) - 3(x + 5) = 3(4x - 10) + 2$
29. $2(3x - 4) + 5 = -9 - 4(5x + 2)$

In Exercises 30 through 41, find the discriminant for the given second degree equation (see Example 5).

30. $6x^2 + 5x - 4 = 0$
31. $2x^2 + 3x - 2 = 0$
32. $x^2 - 6x + 9 = 0$
33. $16x^2 - 40x + 25 = 0$
34. $20x^2 + 13x - 21 = 0$
35. $-9x^2 + 18x - 5 = 0$
36. $x^2 - 4 = 0$ (Hint: $b = 0$)
37. $9x^2 - 16 = 0$
38. $4x^2 + 12x + 9 = 0$
39. $9x^2 + 12x + 4 = 0$
40. $2x^2 + 11x + 12 = 0$
41. $2x^2 + 10x + 12 = 0$

In Exercises 42 through 53, complete the solutions for any of the equations in Exercises 30 through 41 that have solutions (see Example 6).

In Exercises 54 through 61, solve the given quadratic equation, if possible (see Example 7).

54. $-6x = -(9x^2 + 1)$

55. $x^2 = 5(2x - 5)$

56. $6x^2 = 2 - x$

57. $4x^2 = 3 - 11x$

58. $2x^2 - 1 = 2(x - 1)$

59. $3x(x + 8) = 2(x - 20) + 5$

60. $2x(x + 3) = -4x(x - 3) - 2(x - 21)$

61. $x(2x + 1) - 2(5x - 4) = 3(x - 4)$

62. If the discriminant $b^2 - 4ac$ in the quadratic formula is zero, we have seen that the second degree equation, $ax^2 + bx + c = 0$, has one root. Using the quadratic formula, verify that this root is given by the following equation:

$$x = \frac{-b}{2a}$$

63. Write a BASIC program that accepts as input the values of a, b and c from a second degree equation and determines the solution or solutions of the equation, if a solution exists. (Hint: you must eliminate the possibility that $b^2 - 4ac < 0$ before attempting to apply the quadratic formula.)

2.3 Factoring Second Degree Polynomials

Solving second degree, or **quadratic**, equations by using the quadratic formula is both useful and reliable. This formula enables us to calculate the solution or solutions when they exist and to identify equations that have no real number solution. In some instances, however, the calculations can become formidable. Consider, for example, the quadratic equation: $7x^2 - 21x - 196 = 0$. Here, $a = 7$, $b = -21$, $c = -196$; so using the quadratic formula to solve this equation involves calculating the discriminant

$$b^2 - 4ac = (-21)^2 - 4(7)(-196)$$

Once we find this number, its square root will have to be evaluated! Factoring quadratic expressions provides another, and sometimes simpler, approach to solving quadratic equations.

Factors of Integers

When two or more integers are multiplied, the answer is referred to as the **product** of the given numbers. The numbers that were multiplied to get the product are called **factors** of the product. For example, if we multiply $7 \cdot 8$ to get 56, then 56 is the *product* and 7 and 8 are *factors* of 56. Note that 7 and 8 are not the *only* factors of 56—56 can also be written as $2 \cdot 28$ or $4 \cdot 14$. This means that 2, 28, 4, and 14 are also factors of 56. The general definition of a factor is as follows:

FACTORS

If I is an integer, the integer a is a factor of I if there is another integer b with the property that $I = a \cdot b$.

Another way of stating that a is a factor of I is to say that I is exactly divisible by a (that is, a "goes into" I exactly, with no remainder).

EXAMPLE 1

Find *all* the positive integer factors of each of the following numbers:

a. 28 **b.** 42 **c.** 392 **d.** 13

Solution

a. Since every number is exactly divisible by 1 and itself, 1 and 28 are factors of 28. The other factors are 2, 14, 4, and 7 since $28 = 2 \cdot 14$ and $28 = 4 \cdot 7$.

b. To find all the factors of 42, write it as a product of two positive integers in as many ways as possible:

$$42 = 6 \cdot 7$$
$$42 = 2 \cdot 21$$
$$42 = 3 \cdot 14$$
$$42 = 1 \cdot 42$$

The factors of 42 are 1, 2, 3, 6, 7, 14, 21, and 42.

c. $392 = 2 \cdot 196$
$392 = 4 \cdot 98$
$392 = 7 \cdot 56$
$392 = 8 \cdot 49$
$392 = 14 \cdot 28$
$392 = 1 \cdot 392$

The factors of 392 are 1, 2, 4, 7, 8, 14, 28, 49, 56, 98, 196, and 392.

d. 13 is evenly divisible only by 1 and itself. So 1 and 13 are the *only* factors of 13. Such a number, whose only factors are 1 and itself, is called a **prime** number.

Factoring Quadratic Expressions: Finding a Common Factor

A quadratic expression, or **quadratic polynomial**, is simply an algebraic expression of the form $ax^2 + bx + c$ (x can be replaced by *any* variable name). It looks like the left side of a quadratic equation in standard form. Factoring quadratic polynomials (writing them as products of simpler expressions) is useful in several algebraic applications, among them finding solutions of quadratic equations.

In factoring quadratic polynomials, we shall try to do one or both of the following:

1. Write the quadratic polynomial as a constant times a quadratic; and/or
2. write the quadratic polynomial as the product of two first degree expressions.

Let's consider the first of these tasks.

It is sometimes necessary to determine whether two integers have any **common factors**—any numbers that are factors of both. For example, in applications involving reducing fractions (Section 2.4), we are interested in finding the **greatest common factor**, abbreviated **GCF**, of two integers. Before applying these concepts to polynomials, let's examine them in a numerical example.

EXAMPLE 2

Find the GCF of each of the following pairs of integers:

a. 42 and 392 **b.** 28 and 392 **c.** 13 and 28

Solution

Use the results of Example 1 to list the factors of each number.

a. The factors are

42: 1, 2, 3, 6, 7, 14, 21, 42
392: 1, 2, 4, 7, 8, 14, 28, 49, 56, 98, 196, 392

The largest number that appears in *both* lists is the GCF.

GCF = 14

b. The factors are

28: 1, 2, 4, 7, 14, 28
392: 1, 2, 4, 7, 8, 14, 28, 49, 56, 98, 196, 392
GCF = 28

c. The factors are

13: 1, 13
28: 1, 2, 4, 7, 14, 28

The *only* common factor is 1, so

GCF = 1

Numbers like 13 and 28, whose GCF is 1 are said to be **relatively prime**.

Now let us apply these concepts to the task of factoring a quadratic polynomial of the form $ax^2 + bx + c$. The "pieces" (ax^2, bx and c) into which this expression is divided by the + signs are called the **terms** of the polynomial. The constants a, b and c are the **numerical coefficients** of these terms. The first step in factoring a quadratic polynomial is to find the GCF of the numerical coefficients. If it is not 1, we then "factor it out." An example will illustrate this process.

EXAMPLE 3

Begin the factoring process for each of the following quadratic polynomials by finding the GCF of the numerical coefficients and "factoring it out" when possible.

a. $7x^2 - 21x - 196$ **b.** $3x^2 + 27x + 60$
c. $3x^2 + 10x - 8$ **d.** $12x^2 - 6x - 18$

Solution

a. $7x^2 - 21x - 196$

factors of 7: 1, 7
factors of 21: 1, 3, 7, 21
factors of 196: 1, 2, 4, 7, 14, 28, 49, 98, 196

The GCF = 7, and $7 = 7 \cdot 1$, $21 = 7 \cdot 3$, $196 = 7 \cdot 28$. So we can write

$$\begin{aligned} 7x^2 - 21x - 196 &= (7 \cdot 1)x^2 - (7 \cdot 3)x - (7 \cdot 28) \\ &= 7 \cdot (1x^2) + 7 \cdot (-3x) + 7 \cdot (-28) \end{aligned}$$

"Factoring out" the GCF gives

$$7x^2 - 21x - 196 = 7(x^2 - 3x - 28)$$

(Note that this is a reverse application of the distributive property.)

b. $3x^2 + 27x + 60$

factors of 3: 1, 3

Let's analyze this result. Since 1 and 3 are the *only* factors of 3 (3 is *prime*), they are also the only factors 3 can possibly have in common with 27 and 60 (the other numerical coefficients). So we need only test 27 and 60 for divisibility by 3. (Note: the same considerations apply to the factor 7 in part a.) Since $27 = 3 \cdot 9$ and $60 = 3 \cdot 20$, 3 *is* a common factor of 27 and 60 and hence the desired GCF. This gives us

$$3x^2 + 27x + 60 = 3(x^2 + 9x + 20)$$

c. $3x^2 + 10x - 8$

As in part b, 3 is prime, so we must test 10 and 8 for divisibility by 3. Since 3 is a factor of neither 10 nor 8, the GCF is 1, and we cannot factor out a common factor.

d. $12x^2 - 6x - 18$

factors of 12: 1, 2, 3, 4, 6, 12

factors of 6: 1, 2, 3, 6

factors of 18: 1, 2, 3, 6, 9, 18

The GCF = 6, so

$$12x^2 - 6x - 18 = 6(2x^2 - x - 3)$$

Sometimes we must deal with an expression that is not itself a quadratic, but can be written as the product of a quadratic and another single-term, or **monomial**, factor. Consider, for example, the expression $30x^4 + 35x^3 - 15x^2$. This is clearly *not* a quadratic, but we shall attempt to simplify it by using the following three-step approach:

1. Find the GCF of the numerical coefficients.
2. Extract any variable that appears in *all* terms and give it the *smallest* exponent it has in the given expression.

3. Form the product of the GCF of the numerical coefficients with all variables (raised to appropriate powers) found in step 2; "factor out" this term from the original polynomial.

We apply this method to $30x^4 + 35x^3 - 15x^2$.

1. The GCF of 30, 35 and 15 is 5.
2. The only variable that appears in all three terms is x; its smallest exponent is 2.
3. "Factor out" $5x^2$:

$$30x^4 + 35x^3 - 15x^2 = 5x^2(6x^2 + 7x - 3)$$

The expression we factored out (in this case, $5x^2$) is referred to as a **common monomial factor**.

EXAMPLE 4

For each of the following polynomials, factor out any common monomial factor:

a. $6xy^3z^2 - 21xy^2z^2 + 15xyz^2$ **b.** $x^3y^3 - 9xy^3$
c. $3a^2b^3x^2 + 2xy^2 - 11$

Solution

a. $6xy^3z^2 - 21xy^2z^2 + 15xyz^2$

1. The GCF of 6, 21 and 15 is 3.
2. The variables x, y and z appear in all three terms with smallest exponents 1, 1 and 2 respectively.
3. Factoring out $3xyz^2$ gives
$6xy^3z^2 - 21xy^2z^2 + 15xyz^2 = 3xyz^2(2y^2 - 7y + 5)$

b. $x^3y^3 - 9xy^3$

1. The GCF of 1 and 9 is 1.
2. The variables x and y appear in both terms, with smallest exponents of 1 and 3.
3. Factoring out $1xy^3$ (or simply xy^3) gives:
$x^3y^3 - 9xy^3 = xy^3(x^2 - 9)$

c. $3a^2b^3x^2 + 2xy^2 - 11$

1. The GCF of 3, 2 and 11 is 1.
2. *No* variable appears in all three terms.

Therefore, there is no common monomial factor.

Complete Factoring of a Quadratic Polynomial

Finding a common monomial factor, if one exists, is just the first step in factoring a quadratic polynomial. Completing the task requires a second step—writing the quadratic polynomial as the product of two first degree expressions, if possible. To accomplish this, first look at a common algorithm (method) for *multiplying* two first degree expressions.

Suppose we wish to multiply $(x + 4) \cdot (2x - 3)$. We can apply the distributive property as follows:

$$\begin{aligned}(x + 4)(2x - 3) &= x(2x - 3) + 4(2x - 3) \\ &= x(2x) + x(-3) + 4(2x) + 4(-3)\end{aligned}$$

If we assign the names *first term*, *last term*, *inner term* and *outer term* as follows, then the multiplication method can be called by the acronym **FOIL**.

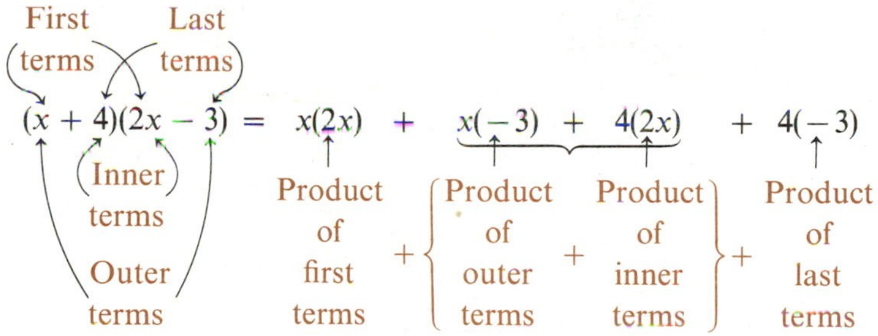

First terms, Last terms, Inner terms, Outer terms

$$(x + 4)(2x - 3) = x(2x) + x(-3) + 4(2x) + 4(-3)$$

Product of first terms + {Product of outer terms + Product of inner terms} + Product of last terms

$$\begin{aligned}&= 2x^2 + (-3x + 8x) - 12 \\ &= 2x^2 + 5x - 12\end{aligned}$$

THE FOIL METHOD

The result of multiplying two first degree expressions by the FOIL method is a three-term polynomial.

1. The *first term* is the product of the first terms of the individual factors.
2. The *second term* is obtained as follows: add the product of the outer terms to the product of the inner terms.
3. The *third term* is the product of the last (that is, second) terms of the individual factors.

EXAMPLE 5

Use the FOIL method to find the following products:

a. $(x + 3)(x - 4)$ **b.** $(3x + 4)(6x + 8)$ **c.** $(5x + 2)(5x - 2)$

Solution

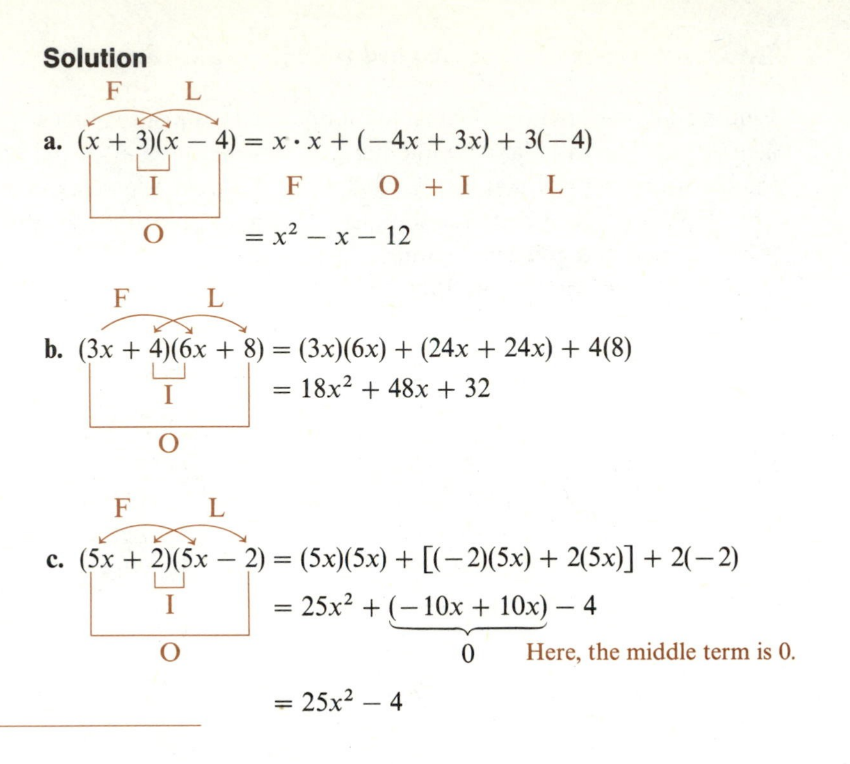

To factor a quadratic polynomial, we use the FOIL method in reverse. Given a three-term quadratic, we want to find two *linear* (first degree) factors. Remember to check first for a common monomial factor.

Consider, for example, the quadratic polynomial $7x^2 - 21x - 196$ of Example 3. We have already found the common monomial factor (the number 7, since no variable appears in all three terms). So we have

$$7x^2 - 21x - 196 = 7(x^2 - 3x - 28)$$

To complete the process, we must look for factors of the polynomial in parentheses, $x^2 - 3x - 28$. We want to fill in this outline:

$$(\square \overset{+}{\underset{-}{\text{or}}} \triangle)(\bigcirc \overset{+}{\underset{-}{\text{or}}} \square) = x^2 - 3x - 28$$

The FOIL method indicates that x^2 must be the product of the first terms, so we can begin to fill in the outline (since $x^2 = x \cdot x$).

$$(x \overset{+}{\underset{-}{\text{or}}} \triangle)(x \overset{+}{\underset{-}{\text{or}}} \square) = x^2 - 3x - 28$$

Further analysis of the FOIL method yields $\triangle \times \square = 28$. That is, the remaining terms must be pairs of factors of 28. The candidates are 1 and 28, 2 and 14, 4 and 7 (with the factors of each pair having opposite signs since we are actually dealing with -28). The pair we choose will be determined by the fact that the middle term of the product must be $-3x$. Trial and error shows that 4 and -7 are the correct choices for last terms (demonstrate this for yourself). So

$$x^2 - 3x - 28 = (x + 4)(x - 7)$$

and

$$7x^2 - 21x - 196 = 7(x + 4)(x - 7)$$

EXAMPLE 6

Factor the following quadratic polynomials completely:

a. $3x^2 + 27x + 60$ **b.** $3x^2 + 10x - 8$

c. $6xy^3z^2 - 21xy^2z^2 + 15xyz^2$ **d.** $x^3y^3 - 9xy^3$

Solution

a. We follow the two-step procedure described earlier to factor polynomials.

1. The common monomial factor in $3x^2 + 27x + 60$ is 3 (see Example 3).

$$3x^2 + 27x + 60 = 3(x^2 + 9x + 20)$$

2. Using the reverse of the FOIL method to factor $x^2 + 9x + 20$, we have

$$\begin{aligned} 3x^2 + 27x + 60 &= 3(x^2 + 9x + 20) \\ &= 3(x + \_\_)(x + \_\_) \end{aligned}$$

The product of the first terms must be x^2.

We need factors of 20 to fill these two blank spots:

$$20 = 1 \times 20 = 2 \times 10 = 4 \times 5$$

Using 4 and 5 as factors gives the desired middle term. So,

$$\begin{aligned} 3x^2 + 27x + 60 &= 3(x^2 + 9x + 20) \\ &= 3(x + 4)(x + 5) \end{aligned}$$

b. For $3x^2 + 10x - 8$, there is *no* common factor (see Example 3); so we proceed to step 2. The first terms of the **binomials** (two-term expressions) must be factors of $3x^2$. The only candidates are $3x$ and $1x$. So

$$3x^2 + 10x - 8 = (3x + \_\_)(x + \_\_)$$

The factors of 8 are $8 = 1 \times 8 = 2 \times 4$. Since our third term is -8, we must choose *opposite* signs for the last terms. The correct factorization is

$$3x^2 + 10x - 8 = (3x - 2)(x + 4).$$

c. $6xy^3z^2 - 21xy^2z^2 + 15xyz^2 = 3xyz^2(2y^2 - 7y + 5)$ See Example 4.

$= 3xyz^2(2y + \_\_)(y + \_\_)$ Since $2y^2 = 2y \cdot y$

$= 3xyz^2(2y - 5)(y - 1)$ Since $5 = 5 \cdot 1$

The choice of two negative signs here is determined by the facts that they have to be the same (the third term of the quadratic is positive) and that the middle term of the quadratic is negative.

d. $x^3y^3 - 9xy^3 = xy^3(x^2 - 9)$ See Example 4.

$= xy^3(x^2 + 0x - 9)$ (We "filled in" a middle term.)

$= xy^3(x + \_\_)(x + \_\_)$ $[x \cdot x = x^2]$

$= xy^3(x + 3)(x - 3)$ $9 = 3 \cdot 3$ and the signs must be opposite.

The reader should verify that each of the above factorizations is correct by performing the multiplications to arrive at the original quadratic.

Part d of Example 6 illustrates a special type of quadratic. After extracting the common monomial factor, we were left with the problem of factoring $x^2 - 9$. This quadratic is missing an x term; also, the first and last terms are perfect squares—$x^2 = (x)^2$ and $9 = 3^2$. A quadratic of this form $(a^2 - b^2)$ is often referred to as the *difference of two squares* and can be factored according to the following rule:

$$a^2 - b^2 = (a + b)(a - b)$$

EXAMPLE 7

Factor each of the following:

a. $y^2 - 100$ **b.** $5x^2 - 80$ **c.** $16x^2 - 100$

Solution

a. $y^2 - 100 = (y)^2 - (10)^2 = (y + 10)(y - 10)$

b. $5x^2 - 80 = 5(x^2 - 16)$ Be sure to watch for a common monomial factor!

$= 5[(x)^2 - (4)^2]$

$= 5(x + 4)(x - 4)$

c. $16x^2 - 100 = 4(4x^2 - 25)$

$= 4[(2x)^2 - (5)^2]$

$= 4(2x + 5)(2x - 5)$

Using Factoring to Solve Quadratic Equations

Factoring quadratic expressions provides us with an alternative tool for solving quadratic equations. In order to use this tool, we first make an observation. Suppose we are given the fact that the product of two numbers is zero. Then it is reasonable to conclude (in fact, it is a property of the real number system) that *at least one* of the numbers must be zero.

Apply this observation to the problem, mentioned at the beginning of this section, of solving the quadratic equation

$$7x^2 - 21x - 196 = 0$$

We have already factored the polynomial on the left side of this equation as follows:

$$7x^2 - 21x - 196 = 7(x + 4)(x - 7)$$

For the original equation, then, we may substitute

$$7(x + 4)(x - 7) = 0$$

The left side represents a product of three factors. We are looking for values of x that make this product equal to zero. But this implies that at least one of the factors must be 0. Since $7 \neq 0$, we must have

$$x + 4 = 0 \qquad \text{or} \qquad x - 7 = 0$$

That is,

$$x = -4 \qquad \text{or} \qquad x = 7$$

These are the *solutions* of the original quadratic equation. You can verify this by substituting them in the original equation.

The use of factoring to solve a quadratic equation involves the following steps:

SOLVING QUADRATIC EQUATIONS BY FACTORING

1. Write the equation in the form $ax^2 + bx + c = 0$.
2. Factor the left side of the equation, if possible; if not, try the quadratic formula.
3. Set each factor equal to 0, and solve the resulting linear equations.
4. Check the results in the *original* quadratic equation.

EXAMPLE 8

The owner of the local Computer Heaven store has determined that, if he sells c home computers per day, his daily profit will be given by the formula

$$\text{profit} = c^2 + 5c - 50$$

a. How many computers must he sell in a day to "break even" (profit = 0)?
b. How many computers must he sell per day to earn a daily profit of $250?

Solution

a. To determine the break-even point, we must solve the quadratic equation

$$c^2 + 5c - 50 = 0$$

We shall follow the steps outlined above.

1. $c^2 + 5c - 50 = 0$
2. $(c + 10)(c - 5) = 0$
3. $c + 10 = 0$; $c + 10 - 10 = 0 - 10$; $c = -10$ | $c - 5 = 0$; $c - 5 + 5 = 0 + 5$; $c = 5$

$$\begin{aligned} c + 10 &= 0 \\ c + 10 - 10 &= 0 - 10 \\ c &= -10 \end{aligned} \qquad \Bigg| \qquad \begin{aligned} c - 5 &= 0 \\ c - 5 + 5 &= 0 + 5 \\ c &= 5 \end{aligned}$$

Since we are talking about computer sales, we want only the *positive* solution (he can't sell -10 computers). Therefore, the store must sell five computers to break even.

4. **Check:**

$$\begin{aligned} (5)^2 + 5(5) - 50 &= 0 \\ 25 + 25 - 50 &= 0 \\ 50 - 50 &= 0 \\ 0 &= 0 \quad \checkmark \end{aligned}$$

b. To find the sales required to earn $250 profit we must solve the equation

$$c^2 + 5c - 50 = 250$$

1. $c^2 + 5c - 50 - 250 = 250 - 250$

$$c^2 + 5c - 300 = 0$$

2. $(c + 20)(c - 15) = 0$
3.

$$\begin{aligned} c + 20 &= 0 \\ c + 20 - 20 &= 0 - 20 \\ c &= -20 \end{aligned} \qquad \Bigg| \qquad \begin{aligned} c - 15 &= 0 \\ c - 15 + 15 &= 0 + 15 \\ c &= 15 \end{aligned}$$

Again, we use the positive solution. He must sell 15 computers.

4. **Check:** $(15)^2 + 5(15) - 50 = 250$

$$225 + 75 - 50 = 250$$

$$300 - 50 = 250$$

$$250 = 250 \quad \checkmark$$

EXERCISES FOR SECTION 2.3

In Exercises 1 through 10, find the greatest common factor (GCF) of the given group of numbers (see Examples 1 and 2).

1. 4, 18
2. 18, 25
3. 60, 108
4. 11, 35
5. 7, 21, 63
6. 12, 108, 144
7. 32, 160, 192
8. 54, 60, 108
9. 31, 63, 108
10. 63, 84, 189

In Exercises 11 through 24, factor out any common monomial factor from the given expression (see Examples 3 and 4).

11. $40x^2 - 250$
12. $6a^2b - 21ac^2 + 30$
13. $3y^2 - 7y - 20$
14. $7a^2y^3 - 63a^2x^2$
15. $15x^3y^2 + 30y^2z$
16. $9x^2 + 6x - 8$
17. $25a^2b + 4ac^2$
18. $-10a^3x - 9bx^3$
19. $4x^2y + 48xy - 112y$
20. $6xy^2 - 12xy + 6x$
21. $15x^2y^2z^2 - 30xy^2z + 3y^2z$
22. $a^3b^2c + 5a^2bc^2 - 4a^2c$
23. $50x^3y^2 + 140x^2y^2 + 98xy^2$
24. $-34x^2y^2z^3 - 51xy^2z^2 + 17x^2z^2$

In Exercises 25 through 36, use the FOIL method to find the indicated products (see Example 5).

25. $(x - 1)(x + 5)$
26. $(x + 2)(x - 3)$
27. $(2x + 1)(3x - 2)$
28. $(3x + 1)(2x - 2)$
29. $(7x + 1)(7x - 1)$
30. $(5x - 4)^2$
31. $(2x + 3)^2$
32. $(4x + 9)(4x - 9)$
33. $(-6x + 7)(x - 5)$
34. $(-4x + 9)(2x - 5)$
35. $(3x + 2)(3x - 2)$
36. $(6x + 1)(5x + 3)$

In Exercises 37 through 50, factor each expression completely (see Examples 6 and 7).

37. $4x^2 - 25$
38. $x^2 - x - 6$
39. $x^2 + 6x + 5$
40. $25x^2 - 36$
41. $3y^2 - 7y - 20$
42. $27x^2 + 18x - 24$

43. $30x^2 - 5x - 10$

44. $27x^2 - 243$

45. $4x^2 + 48x - 112$

46. $6y^2 - 12y + 6$

47. $100x^2 - 16$

48. $12x^2 + 24x - 96$

49. $9x^2y^2 + 30xy^2 + 25y^2$

50. $35a^2x^2 + 31a^2x + 6a^2$

In Exercises 51 through 64, use factoring to solve the given quadratic equation (see Example 8).

51. $6x^2 - x - 2 = 0$

52. $3y^2 - 7y - 20 = 0$

53. $36x^2 - 25 = 0$

54. $40x^2 - 250 = 0$

55. $3x^2 + 12x = 15$

56. $9x^2 = 30x - 25$

57. $7x^2 = 112$

58. $4(x^2 + 2x) = 4(2x + 16)$

59. $4(x^2 + 2x) = 32$

60. $2x(x + 1) = 40$

61. $6(x^2 - 2) = -x$

62. $3x^2 = 10x + 25$

63. $2x^2 = 13x + 7$

64. $(2x + 1)^2 = 9$

65. The concepts of this section can be used to do a sort of "mental" arithmetic. Suppose, for example, we wish to calculate $104 \cdot 96$. First, we notice that $104 = 100 + 4$ and $96 = 100 - 4$. So we can find $104 \cdot 96$ if we can calculate $(100 + 4)(100 - 4)$. Compare this with Example 7:

$$\begin{aligned}(100 + 4)(100 - 4) &= (100)^2 - (4)^2 \\ &= 10000 - 16 \\ &= 9984\end{aligned}$$

So $104 \cdot 96 = 9984$.
Use this short-cut to perform the following calculations:
a. $101 \cdot 99$ **b.** $52 \cdot 48$

2.4 Algebraic Fractions

A rational number (a numeric fraction) is any number that can be expressed as the quotient of two integers, where the divisor, which is the number on the bottom, or the **denominator**, is nonzero. An **algebraic fraction**, or **rational expression**, is an expression that is the quotient of two algebraic expressions. In this section, we shall study the "arithmetic" of algebraic fractions—how to add, subtract, multiply, divide, and reduce them. In all cases, we shall assume that any value of a variable that would cause a denominator to become zero is excluded, since division by zero is an undefined operation (see Section 1.1.).

Reducing Algebraic Fractions

Let's consider first how one would go about "reducing" a numeric fraction—for example, $\frac{32}{72}$. To reduce such a fraction, we divide the numerator (the top number) and the denominator (the bottom number) by the *same* nonzero factor, if we can find such a number. The resulting fraction is *equivalent* to (represents the same number as) the original fraction. To save time and effort, we try to divide by the GCF of the numerator and denominator. If the GCF is one, the fraction cannot be reduced further; otherwise, we divide both numbers by the GCF.

The task of reducing $\frac{32}{72}$ to *lowest terms* (so that the new numerator and denominator have no common factors), then, involves two steps:

1. Find the GCF of the numerator and denominator.

   factors of 32: 1, 2, 4, 8, 16, 32

   factors of 72: 1, 2, 3, 4, 6, 8, 9, 12, 18, 24, 36, 72

   The GCF is 8.
2. Divide numerator and denominator by the GCF to obtain an equivalent fraction.

$$\frac{32}{72} = \frac{32/8}{72/8} = \frac{4}{9}$$

The fraction is now reduced to lowest terms since 4 and 9 have no common factors.

Exactly the same procedure is followed to reduce algebraic fractions:

REDUCING ALGEBRAIC FRACTIONS

1. Factor the numerator and denominator, if possible, to find the GCF.
2. Divide both numerator and denominator by the GCF; this is often referred to as "canceling" common factors.

EXAMPLE 1

Reduce each of the following algebraic fractions to lowest terms:

a. $\dfrac{6x^4y^7}{27x^2y^8}$ **b.** $\dfrac{x^3 + 7x^2}{x^3 + 9x^2 + 14x}$ **c.** $\dfrac{x^2 - 64}{3x - 24}$ **d.** $\dfrac{3x^2 - 7x + 4}{x^2 - 1}$

Solution

a. $\dfrac{6x^4y^7}{27x^2y^8}$

We shall use the two-step procedure for reducing algebraic fractions.

1. Here, the GCF is $3x^2y^7$.

2. $$\frac{\text{numerator}}{\text{GCF}} = \frac{6x^4y^7}{3x^2y^7} = 2x^2$$

$$\frac{\text{denominator}}{\text{GCF}} = \frac{27x^2y^8}{3x^2y^7} = 9y$$

This gives the final result

$$\frac{6x^4y^7}{27x^2y^8} = \frac{2x^2}{9y}$$

b. $$\frac{x^3 + 7x^2}{x^3 + 9x^2 + 14x} = \frac{x^2(x + 7)}{x(x^2 + 9x + 14)}$$ Factor out common monomial factors.

$$= \frac{x^2(x + 7)}{x(x + 2)(x + 7)}$$ Complete factorization of denominator.

$$= \frac{x}{x + 2}$$ Now "cancel" the common factors; GCF $= x(x + 7)$.

since $\dfrac{x^2(x + 7)}{x(x + 7)} = x$ and $\dfrac{x(x + 2)(x + 7)}{x(x + 7)} = x + 2$.

c. $$\frac{x^2 - 64}{3x - 24} = \frac{(x)^2 - (8)^2}{3(x - 8)}$$

$$= \frac{(x + 8)\cancel{(x - 8)}}{3\cancel{(x - 8)}}$$ Factored forms

$$= \frac{x + 8}{3}$$ Cancel common factor of $x - 8$.

Note that dividing numerator and denominator by their GCF is equivalent to "canceling" (crossing out or eliminating) any factors the numerator and denominator have in common.

d. $$\frac{3x^2 - 7x + 4}{x^2 - 1} = \frac{(3x - 4)\cancel{(x - 1)}}{(x + 1)\cancel{(x - 1)}}$$ Factor completely.

$$= \frac{3x - 4}{x + 1}$$ Cancel common factor of $x - 1$.

Multiplying and Dividing Algebraic Fractions

The rule for multiplying numeric fractions is:

MULTIPLYING NUMERIC FRACTIONS

$$\frac{a}{b} \cdot \frac{c}{d} = \frac{a \cdot c}{b \cdot d}$$

For example,

$$\frac{8}{9} \cdot \frac{6}{7} = \frac{8 \cdot 6}{9 \cdot 7} = \frac{48}{63} \left(= \frac{16}{21}, \text{ if we reduce the product} \right)$$

A similar rule applies for multiplication of algebraic fractions: *the product of two rational expressions is defined to be the product of the numerators over the product of the denominators.* With algebraic fractions, it is advisable (if we want to reduce the product to lowest terms) to factor the numerators and denominators before beginning the multiplication.

EXAMPLE 2

Perform the following multiplications, reducing all products to lowest terms.

a. $\frac{7x}{5} \cdot \frac{3}{x^3}$

b. $\frac{5x - 15}{x + 1} \cdot \frac{x^2 - 1}{10x + 20}$

c. $\frac{5x - 10}{x + 2} \cdot \frac{6x + 12}{7x - 14}$

d. $\frac{x^2 - 7x + 12}{x^2 - 16} \cdot \frac{2x + 8}{x^2 - 2x - 3}$

Solution

a. $\frac{7x}{5} \cdot \frac{3}{x^3} = \frac{(7x)(3)}{5(x^3)} = \frac{21x}{5x^3} = \frac{21}{5x^2}$ Divide numerator and denominator by x.

b.
$$\frac{5x - 15}{x + 1} \cdot \frac{x^2 - 1}{10x + 20} = \frac{5(x - 3)}{x + 1} \cdot \frac{(x + 1)(x - 1)}{10(x + 2)}$$

$$= \frac{\cancel{5}(x - 3)\cancel{(x + 1)}(x - 1)}{\underset{2}{\cancel{10}}\cancel{(x + 1)}(x + 2)}$$

$$= \frac{(x - 3)(x - 1)}{2(x + 2)}$$

Cancel common factors; GCF $= 5(x + 1)$

c. $\dfrac{5x-10}{x+2}\cdot\dfrac{6x+12}{7x-14}=\dfrac{5(x-2)}{x+2}\cdot\dfrac{6(x+2)}{7(x-2)}$

$$=\frac{(5)(6)\cancel{(x-2)}\cancel{(x+2)}}{7\cancel{(x+2)}\cancel{(x-2)}} \qquad \text{GCF} = (x+2)(x-2)$$

$$=\frac{30}{7}$$

d. $\dfrac{x^2-7x+12}{x^2-16}\cdot\dfrac{2x+8}{x^2-2x-3}$

$$=\frac{(x-4)(x-3)}{(x-4)(x+4)}\cdot\frac{2(x+4)}{(x-3)(x+1)}$$

$$=\frac{2\cancel{(x-4)}\cancel{(x-3)}\cancel{(x+4)}}{\cancel{(x-4)}\cancel{(x+4)}\cancel{(x-3)}(x+1)} \qquad \text{GCF} = (x-4)(x-3)(x+4)$$

$$=\frac{2}{x+1}$$

Division of algebraic fractions is handled in the same way as division of numeric fractions: the problem is converted to a multiplication problem by inverting (taking the reciprocal of) the divisor. Consider, for example, the division problem $\frac{4}{5} \div \frac{7}{10}$. We can write this as a complex fraction, a fraction whose numerator or denominator contains a fraction.

$$\frac{\frac{4}{5}}{\frac{7}{10}}$$

Suppose we multiply the numerator and denominator of *this* fraction by $\frac{10}{7}$; the new fraction will be equivalent to the old one (and thus to the original division problem).

$$\frac{4}{5} \div \frac{7}{10} = \frac{\frac{4}{5}}{\frac{7}{10}}\cdot\frac{\frac{10}{7}}{\frac{10}{7}} = \frac{\frac{4}{5}\cdot\frac{10}{7}}{\frac{7}{10}\cdot\frac{10}{7}}$$

Examine the denominator of this new fraction:

$$\frac{7}{10}\cdot\frac{10}{7}=\frac{7\cdot 10}{10\cdot 7}=\frac{70}{70}=1$$

So

$$\frac{4}{5}\div\frac{7}{10}=\frac{\frac{4}{5}}{\frac{7}{10}}=\frac{4}{5}\;\frac{10}{7}=\frac{4\cdot 10}{5\cdot 7}=\frac{40}{35}=\frac{8}{7}$$

↑ This fraction ($\frac{10}{7}$) is the *reciprocal* of the original divisor.

Similarly, any problem involving division of algebraic fractions can be solved by converting it to a multiplication problem. That is,

DIVISION OF ALGEBRAIC FRACTIONS

$$\frac{a}{b} \div \frac{c}{d} = \frac{a}{b} \cdot \frac{d}{c}$$

EXAMPLE 3

Perform the following divisions:

a. $\frac{x^2}{yz^2} \div \frac{x}{y^2z}$ **b.** $\frac{18x^2 - 3x - 6}{2x^4 + 3x^3 - 5x^2} \div \frac{12x + 6}{2x^4 + 5x^3}$

Solution

a.
$$\frac{x^2}{yz^2} \div \frac{x}{y^2z} = \frac{x^2}{yz^2} \cdot \frac{y^2z}{x}$$
$$= \frac{x^2y^2z}{xyz^2} = \frac{xy}{z}$$

b.
$$\frac{18x^2 - 3x - 6}{2x^4 + 3x^3 - 5x^2} \div \frac{12x + 6}{2x^4 + 5x^3} = \frac{18x^2 - 3x - 6}{2x^4 + 3x^3 - 5x^2} \cdot \frac{2x^4 + 5x^3}{12x + 6}$$
$$= \frac{3(6x^2 - x - 2)}{x^2(2x^2 + 3x - 5)} \cdot \frac{x^3(2x + 5)}{6(2x + 1)}$$
$$= \frac{3(3x - 2)(2x + 1)}{x^2(2x + 5)(x - 1)} \cdot \frac{x^3(2x + 5)}{6(2x + 1)}$$
$$= \frac{3x^3(3x - 2)\cancel{(2x + 1)}\cancel{(2x + 5)}}{6x^2\cancel{(2x + 5)}(x - 1)\cancel{(2x + 1)}}$$
$$= \frac{x(3x - 2)}{2(x - 1)}$$

Adding and Subtracting Algebraic Fractions

Adding fractions is a bit more complicated than multiplying them because the method most commonly used requires that the fractions added have the same denominator. Suppose, for example, we wish to add the fractions $\frac{7}{15}$ and $\frac{11}{18}$. The procedure we shall follow is this three-step one:

1. Find a number that is a multiple of both denominators—in this case, a number that has both 15 and 18 as factors. The smallest such number, called the **least common multiple** or **LCM**, is 180 ($180 = 12 \cdot 15$ and $180 = 10 \cdot 18$).
2. Convert each of the original fractions to a fraction with 180 (the LCM of the original denominators) as denominator.

$$\frac{7}{15} = \frac{7}{15} \cdot \frac{12}{12} = \frac{84}{180}$$ Multiply top and bottom by $\frac{180}{15}$.

$$\frac{11}{18} = \frac{11}{18} \cdot \frac{10}{10} = \frac{110}{180}$$ We multiplied 18 by 10; so we multiply 11 by 10 also.

3. The sum of the original fractions is

$$\frac{\text{sum of new numerators}}{\text{new denominator}}$$

That is,

$$\frac{7}{15} + \frac{11}{18} = \frac{84}{180} + \frac{110}{180} = \frac{84 + 110}{180}$$

$$= \frac{194}{180} = \frac{97}{90}$$ Reduced to lowest terms

Now we apply a similar method to addition of algebraic fractions. Consider, for example, the sum

$$\frac{3}{x^2 + 6x + 9} + \frac{2x}{x^2 + 8x + 15}$$

We find the sum as follows:

1. Factor both numerators and denominators completely.

$$\frac{3}{x^2 + 6x + 9} + \frac{2x}{x^2 + 8x + 15}$$

$$= \frac{3}{(x + 3)(x + 3)} + \frac{2x}{(x + 5)(x + 3)}$$

$$= \frac{3}{(x + 3)^2} + \frac{2x}{(x + 5)(x + 3)}$$

2. Find the LCM of the denominators by forming the product of all factors that appear in any denominator; give each one the *largest* exponent it has in a denominator. Here, the LCM is $(x + 3)^2(x + 5)$.

3. Convert each fraction to a fraction with the LCM as denominator by multiplying both top and bottom by the same quantity.

$$\frac{3}{(x+3)^2} = \frac{3}{(x+3)^2} \cdot \frac{x+5}{x+5} = \frac{3(x+5)}{(x+3)^2(x+5)}$$

$$\frac{2x}{(x+5)(x+3)} = \frac{2x}{(x+5)(x+3)} \cdot \frac{x+3}{x+3}$$

$$= \frac{2x(x+3)}{(x+3)^2(x+5)}$$

4. Finally, add the resulting fractions by adding the numerators and using the LCM as denominator. Simplify the numerator, and reduce the sum if possible.

$$\frac{3}{x^2+6x+9} + \frac{2x}{x^2+8x+15}$$

$$= \frac{3}{(x+3)^2} + \frac{2x}{(x+3)(x+5)}$$

$$= \frac{3(x+5)}{(x+3)^2(x+5)} + \frac{2x(x+3)}{(x+3)^2(x+5)}$$

$$= \frac{3(x+5)+2x(x+3)}{(x+3)^2(x+5)}$$

$$= \frac{3x+15+2x^2+6x}{(x+3)^2(x+5)}$$ Remove parentheses in the numerator.

$$= \frac{2x^2+9x+15}{(x+3)^2(x+5)}$$ Combine like terms in the numerator.

This sum cannot be reduced since the numerator cannot be factored.

EXAMPLE 4

Perform the indicated operations.

a. $\frac{4}{x^3y} + \frac{3}{x^2y^2z}$ **b.** $\frac{1}{y^2+2y} + \frac{1}{2y+4}$ **c.** $\frac{2x+3}{x^2+3x+2} - \frac{x+1}{2x^2+7x+6}$

Solution

a. $\frac{4}{x^3y} + \frac{3}{x^2y^2z}$

1. Factoring is not necessary here.

2. LCM $= x^3y^2z$

3. $\dfrac{4}{x^3y} = \dfrac{4}{x^3y} \cdot \dfrac{yz}{yz} = \dfrac{4yz}{x^3y^2z}$

$\dfrac{3}{x^2y^2z} = \dfrac{3}{x^2y^2z} \cdot \dfrac{x}{x} = \dfrac{3x}{x^3y^2z}$

4. $\dfrac{4}{x^3y} + \dfrac{3}{x^2y^2z} = \dfrac{4yz}{x^3y^2z} + \dfrac{3x}{x^3y^2z} = \dfrac{4yz + 3x}{x^3y^2z}$

b. $\dfrac{1}{y^2 + 2y} + \dfrac{1}{2y + 4}$

1. $\dfrac{1}{y^2 + 2y} = \dfrac{1}{y(y + 2)}$

$\dfrac{1}{2y + 4} = \dfrac{1}{2(y + 2)}$

2. LCM $= 2y(y + 2)$

3. $\dfrac{1}{y(y + 2)} = \dfrac{1}{y(y + 2)} \cdot \dfrac{2}{2} = \dfrac{2}{2y(y + 2)}$

$\dfrac{1}{2(y + 2)} = \dfrac{1}{2(y + 2)} \cdot \dfrac{y}{y} = \dfrac{y}{2y(y + 2)}$

4. $\dfrac{1}{y^2 + 2y} + \dfrac{1}{2y + 4} = \dfrac{2}{2y(y + 2)} + \dfrac{y}{2y(y + 2)}$

$= \dfrac{2 + y}{2y(y + 2)} = \dfrac{\cancel{y + 2}}{2y\cancel{(y + 2)}} = \dfrac{1}{2y}$

c. $\dfrac{2x + 3}{x^2 + 3x + 2} - \dfrac{x + 1}{2x^2 + 7x + 6}$

Note that to *subtract* fractions, we simply subtract the second numerator from the first in step 4.

1. $\dfrac{2x + 3}{x^2 + 3x + 2} = \dfrac{2x + 3}{(x + 1)(x + 2)}$

$\dfrac{x + 1}{2x^2 + 7x + 6} = \dfrac{x + 1}{(2x + 3)(x + 2)}$

2. LCM $= (x+1)(x+2)(2x+3)$

3. $$\frac{2x+3}{(x+1)(x+2)} = \frac{2x+3}{(x+1)(x+2)} \cdot \frac{2x+3}{2x+3} = \frac{(2x+3)(2x+3)}{(x+1)(x+2)(2x+3)}$$

$$\frac{x+1}{(2x+3)(x+2)} = \frac{x+1}{(2x+3)(x+2)} \cdot \frac{x+1}{x+1} = \frac{(x+1)(x+1)}{(x+1)(x+2)(2x+3)}$$

4. $$\frac{2x+3}{x^2+3x+2} - \frac{x+1}{2x^2+7x+6} = \frac{(2x+3)(2x+3)}{(x+1)(x+2)(2x+3)} - \frac{(x+1)(x+1)}{(x+1)(x+2)(2x+3)}$$

$$= \frac{(2x+3)(2x+3) - (x+1)(x+1)}{(x+1)(x+2)(2x+3)}$$

$$= \frac{4x^2+12x+9-(x^2+2x+1)}{(x+1)(x+2)(2x+3)}$$

$$= \frac{4x^2+12x+9-x^2-2x-1}{(x+1)(x+2)(2x+3)}$$

$$= \frac{3x^2+10x+8}{(x+1)(x+2)(2x+3)}$$

$$= \frac{(3x+4)\cancel{(x+2)}}{(x+1)\cancel{(x+2)}(2x+3)} = \frac{3x+4}{(x+1)(2x+3)}$$

EXERCISES FOR SECTION 2.4

In Exercises 1 through 20, reduce each fraction to lowest terms (see Example 1).

1. $\dfrac{2xy^2}{3x^2y}$ **2.** $\dfrac{5a^2b^3}{a^3b^2}$ **3.** $\dfrac{21x^2y^3}{14y^2}$

4. $\dfrac{12x^3y^2z}{30x^2y^2z^2}$ **5.** $\dfrac{3x^2}{x^2+x}$ **6.** $\dfrac{y^3+4y^2}{2y}$

7. $\dfrac{2x-2y}{6xy}$ **8.** $\dfrac{3x^2-9x}{xy}$ **9.** $\dfrac{5x^2+15x}{-5x}$

10. $\dfrac{x+3}{x^2-9}$ **11.** $\dfrac{5x^2-20}{2x^2+4x}$ **12.** $\dfrac{5x^2-20}{x^3-2x^2}$

13. $\dfrac{x^2+3x}{x^2-3x}$ **14.** $\dfrac{x^3-3x^2}{x+3}$ **15.** $\dfrac{2x^2+5x-3}{x^2-9}$

16. $\dfrac{2x^2 - 9x + 4}{36x^2 - 9}$

17. $\dfrac{7x^2 + 42x + 49}{7x^3 + 21x^2}$

18. $\dfrac{3x^2 - 7x + 4}{5x^2 - 5}$

19. $\dfrac{9x^2 - 16}{3x^2 - 7x + 4}$

20. $\dfrac{9x^3 - 12x^2 + 3x}{18x^2 - 12x - 6}$

In Exercises 21 through 40, perform the indicated operation, expressing the answer in lowest terms (see Examples 2 and 3).

21. $\dfrac{2y}{3x} \cdot \dfrac{3x^2y}{2}$

22. $\dfrac{x^2}{y^2} \cdot \dfrac{y^2z}{2xy}$

23. $\dfrac{6x}{5} \cdot \dfrac{10}{x^3}$

24. $\dfrac{12x^3y^2}{-7} \cdot \dfrac{4}{-3x^2y}$

25. $\dfrac{2x + 4}{3x - 3} \cdot \dfrac{9x - 9}{4x - 8}$

26. $\dfrac{6x^2 + 9x}{2x - 4} \cdot \dfrac{x - 2}{3x}$

27. $\dfrac{x^2 - 2x - 8}{x^2 - 4} \cdot \dfrac{x^2 + 3x - 10}{x^2 + x - 20}$

28. $\dfrac{3x^2 - 3x - 6}{5x^2 + 7x - 6} \cdot \dfrac{7x + 14}{x^2 - x - 2}$

29. $\dfrac{6x^2 + 13x + 6}{6x - 12} \cdot \dfrac{3x^2 - 3x - 6}{4x^2 + 12x + 9}$

30. $\dfrac{x^2 + 1}{3x + 4} \cdot \dfrac{9x^2 - 16}{3x^3 - 4x^2}$

31. $\dfrac{6y}{4x} \div \dfrac{3x^2y}{2}$

32. $\dfrac{z^2}{x} \div \dfrac{y^2z}{3xy}$

33. $\dfrac{3x}{4y^2z} \div \dfrac{6x^3}{5z^2}$

34. $\dfrac{-6x^3}{5z^2} \div \dfrac{3x}{4y^2z}$

35. $\dfrac{3x^2 - 9x}{15x^3} \div \dfrac{3x}{5x^2}$

36. $\dfrac{x^2 - 9}{4} \div \dfrac{x^2 + 6x + 9}{6}$

37. $\dfrac{a^2 - a}{3b^2 - 12} \div \dfrac{2a^2}{b + 2}$

38. $\dfrac{15xy - 5y}{3x + 1} \div \dfrac{27x^2y^2 - 3y^2}{3x^2 - 2x - 1}$

39. $\dfrac{16y^2 - 4}{4y^2 - 6y + 2} \div \dfrac{4y^2 + 6y + 2}{y - 1}$

40. $\dfrac{3x^2 + 2x - 1}{x^2 - 1} \div \dfrac{6x^2 - 8x + 2}{x^2 - 2x + 1}$

In Exercises 41 through 60, perform the indicated operation, expressing the answer in lowest terms (see Example 4).

41. $\dfrac{2x}{3y} + \dfrac{5}{3y}$

42. $\dfrac{3}{z} + \dfrac{7y}{z}$

43. $\dfrac{2x}{x - 3} - \dfrac{6}{x - 3}$

44. $\dfrac{3y}{y + 6} + \dfrac{18}{y + 6}$

45. $\frac{3}{x} + \frac{3}{4x}$

46. $\frac{7x}{y} - \frac{2x}{3y}$

47. $\frac{4}{x^2 + 4x + 4} + \frac{x - 2}{x + 2}$

48. $\frac{x}{x^2 + 7x + 6} - \frac{3}{x + 6}$

49. $\frac{x + 5}{x - 7} + \frac{x + 7}{x - 5}$

50. $\frac{x - 4}{x + 6} - \frac{x - 6}{x + 4}$

51. $\frac{4}{x^2 + 4x + 4} - \frac{x - 2}{2x + 4}$

52. $\frac{x^2 + 1}{x^2 - 1} - \frac{x}{x - 1}$

53. $\frac{x}{x^2 - 25} + \frac{5}{x^2 + 10x + 25}$

54. $\frac{x}{2x + 8} - \frac{x}{2x - 8}$

55. $\frac{x + 1}{x^2 - x - 2} - \frac{x - 2}{x^2 + x - 6}$

56. $\frac{x + 1}{x^2 - x - 2} + \frac{x - 2}{x^2 + x - 6}$

57. $\frac{4}{x^2 + 2x + 4} + \frac{x + 2}{x - 2}$

58. $\frac{y + 3}{y - 9} + \frac{3}{y^2 - 3y + 9}$

59. $x^2 + x + \frac{1}{x}$

$\left(\text{Hint: write } x^2 \text{ as } \frac{x^2}{1} \text{ and } x \text{ as } \frac{x}{1}.\right)$

60. $y^3 - y^2 + \frac{2}{y^2}$

2.5 Solving Equations That Contain Fractions

One reason for studying algebraic fractions is that many real-life situations give rise to equations that contain fractions. Imagine, for example, the following situation: Tim has borrowed his father's new foreign sports car and is driving along a state highway with a posted speed limit of 55 miles per hour. He is stopped by a police officer, who says that Tim has been speeding. Unfortunately, Tim isn't sure whether or not this is true because the car's speedometer registered 90 *kilometers* per hour, and his brain knows only *miles* per hour. Tim would know whether or not to claim innocence if he could convert kilometers per hour to miles per hour.

Luckily for Tim, this problem is easy to solve, given the fact that 1 mile is approximately equal to 1.6 kilometers. Our problem is to convert 90 kilometers to some number of miles. We'll do this by setting up a **proportion,** an equation of the form

$$\frac{a}{b} = \frac{c}{d}$$

In this case, we'll use

$$\frac{\text{miles}}{\text{kilometers}} = \frac{\text{miles}}{\text{kilometers}}$$

or

$$\frac{1 \text{ mile}}{1.6 \text{ kilometers}} = \frac{x \text{ miles}}{90 \text{ kilometers}}$$

In doing this, we must be aware that, when a, b, c, and d are numbers representing measurements, we must keep like units of measure in the same relative positions. Once we have set up the proportion, we can then omit the units of measure. Our equation becomes

$$\frac{1}{1.6} = \frac{x}{90}$$

This is the simplest type of fractional equation to solve. We use the **principle of equality of fractions**:

EQUALITY OF FRACTIONS

The fractions a/b and c/d are equal, that is

$$\frac{a}{b} = \frac{c}{d}$$

if and only if $a \cdot d = b \cdot c$.

Many algebra textbooks refer to this technique as *cross-multiplication.*

Let us apply the principle of equality of fractions to Tim's problem.

$$\frac{1}{1.6} = \frac{x}{90}$$

$$1.6 \cdot x = 1 \cdot 90$$

$$x = 90/1.6$$

$$x = 56.25$$

That is, Tim's speed was 56.25 miles per hour—he *was* exceeding the speed limit.

EXAMPLE 1

Use the principle of equality of fractions (cross-multiplication) to solve any of the following equations that have solutions:

a. $\frac{3}{4} = \frac{x}{42}$ **b.** $\frac{3}{x} = \frac{5}{13}$ **c.** $\frac{5}{2y + 1} = \frac{2}{y + 3}$ **d.** $\frac{x}{x - 3} = \frac{3}{x - 3}$

Solution

a. $\frac{3}{4} \times \frac{x}{42}$

$4x = 3 \cdot 42$ Cross-multiply

$4x = 126$

$x = 126/4$ Divide both sides by 4.

$x = 31.5$

Check

$$\frac{3}{4} = \frac{31.5}{42}$$

$$0.75 = 0.75 \checkmark$$

b. $\frac{3}{x} \times \frac{5}{13}$

$5x = 3 \cdot 13$

$5x = 39$

$x = 7.8$

Check

$$\frac{3}{7.8} = \frac{5}{13}$$

$$3 \cdot 13 = 5 \cdot 7.8$$

$$39 = 39 \checkmark$$

c. $\frac{5}{2y + 1} \times \frac{2}{y + 3}$

$5(y + 3) = 2(2y + 1)$

$5y + 15 = 4y + 2$

$5y = 4y - 13$

$y = -13$

Check

$$\frac{5}{2(-13)+1} = \frac{2}{-13+3}$$

$$\frac{5}{-25} = \frac{2}{-10}$$

$$\frac{-1}{5} = \frac{-1}{5} \quad \checkmark$$

d.

$$\frac{x}{x-3} = \frac{3}{x-3}$$

$$x(x-3) = 3(x-3)$$

$$x^2 - 3x = 3x - 9$$

$$x^2 - 3x - 3x + 9 = 3x - 9 - 3x + 9$$

$$x^2 - 6x + 9 = 0$$

$$(x-3)(x-3) = 0 \qquad \text{Solve by factoring.}$$

$$x - 3 = 0$$

$$x = 3$$

Check

$$\frac{3}{3-3} = \frac{3}{3-3}$$

$$\frac{3}{0} = \frac{3}{0}$$

Division by zero is impossible, so the "solution" is *invalid.*

This equation has *no solution.*

Example 1d illustrates a potential problem in dealing with equations that involve fractions. It is possible to arrive at a "solution" that seems correct but that, when substituted in the original equation, causes a denominator to equal zero. For this reason, *it is essential to check all solutions* of equations containing fractions *in the original equation.*

Not all algebraic equations that contain fractions are quite this simple to deal with. Suppose, for example, that we want to solve the equation:

$$\frac{x}{5} + \frac{x-2}{3} = 2$$

One would suspect that this equation would be easier to solve if we could "get rid of" the fractions—that is, substitute an equivalent equation containing no fractions. We begin by finding the least common multiple of the denominators of *all* the fractions in the equation. In this case, the denominators are 5 and 3; their LCM is 15. Next, we multiply both sides of the equation by the LCM of the denominators. This gives us

$$15\left(\frac{x}{5} + \frac{x-2}{3}\right) = 15 \cdot 2$$

$$15\left(\frac{x}{5}\right) + 15\left(\frac{x-2}{3}\right) = 15 \cdot 2$$

(Recall that the distributive property says that we must multiply *each* term on the left by 15.) Now we solve the resulting equation:

$$\begin{aligned} 3x + 5(x-2) &= 30 \\ 3x + 5x - 10 &= 30 \\ 8x - 10 &= 30 \\ 8x &= 40 \\ x &= 5 \end{aligned}$$

Finally, we check the solution in the original equation:

$$\begin{aligned} 5/5 + (5-2)/3 &= 2 \\ 1 + 3/3 &= 2 \\ 1 + 1 &= 2 \\ 2 &= 2 \quad \checkmark \end{aligned}$$

The solution checks, so $x = 5$ is the solution of the original equation.

Let us pause to summarize the method we have just used. To solve an equation containing fractions, we follow this four-step procedure:

SOLVING AN EQUATION THAT CONTAINS FRACTIONS

1. Find the least common multiple of all the denominators (this may require factoring some denominators).
2. Multiply both sides of the equation (i.e., *every term*) by the LCM of the denominators.
3. Solve the resulting equation.
4. Check all potential solutions in the *original* equation.

EXAMPLE 2

Solve the following equations:

a. $\dfrac{1}{x} + \dfrac{3}{x^2} = 2$ **b.** $\dfrac{2x}{x+2} - \dfrac{2}{x^2-4} = \dfrac{4}{x-2}$

Solution

a. Using the four steps for solving equations that contain fractions, we have

1. $\frac{1}{x} + \frac{3}{x^2} = 2$ Denominators: x, x^2; LCM $= x^2$

2. $\frac{x^2}{1} \cdot \frac{1}{x} + \frac{x^2}{1} \cdot \frac{3}{x^2} = x^2 \cdot 2$

$$\frac{x^2}{x} + \frac{3x^2}{x^2} = 2x^2$$

$$x + 3 = 2x^2$$

3. $2x^2 = x + 3$

$$2x^2 - x - 3 = 0$$

$$(2x - 3)(x + 1) = 0 \quad \text{Solve by factoring.}$$

$2x - 3 = 0$	$x + 1 = 0$
$2x = 3$	$x = -1$
$x = \frac{3}{2}$	

The possible solutions are

$$x = \tfrac{3}{2} \quad \text{and} \quad x = -1$$

4. **Check**

for $x = \frac{3}{2}$	for $x = -1$
$\frac{1}{\frac{3}{2}} + \frac{3}{(\frac{3}{2})^2} = 2$	$\frac{1}{-1} + \frac{3}{(-1)^2} = 2$
$\frac{2}{3} + \frac{3}{\frac{9}{4}} = 2$	$-1 + \frac{3}{1} = 2$
$\frac{2}{3} + \frac{4}{3} = 2$	$-1 + 3 = 2$
$\frac{6}{3} = 2$	$2 = 2$ ✓
$2 = 2$ ✓	

Both solutions check, so the solutions are $x = \frac{3}{2}$ and $x = -1$.

b. $\frac{2x}{x + 2} - \frac{2}{x^2 - 4} = \frac{4}{x - 2}$

1. The denominators are $x + 2$, $x^2 - 4$, and $x - 2$.
   We factor $x^2 - 4$: $x^2 - 4 = (x + 2)(x - 2)$
   LCM $= (x + 2)(x - 2)$

2. $$\frac{(x+2)(x-2)}{1} \cdot \frac{2x}{x+2} - \frac{(x+2)(x-2)}{1} \cdot \frac{2}{(x+2)(x-2)} = \frac{(x+2)(x-2)}{1} \cdot \frac{4}{x-2}$$

$$\frac{2x\cancel{(x+2)}(x-2)}{\cancel{x+2}} - \frac{2\cancel{(x+2)}\cancel{(x-2)}}{\cancel{(x+2)}\cancel{(x-2)}} = \frac{4(x+2)\cancel{(x-2)}}{\cancel{x-2}}$$

$$2x(x-2) - 2 = 4(x+2)$$

3. $$2x^2 - 4x - 2 = 4x + 8$$
$$2x^2 - 8x - 10 = 0$$
$$2(x^2 - 4x - 5) = 0$$
$$2(x - 5)(x + 1) = 0$$

$x - 5 = 0$	$x + 1 = 0$
$x = 5$	$x = -1$

4. **Check**

$x = 5$

$$\frac{2(5)}{5+2} - \frac{2}{5^2 - 4} = \frac{4}{5-2}$$
$$\frac{10}{7} - \frac{2}{21} = \frac{4}{3}$$
$$\frac{30}{21} - \frac{2}{21} = \frac{28}{21}$$
$$\frac{28}{21} = \frac{28}{21} \checkmark$$

$x = -1$

$$\frac{2(-1)}{-1+2} - \frac{2}{(-1)^2 - 4} = \frac{4}{-1-2}$$
$$\frac{-2}{1} - \frac{2}{-3} = \frac{4}{-3}$$
$$-2 + \frac{2}{3} = \frac{-4}{3}$$
$$\frac{-4}{3} = \frac{-4}{3} \checkmark$$

The solutions are $x = 5$ and $x = -1$.

EXAMPLE 3

Solve the following equation:

$$\frac{x}{x-5} - \frac{5}{x+5} = \frac{10x}{x^2 - 25}$$

Solution

1. The denominators are $x - 5$, $x + 5$, and $x^2 - 25$.
   We factor $x^2 - 25$: $x^2 - 25 = (x + 5)(x - 5)$
   LCM $= (x + 5)(x - 5)$

2. $$\frac{(x+5)(x-5)}{1} \cdot \frac{x}{x-5} - \frac{(x+5)(x-5)}{1} \cdot \frac{5}{x+5}$$

$$= \frac{(x+5)(x-5)}{1} \cdot \frac{10x}{(x+5)(x-5)}$$

$$\frac{x(x+5)\cancel{(x-5)}}{\cancel{x-5}} - \frac{5\cancel{(x+5)}(x-5)}{\cancel{x+5}} = \frac{10x\cancel{(x+5)}\cancel{(x-5)}}{\cancel{(x+5)}\cancel{(x-5)}}$$

$$x(x+5) - 5(x-5) = 10x$$

3. $$x^2 + 5x - 5x + 25 = 10x$$
$$x^2 + 25 = 10x$$
$$x^2 - 10x + 25 = 0$$

   We now use the quadratic formula, with $a = 1$, $b = -10$, and $c = 25$;

$$\text{discriminant} = (-10)^2 - 4(1)(25) = 100 - 100 = 0.$$

   There is *one* potential solution:

$$x = \frac{-(-10) + 0}{2(1)} = \frac{10}{2} = 5$$

4. **Check**

$$\frac{5}{5-5} - \frac{5}{5+5} = \frac{10(5)}{(5)^2 - 25}$$

$$\frac{5}{0} - \frac{5}{10} = \frac{50}{0}$$

   Since we cannot divide by zero, $x = 5$ is *not* a solution. The given equation has *no solution.*

A final type of application of fractional equations, in "rearranging" mathematical formulas, is illustrated by our last example.

EXAMPLE 4

Electromagnetic waves originate in the oscillation of an electric charge. The charge attracts or repels other charges with a force given by the formula $F = Qq/kR^2$, where q and Q are electrical charges, R is a distance, and k is a constant.

a. Solve this equation for the constant k.

b. Solve the equation for q.

Solution

a. First, we rewrite the equation as

$$\frac{F}{1} = \frac{Qq}{kR^2}$$

Now, we cross-multiply.

$$F(kR^2) = 1(Qq)$$
$$FkR^2 = Qq$$

Finally, dividing both sides by FR^2 to "isolate" k gives

$$\frac{FkR^2}{FR^2} = \frac{Qq}{FR^2}$$
$$k = Qq/FR^2$$

b. From the results of cross-multiplication above,

$$FkR^2 = Qq \qquad \text{or} \qquad Qq = FkR^2$$

Dividing both sides by Q gives

$$q = FkR^2/Q$$

EXERCISES FOR SECTION 2.5

In Exercises 1 through 12, use cross-multiplication to solve any equations that have solutions (see Example 1). *Be sure to check all solutions.*

1. $\frac{2}{x} = \frac{4}{5}$

2. $\frac{3}{10} = \frac{15}{x}$

3. $\frac{20}{x} = \frac{5}{4}$

4. $\frac{x}{45} = \frac{2}{3}$

5. $\frac{2}{y} = \frac{y}{2}$

6. $\frac{x}{5} = \frac{5}{x}$

7. $\dfrac{x}{x-3} = \dfrac{5}{x-3}$
8. $\dfrac{4}{y+2} = \dfrac{y}{y+2}$
9. $\dfrac{x}{x+1} = \dfrac{5x}{4}$

10. $\dfrac{y}{2y+3} = \dfrac{3y}{5}$
11. $\dfrac{x}{x-5} = \dfrac{1}{x-5}$
12. $\dfrac{1}{2x-3} = \dfrac{2x}{2x-3}$

In Exercises 13 through 30, solve the given equation, if possible, checking *all* solutions in the *original* equation (see Examples 2 and 3).

13. $\dfrac{3}{x} + \dfrac{1}{2} = 5$
14. $\dfrac{2}{3} + \dfrac{4}{x} = 1$

15. $\dfrac{4}{x} - \dfrac{2}{3} = 1$
16. $\dfrac{3}{x} - \dfrac{1}{2} = 4$

17. $\dfrac{x}{x+2} + \dfrac{1}{2} = 1$
18. $\dfrac{1}{3} - \dfrac{2x}{3x-1} = 1$

19. $\dfrac{x+3}{2} = \dfrac{x-2}{3} + \dfrac{1}{2}$
20. $\dfrac{2x+3}{5} - \dfrac{x+2}{2} = \dfrac{7}{10}$

21. $\dfrac{5}{2y-3} = -2 - \dfrac{1}{y}$
22. $\dfrac{x}{x+2} = 5 - \dfrac{2}{x+2}$

23. $\dfrac{x}{x-5} = 4 + \dfrac{5}{x-5}$
24. $\dfrac{2x-1}{3x} + \dfrac{x}{4} = 1$

25. $y + \dfrac{2}{y} = \dfrac{11}{y}$
26. $2x - \dfrac{3}{x} = \dfrac{5}{x}$

27. $\dfrac{x}{x+2} + 2 = \dfrac{6-x}{x^2-4}$
28. $\dfrac{4}{x} + \dfrac{7}{x+5} = \dfrac{3}{x-1}$

29. $\dfrac{4}{5y^2-3y} - \dfrac{1}{y} = \dfrac{2y}{5y-3}$
30. $\dfrac{4}{3y-1} - \dfrac{2}{y+1} = \dfrac{1}{y}$

In Exercises 31 through 40, solve each equation for the indicated variable (see Example 4).

31. $F = \dfrac{kmM}{d^2}$ for m
32. $F = \dfrac{kmM}{d^2}$ for d

33. $\dfrac{C}{F-32} = \dfrac{5}{9}$ for C
34. $\dfrac{C}{F-32} = \dfrac{5}{9}$ for F

35. $\dfrac{1}{x} = \dfrac{1}{y} + \dfrac{1}{z}$ for y
36. $\dfrac{1}{x} = \dfrac{1}{y} - \dfrac{1}{z}$ for z

37. $\frac{1}{x} = \frac{1}{y} + \frac{1}{z}$ for x

38. $\frac{1}{x} = \frac{1}{y} - \frac{1}{z}$ for x

39. $Ft = \frac{wx}{g} - \frac{wy}{g}$ for x

40. $Ft = \frac{wx}{g} - \frac{wy}{g}$ for w

2.6 Rectangular Coordinates

Graphing First Degree Equations and Inequalities

In many applications, especially in business, people make use of tables to help organize and interpret information. A table or chart of some kind can often present information in a format that is easy to visualize and analyze. We shall discuss the versatility of tables in more detail and from a slightly different point of view in Chapter 7. Right now, let's consider a simple example.

Suppose Video Toys Unlimited wants to analyze its sales of a particular video game for the year 1984. A table like the one in Table 2.1 might be helpful.

TABLE 2.1

Month	Units Sold
January (1)	175
February (2)	250
March (3)	325
April (4)	1200
May (5)	990
June (6)	760
July (7)	850
August (8)	975
September (9)	1100
October (10)	1250
November (11)	1795
December (12)	2550

If VTU's sales manager had to make a presentation about the sales of this item to top-level management, she would probably prefer to present this information in graphic form. She might do so by preparing a **bar chart** like the one in Figure 2.1.

FIGURE 2.1

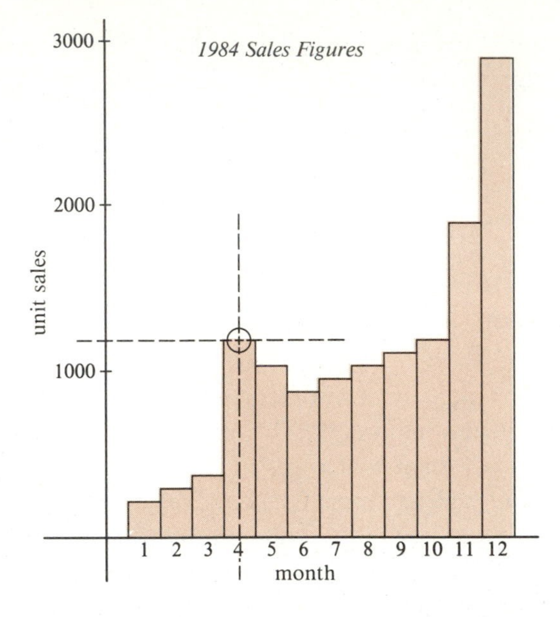

This bar chart, or *graph*, represents the same information that is in the table in a pictorial format. It uses a typical graph format: months (one of the *variables* in this example) are represented along a horizontal line, and unit sales figures (the other variable) are represented along a vertical line. To plot the sales figure for a given month, we draw an imaginary vertical line through the month number (for example, 4 for April) and an imaginary horizontal line through the sales figure (1200 for that month). The point where these lines meet (circled in Figure 2.1) represents the sales figures for that month. In a bar chart, a vertical bar is then drawn from this point to the horizontal line representing the months.

The Rectangular Coordinate System

In mathematics, we are more often involved in drawing **line graphs** (either straight or curved) to illustrate pictorially the relationship between two variables. As in the bar graph example, we begin by drawing intersecting vertical and horizontal number lines to represent the variables. These number lines, called the **axes**, determine a **rectangular coordinate system**. We usually refer to the vertical number line as the ***y*-axis** and the horizontal line as the ***x*-axis**. The point of intersection of the two axes is called the **origin** (Figure 2.2).

FIGURE 2.2

Rectangular coordinates.

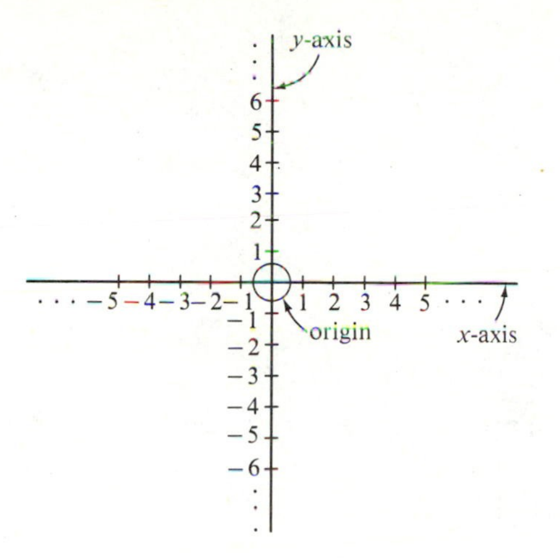

We choose the origin so that it is the zero point of both axes. We also assume that the positive direction for the x-axis is to the right and for the y-axis is upward. It is then possible to determine uniquely any point in the plane by specifying its distance from the origin along the x-axis, called the **x-coordinate**, and its distance from the origin along the y-axis, called the **y-coordinate**. (Note that if we travel to the left on the x-axis, the x-coordinate is negative; if we travel downward along the y-axis, the y-coordinate is negative.) A point is identified then, by an **ordered pair** of numbers. For example, the ordered pair $(3, -2)$ specifies a point reached by going three units to the right on the x-axis and two units down along the y-axis. We specify *ordered* pair to indicate that the points $(3, -2)$ and $(-2, 3)$ are different. An example will illustrate the process of locating points by means of ordered pairs.

EXAMPLE 1

On a rectangular coordinate system, locate the following points:

a. $(3, -2)$ **b.** $(-2, 3)$ **c.** $(0, 0)$ **d.** $(0, 4)$
e. $(4, 0)$ **f.** $(-3, -5)$ **g.** $(5, 3)$ **h.** $(5, -3)$

Solution

a. We locate $(3, -2)$ by moving three units to the right along the x-axis and then two units down along the y-axis, as the arrows in Figure 2.3 indicate.

FIGURE 2.3

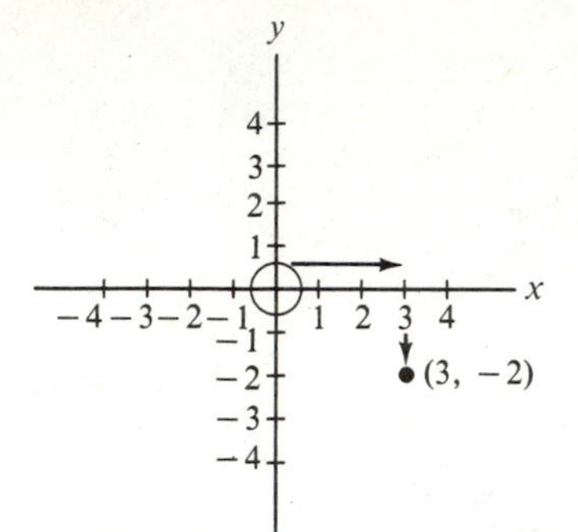

We plot the remaining points on a single rectangular coordinate system (Figure 2.4).

FIGURE 2.4

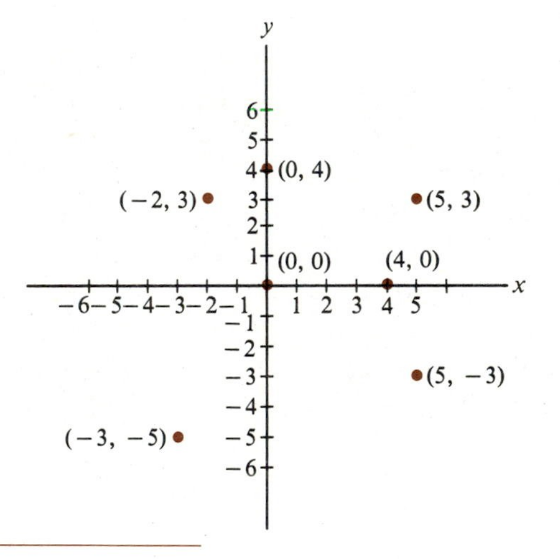

Note that the point (0, 0) (Example 1c) is the *origin* of the rectangular coordinate system.

Graphing First Degree Equations in Two Unknowns

As we pointed out in our bar chart example, rectangular coordinate systems can provide a convenient means of representing relationships pictorially. One such relationship is an equation involving two unknowns. For example, suppose a toy manufacturer produces dolls and doll strollers. Suppose also that the number of dolls produced each month should be four more than twice the number of strollers. If y represents the number of dolls and x represents the number of strollers, this relationship can be expressed by the following equation:

$$y = 2x + 4$$

Both x and y in this equation are variables, but a particular value of y is uniquely determined by the corresponding value of x. Any pair of x and y values that together make this equation a true statement is a *solution* of the equation. Unfortunately, we cannot list all such pairs, since there are infinitely many. For example,

when $x = 1$, $y = 2(1) + 4 = 6$;
when $x = 5$, $y = 2(5) + 4 = 14$;
when $x = 0$, $y = 2(0) + 4$; etc.

In table form, we have

x	y
1	6
5	14
0	4

Let us now plot these points on a rectangular coordinate system (Figure 2.5).

FIGURE 2.5

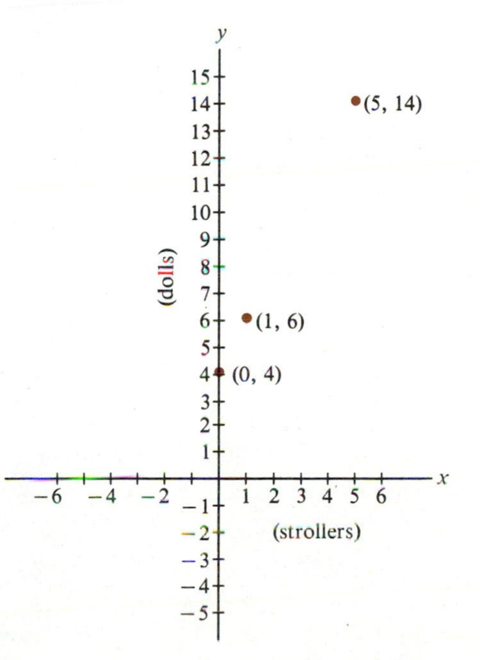

Notice that the three points lie on a straight line. We might conclude that *all* the points whose x and y coordinates satisfy the equation $y = 2x + 4$ lie on this straight line. This conclusion is correct; and we call this pictorial representation of the solution the **graph** of the equation. Thus, the straight line pictured in Figure 2.6 is the graph of $y = 2x + 4$.

FIGURE 2.6

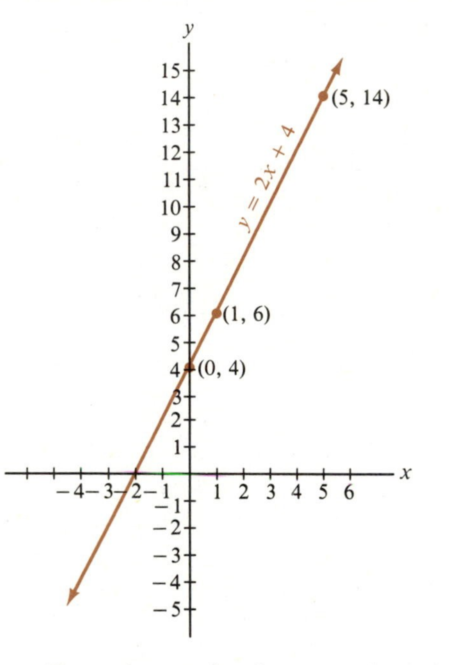

Since the graph of an equation of the form $y = ax + b$ is *always* a straight line, we call such an equation a **linear equation**.

EXAMPLE 2

Graph each of the following linear equations:

a. $y = -2x + 1$ **b.** $y - 3x = 0$

Solution

a. $y = -2x + 1$

We begin by choosing some values for x and finding the corresponding values of y.

x	y	
-2	5	$y = -2(-2) + 1 = 4 + 1 = 5$
0	1	$y = -2(0) + 1 = 0 + 1 = 1$
1	-1	$y = -2(1) + 1 = -2 + 1 = -1$
2	-3	$y = -2(2) + 1 = -4 + 1 = -3$

Now we plot these points and draw the straight line that contains all of them (Figure 2.7).

FIGURE 2.7

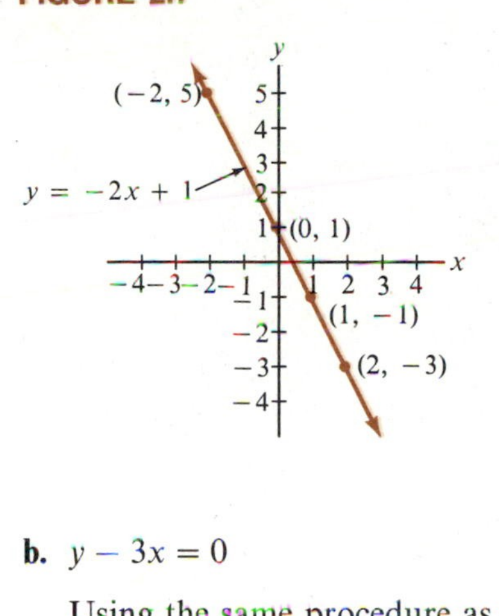

b. $y - 3x = 0$

Using the same procedure as above, we arrive at the graph of Figure 2.8.

x	y
-2	-6
-1	-3
0	0
1	3
2	6

$$y - 3(-2) = 0 \rightarrow y + 6 = 0 \rightarrow y = -6$$
$$y - 3(-1) = 0 \rightarrow y + 3 = 0 \rightarrow y = -3$$
$$y - 3(0) = 0 \rightarrow y - 0 = 0 \rightarrow y = 0$$
$$y - 3(1) = 0 \rightarrow y - 3 = 0 \rightarrow y = 3$$
$$y - 3(2) = 0 \rightarrow y - 6 = 0 \rightarrow y = 6$$

FIGURE 2.8

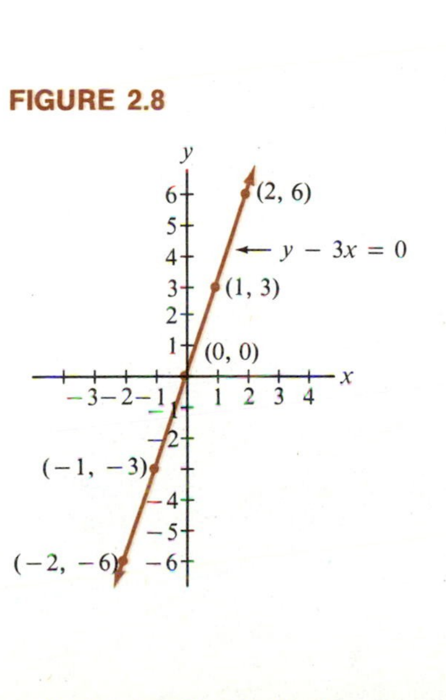

Sometimes it is useful to apply the methods of Section 2.2 to linear equations in two variables, thus obtaining an equivalent equation that expresses y in terms of x. This process is illustrated by the next example.

EXAMPLE 3

Graph the equation $2y + 4x = 6$.

Solution

It will be easier to construct a table if we first solve for y in terms of x as follows:

$2y + 4x = 6$	Original equation
$2y + 4x - 4x = -4x + 6$	Add $-4x$ to both sides.
$2y = -4x + 6$	Combine like terms.
$\frac{1}{2} \cdot 2y = \frac{1}{2}(-4x + 6)$	Multiply both sides by $\frac{1}{2}$.
$y = -2x + 3$	Equivalent equation, expressing y in terms of x.

Now we can construct a table for $y = -2x + 3$:

x	y	
-2	7	$y = -2(-2) + 3 = 4 + 3 = 7$
-1	5	$y = -2(-1) + 3 = 2 + 3 = 5$
0	3	$y = -2(0) + 3 = 0 + 3 = 3$
1	1	$y = -2(1) + 3 = -2 + 3 = 1$
2	-1	$y = -2(2) + 3 = -4 + 3 = -1$

Finally, we draw the graph (Figure 2.9).

FIGURE 2.9

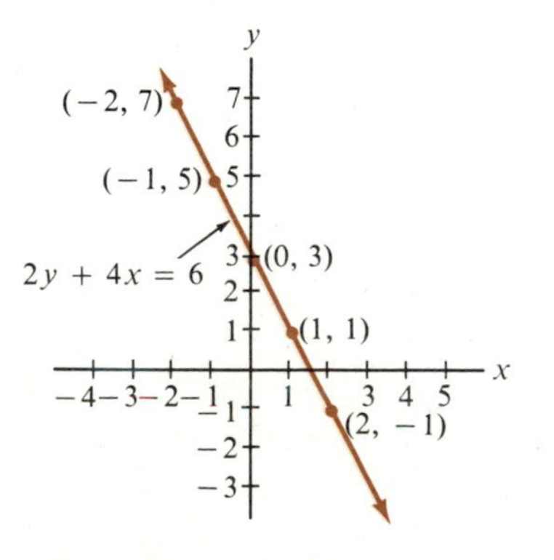

When, as in Example 3, we solve a linear equation in two variables, x and y, for y in terms of x, we say the equation is in **slope-intercept form**. We write the general slope-intercept form of a linear equation as

$$y = mx + b$$

where m is called the **slope** of the line and b is the **y-intercept**, the point where the graph of the line crosses the y-axis.

For example, the slope-intercept form of the equation of Example 3 is

$$y = -2x + 3$$

The slope of this line is -2 and the y-intercept is 3. Notice that the graph of this line crosses the y-axis at the point (0, 3). Notice also that this graph slopes *downward* as we go from left to right along the x-axis. This is true of the graph of any line whose slope is negative. If the slope is positive, the graph slopes *upward* as we go from left to right along the x-axis.

EXAMPLE 4

For each of the lines in Example 2 determine the slope and the y-intercept.

Solution

a. $y = -2x + 1$

This equation is already in slope-intercept form.

$$\text{slope} = -2$$
$$y\text{-intercept} = 1$$

b. $y - 3x = 0$

This equation is *not* in slope-intercept form, so we must first solve for y.

$$y - 3x = 0$$
$$y - 3x + 3x = 0 + 3x$$
$$y = 3x + 0$$ This is the slope-intercept form.
$$\text{slope} = 3$$
$$y\text{-intercept} = 0$$

Because the y-intercept is 0, this line passes through the origin.

Graphing Linear Inequalities

Suppose we want to change the restrictions in our doll and stroller manufacturing example to the following: the number of dolls produced should be *no more than* twice the number of strollers plus 4. Then, we would have to substitute for the *equation*

$$y = 2x + 4$$

(recall that y is the number of dolls and x is the number of strollers) the **inequality**

$$y \leq 2x + 4$$

where $\leq$ is read *is less than or equal to.* Now, consider the problem of *graphing* this inequality. We want to picture all points whose y-coordinates are less than or equal to twice their x-coordinates plus 4. We begin by graphing the linear equation $y = 2x + 4$ (Figure 2.10).

FIGURE 2.10

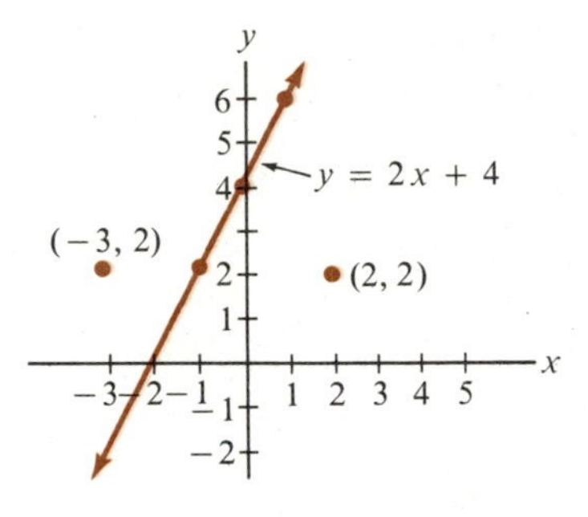

The line representing $y = 2x + 4$ can be thought of as dividing the plane into two halves; points to the left of the line constitute one half, and points to the right make up the other half.

If any point in one half-plane satisfies the inequality

$$y < 2x + 4$$

then *all* the points in that half-plane satisfy the inequality. Also, if any point in a half-plane fails to satisfy the inequality, then *no* point in that half-plane satisfies the inequality.

In order to graph the inequality, it will suffice to test two points—one to the left of the line and one to the right. Suppose we choose $(-3, 2)$ and $(2, 2)$, and construct a table:

x	y	$2x + 4$	
-3	2	-2	$2 > -2$, so $y > 2x + 4$
2	2	8	$2 < 8$, so $y < 2x + 4$

The table indicates that the point (2, 2) is in the correct half-plane. We graph the inequality in Figure 2.11, shading the half-plane that forms the solution of the inequality.

FIGURE 2.11

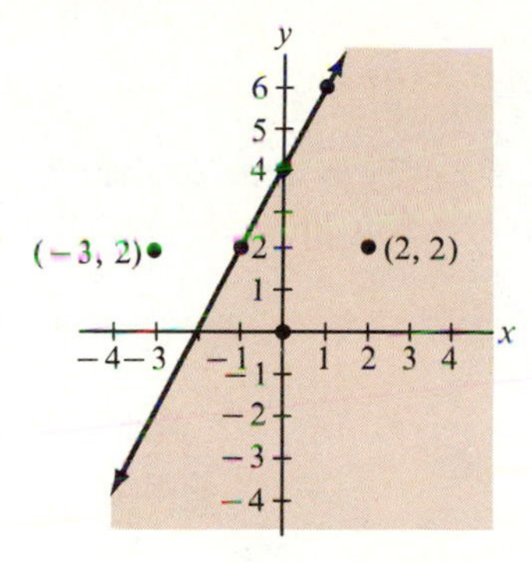

EXAMPLE 5

Graph the following linear inequalities:

a. $y \leq -2x + 1$ **b.** $y - 3x \geq 0$

Solution

a. $y \leq -2x + 1$

First, we graph the equation $y = -2x + 1$ (Figure 2.12) (compare Example 2a), and then we test a point in each half-plane, for example, $(-2, 1)$ and $(1, 1)$.

x	y	$-2x + 1$	
-2	1	5	$(1 < 5$ so $y < -2x + 1)$
1	1	-1	$(1 > -1$ so $y > -2x + 1)$

FIGURE 2.12

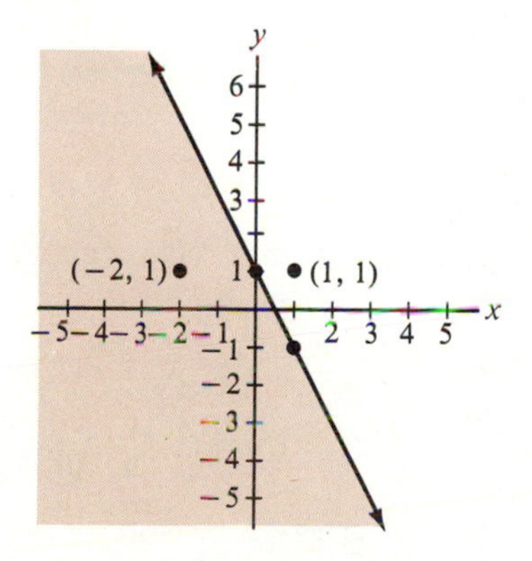

The point $(-2, 1)$ satisfies the inequality, so we shade the half-plane containing that point.

b. $y - 3x \geq 0$

We begin by graphing the equation $y - 3x = 0$, or equivalently $y = 3x$ (see Examples 4 and 2b). Next we test a point on each side of the line, for example $(-1, 1)$ and $(2, 0)$.

x	y	$y - 3x$
-1	1	$4\ (>0)$
2	0	$-6\ (<0)$

Since $y - 3x > 0$ for the point $(-1, 1)$, this point lies in the correct half-plane. We shade this half-plane for the graph of the inequality (Figure 2.13).

FIGURE 2.13

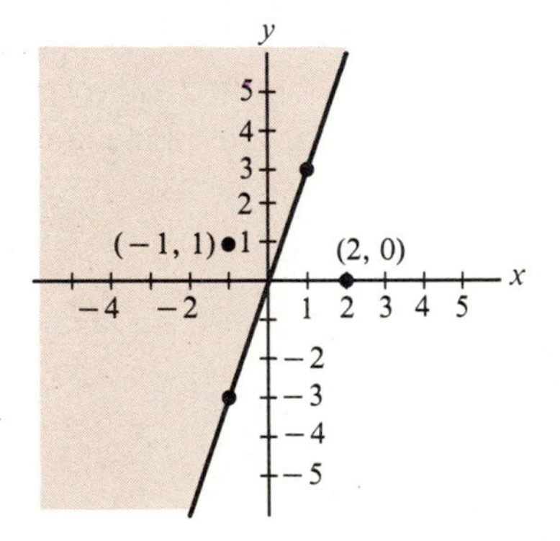

EXERCISES FOR SECTION 2.6

In Exercises 1 through 12, plot the given point on a rectangular coordinate system (see Example 1).

1. $(2, 3)$ **2.** $(3, 2)$ **3.** $(0, 0)$

4. $(5, 0)$ **5.** $(0, 5)$ **6.** $(-1, 6)$

7. $(6, -1)$ **8.** $(-5, 0)$ **9.** $(0, -5)$

10. $(-4, -6)$ **11.** $(-6, -4)$ **12.** $(3, -3)$

In Exercises 13 through 26, (a) find the slope and y-intercept of the given linear equation, and (b) graph the equation. If necessary, first solve for y in terms of x (see Examples 2–4).

13. $y = 2x - 5$ **14.** $y = -2x + 5$ **15.** $y = -2x - 5$
16. $y = 2x + 5$ **17.** $y = -3x - 4$ **18.** $y = 3x - 4$
19. $3x + y = 4$ **20.** $y - 3x = 4$ **21.** $3x + 4y = 8$
22. $3y - 4x = 9$ **23.** $x = 2y$ **24.** $x = -2y$
25. $3y + 2x = 0$ **26.** $3y - 2x = 0$

In Exercises 27 through 38, graph the given linear inequality (see Example 5).

27. $y \geq 2x - 5$ **28.** $y \leq -2x + 5$ **29.** $y \leq -2x - 5$
30. $y \geq 2x + 5$ **31.** $y \leq -3x - 4$ **32.** $y \geq 3x - 4$
33. $y \geq -\frac{3}{4}x + 2$ **34.** $y \leq \frac{4}{3}x + 3$ **35.** $y \leq \frac{1}{2}x$
36. $y \geq -\frac{2}{3}x$ **37.** $y \leq 2$ **38.** $y \leq x$

39. The relationship between the Fahrenheit and Celsius temperature scales is given by the linear equation

$$F = \tfrac{9}{5}C + 32$$

where F represents Fahrenheit temperature and C represents Celsius temperature.

a. Find the Fahrenheit temperature that is equivalent to 100° Celsius.
b. Find the Fahrenheit temperature equivalent to −40° Celsius.
c. Graph this equation with Celsius temperature (C) on the horizontal axis and Fahrenheit temperature (F) on the vertical axis.

40. Cosmetologists use the breaking strength of a human hair when stretched as an indicator in studying healthy human hair. Suppose that the force (F), in grams, needed to break a hair is related to the diameter (d), in microns (0.0001 cm), by the following linear equation:

$$F = 0.8d - 18$$

a. Complete the table below by calculating the force needed to break a healthy hair with each of the given diameters:

d	F
35	
45	
55	
65	

b. Use the results of part a to graph the equation, using d for the horizontal axis and F for the vertical axis.

2.7 Graphing Second Degree Equations

Although there are many types of second degree equations in two variables, we shall consider here only one type, equations of the form

$$y = ax^2 + bx + c$$

One approach to graphing such an equation is to construct a table as we did for linear equations. Suppose, for example, that we wish to graph the equation $y = x^2 + 2x - 3$. We begin by constructing a table.

x	y	
-3	0	$y = (-3)^2 + 2(-3) - 3 = 9 - 6 - 3 = 0$
-2	-3	$y = (-2)^2 + 2(-2) - 3 = 4 - 4 - 3 = -3$
-1	-4	$y = (-1)^2 + 2(-1) - 3 = 1 - 2 - 3 = -4$
0	-3	$y = (0)^2 + 2(0) - 3 = -3$
1	0	$y = (1)^2 + 2(1) - 3 = 1 + 2 - 3 = 0$
2	5	$y = (2)^2 + 2(2) - 3 = 4 + 4 - 3 = 5$

Next, we plot these points on a rectangular coordinate system (Figure 2.14):

FIGURE 2.14

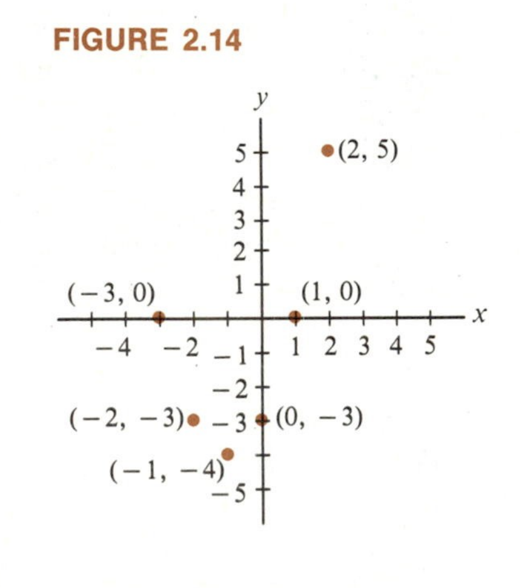

From the positions of these points it is obvious that the graph is *not* a straight line. In fact, it will be a curved line, specifically a curve called a **parabola.** A parabola has certain characteristics.

CHARACTERISTICS OF A PARABOLA

1. A parabola has a **vertex**, that is, a highest or lowest point. (In the graph of Figure 2.14, the point $(-1, -4)$ looks as if it might be the vertex.)
2. If the equation of the parabola is in the form $y = ax^2 + bx + c$, the coordinates of the vertex are given by:

$$\left(-\frac{b}{2a}, -\frac{b^2 - 4ac}{4a}\right)$$

3. If we draw the vertical line that passes through the vertex, the parabola is *symmetrical* about that line. That is, if the graph were folded over this vertical line, called the **axis of symmetry**, the two parts of the graph on either side would coincide.

Let's use these characteristics of a parabola to complete the graph of

$$y = x^2 + 2x - 3$$

The coordinates of the vertex are:

$$-\frac{b}{2a} = \frac{-2}{2(1)} = \frac{-2}{2} = -1$$

and

$$-\frac{b^2 - 4ac}{4a} = -\frac{(2)^2 - 4(1)(-3)}{4(1)} = -\frac{4 + 12}{4} = -\frac{16}{4} = -4$$

Therefore, the vertex *is* the point $(-1, -4)$, as we suspected. Knowing the exact location of the vertex allows us to complete the graph (Figure 2.15).

FIGURE 2.15

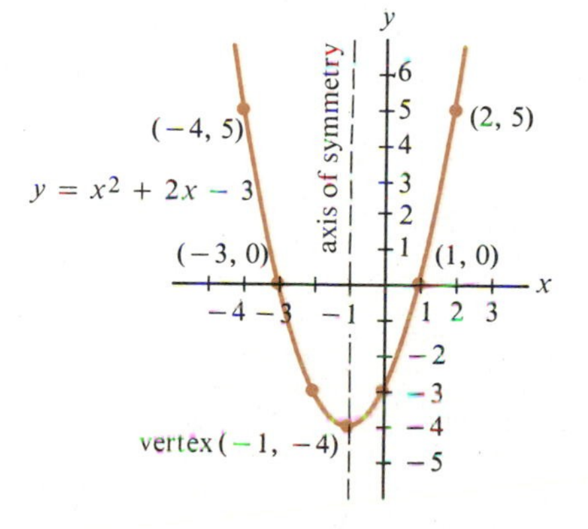

Note that the symmetry of the graph about the vertical line through the point $(-1, -4)$ allowed us to add the point $(-4, 5)$ to those previously plotted. When the vertex is the lowest point of the parabola, as in this example, we say that the graph is **concave up** (that is, it opens upward). If the vertex is the highest point, we say the parabola is **concave down** (it opens downward; see Example 1a).

EXAMPLE 1

Graph each of the following second degree equations:

a. $y = -x^2 - 4x + 3$ **b.** $y = 2x^2 + 1$

Solution

a. $y = -x^2 - 4x + 3$

First, we find the coordinates of the vertex. Here, $a = -1$, $b = -4$, and $c = 3$.

$$-\frac{b}{2a} = -\frac{-4}{2(-1)} = -\frac{-4}{-2} = -2$$

$$-\frac{b^2 - 4ac}{4a} = -\frac{(-4)^2 - 4(-1)(3)}{4(-1)} = -\frac{16 + 12}{-4}$$

$$= -\frac{28}{-4} = -(-7) = 7$$

The vertex is $(-2, 7)$. Now we construct a table of points, plot them, and draw a smooth curve through the plotted points (Figure 2.16).

FIGURE 2.16

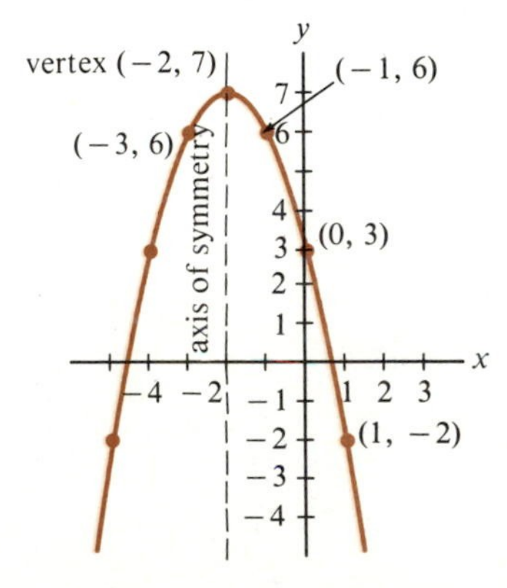

x	y
-3	6
-2	7
-1	6
0	3
1	-2

$y = -(-3)^2 - 4(-3) + 3 = -9 + 12 + 3 = 6$
$y = 7$ (This is the vertex.)
$y = -(-1)^2 - 4(-1) + 3 = -1 + 4 + 3 = 6$
$y = -(0)^2 - 4(0) + 3 = 3$
$y = -(1)^2 - 4(1) + 3 = -1 - 4 + 3 = -2$

Note that this graph is concave down, and the vertex gives the highest point.

b. $y = 2x^2 + 1$ or equivalently, $y = 2x^2 + 0x + 1$

Here $a = 2$, $b = 0$, and $c = 1$.

$$-\frac{b}{2a} = -\frac{0}{2(2)} = 0$$

$$-\frac{b^2 - 4ac}{4a} = -\frac{0^2 - 4(2)(1)}{4(2)} = -\frac{-8}{8} = -(-1) = 1$$

The point (0, 1) is the vertex. We construct the table:

x	y
-2	9
-1	3
0	1
1	3
2	9

$y = 2(-2)^2 + 1 = 2(4) + 1 = 9$
$y = 2(-1)^2 + 1 = 2(1) + 1 = 3$
$y = 2(0)^2 + 1 = 1$
$y = 2(1)^2 + 1 = 2(1) + 1 = 3$
$y = 2(2)^2 + 1 = 2(4) + 1 = 9$

This yields the graph of Figure 2.17, which is symmetric about the y-axis.

FIGURE 2.17

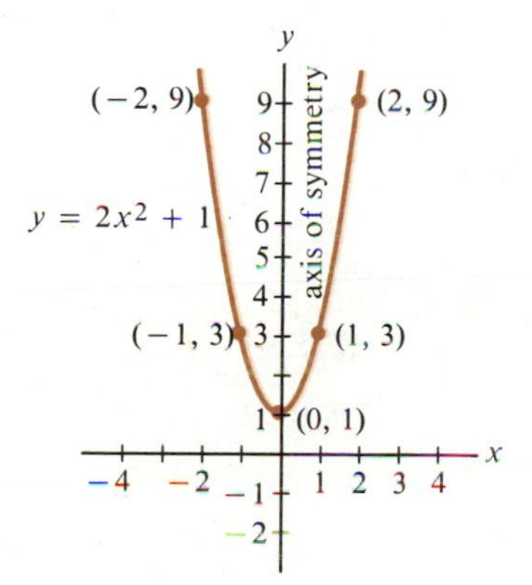

EXAMPLE 2
Graph the equation $y - x^2 + 2x = 3$.

Solution
First, we use the techniques of Section 2.2 to write an equivalent equation of the form

$$y = ax^2 + bx + c$$

We have

$y - x^2 + 2x = 3$	Original equation
$y - x^2 + x^2 + 2x = x^2 + 3$	Add x^2 to both sides.
$y + 2x = x^2 + 3$	$-x^2 + x^2 = 0$
$y + 2x - 2x = x^2 - 2x + 3$	Add $-2x$ to both sides.
$y = x^2 - 2x + 3$	The form we need.

Now, we find the vertex of the graph. Here, $a = 1, b = -2, c = 3$.

$$-\frac{b}{2a} = -\frac{(-2)}{2(1)} = -(-1) = 1$$

$$-\frac{b^2 - 4ac}{4a} = -\frac{(-2)^2 - 4(1)(3)}{4(1)} = -\frac{4 - 12}{4}$$

$$= -\frac{-8}{4} = -(-2) = 2$$

The vertex is (1, 2).

Finally, we construct a table and draw the graph (Figure 2.18):

FIGURE 2.18

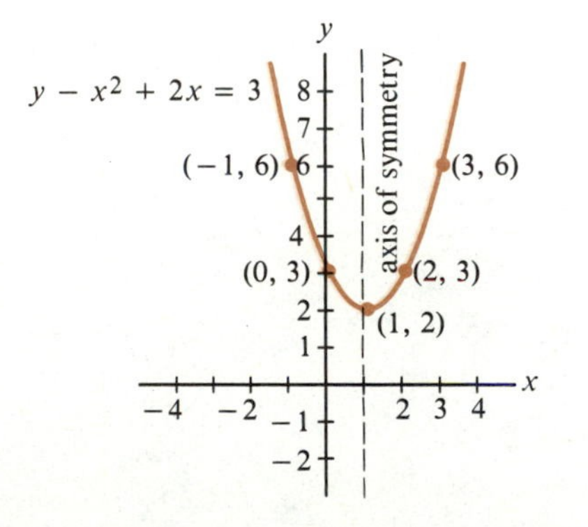

x	y	
3	6	$y = (3)^2 - 2(3) + 3 = 9 - 6 + 3 = 6$
2	3	$y = (2)^2 - 2(2) + 3 = 4 - 4 + 3 = 3$
1	2	$y = (1)^2 - 2(1) + 3 = 1 - 2 + 3 = 2$
0	3	$y = 0^2 - 2(0) + 3 = 3$
-1	6	$y = (-1)^2 - 2(-1) + 3 = 1 + 2 + 3 = 6$

An examination of the coefficients of x^2 in Examples 1 and 2 leads to the following observation: In graphing the equation $y = ax^2 + bx + c$, the graph is concave up when a is positive and concave down when a is negative. In other words, the vertex is the lowest point of the graph when a is positive and the highest point when a is negative.

EXERCISES FOR SECTION 2.7

In Exercises 1 through 12, find the vertex of the graph of the given quadratic equation, and then draw the graph (a parabola). Indicate whether the graph is concave up or concave down (see Example 1).

1. $y = x^2 - 6x + 9$

2. $y = -x^2 + 6x - 9$

3. $y = -x^2 + 9$

4. $y = x^2 - 4$

5. $y = x^2 - 6x + 7$

6. $y = -3x^2 + 6x + 4$

7. $y = 4x^2 - 9$

8. $y = 9x^2 - 4$

9. $y = -2x^2 - 8x + 1$

10. $y = -4x^2 + 12x + 9$

11. $y = 4x^2 + 12x + 9$

12. $y = 2x^2 + 4x - 3$

In Exercises 13 through 22, transform the given equation into one of the form $y = ax^2 + bx + c$, and then draw the graph (see Example 2).

13. $x^2 - 8x = y - 16$

14. $4x^2 + y = 9$

15. $y - 5 = x^2 - 2x$

16. $2x + 5 = x^2 + y$

17. $y + 3 = 4x(x + 3)$

18. $4x^2 + y = 3(4x - 1)$

19. $y - 5 = -x(x + 4)$

20. $y - 5 = x(x + 4)$

21. $-4x(x - 2) = 7 - y$

22. $4x(x + 2) = -7 - y$

2.8 Functions

In mathematics textbooks, there are many different approaches to the topic we are about to discuss. We shall adopt an approach that, in our opinion, is especially well suited to the study of the function concept in the context of computer programming. We shall define a **function** F on a subset A of the real numbers to be a rule that assigns to each element x of A a *unique* real number, called the **image** of x and denoted by $F(x)$. The number x in parentheses is called the **argument**; the set A is called the **domain** of the function; and the set of all images is called the **range** of the function.

As a simple example of a function, let us consider the following: Our *rule* will be

To each real number, assign its double.

Since mathematicians generally prefer to deal with formulas or equations, we can redefine the function above as follows:

For each real number x, $F(x) = 2x$.

To verify that this is a function, we must satisfy ourselves that every real number has a *unique* double. In other words, if a and b are not equal, then $2a$ and $2b$ are not equal. This is true, since $2a = 2b$ *if and only if* $a = b$.

The domain of this function is the set of all real numbers; the range, or set of images, is also the set of all real numbers since every real number is the double of some other real number.

It is often useful to graph a function defined on a subset of the real numbers. To do this, we simply use the horizontal axis to denote the subset of the domain and the vertical axis to denote the images. The graph of $F(x) = 2x$ is illustrated in Figure 2.19.

x	$F(x)$
−1	−2
0	0
1	2
2	4

FIGURE 2.19

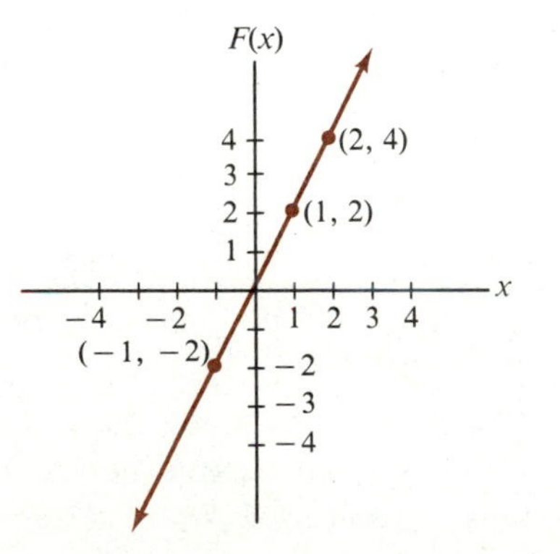

To find $F(x)$ for a specific value of x—for example to find $F(1)$—we substitute the value (here 1) for x in the equation defining the function F. Therefore $F(1) = 2 \cdot 1 = 2$, $F(2) = 2 \cdot 2 = 4$, and so on.

Graphing a function, like graphing an equation, gives us a picture of the function. It can be helpful in determining the domain and range of the function. Graphing can also be a useful tool in determining whether or not a given equation defines a function. If there is a vertical line (*any* vertical line) that intersects the graph in more than one point, then there are two $F(x)$ values—that is, two images—that correspond to the same value of x. This x value, then, does not have a unique image; so the graph is *not* the graph of a function. If no such vertical line exists then every x value must have a unique image and the graph *is* the graph of a function.

EXAMPLE 1

Graph each of the following equations. Use the vertical line test to determine whether or not the equation defines a function. If it does, determine the domain and range of the function.

a. $F(x) = x^2$ **b.** $G(x) = -3$ **c.** $D(x) = \pm\sqrt{x}$

Solution

a. $F(x) = x^2$

The graph is a parabola with vertex at the origin (0, 0) (Figure 2.20). This equation *does* define a function, since no vertical line intersects the graph in more than one point. The *domain* is the set of all real numbers.

The *range* is the set of nonnegative real numbers since the graph never goes below the x-axis; that is, $F(x)$ has no negative values.

FIGURE 2.20

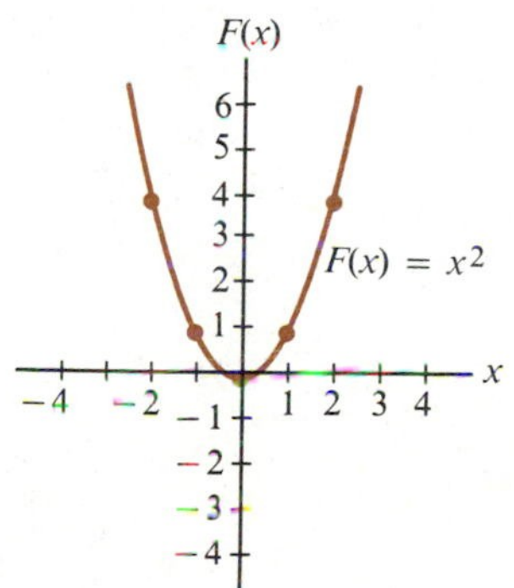

x	$F(x)$	
-2	4	$F(-2) = (-2)^2$
-1	1	$F(-1) = (-1)^2$
0	0	$F(0) = 0^2$
1	1	$F(1) = 1^2$
2	4	$F(2) = 2^2$

b. $G(x) = -3$

This rule states that the image of every real number is the number -3. The graph of this equation is shown in Figure 2.21. This is clearly the graph of a function, since every vertical line intersects it in exactly one point. The *domain* is the set of all real numbers. The *range* is the set $\{-3\}$ since -3 is the *only* image.

FIGURE 2.21

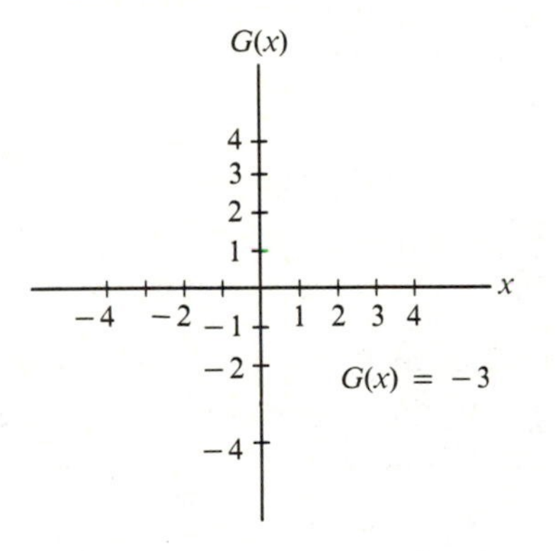

c. $D(x) = \pm\sqrt{x}$

This rule associates with each real number *two* values, when they exist—the positive square root and the negative square root. It does *not* define a function since, for example, 9 has two potential images, $+3$ and -3. Let's try to graph this equation. Note that $D(x)$ is not defined for negative values of x since the square root of a negative number is not a real number.

x	$D(x)$
0	0
4	$+2$ and -2
9	$+3$ and -3
16	$+4$ and -4
25	$+5$ and -5

FIGURE 2.22

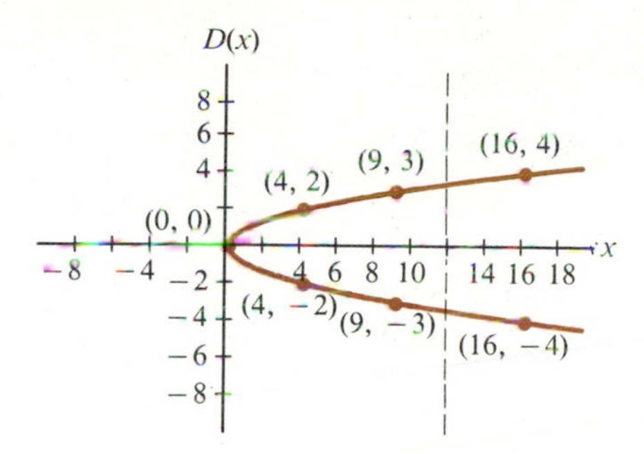

Note that the graph of $D(x)$ (Figure 2.22) *fails* the vertical line test since many vertical lines—all those drawn to the right of the $D(x)$ axis—intersect the curve in two points. The figure illustrates one such line.

In a programming context, it is often useful for the programmer to be able to identify a particular formula as a function. In Section 2.4, we encountered a toy manufacturer who produced, among other things, dolls and doll strollers. The number of dolls manufactured in any given month was related to the number of strollers manufactured by the equation:

$$y = 2x + 4$$

where y denoted dolls and x denoted strollers.

This relationship may also be described as follows: The number of dolls is a *function* of the number of strollers. In symbols, we write

$$D(x) = 2x + 4$$

where x is still the number of strollers and $D(x)$ is the number of dolls. The use of the function notation emphasizes the fact that the number of dolls produced *depends on* (is a function of) the number of strollers.

In BASIC, such a programmer-defined function would be defined in a statement like the following:

```
200 DEF FND(X) = 2*X + 4
```

Here, D is the name of the function, and X, in parentheses, is the *argument* of the function. The expression to the right of the equal sign is the formula that defines the rule for this particular function. If the program later refers to a value like FND(600), the value returned will be the *image*, under the function D, of 600. That is:

```
FND(600) = 2*600 + 4 = 1204
```

It is important that a programmer understand the concept of a function in order to use it in a program. The BASIC interpreter will, for example, reject a function definition if the rule used doesn't really produce a functional relationship.

EXAMPLE 2

Write BASIC statements for each of the equations of Example 2 in Section 2.6 and of Example 1 in Section 2.7, defining y as a function of x if the rule would actually define a function.

Solution

Examining the graphs verifies that all four equations do define y as a function of x.

Example 2 of Section 2.6

a. $y = -2x + 1$

BASIC statement: `DEF FNF(X) = -2*X+1`

b. $y - 3x = 0$

This equation must first be written as $y = 3x$

BASIC statement: `DEF FNF(X) = 3*X`

Example 1 of Section 2.7

a. $y = -x^2 - 4x + 3$

BASIC statement: `DEF FNG(X) = -(X^2) - 4*X + 3`

b. $y = 2x^2 + 1$

BASIC statement: `DEF FNG(X) = 2*X^2 + 1`

BASIC, like most other programming languages, also has some **intrinsic**, or **library, functions**. These functions need not be defined by the programmer in order to be used in a program. They include, for the mathematically inclined, the trigonometric functions (sine, cosine, tangent, etc.), the logarithmic and exponential functions, and a function for choosing random numbers, among others.

BASIC also has a library function called the **greatest integer function**. Its name in BASIC is INT. When applied to a specific argument, INT returns as the image the *greatest integer that is less than or equal to* the argument.

EXAMPLE 3

Determine the value of each of the following:

a. `INT(7.3)` **b.** `INT(8.9)` **c.** `INT(23)` **d.** `INT(-4.32)`

Solution

a. The greatest integer less than or equal to 7.3 is 7, so

```
INT(7.3) = 7
```

b. The greatest integer less than or equal to 8.9 is 8, so

```
INT(8.9) = 8
```

Notice that INT does *not* round off.

c. Since 23 is an integer, it is the greatest integer less than or equal to itself:

```
INT(23) = 23
```

d. Be careful here. The greatest integer *less than* or equal to -4.32 is -5. You might be tempted to choose -4, but -4 is *greater than* -4.32.

```
INT(-4.32) = -5.
```

EXAMPLE 4

Draw the graph of the greatest integer function.

Solution

The easiest way to approach this problem is to work with intervals and attempt to establish a pattern. For example, for numbers between 0 and 1 (not including 1, but including 0), the INT function yields an image of 0.

When the value of X reaches 1, INT(X) jumps to 1 and stays there until X = 2. The pattern seems to go in steps as in Figure 2.23.

FIGURE 2.23

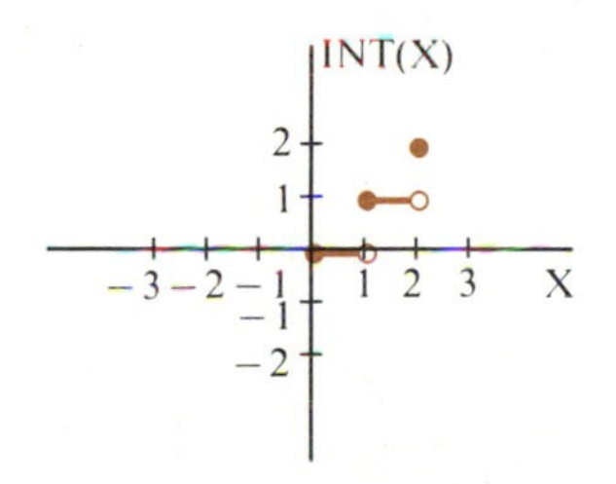

Completing the pattern gives the graph of Figure 2.24.

FIGURE 2.24

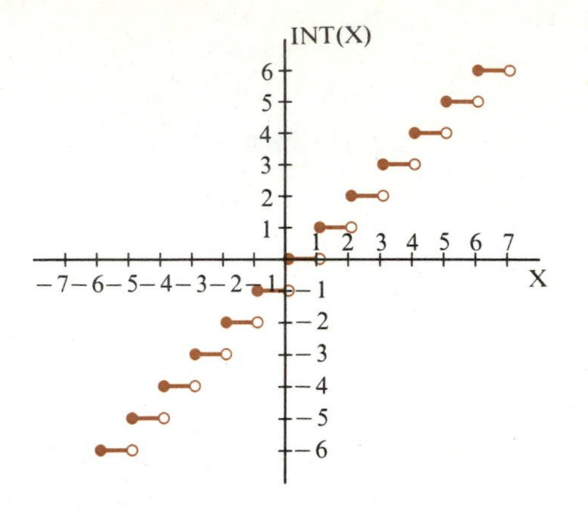

EXERCISES FOR SECTION 2.8

In Exercises 1 through 6, use the vertical line test to determine whether the given graph is the graph of a function (see Example 1).

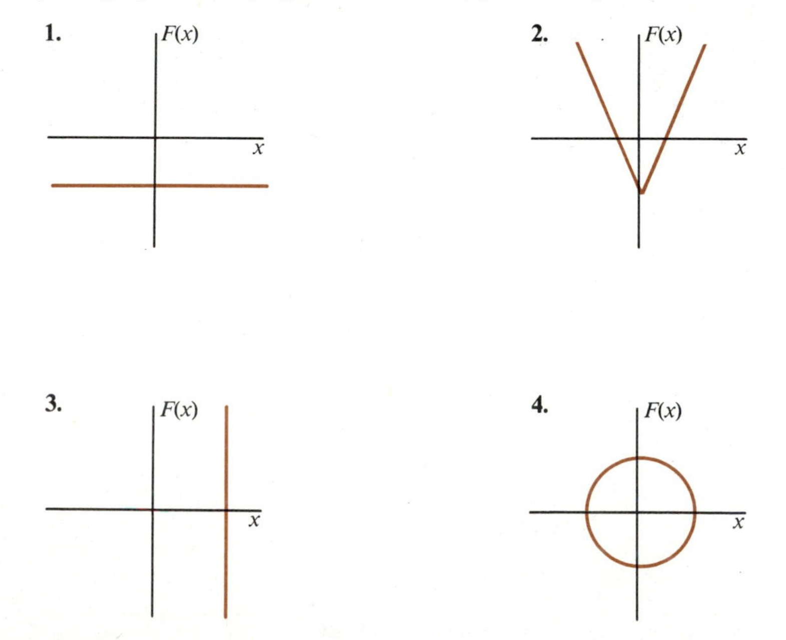

5. 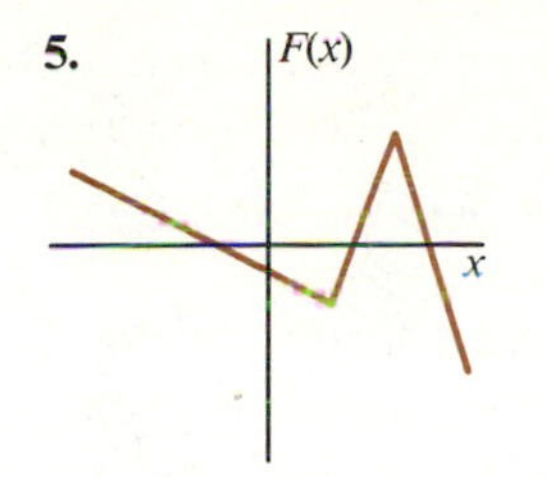

6. 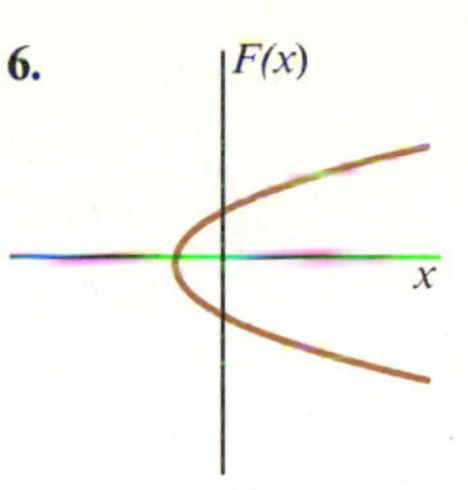

In Exercises 7 through 14, graph the given function and determine its domain and range (see Example 1).

7. $F(x) = -x^2$
8. $F(x) = 3x - 1$
9. $F(x) = x$
10. $Q(x) = x^2 - 2x + 4$
11. $P(x) = -x^2 + 4x + 4$
12. $H(x) = \sqrt{x - 1}$ (Use the principal square root here.)
13. $G(x) = 4$
14. $F(x) = 2x + 3$

In Exercises 15 through 22 write a BASIC statement to define each of the functions in Exercises 7 through 14 (see Example 2).

23. We define a relationship M on the set of integers as follows: M assigns to an integer n the remainder after dividing n by 3. For example, $M(7) = 1$, since $7 \div 3 = 2$, with a remainder of 1.
   a. Determine: $M(0)$, $M(2)$, $M(3)$, $M(8)$, $M(9)$, $M(10)$
   b. Does the rule for M define a function? If so, what is the range?
24. A function that is available in both BASIC and Pascal is the ABS, or **absolute value,** function. This function is defined on the set of real numbers as follows:

$$\text{ABS(X)} = \begin{cases} \text{X}, & \text{if X is positive} \\ 0, & \text{if X is 0} \\ -\text{X}, & \text{if X is negative} \end{cases}$$

   a. Draw the graph of this function.
   b. Determine the domain and range of the absolute value function.

2.9 Chapter Review

This chapter reviewed the basic concepts of algebra, concentrating on those most likely to be encountered in computer programming applications. It began with a review of exponents and the properties of exponents in Section 2.1. Methods of solving first and second degree equations were developed in Sections 2.2 and 2.3, including solution of second degree equations both by factoring and by using the quadratic formula. The first part of the chapter concluded with a discussion of algebraic fractions in Sections 2.4 and 2.5.

Attention then turned to rectangular coordinate systems and graphing in Sections 2.6 and 2.7. Techniques for graphing both linear and quadratic equations were discussed, along with the graphing of linear inequalities. The chapter concluded with an explanation of the concept of a function in algebra and examples of the use of functions in BASIC programming.

VOCABULARY

exponent
base
x^a, for any integer a
x^{-a}, for $x \neq 0$, a any integer
square root
principal root
nth root ($n > 1$)
$x^{1/n}$ ($n \neq 0$)
$x^{m/n}$ ($m > 0, n \neq 0$)
equation
solution of an equation
equivalent equations
quadratic formula
discriminant
quadratic equation
product
factor
prime
quadratic polynomial
common factors
greatest common factor (GCF)
relatively prime
terms of a polynomial
numerical coefficient
FOIL method
algebraic fraction
equivalent fractions
least common multiple (LCM)
principle of equality of fractions
cross-multiplication
rectangular coordinate system
x-axis
y-axis
origin
x-coordinate
y-coordinate
ordered pair
graph of an equation
linear equation
slope
y-intercept
parabola
vertex of a parabola
axis of symmetry
concave up
concave down
function
image
argument
domain
range
vertical line test

CHAPTER TEST

Simplify any of the following expressions that can be simplified:

1. $(2^3)^2$

2. $\left(\frac{1}{3}\right)^{-1}$

3. $8^{2/3}$

4. $(3a^2b^3)^{-2}$

5. $\left(\frac{x^2y}{3z}\right)^3$

6. $\left(\frac{a^{-1}b^2c^{-2}}{abc^3}\right)^3$

7. $\left(\frac{a^{-1}b^2c^{-2}}{abc^3}\right)^{-3}$

8. $(x^4y^3z^{-6})^{2/3}$

Find the solution of each of the following linear equations if a *unique* solution exists. If not, state whether the equation has no solution or infinitely many solutions.

9. $3x - 2 = 5$

10. $5x - 3 = 7(x + 1)$

11. $-3(2x + 1) + 2(5x + 1) = 4x + 1$

12. $5(x - 3) + 2x = 3(x - 7)$

13. $2(x - 2) = 4(2x + 5) + 3(-2x - 8)$

Using the quadratic formula, solve each of the following quadratic equations that can be solved.

14. $4x^2 - 5x + 1 = 0$

15. $x^2 = 2x - 5$

16. $5x^2 - 5(2x - 1) = 0$

17. $3x^2 + 7x - 6 = 0$

18. $4x^2 = 16$

Factor each of the following polynomials completely.

19. $x^2 + 4x - 5$

20. $36x^2y^2 + 12xy^2 - 8y^2$

21. $12x^2 - 27$

22. $25x^2 + 20x + 4$

23. $6x^3y^2z + 3x^2y^2z - 2xy^2z$

Use factoring to solve each of the following quadratic equations.

24. $4x^2 - 5x + 1 = 0$

25. $3(x^2 - 2) = -7x$

26. $36x^2 = 49$

27. $9x^2 + 6x + 1 = 0$

Perform the given operation and reduce each answer to lowest terms.

28. $\frac{ab}{c^2d} \cdot \frac{cd}{a^2b}$

29. $\dfrac{12x + 12}{4x + 12} \cdot \dfrac{x^2 + x - 6}{3x + 3}$

30. $\dfrac{ab}{c^2d} \div \dfrac{cd}{a^2b}$

31. $\dfrac{x + 9}{x^2 - 9} \div \dfrac{x^2 + 10x + 9}{x^2 + 6x + 9}$

32. $\dfrac{4x}{x - 5} - \dfrac{20}{x - 5}$

33. $\dfrac{5}{x^2y^2} + \dfrac{3}{xy^3}$

34. $\dfrac{3x}{x^2 - y^2} - \dfrac{y}{x + y}$

35. $\dfrac{x}{x^2 - 9} + \dfrac{2x}{x^2 - 9x + 18}$

Solve all of the following equations that have solutions. Be sure to *check your answers.*

36. $\dfrac{3}{2x} - \dfrac{5}{7} = 1$

37. $\dfrac{x}{x + 3} + \dfrac{12}{x - 9} = \dfrac{-2}{x + 3}$

Graph each of the following.

38. $y = 2x - 1$

39. $-3y = x + 6$

40. $x - 2y = 0$

41. $y \leq 2x - 1$

42. $y \geq \frac{1}{2}x$

43. $y = x^2 + 4x - 5$

44. $y = -3x^2 - 6x + 7$

State whether each of the following is the graph of a function. If it is, indicate the domain and range of the function.

45.

46.

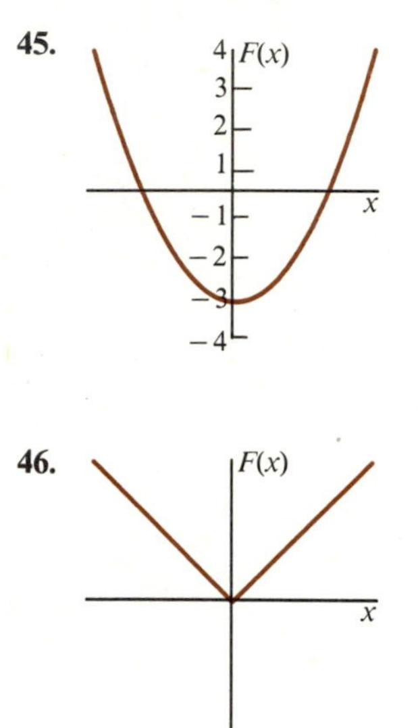

47.

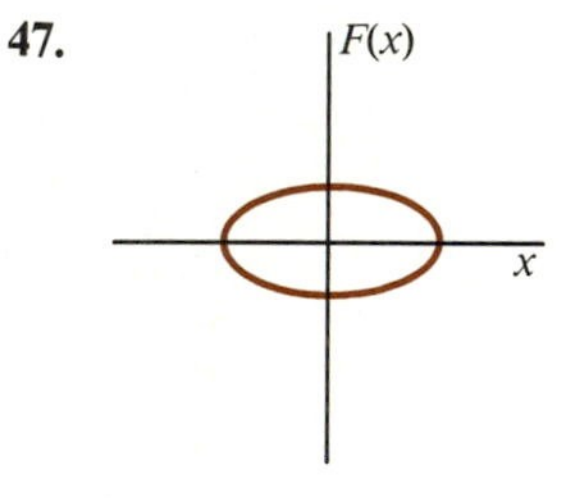

48.

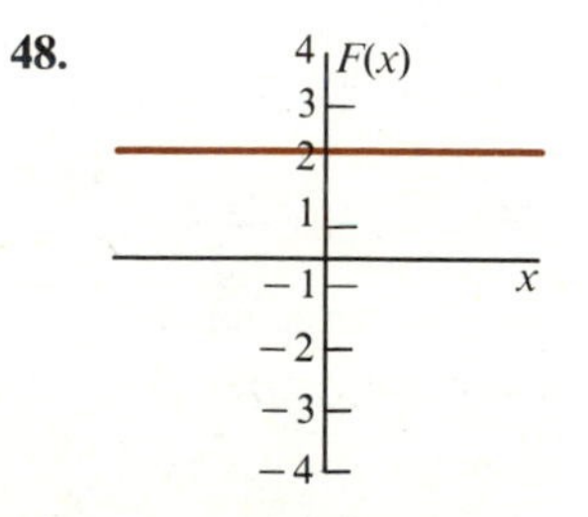

CHAPTER

3

From the ancient Chinese abacus to the seventeenth century adding machine to the twentieth century devices of slide rule, calculator, and microcomputer, humankind seems in constant quest to ease and speed data storage and calculations.

Electronic computers use a different system of number representation than their human inventors; it is called the binary system of numeration. We study that system, as well as other computer-related systems in this chapter.

Number Bases and the Computer

3.1 The Binary System

Probably because we humans have ten fingers and ten toes, we have usually found the decimal number system to be best suited to our needs in solving problems that require counting or calculations. The earliest computers also solved arithmetic problems using the decimal, or base ten, system of numeration. This was soon found, however, to be impractical largely because of the amount of storage space it required. Computers store information essentially in terms of the presence or absence of an electrical charge (that is, as either an *on* or *off* condition). This fact led designers to choose a number system for use in computer design that more clearly reflects the way computers store and manipulate data. This system is called the **binary**, or **base two, number system**. We will study the binary number system in the first half of this chapter. First, we review the decimal system of numeration.

Expansion and Positional Notation

The decimal number system is a base ten system that uses ten symbols to represent numbers. These are the digits 0, 1, 2, 3, 4, 5, 6, 7, 8, and 9. Any decimal number can be represented uniquely in terms of these digits. The meaning of a specific digit in a number depends on its position in the decimal representation of the number. Thus the concept of **place value** is important in this number system.

For example, the number 2345 is read as "*two* thousand *three* hundred forty-five" and the number 3245 is read as "*three* thousand *two* hundred forty-five." Although these numbers are constructed using the same four digits, changing the positions of the digits 2 and 3 also changes the value of the number. The place value nature of the system assigns a *value* to each position.

Below, we write the expanded form of 2345:

$$2345 = (2 \cdot 1000) + (3 \cdot 100) + (4 \cdot 10) + (5 \cdot 1)$$

The equivalent form of this, using exponents, is preferred:

$$2345 = (2 \cdot 10^3) + (3 \cdot 10^2) + (4 \cdot 10^1) + (5 \cdot 10^0)$$

(Recall that 10 raised to the zero power is 1.)

EXAMPLE 1

Write each of the following in expanded form:

a. 3245 **b.** 678 **c.** 50401

Solution

a. $3245 = (3 \cdot 10^3) + (2 \cdot 10^2) + (4 \cdot 10^1) + (5 \cdot 10^0)$

b. $678 = (6 \cdot 10^2) + (7 \cdot 10^1) + (8 \cdot 10^0)$

c. $50401 = (5 \cdot 10^4) + (0 \cdot 10^3) + (4 \cdot 10^2) + (0 \cdot 10^1) + (1 \cdot 10^0)$

We can use the notion of expansion for decimal numbers with fractional parts, that is, numbers involving a decimal point. The values of the positions to the right of the decimal point are

tenths	hundredths	thousandths	ten-thousandths
0.1	0.01	0.001	0.0001
10^{-1}	10^{-2}	10^{-3}	10^{-4}

EXAMPLE 2

Write each of the following in expanded form:

a. 32.46 **b.** 65.082

Solution

a. $32.46 = (3 \cdot 10^1) + (2 \cdot 10^0) + (4 \cdot 10^{-1}) + (6 \cdot 10^{-2})$

b. $65.082 = (6 \cdot 10^1) + (5 \cdot 10^0) + (0 \cdot 10^{-1}) + (8 \cdot 10^{-2}) + (2 \cdot 10^{-3})$

Binary Numbers and Positional Notation

Just as the base ten system is composed of ten digits, the base two or binary system is composed of two digits, 0 and 1. Recall that the binary system was applied to computers because of the "off and on" nature of electrical current. The 0 and 1 correspond to *off* and *on*, respectively.

Any number, then, can be represented by a binary numeral as a sequence of zeros and ones, with the value of a particular binary digit (or **bit**) determined by its position in the binary numeral. We use a subscript 2 to denote that the numeral is written in binary form. We can express 101_2, for example, in expanded form as

$$101_2 = (1 \cdot 2^2) + (0 \cdot 2^1) + (1 \cdot 2^0)$$

EXAMPLE 3

Write the expansions of the following base two numbers:

a. 1111_2 **b.** 10100_2

Solution

a. $1111_2 = (1 \cdot 2^3) + (1 \cdot 2^2) + (1 \cdot 2^1) + (1 \cdot 2^0)$

b. $10100_2 = (1 \cdot 2^4) + (0 \cdot 2^3) + (1 \cdot 2^2) + (0 \cdot 2^1) + (0 \cdot 2^0)$

We use this expansion method to convert from another base to base ten. To facilitate conversions, we present some powers of two in Table 3.1.

TABLE 3.1

Powers of two.

Power of 2	Value	Power of 2	Value
2^0	1	2^6	64
2^1	2	2^7	128
2^2	4	2^8	256
2^3	8	2^9	512
2^4	16	2^{10}	1024
2^5	32	2^{11}	2048

We convert to the decimal from the binary system in the next example.

EXAMPLE 4

Evaluate each binary numeral by converting it to base ten.

a. 101_2 **b.** 10100_2 **c.** 1101100101_2

Solution

a. $101_2 = (1 \cdot 2^2) + (0 \cdot 2^1) + (1 \cdot 2^0)$
$= 4 + 0 + 1 = 5$

b. $10100_2 = (1 \cdot 2^4) + (0 \cdot 2^3) + (1 \cdot 2^2) + (0 \cdot 2^1) + (0 \cdot 2^0)$
$= 16 + 0 + 4 + 0 + 0 = 20$

c. $1101100101_2 = (1 \cdot 2^9) + (1 \cdot 2^8) + (0 \cdot 2^7) + (1 \cdot 2^6) + (1 \cdot 2^5)$
$+ (0 \cdot 2^4) + (0 \cdot 2^3) + (1 \cdot 2^2) + (0 \cdot 2^1) + (1 \cdot 2^0)$
$= 512 + 256 + 0 + 64 + 32 + 0 + 0 + 4 + 0 + 1 = 869$

Binary Fractions

As in the decimal system, binary numerals can have fractional parts, represented by digits to the right of a point, now called the **binary point**. These digits have a meaning similar to those in the decimal system. For example,

$$101.1_2 = (1 \cdot 2^2) + (0 \cdot 2^1) + (1 \cdot 2^0) + (1 \cdot 2^{-1})$$

Again, a table of values might prove useful for complicated conversions (Table 3.2)

TABLE 3.2

Negative powers of two.

Power of 2	Fraction	Value
2^{-1}	$\frac{1}{2}$	0.5
2^{-2}	$\frac{1}{4}$	0.25
2^{-3}	$\frac{1}{8}$	0.125
2^{-4}	$\frac{1}{16}$	0.0625
2^{-5}	$\frac{1}{32}$	0.03125

We conclude this section with two more examples of converting base two numerals to base ten numerals.

EXAMPLE 5

Convert 11.011_2 to base ten.

Solution

$$\begin{aligned} 11.011_2 &= (1 \cdot 2^1) + (1 \cdot 2^0) + (0 \cdot 2^{-1}) + (1 \cdot 2^{-2}) + (1 \cdot 2^{-3}) \\ &= 2 + 1 + 0 + 0.25 + 0.125 \\ &= 3.375 \end{aligned}$$

EXAMPLE 6

Convert 0.11001_2 to a decimal numeral.

Solution

$$\begin{aligned} 0.11001_2 &= (1 \cdot 2^{-1}) + (1 \cdot 2^{-2}) + (0 \cdot 2^{-3}) + (0 \cdot 2^{-4}) + (1 \cdot 2^{-5}) \\ &= 0.5 + 0.25 + 0 + 0 + 0.03125 \\ &= 0.78125 \end{aligned}$$

EXERCISES FOR SECTION 3.1

In Exercises 1 through 12, write each base ten numeral in expanded form (see Examples 1 and 2).

1. 289	**2.** 1256	**3.** 812	**4.** 2990
5. 50238	**6.** 216.5	**7.** 900.65	**8.** 12.456
9. 1002.01	**10.** 9.0023	**11.** 0.0024	**12.** 909.1206

In Exercises 13 through 21, convert each binary numeral to base ten by first writing in expanded form (see Examples 3 and 4).

13. 1101_2	**14.** 1011_2	**15.** 1001_2
16. 10101_2	**17.** 10111_2	**18.** 10011_2
19. 1010001_2	**20.** 1001111_2	**21.** 111101_2

In Exercises 22 through 30, convert each binary numeral to a base ten numeral (see Examples 5 and 6).

22. 10.01_2	**23.** 10.11_2	**24.** 101.011_2
25. 1.0111_2	**26.** 110.0001_2	**27.** 0.11011_2
28. 0.00001_2	**29.** 10.00011_2	**30.** 11.010101_2

3.2 Decimal-to-Binary Conversion

We have already stated that calculations with and representations of numbers in a computer are generally related in some way to their representation in the binary number system; it is useful, then, to be able to make the conversion from the decimal system to the binary system. It is this latter form that computers can "understand."

Two methods of making this conversion are explained in the remainder of this section. In each case, the problem is to convert the (decimal) numeral 61 to its binary equivalent.

First Method

Look again at the powers of two in Tables 3.1 and 3.2. In order to convert 61 to its binary equivalent, we search for the highest power of two that is less than or equal to 61. That number is 32 or 2^5. The next step is to subtract this number from 61 (that is, $61 - 32 = 29$) and repeat the process with this new number, 29. The largest power of two that is less than or equal to 29 is 16 (or 2^4). Subtracting 16 from 29 yields 13 and the process is repeated with 13. We depict this method below:

$$\begin{array}{rl}
61 & \\
-32 & \text{32 is the largest power of 2 less than or equal to 61.} \\
\overline{29} & \text{Subtract it from 61. } 61 = 2^5 + 29 \\
-16 & \text{16 is the largest power of 2 less than or equal to 29.} \\
\overline{13} & \text{Subtract it from 29. } 29 = 2^4 + 13 \\
-8 & \text{8 is the largest power of 2 less than or equal to 13.} \\
\overline{5} & \text{Subtract it from 13. } 13 = 2^3 + 5 \\
-4 & \text{4 is the largest power of 2 less than or equal to 5.} \\
\overline{1} & \text{Subtract it from 5. } 5 = 2^2 + 1 \\
-1 & \text{1 is the largest power of 2 less than or equal to 1.} \\
\overline{0} & \text{Subtract it from 1. } 1 = 2^0 + 0
\end{array}$$

Therefore, $61 = 2^5 + 2^4 + 2^3 + 2^2 + 2^0$. The expanded form of 61 is, then,

$$(1 \cdot 2^5) + (1 \cdot 2^4) + (1 \cdot 2^3) + (1 \cdot 2^2) + (0 \cdot 2^1) + (1 \cdot 2^0)$$

We write

$$61 = 111101_2$$

Note the importance of "supplying" a zero in the position of 2^1. We examine an example before going on to the second method.

EXAMPLE 1

Convert 872 to its binary representation.

Solution

The largest power of two that is less than or equal to 872 is $512 = 2^9$ (Table 3.1). We subtract it and continue the procedure:

$$\begin{array}{rl}
872 & \\
-512 & 512 = 2^9 \\
\overline{360} & \\
-256 & 256 = 2^8 \\
\overline{104} & \\
-64 & 64 = 2^6 \\
\overline{40} & \\
-32 & 32 = 2^5 \\
\overline{8} & \\
-8 & 8 = 2^3 \\
\overline{0} & \leftarrow \text{Indicator to STOP}
\end{array}$$

Our final result is

$$872 = 1101101000_2$$

Second Method

The second method, less obviously related to the place value system of notation, involves repeated division by two. More specifically, we divide the original decimal numeral by two and note the remainder. We take the result of that division (temporarily ignoring the remainder) and divide by two, again noting the remainder. This is repeated until the division leaves a zero. We demonstrate this algorithm in Example 2.

EXAMPLE 2

Use the second method to convert 61 to its base two equivalent.

Solution

After each division, we write the quotient *under* the dividend and the remainder to the right.

Division	Remainder	
2) 61		
2) 30	Remainder = 1	↑
2) 15	Remainder = 0	
2) 7	Remainder = 1	read
2) 3	Remainder = 1	up
2) 1	Remainder = 1	
0	Remainder = 1	

We now form the binary representation by listing the division remainders in reverse order. The binary representation of 61 is, therefore, 111101_2.

EXAMPLE 3

Convert 455 to a base two numeral.

Solution

Division	Remainder
2) 455	
2) 227	Remainder = 1
2) 113	Remainder = 1
2) 56	Remainder = 1
2) 28	Remainder = 0
2) 14	Remainder = 0
2) 7	Remainder = 0
2) 3	Remainder = 1
2) 1	Remainder = 1
0	Remainder = 1

Thus,

$$455 = 111000111_2$$

To see why this method works, look at Example 2 in another way.

$$\begin{array}{rll}
2\,\underline{)\,61} & & \\
2\,\underline{)\,30} & \text{Remainder} = 1 & 61 = 2 \times 30 + 1 \\
2\,\underline{)\,15} & \text{Remainder} = 0 & 30 = 2 \times 15 + 0 \\
2\,\underline{)\,7} & \text{Remainder} = 1 & 15 = 2 \times 7 + 1 \\
2\,\underline{)\,3} & \text{Remainder} = 1 & 7 = 2 \times 3 + 1 \\
2\,\underline{)\,1} & \text{Remainder} = 1 & 3 = 2 \times 1 + 1 \\
0 & \text{Remainder} = 1 & 1 = 2 \times 0 + 1
\end{array}$$

Now start with $61 = 2 \cdot 30 + 1$ and make substitutions:

$$\begin{aligned}
61 &= 2 \cdot 30 + 1 \\
&= 2 \cdot (2 \cdot 15 + 0) + 1 = 2^2 \cdot 15 + 2^1 \cdot 0 + 1 \\
&= 2^2 \cdot (2 \cdot 7 + 1) + 2^1 \cdot 0 + 1 \\
&= 2^3 \cdot 7 + 2^2 \cdot 1 + 2^1 \cdot 0 + 1 \\
&= 2^3 \cdot (2 \cdot 3 + 1) + 2^2 + 0 \cdot 2^1 + 2^0 \\
&= 2^4 \cdot 3 + 2^3 + 2^2 + 0 \cdot 2^1 + 2^0 \\
&= 2^4 \cdot (2 \cdot 1 + 1) + 2^3 + 2^2 + 0 \cdot 2^1 + 2^0 \\
61 &= 2^5 + 2^4 + 2^3 + 2^2 + 0 \cdot 2^1 + 2^0
\end{aligned}$$

or

$$61 = (1 \cdot 2^5) + (1 \cdot 2^4) + (1 \cdot 2^3) + (1 \cdot 2^2) + (0 \cdot 2^1) + (1 \cdot 2^0)$$

This gives us

$$61 = 111101_2.$$

Decimal-to-Binary Conversion: Fractions

To convert decimal fractions to binary fractions, we use a process similar to the second method above but using successive multiplications instead of divisions. We begin by multiplying the given decimal number by two. The whole number part of this product becomes the first digit after the binary point. We then multiply the fractional part of this product by two. The next binary digit will be the whole number part of this new product. The process is continued until we arrive at a product whose fractional part is zero or until either an acceptable degree of precision or a sequence of repeating digits is reached. The next two examples illustrate conversion to binary from decimal systems for numbers between zero and one.

EXAMPLE 4

Convert 0.71875 to a binary numeral.

Solution

Number	Number × 2	Whole Number Part
0.71875	1.43750	1
0.4375	0.8750	0
0.875	1.750	1
0.750	1.50	1
0.50	1.0	1
0 Done!		

Thus, $0.71875 = 0.10111_2$.

EXAMPLE 5

Convert 0.45 to a binary numeral.

Solution

Notice that in the sequence below, a fractional part of zero is never reached. We stop after a sequence of repeating digits (1100) is reached:

Number	Number × 2	Whole Number Part
0.45	0.90	0
0.90	1.80	1
0.80	1.60	1 ⎤
0.60	1.20	1 ⎥
0.20	0.40	0 ⎥
0.40	0.80	0 ⎦
0.80	1.60	1 ⎤
0.60	1.20	1 ⎥
0.20	0.40	0 ⎥
0.40	0.80	0 ⎦
0.80	1.60	1
0.60	. . .	

So, $0.45 = 0.0111001100_2$.

The final example of this section deals with converting a mixed number greater than one (a base ten numeral with something on each side of the decimal point) to the binary system.

EXAMPLE 6

Convert 12.58 to its base two equivalent.

Solution

We use the first method described to first convert 12 to binary.

$$\begin{array}{rl} 12 & \\ -8 & 8 = 2^3 \\ \hline 4 & \\ -4 & 4 = 2^2 \\ \hline 0 & \end{array}$$

So, $12 = 1100_2$.

Next, we must convert 0.58 to binary:

Number	Number × 2	Whole Part
0.58	1.16	1
0.16	0.32	0
0.32	0.64	0
0.64	1.28	1
0.28	0.56	0
0.56	1.12	1
0.12	0.24	0
0.24	0.48	0
0.48	0.96	0
0.96	1.92	1
0.92	1.84	1
etc.		

Ending with this precision, we have $0.58 = 0.10010100011_2$.

Finally, we combine the portions to the right and left of the binary point: $12.58 = 1100.10010100011_2$.

EXERCISES FOR SECTION 3.2

In Exercises 1 through 12, convert each decimal number to its binary equivalent (see Examples 1–3).

1. 28
2. 72
3. 871
4. 442
5. 19
6. 120
7. 321
8. 252
9. 255
10. 256
11. 257
12. 290

In Exercises 13 through 24, convert each decimal to its binary equivalent (see Examples 4 and 5).

13. 0.5625 **14.** 0.375 **15.** 0.875
16. 0.6875 **17.** 0.53125 **18.** 0.78125
19. 0.046875 **20.** 0.921875 **21.** 0.35
22. 0.90 **23.** 0.1 **24.** 0.99

In Exercises 25 through 30, convert each base ten number to a base two number (see Example 6).

25. 28.35 **26.** 72.90 **27.** 72.1
28. 19.375 **29.** 14.5 **30.** 10.34

Exercises 31 through 33 involve writing and interpreting BASIC programs.

31. Consider the following BASIC program to convert a positive integer from base ten to base two:

```
 10  LET Z = 0
 20  DIM D(30)
 30  PRINT "INPUT A BASE TEN POSITIVE INTEGER."
 40  INPUT X
 50  FOR I = 30 TO 0 STEP -1
 60    IF X < (2^I) THEN D(I) = 0: GOTO 800
 70    LET T = INT(2^I)
 80    LET Z = Z + 1
 90    LET X = X - T
100    LET D(I) = 1
110    IF Z = 1 THEN Q = I
800  NEXT I
900  PRINT "THE BASE TWO EQUIVALENT IS"
910  FOR I = Q TO 0 STEP -1
920    IF D(I) = 1 THEN PRINT "1" ;
930    IF D(I) = 0 THEN PRINT "0" ;
940  NEXT I
950  PRINT
999  END
```

a. Refine the program. Be sure to document it with REM statements.
b. What role does the variable Z play in the program?
c. What limitations exist, if any, on the base ten number input by the user?
d. Which text method of conversion does the program employ, the first or the second?

32. Using the program of Exercise 31 as a guide, write a different program to convert base ten numbers between zero and one to their binary equivalents.

33. In Example 5 we saw that the terminating decimal number 0.45 had an infinite (nonterminating) binary representation. Thus, many numbers can be represented only approximately in the computer's memory. (This fact is also one of the reasons why the associative and commutative rules of addition and multiplication do not always work on a computer. See Chapter 1 for a further discussion of this situation.)

    a. On a microcomputer run the following BASIC mini-program:

    ```
    10  PRINT  654.321 - 654
    ```

    Explain the result. Do you think all computers will give the same result?

    b. Rework part a for this statement:

    ```
    10  PRINT  765.4321 - 765
    ```

3.3 Binary Arithmetic

In order to perform computations in the binary number system, it is helpful to examine closely the methods we use to perform similar calculations in the decimal system. We can then use our observations about what we do in base ten to develop algorithms for base two.

Addition

In base ten, if we want to add the decimal numbers 172 and 354, we begin by writing them one under the other, putting digits with the same place value in the same column. Our addition problem then looks like this:

$$\begin{array}{r} 172 \\ +354 \\ \hline \end{array}$$

The next step is to add digits that are in the same column, writing the sums underneath. We obtain

$$\begin{array}{r} 1 \\ +3 \\ \hline 4 \end{array} \qquad \begin{array}{r} 7 \\ +5 \\ \hline 12 \end{array} \qquad \begin{array}{r} 2 \\ +4 \\ \hline 6 \end{array}$$

The 12 in the middle column is not exactly what we want, however. So we use the fact that the 12, representing twelve *tens*, means one hundred and two tens;

we *carry* the one to the hundreds column and our addition problem looks more familiar:

$$\begin{array}{r} {\scriptstyle 1} \\ 172 \\ +354 \\ \hline 526 \end{array}$$

Of course, a decimal addition problem may involve more than one "carry" as in the following:

$$\begin{array}{r} {\scriptstyle 1\,1\,1} \\ 6825 \\ +7489 \\ \hline 14314 \end{array}$$

Addition in base two involves the same two-step algorithm:

1. Find the sum of each pair of digits, and
2. carry as necessary.

The first of these is simpler than for base ten addition since there are four combinations of digits to add. The only base two addition facts we must memorize are: $0 + 0 = 0$; $0 + 1 = 1$; $1 + 0 = 1$; $1 + 1 = 10_2$.

To add two bits (binary digits) the following table may be useful:

+	0	1
0	0	1
1	1	10

Digits are then carried as in the decimal system; two examples follow:

EXAMPLE 1

Add $101_2 + 11_2$.

Solution

$$\begin{array}{r} {\scriptstyle 1} \\ 101_2 \\ +11_2 \\ \hline 0 \end{array}$$

Notice that $1 + 1 = 10$ in base two. We put down the 0 and carry the 1.

We complete the addition of the numbers as follows:

$$\begin{array}{r} {\scriptstyle 1\,1} \\ 101_2 \\ +11_2 \\ \hline 1000_2 \end{array}$$

One way to check the result is to convert each number to base ten:

$$\begin{array}{rcr} 101_2 & \rightarrow & 5 \\ +11_2 & \rightarrow & +3 \\ \hline 1000_2 & \rightarrow & 8 \ \checkmark \end{array} \qquad \text{It checks!}$$

EXAMPLE 2

Add $11011_2 + 1111_2$.

Solution

$$\begin{array}{r} {\scriptstyle 1111} \\ 11011_2 \\ +1111_2 \\ \hline 101010_2 \end{array}$$

Check

$$\begin{array}{rcr} 11011_2 & \rightarrow & 27 \\ +1111_2 & \rightarrow & +15 \\ \hline 101010_2 & \rightarrow & 42 \ \checkmark \end{array}$$

Complements

Before we examine subtraction in base two, it will be helpful to discuss the notion of the **complement** of a number. In base ten, the **nines complement** of a number is obtained by subtracting each digit of the number from nine:

number:	3608
nines complement:	6391

EXAMPLE 3

Find the nines complement of each of the following numbers:

a. 172 **b.** 9023

Solution

a. Subtracting each digit of 172 from 9 yields 827

b. The nines complement of 9023 is 0976 or simply 976.

A computer can use the notion of complement to subtract without needing to "borrow." The technique is as follows: To find $x - y$ by the **nines complement**

method of subtraction

1. determine the nines complement of y. Call it y',
2. find $x + y'$;
3. "shift"—that is, move the leftmost digit from step 2 to the rightmost position and then add.

It is shown in Exercise 44 that the nines complement method is equivalent to the customary "borrowing" method of subtraction. The two methods are compared below to find $9763 - 3608$:

Regular subtraction

$$\begin{array}{r} {\scriptstyle 5\,13} \leftarrow \text{"borrow"} \\ 97\not{6}\not{3} \\ -3608 \\ \hline 6155 \end{array}$$

Nines complement subtraction

The nines complement of 3608 is 6391.

$$\begin{array}{r} 9763 \\ +6391 \\ \hline 16154 \\ \llcorner\!\!\to +1 \\ \hline 6155 \end{array} \quad \text{Answer}$$

EXAMPLE 4

Use the nines complement method to find $2456 - 827$.

Solution

First, we find the nines complement of 0827. Notice that we fill in a leading zero because 2456 has one more digit than 827. The nines complement of 0827 is 9172. Now we *add* 9172 to 2456 and "shift" as below:

$$\begin{array}{r} 2456 \\ +9172 \\ \hline 11628 \\ \llcorner\!\!\to +1 \\ \hline 1629 \end{array} \quad \text{Answer}$$

A big advantage of this method is that it can be adapted to any base. In the binary system, we use the technique as described above except instead of finding the nines complement, we find the *ones complement* of a number. We illustrate the subtraction method with the next example.

EXAMPLE 5

Find $110011_2 - 10111_2$.

Solution

We find the *ones* complement of 010111 (the leading zero is filled because 110011 has six digits) by subtracting each digit from one. This is equivalent to exchanging ones and zeros.

$$010111_2 \rightarrow 101000_2 \qquad \text{The ones complement}$$

Next, we add and shift:

$$\begin{array}{r} 110011_2 \\ +101000_2 \\ \hline 1011011_2 \\ \llcorner\!\rightarrow +1 \\ \hline 11100_2 \end{array}$$

Check

We check the problem by converting each number to base ten and then subtracting:

$$\begin{array}{rcr} 110011_2 & \rightarrow & 51 \\ -10111_2 & \rightarrow & -23 \\ \hline 11100_2 & \rightarrow & 28 \; \checkmark \end{array}$$

Another algorithm for binary subtraction involves using the twos complement of the number to be subtracted. The **twos complement** of a binary number is found as follows:

1. Replace all zeros in the number with ones and all ones with zeros (i.e., find the ones complement).
2. Add one to the result of the first step (this replaces the "end-around-carry").

We illustrate the twos complement method for the subtraction of Example 5:

$$110011_2 - 10111_2$$

As before, we begin by adding a leading zero to 10111_2, so that it has the same number of digits as 110011_2. Next, we find the twos complement:

$$010111_2 \rightarrow 101000_2 + 1_2 = 101001_2 \qquad \text{Twos complement}$$

We then add:

$$\begin{array}{r} {\scriptstyle 1\,1} \\ 110011_2 \\ +101001_2 \\ \hline 1011100_2 \end{array} \xrightarrow[\text{first digit}]{\text{drop}} 011100_2$$

Since the resulting sum has more digits than either of the original numbers, we drop the high-order (leftmost) bit. The answer is $110011_2 - 10111_2 = 11100_2$.

Note that this result is the same as the answer we obtained in Example 5 using the ones complement.

The twos complement method lends itself to solving subtraction problems in which the difference is negative as well. It is also more directly related than the ones complement to the way computers do arithmetic, since most computers *store* negative numbers in complement form. We'll illustrate the process first with a base ten subtraction example.

Suppose we want to perform the following subtraction: 425 − 591. We begin by finding the *tens complement* of 591:

1. Find the nines complement by subtracting each digit from 9. It is *408*.
2. Add 1 to the nines complement. The result is *409*, the tens complement of 591.

Now perform the addition:

$$\begin{array}{r} \overset{1}{4}25 \\ +409 \\ \hline 834 \end{array}$$

Notice that there is *no* high-order digit to drop; notice also that this is *not* the correct answer. *Whenever there is no high-order digit to drop in a subtraction problem, the difference is negative.* We find the "real" answer by taking the tens complement of 834 and prefixing it with a minus sign:

$$\begin{array}{rl} 834 \rightarrow 165 & \text{Nines complement} \\ +1 & \\ \hline 166 & \text{Tens complement} \end{array}$$

The answer is 425 − 591 = −166.
Check: 591 + (−166) = 425

EXAMPLE 6

Use the twos complement method to perform the following subtractions:

a. $10011_2 - 1010_2$ **b.** $101101_2 - 110001_2$ **c.** $10111_2 - 110011_2$

Solution

a. 1. Find the twos complement of 01010_2. Notice that we added a leading zero.

$$\begin{array}{rl} 01010_2 \rightarrow 1010\overset{1}{1}_2 & \text{Ones complement} \\ +1_2 & \\ \hline 10110_2 & \text{Twos complement} \end{array}$$

2. Now add:

$$\begin{array}{r} {\scriptstyle 11} \\ 10011_2 \\ +10110_2 \\ \hline 101001_2 \end{array}$$

3. Finally, drop the leftmost digit, the high-order bit. The answer is

$$10011_2 - 1010_2 = 1001_2$$

b. $101101_2 - 110001_2$

1. $110001_2 \rightarrow 001110_2$ Ones complement

$$\begin{array}{r} 001110_2 \\ +1_2 \\ \hline 001111_2 \end{array}$$

Twos complement

2.
$$\begin{array}{r} {\scriptstyle 1111} \\ 101101_2 \\ +001111_2 \\ \hline 111100_2 \end{array}$$

3. There is no high-order bit to drop (because the answer has the same number of digits as the number we're subtracting from), so the answer must be negative. We find the twos complement of 111100_2.

$$111100_2 \rightarrow \begin{array}{r} {\scriptstyle 11} \\ 000011_2 \\ +1_2 \\ \hline 000100_2 \end{array}$$

Ones complement

Twos complement

The answer is $101101_2 - 110001_2 = -100_2$

Check

We check this result by converting to base ten.

$$\begin{array}{rcr} 101101_2 & \rightarrow & 45 \\ -110001_2 & \rightarrow & -49 \\ \hline -100_2 & \rightarrow & -4 \end{array}$$

c. $10111_2 - 110011_2$

1. $110011_2 \rightarrow 001100_2$

$$\begin{array}{r} 001100_2 \\ +1_2 \\ \hline 001101_2 \end{array}$$

2.
$$\begin{array}{r} {\scriptstyle 1111} \\ 010111_2 \\ +001101_2 \\ \hline 100100_2 \end{array}$$

Notice that we give *both* numbers the same number of digits as in the *larger* of the original numbers.

3. $100100_2 \rightarrow \overset{1\,1}{011011_2}$

$$\begin{array}{r} 011011_2 \\ +1_2 \\ \hline 011100_2 \end{array}$$

The answer is $10111_2 - 110011_2 = -11100_2$
The reader should verify this answer by checking it in base ten.

Multiplication

Multiplication in the binary system is best described in terms of the process of decimal multiplication. The binary multiplication table, however, is extremely simple!

×	0	1
0	0	0
1	0	1

We display binary multiplication in the examples below. Notice in each case that we must maintain the alignment of the partial products, just as in decimal multiplication.

EXAMPLE 7

Multiply $110_2 \cdot 101_2$.

Solution

$$\begin{array}{r} 110_2 \\ \times\,101_2 \\ \hline 110 \\ 000 \\ 110 \\ \hline 11110_2 \end{array}$$

Check

We can check by converting to base ten and then multiplying:

$$\begin{array}{rcr} 110_2 & \rightarrow & 6 \\ \times\,101_2 & \rightarrow & \times\,5 \\ \hline 11110_2 & \rightarrow & 30 \end{array}$$

EXAMPLE 8

Find the product of 11011_2 with 1011_2.

Solution

$$
\begin{array}{r}
11011_2 \\
\times\ 1011_2 \\
\hline
11011 \\
11011 \\
00000 \\
11011 \\
\hline
100101001_2
\end{array}
$$

Check

$$
\begin{array}{rcr}
11011_2 & \rightarrow & 27 \\
\times\ 1011_2 & \rightarrow & \times\ 11 \\
\hline
100101001_2 & \rightarrow & 297
\end{array}
$$

Division

The algorithm for division in the binary system can also be mimicked from the decimal system. To divide 10101_2 by 11_2, for example, we proceed as follows:

$$
\begin{array}{r}
1 \\
11_2\,\overline{)\,10101_2} \\
-11 \\
\hline
10
\end{array}
\qquad
\begin{array}{l}
\text{11 "goes into" 101 one time.} \\
\text{Multiply and subtract.}
\end{array}
$$

The process is continued until the remainder after subtracting and "bringing down" is less than the *divisor* 11_2.

$$
\begin{array}{r}
111_2 \\
11_2\,\overline{)\,10101_2} \\
-11 \\
\hline
100 \\
-11 \\
\hline
11 \\
-11 \\
\hline
0
\end{array}
\qquad \text{Remainder} = 0
$$

The resulting *quotient* is 111_2.

We complete this section with another example of binary division.

EXAMPLE 9

Divide 1110100_2 by 1011_2.

Solution

The reader should verify each of the steps below.

$$\begin{array}{r} 1010_2 \\ 1011_2 \overline{) \, 1110100_2} \\ -1011 \\ \hline 1110 \\ -1011 \\ \hline 110 \\ -0 \\ \hline 110 \end{array}$$

Thus, when 1110100_2 is divided by 1011_2, the result is 1010_2 with remainder 110_2.

We leave the check to the reader. The method of checking is to convert each number (divisor, dividend, quotient, and remainder) to its decimal equivalent.

EXERCISES FOR SECTION 3.3

In Exercises 1 through 10 find the indicated sum (see Examples 1 and 2).

1. $101_2 + 10_2$
2. $110_2 + 11_2$
3. $111_2 + 11_2$
4. $1011_2 + 100_2$
5. $1101_2 + 101_2$
6. $1011_2 + 111_2$
7. $10110_2 + 1001_2$
8. $101101_2 + 11011_2$
9. $101010_2 + 101011_2$
10. $1101111_2 + 1011011_2$

In Exercises 11 through 15, find the nines complement of each base ten number (see Example 3).

11. 2582
12. 723
13. 1002
14. 3891
15. 12392

In Exercises 16 through 20, use the complement method of subtraction to find each indicated base ten difference (see Example 4).

16. 9763 – 3605
17. 1382 – 1234
18. 762 – 128
19. 3056 – 239
20. 2000 – 289

In Exercises 21 through 25, find the ones complement and the twos complement of each binary number (see Examples 5 and 6).

21. 1011_2 **22.** 11011_2 **23.** 1001101_2
24. 1101111_2 **25.** 10100011001_2

In Exercises 26 through 31, use the ones complement method and the twos complement method to find each indicated binary difference (see Examples 5 and 6).

26. $110011_2 - 10011_2$ **27.** $110011_2 - 101110_2$
28. $101010_2 - 11011_2$ **29.** $101010101_2 - 100111_2$
30. $1010111_2 - 11111_2$ **31.** $1011000_2 - 111111_2$

In Exercises 32 through 37, find the indicated binary product (see Examples 7 and 8).

32. $110_2 \cdot 11_2$ **33.** $101_2 \cdot 110_2$
34. $110_2 \cdot 110_2$ **35.** $11011_2 \cdot 1101_2$
36. $110010_2 \cdot 1011_2$ **37.** $110110_2 \cdot 101011_2$

In Exercises 38 through 43, find the indicated binary quotients and remainders (see Example 9).

38. $1110100_2 \div 101_2$ **39.** $110110_2 \div 1001_2$
40. $11011010_2 \div 11011_2$ **41.** $101101_2 \div 101_2$
42. $100001_2 \div 1111_2$ **43.** $100101111_2 \div 1001010_2$

44. The method of subtraction by complements uses the shift or "end-around-carry" technique. Its mathematical justification can be outlined as follows:
We assume here that x is a four-digit number, but the verification can easily be adapted for any subtraction of an integer from a larger integer. Verify each statement below using $x = 2357$ and $y = 1455$.
a. The nines complement of y, y', is $9999 - y$.
b. $x + y'$ is equal to $x + 9999 - y$.
c. $x + y' - 10000 + 1$ is $x + 9999 - y - 10000 + 1$. (This is the end-around-carry step.)
d. $x - y$ is the same as the result of the end-around-carry, $x + 9999 - y - 10000 + 1$.

45. Rework Exercise 44 for the case when x is a five-digit number.

3.4 The Octal Number System

The binary number system, with its ability to represent numbers in terms of the digits 0 and 1, is ideally suited for storing, manipulating, and retrieving information in electronic computers. For the programmer, however, it can be very cumbersome since it requires many zeros and ones to represent very large or very small quantities. The **octal** (base eight) and **hexadecimal** (base sixteen) number systems are easier for the programmer to work with. Because of their special relationship to the binary system (which will be examined later in this chapter) they are also easily adapted for use in computers. In this section, we shall describe the octal number system and its arithmetic.

The Octal/Decimal Relationship

The octal or base eight number system, like the decimal and binary systems, uses the concepts of place value and positional notation. Place values, of course, are calculated as multiples of powers of eight. We use eight digits in the octal system: 0, 1, 2, 3, 4, 5, 6, and 7.

To convert a number like 314_8 to a decimal numeral, we simply expand it realizing that the 3 represents $3 \cdot 8^2$, the 1 means $1 \cdot 8^1$, and the 4 means $4 \cdot 8^0$. As we proceed with examples, the listing of powers of eight in Table 3.3 will prove useful.

TABLE 3.3
Powers of eight.

n	8^n	n	8^n
4	4096	0	1
3	512	−1	0.125
2	64	−2	0.015625
1	8	−3	0.001953125

EXAMPLE 1
Convert each of the following to its decimal equivalent:

a. 314_8 **b.** 2071_8

Solution

a. $314_8 = (3 \cdot 8^2) + (1 \cdot 8^1) + (4 \cdot 8^0)$
$= (3 \cdot 64) + (1 \cdot 8) + (4 \cdot 1)$
$= 192 + 8 + 4 = 204$

b. $2071_8 \quad (2 \cdot 8^3) + (0 \cdot 8^2) + (7 \cdot 8^1) + (1 \cdot 8^0)$
$= (2 \cdot 512) + (0 \cdot 64) + (7 \cdot 8) + (1 \cdot 1)$
$= 1024 + 0 + 56 + 1 = 1081$

EXAMPLE 2

Convert each of the following to its base ten equivalent:

a. 16.5_8 **b.** 0.13_8

Solution

a. $16.5_8 = (1 \cdot 8^1) + (6 \cdot 8^0) + (5 \cdot 8^{-1})$
$= (1 \cdot 8) + (6 \cdot 1) + (5 \cdot 0.125)$
$= 8 + 6 + 0.625 = 14.625$

b. $0.13_8 = (1 \cdot 8^{-1}) + (3 \cdot 8^{-2})$
$= 0.125 + 0.046875$
$= 0.171875$

We can accomplish conversion *from* decimal *to* octal using either of the methods described in Section 3.2. We choose the second method below and convert a number like 196 to base eight by repeated division by eight. Recall that after each division we keep track of the remainder. Again, we repeat the division by eight until there is a zero quotient.

EXAMPLE 3

Convert each of the following to its octal equivalent:

a. 196 **b.** 2437

Solution

a. We begin by dividing by 8:

$$\begin{array}{rl} 8\,\overline{)\,196} & \\ 8\,\overline{)\,24} & \text{Remainder} = 4 \\ 8\,\overline{)\,3} & \text{Remainder} = 0 \\ 0 & \text{Remainder} = 3 \end{array}$$

Reading from the bottom up, we have

$$196 = 304_8$$

b. $8\,\overline{)\,2437}$

$8\,\overline{)\,304}$	Remainder = 5
$8\,\overline{)\,38}$	Remainder = 0
$8\,\overline{)\,4}$	Remainder = 6
0	Remainder = 4

So,

$$2437 = 4605_8$$

Arithmetic in Base Eight: Addition

Arithmetic operations in the octal number system can be performed using the same algorithms that apply to decimal arithmetic. Since the base of our number system is eight, a base eight table of addition facts will be useful (Table 3.4).

TABLE 3.4

Octal addition table.

+	0	1	2	3	4	5	6	7
0	0	1	2	3	4	5	6	7
1	1	2	3	4	5	6	7	10
2	2	3	4	5	6	7	10	11
3	3	4	5	6	7	10	11	12
4	4	5	6	7	10	11	12	13
5	5	6	7	10	11	12	13	14
6	6	7	10	11	12	13	14	15
7	7	10	11	12	13	14	15	16

EXAMPLE 4

Perform the indicated addition:

$$\begin{array}{r} 4506_8 \\ +3675_8 \\ \hline \end{array}$$

Solution

Below, we "spread out" the addition problem in order to see how the "carrying" is done in base eight.

$$\begin{array}{r} 4 \\ +3 \\ \hline 7 \end{array} \quad \begin{array}{r} 5 \\ +6 \\ \hline 13 \end{array} \quad \begin{array}{r} 0 \\ +7 \\ \hline 7 \end{array} \quad \begin{array}{r} 6 \\ +5 \\ \hline 13 \end{array}$$

The "1" part of each "13" must be carried. Notice, though, that when we carry a "1" to "7," the result is 10_8, and that causes another carry. The result is illustrated below:

$$
\begin{array}{r}
{\scriptstyle 1\,1\,1} \\
4506_8 \\
+\,3675_8 \\
\hline
10403_8
\end{array}
$$

EXAMPLE 5

Add the following three octal numbers:

$$
\begin{array}{r}
1217_8 \\
2335_8 \\
+\,305_8 \\
\hline
\end{array}
$$

Solution

Take a look at the rightmost column of octal digits. Although we *think* that $7 + 5 + 5$ is seventeen, we must *write* seventeen in base eight as 21. So, we put down the 1 and carry the 2. The reader is urged to verify the other "carries" in the finished addition below:

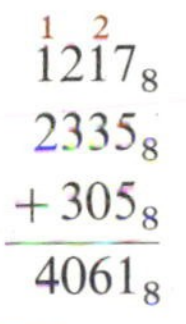

Subtraction

In Section 3.2, we saw the complement method of subtraction used. In this section, we shall subtract in base eight using the idea of **sevens complement**. To construct the sevens complement of a base eight number, we subtract each digit from seven. For example, consider 2654_8. Subtracting each digit from seven yields

number:	2654_8
sevens complement:	5123_8

EXAMPLE 6

Subtract 7654_8 from 10443_8.

Solution

First, we must find the sevens complement of 07654. The leading zero has been filled in because there are five octal digits in the minuend; the sevens complement of 07654 is 70123. Next, we perform the subtraction (actually, we *add* the complement) and the end-around-carry:

$$\begin{array}{r} 10443_8 \\ +70123_8 \\ \hline 100566 \\ \hookrightarrow +1 \\ \hline 567_8 \end{array}$$

Multiplication and Division

Multiplication and division in base eight use the same algorithms as we used in the binary and decimal systems. We begin by displaying the octal multiplication table in Table 3.5.

TABLE 3.5

The octal multiplication table.

×	0	1	2	3	4	5	6	7
0	0	0	0	0	0	0	0	0
1	0	1	2	3	4	5	6	7
2	0	2	4	6	10	12	14	16
3	0	3	6	11	14	17	22	25
4	0	4	10	14	20	24	30	34
5	0	5	12	17	24	31	36	43
6	0	6	14	22	30	36	44	52
7	0	7	16	25	34	43	52	61

EXAMPLE 7

Find the product: $467_8 \times 25_8$.

Solution

$$\begin{array}{rl} {\scriptstyle 11} & \\ {\scriptstyle 44} & \\ 467_8 & \\ \times 25_8 & \\ \hline 3023 & 5 \cdot 7 = 43;\ 5 \cdot 6 = 36 + 4 \text{ carry} = 42;\ 5 \cdot 4 = 24 + 4 \text{ carry} = 30 \\ 1156 & 2 \cdot 7 = 16;\ 2 \cdot 6 = 14 + 1 \text{ carry};\ 2 \cdot 4 = 10 + 1 \text{ carry} = 11 \\ \hline 14603_8 & \end{array}$$

EXAMPLE 8

Divide 14603_8 by 25_8.

Solution

Since we think in base ten, when we try to determine how many times 25_8 goes into 146_8, we convert to base ten in our heads and ask "how many times does 21 go into 102?" The answer is 4 and the process of multiplying by 4, subtracting, and bringing down is displayed below.

$$\begin{array}{r} 4{}_8 \\ 25_8 \,\overline{)\,14603_8} \\ \underline{124} \\ 22 \end{array}$$

The reader is urged to verify each number in the completed division below:

$$\begin{array}{r} 467_8 \\ 25_8 \,\overline{)\,14603_8} \\ \underline{124} \\ 220 \\ \underline{176} \\ 223 \\ \underline{223} \\ 0 \end{array}$$

We conclude this section by mentioning that, for practical purposes, anyone who really had to perform calculations in base eight would probably use octal-to-decimal and decimal-to-octal conversion tables. One advantage in studying arithmetic in other bases, though, is that it provides valuable insight into the actual algorithms used in arithmetic. Such insight is necessary for the person who designs computers or programs computer operating systems.

EXERCISES FOR SECTION 3.4

In Exercises 1 through 15, convert each octal number to its base ten equivalent (see Examples 1 and 2).

1. 205_8 **2.** 315_8 **3.** 17_8

4. 755_8 **5.** 206_8 **6.** 257_8

7. 2033_8
8. 1007_8
9. 4461_8
10. 10205_8
11. 15.6_8
12. 0.23_8
13. 5.55_8
14. 203.71_8
15. 103.103_8

In Exercises 16 through 24, convert each decimal number to its octal equivalent (see Examples 3 and 4).

16. 197
17. 291
18. 2437
19. 255
20. 256
21. 257
22. 5029
23. 1008
24. 8888

In Exercises 25 through 30, add the given base eight numbers (see Example 4).

25. $\begin{array}{r} 3052_8 \\ +1243_8 \\ \hline \end{array}$

26. $\begin{array}{r} 1025_8 \\ +4234_8 \\ \hline \end{array}$

27. $\begin{array}{r} 5007_8 \\ +3254_8 \\ \hline \end{array}$

28. $1213_8 + 2043_8$
29. $1036_8 + 2441_8 + 3073_8$
30. $303_8 + 2074_8 + 7005_8 + 3343_8$

In Exercises 31 through 36 perform the indicated operation (see Example 6).

31. $5672_8 - 2331_8$
32. $4235_8 - 1127_8$
33. $5340_8 - 2451_8$
34. $23506_8 - 2417_8$
35. $4024_8 - 1331_8$
36. $4006_8 - 1667_8$

In Exercises 37 through 42, find the indicated product (see Example 7).

37. $236_8 \times 24_8$
38. $253_8 \times 52_8$
39. $341_8 \times 27_8$
40. $254_8 \times 123_8$
41. $254_8 \times 107_8$
42. $467_8 \times 660_8$

In Exercises 43 through 45, perform the indicated division (see Example 8).

43. Divide 5043_8 by 24_8.
44. Divide 14504_8 by 35_8.
45. Divide 32570_8 by 103_8.

3.5 The Hexadecimal Number System

Like the binary, octal, and decimal systems, the hexadecimal number system uses the concepts of positional notation and place value. With sixteen as the base, the hexadecimal system necessarily *looks* different from the other three number bases we have studied thus far.

The number of digits in the binary system is two (0 and 1); in the octal system we used eight digits (0, 1, 2, 3, 4, 5, 6, and 7); of course, in the decimal system we use the ten digits 0 through 9. In a similar fashion, the base sixteen (hexadecimal) system requires sixteen digits. Since a single character is needed to represent each digit, we use 0 through 9 plus the following six letters:

A	for decimal	10
B	for decimal	11
C	for decimal	12
D	for decimal	13
E	for decimal	14
F	for decimal	15

Hexadecimal-to-Decimal and Decimal-to-Hexadecimal Conversion

We shall use a subscript H to denote the hexadecimal numerals. Conversion from a hexadecimal to a decimal numeral can be done by expanding the hexadecimal numeral in powers of the base (16). For example, $13A_H$ is equivalent to

$$13A_H = (1 \cdot 16^2) + (3 \cdot 16^1) + (10 \cdot 16^0) = 256 + 48 + 10 = 314$$

EXAMPLE 1

Convert $C3D0_H$ to its decimal equivalent.

Solution

$$\begin{aligned} C3D0_H &= (12 \cdot 16^3) + (3 \cdot 16^2) + (13 \cdot 16^1) + (0 \cdot 16^0) \\ &= (12 \cdot 4096) + (3 \cdot 256) + (13 \cdot 16) + (0 \cdot 1) \\ &= 49152 + 768 + 208 + 0 \\ &= 50128 \end{aligned}$$

To convert a number from the decimal system to the hexadecimal system, we use method two of Section 3.2. In the next example, we apply that method, repeated division by 16, to get the desired result.

EXAMPLE 2

Convert each of the following numbers to its hexadecimal equivalent:

a. 58 **b.** 61469

Solution

a. $16\,\overline{)\,58}$

$16\,\overline{)\,3}$ Remainder $= 10 \rightarrow A$

0 Remainder $= 3$

Reading up, $58 = 3A_H$.

b. 16) 61469

16) 3841 Remainder = 13 → D

16) 240 Remainder = 1

16) 15 Remainder = 0

0 Remainder = 15 → F

So,

$$61469 = F01D_H$$

Arithmetic in Base Sixteen: Addition

Arithmetic in the hexadecimal number system follows the same rules as in the binary, octal, and decimal systems. Since we are dealing with some unfamiliar symbols for digits, addition and multiplication tables will be especially helpful.

The addition table for base sixteen appears in Table 3.6. Note that the subscript H has been omitted in all table values. We can use the table to perform hexadecimal addition as the next example illustrates.

EXAMPLE 3

Perform the following hexadecimal additions:

a. $4C3A_H + 8BAD_H$ **b.** $BEAD_H + 90E8_H$

TABLE 3.6

The hexadecimal addition table.

+	0	1	2	3	4	5	6	7	8	9	A	B	C	D	E	F
0	0	1	2	3	4	5	6	7	8	9	A	B	C	D	E	F
1	1	2	3	4	5	6	7	8	9	A	B	C	D	E	F	10
2	2	3	4	5	6	7	8	9	A	B	C	D	E	F	10	11
3	3	4	5	6	7	8	9	A	B	C	D	E	F	10	11	12
4	4	5	6	7	8	9	A	B	C	D	E	F	10	11	12	13
5	5	6	7	8	9	A	B	C	D	E	F	10	11	12	13	14
6	6	7	8	9	A	B	C	D	E	F	10	11	12	13	14	15
7	7	8	9	A	B	C	D	E	F	10	11	12	13	14	15	16
8	8	9	A	B	C	D	E	F	10	11	12	13	14	15	16	17
9	9	A	B	C	D	E	F	10	11	12	13	14	15	16	17	18
A	A	B	C	D	E	F	10	11	12	13	14	15	16	17	18	19
B	B	C	D	E	F	10	11	12	13	14	15	16	17	18	19	1A
C	C	D	E	F	10	11	12	13	14	15	16	17	18	19	1A	1B
D	D	E	F	10	11	12	13	14	15	16	17	18	19	1A	1B	1C
E	E	F	10	11	12	13	14	15	16	17	18	19	1A	1B	1C	1D
F	F	10	11	12	13	14	15	16	17	18	19	1A	1B	1C	1D	1E

Solution

a.
$$\begin{array}{r} \overset{1}{4}\,C\,\overset{1}{3}\,A_H \\ +\ 8\,BAD_H \\ \hline D7E7_H \end{array}$$

b.
$$\begin{array}{r} B\overset{1}{E}\overset{1}{A}D_H \\ +\ 90E8_H \\ \hline 14F95_H \end{array}$$

Subtraction

To subtract $90E8_H$ from $14F95_H$ we proceed as follows:

1. To find the fifteens complement of $90E8_H$, the number to be subtracted, we subtract each digit of 090E8 (note the leading zero) from 15. Then we substitute the result for the original digit.

$$15 - 0 = 15 \rightarrow F$$
$$15 - 9 = 6$$
$$15 - 0 = 15 \rightarrow F$$
$$15 - E \text{ is equivalent to } 15 - 14 = 1$$
$$15 - 8 = 7$$

   So we replace $090E8_H$ by $F6F17_H$.

2. Next we add $14F95_H$ and $F6F17_H$:

$$\begin{array}{r} 1\overset{1}{4}F95_H \\ +F6F17_H \\ \hline 10BEAC_H \end{array}$$

3. The next step is the end-around-carry:

$$\begin{array}{r} 14F95_H \\ +F6F17_H \\ \hline 10BEAC_H \\ \llcorner\!\longrightarrow +1 \\ \hline 0BEAD_H \end{array}$$

We can verify this answer by referring to Example 3.

EXAMPLE 4

Subtract $3C4_H$ from $DE5F_H$.

Solution

1. The fifteens complement of $03C4_H$ (we supply a leading zero because $DE5F_H$ has four digits) is $FC3B_H$.

2. Adding $DE5F_H$ and $FC3B_H$ gives

$$\begin{array}{r} {}^{1}\,{}^{1} \\ DE5F_H \\ +FC3B_H \\ \hline 1DA9A_H \end{array}$$

3. The end-around-carry:

$$\begin{array}{r} DE5F_H \\ +FC3B_H \\ \hline 1DA9A_H \\ \llcorner\!\!\longrightarrow +1 \\ \hline DA9B_H \end{array}$$

Thus $DE5F_H - 3C4_H = DA9B_H$.

Multiplication and Division

We first construct a multiplication table for the hexadecimal digits (Table 3.7). As before, all numbers in the table are understood to be hexadecimal even though the subscript H is omitted. We now use this table to do some multiplication and division examples. We use the algorithms normally used in decimal arithmetic.

TABLE 3.7

The hexadecimal multiplication table.

×	0	1	2	3	4	5	6	7	8	9	A	B	C	D	E	F
0	0	0	0	0	0	0	0	0	0	0	0	0	0	0	0	0
1	0	1	2	3	4	5	6	7	8	9	A	B	C	D	E	F
2	0	2	4	6	8	A	C	E	10	12	14	16	18	1A	1C	1E
3	0	3	6	9	C	F	12	15	18	1B	1E	21	24	27	2A	2D
4	0	4	8	C	10	14	18	1C	20	24	28	2C	30	34	38	3C
5	0	5	A	F	14	19	1E	23	28	2D	32	37	3C	41	46	4B
6	0	6	C	12	18	1E	24	2A	30	36	3C	42	48	4E	54	5A
7	0	7	E	15	1C	23	2A	31	38	3F	46	4D	54	5B	62	69
8	0	8	10	18	20	28	30	38	40	48	50	58	60	68	70	78
9	0	9	12	1B	24	2D	36	3F	48	51	5A	63	6C	75	7E	87
A	0	A	14	1E	28	32	3C	46	50	5A	64	6E	78	82	8C	96
B	0	B	16	21	2C	37	42	4D	58	63	6E	79	84	8F	9A	A5
C	0	C	18	24	30	3C	48	54	60	6C	78	84	90	9C	A8	B4
D	0	D	1A	27	34	41	4E	5B	68	75	82	8F	9C	A9	B6	C3
E	0	E	1C	2A	38	46	54	62	70	7E	8C	9A	A8	B6	C4	D2
F	0	F	1E	2D	3C	4B	5A	69	78	87	96	A5	B4	C3	D2	E1

EXAMPLE 5

Do the following multiplication problems in base sixteen:

a. $2A3_H \cdot 4B_H$ **b.** $40B_H \cdot 23_H$

Solution

a.

$$\begin{array}{r} 2A3_H \\ \times\ 4B_H \\ \hline 1D01 \\ A8C0 \\ \hline C5C1_H \end{array}$$

1. $$\begin{array}{r} 2A3 \\ \times\ \ B \\ \hline 1D01 \end{array} \quad \begin{cases} B \cdot 3 = 21 \\ B \cdot A = 6E + 2 = 70 \\ B \cdot 2 = 16;\ 16 + 7 = 1D \end{cases}$$

2. $$\begin{array}{r} {}^{2} \\ 2A3 \\ \times\ 40 \\ \hline A8C0 \end{array} \quad \begin{cases} 4 \cdot 3 = C \\ 4 \cdot A = 28 \\ 4 \cdot 2 = 8;\ 8 + 2 = A \end{cases}$$

3. $$\begin{array}{r} {}^{1} \\ 1D01 \\ +A8C0 \\ \hline C5C1 \end{array}$$

b.

$$\begin{array}{r} 40B_H \\ \times 23_H \\ \hline C21 \\ 8160 \\ \hline 8D81_H \end{array}$$

EXAMPLE 6

Divide in base sixteen $C5C1_H \div 4B_H$.

Solution

$$\begin{array}{rll} & 2A3_H & \\ 4B_H\,) & C5C1_H & \\ & 96 & (2 \cdot 4B = 96) \\ \hline & 2FC & (C5 - 96 = 2F) \\ & 2EE & (A \cdot 4B = 2EE) \\ \hline & E1 & (2FC - 2EE = E) \\ & E1 & (3 \cdot 4B = E1) \\ \hline & 0 & (E1 - E1 = 0) \end{array}$$

Binary, Octal, and Hexadecimal Number System Conversions

We have completed our detailed examination of the octal and hexadecimal systems of numeration. Their importance to the computer programmer lies in their special relationship to the binary system. This relationship is based on the fact that 8 and 16 are powers of two, the base of the binary system. Specifically, 8 is 2^3 and 16 is 2^4.

Binary-to-Octal Conversion We use the fact that $8 = 2^3$ to develop a method for converting any binary numeral to its base eight equivalent. An example will illustrate the method we shall use. Suppose we wish to convert 11001110_2 to its octal equivalent.

1. We begin by dividing the digits of the binary numeral into groups of three (because $8 = 2^3$), starting at the binary point and working toward the end. If the leftmost group has fewer than three digits, we supply one or two zeros to complete a group of three. Thus we have

$$011\ 001\ 110_2$$

2. Now we convert each group of three binary digits to its equivalent octal digit. (Note that the largest number that can be expressed as a three-digit binary numeral is 111_2, which is decimal 7. Therefore, each three-digit binary group converts to a *single* octal digit.) So we have

$$\underbrace{011}_{3}\ \underbrace{001}_{1}\ \underbrace{110_2}_{6}$$

3. We now have the octal numeral that is equivalent to the original binary numeral.

$$11001110_2 = 316_8$$

This can be checked by converting each of the numerals above to its decimal equivalent, 206.

EXAMPLE 7

Convert each of the following binary numerals to its octal equivalent:

a. 100101101_2 **b.** 1000011_2

Solution

a. $\underbrace{100}_{4}\ \underbrace{101}_{5}\ \underbrace{101_2}_{5_8}$ Group in threes. Convert each group to octal.

$100101101_2 = 455_8$

b. $\underbrace{001}\,\underbrace{000}\,\underbrace{011}_2$ Group in threes.

$1 \quad 0 \quad 3_8$ Convert each group to octal.

$1000011_2 = 103_8$

EXAMPLE 8

Convert 1011001.11001_2 to its base eight equivalent.

Solution

The grouping of threes starts at the binary point and goes in both directions:

Two leading zeros added to complete group of three. $001\ 011\ 001\ .\ 110\ 010$ Zero added to complete group of three.

$$\underbrace{001}_{1}\ \underbrace{011}_{3}\ \underbrace{001}_{1}\ .\ \underbrace{110}_{6}\ \underbrace{010}_{2}$$

Thus $1011001.11001_2 = 131.62_8$.

Check

$$1011001.11001_2 = 2^6 + 2^4 + 2^3 + 1 + 2^{-1} + 2^{-2} + 2^{-5} = 89.78125$$

$$131.62_8 = 8^2 + 3 \cdot 8 + 1 + 6 \cdot 8^{-1} + 2 \cdot 8^{-2}$$

$$= 89.78125 \checkmark$$

Octal-to-Binary Conversion Conversion of numerals from base eight to base two is simply the reverse of binary-to-octal conversion. We convert each octal digit to a three-digit binary numeral, filling in zeros as needed.

EXAMPLE 9

Convert the following base eight numerals to their binary equivalents:

a. 4073_8 **b.** 26.35_8

Solution

a. 4 0 7 3

↓ ↓ ↓ ↓

100 000 111 011

Note that the 0 in the octal numeral becomes a group of three zeros in the binary numeral.

Thus, $4073_8 = 100000111011_2$.

b. 2 6 . 3 5

↓ ↓ ↓ ↓

010 110 . 011 101

We can drop the leading zero in the binary numeral.

$26.35_8 = 10110.011101_2$

Binary-to-Hexadecimal and Hexadecimal-to-Binary Conversion Binary-to-hexadecimal conversion is based on the same principle as binary-to-octal conversion. Since sixteen, the base of the hexadecimal system, is equal to 2^4, we use groups of four digits, starting at the binary point.

EXAMPLE 10

Convert the following binary numerals to hexadecimal:

a. 11001110_2 **b.** 1100001010_2

Solution

a. $\underbrace{1100}\underbrace{1110}$ Groups of four.

$12 \quad 14$ Convert each group to hexadecimal.

$\downarrow \quad \downarrow$

$C \quad E_H$

$11001110_2 = CE_H$

b. $\underbrace{0011}\underbrace{0000}\underbrace{1010}_2$ Groups of four, zeros added where needed.

$3 \quad 0 \quad 10$ Convert each group to hexadecimal.

$\downarrow$

A

$1100001010_2 = 30A_H$

EXAMPLE 11

Convert 1000001.11101_2 to its hexadecimal equivalent.

Solution

$\underbrace{0100}\underbrace{0001}.\underbrace{1110}\underbrace{1000}$ Groups of four in both directions from the binary point.

14

$\downarrow$

$4 \quad 1 \; . \; E \quad 8$

Thus $1000001.11101_2 = 41.E8_H$

Check

$$1000001.11101_2 = 2^6 + 1 + 2^{-1} + 2^{-2} + 2^{-3} + 2^{-5} = 65.90625$$

$$41.E8_H = (4 \cdot 16) + 1 + (14 \cdot 16^{-1}) + (8 \cdot 16^{-2})$$

$$= 65.90625$$

Conversion from the hexadecimal to the binary system requires that each hexadecimal digit be converted to a group of *four* binary digits, adding zeros where necessary.

EXAMPLE 12

Convert the following numerals from the hexadecimal to the binary system.

a. $7F_H$ **b.** $A03_H$

Solution

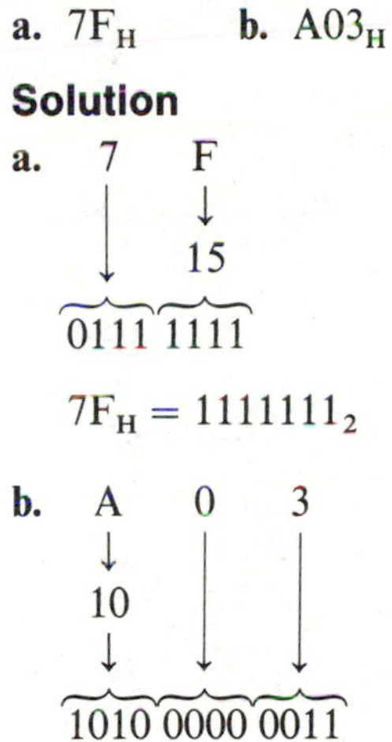

$7F_H = 1111111_2$

Note that we replace the hexadecimal zero by a group of *four* zeros in the binary numeral.

$A03_H = 101000000011_2$

EXAMPLE 13

Convert $48.3C_H$ to its base two equivalent.

Solution

$$\begin{array}{cccc} 4 & 8 & . \; 3 & C_H \\ \downarrow & \downarrow & \downarrow & \downarrow 12 \\ 0100 & 1000 & .\; 0011 & 1100 \end{array}$$

Thus, $48.3C_H = 1001000.001111_2$

Check

$$\begin{aligned} 48.3C_H &= 4 \cdot 16 + 8 + 3 \cdot 16^{-1} + 12 \cdot 16^{-2} \\ &= 64 + 8 + 0.1875 + 0.046875 \\ &= 72.234375 \end{aligned}$$

$$\begin{aligned} 1001000.001111_2 &= 2^6 + 2^3 + 2^{-3} + 2^{-4} + 2^{-5} + 2^{-6} \\ &= 64 + 8 + 0.125 + 0.0625 + 0.03125 + 0.015625 \\ &= 72.234375 \end{aligned}$$

EXERCISES FOR SECTION 3.5

In Exercises 1 through 12 convert each hexadecimal number to its decimal equivalent (see Example 1).

1. $1A3_H$ **2.** $1A4_H$ **3.** $AB9_H$
4. $C3DA_H$ **5.** $A12B_H$ **6.** $B12A_H$
7. $A001_H$ **8.** $D2F_H$ **9.** $BEEF_H$
10. $124AA_H$ **11.** $ABCD_H$ **12.** $A00FF_H$

In Exercises 13 through 21, convert each decimal number to its hexadecimal equivalent (see Example 2).

13. 59 **14.** 68 **15.** 17
16. 161 **17.** 170 **18.** 171
19. 448 **20.** 458 **21.** 459

In Exercises 22 through 35, perform the indicated operations in base sixteen (see Examples 3–6).

22. $4C2B_H + 7A15_H$ **23.** $4C89_H + BBA1_H$ **24.** $8015_H + 29F_H$
25. $347A_H + 8741_H$ **26.** $51C_H - 329_H$ **27.** $2738_H - 15CA_H$
28. $51B6_H - 423_H$ **29.** $ABCD_H - 1EFF_H$ **30.** $2A4_H \cdot 5B_H$
31. $617_H \cdot A3_H$ **32.** $402_H \cdot 2B1_H$ **33.** $814B_H \cdot AA6_H$
34. $C606_H \div 4B_H$ **35.** $62AC_H \div 3C_H$

In Exercises 36 through 43, convert each binary number directly to its octal equivalent (see Examples 7 and 8).

36. 100101_2 **37.** 100111_2 **38.** 1000001_2
39. 1011011_2 **40.** 11101110_2 **41.** 11101.111_2
42. 111100.001_2 **43.** 111100.00010_2

In Exercises 44 through 49, convert each octal number directly to its binary equivalent (see Example 9).

44. 3012_8 **45.** 571_8 **46.** 601_8
47. 35.25_8 **48.** 72.016_8 **49.** 1205.027_8

In Exercises 50 through 57, convert each binary number directly to its hexadecimal equivalent (see Examples 10 and 11).

50. 10110001_2 **51.** 11001011_2 **52.** 1000001_2

53. 1011011_2 **54.** 1000011.10101_2 **55.** 1000011.11101_2

56. 1011.110101_2 **57.** 1100.0010101_2

In Exercises 58 through 63, convert each hexadecimal number directly to its binary equivalent (see Examples 12 and 13).

58. $8F_H$ **59.** 102_H **60.** $A12_H$

61. $B1C_H$ **62.** $46.2B_H$ **63.** $AA.0BC_H$

64. On some popular microcomputers, memory is reserved for specific functions, such as graphics, operating system, and program memory. For example, suppose 400_H to $7FF_H$ is reserved for text and low-resolution graphics. Use hexadecimal subtraction to find out how many locations (bytes) are reserved for text and low-resolution graphics.

65. Suppose BASIC programs can be stored in locations 800_H to BFF_H. How many locations (bytes) is this?

66. Suppose high-resolution graphics reserve locations 4000_H through $5FFF_H$. How many bytes are reserved for high-resolution graphics?

67. *BASIC Exercise.* A student wrote the following BASIC program to convert from base ten to base sixteen:

```
10   LET Z = 0
20   DIM D(10), D$(10)
30   INPUT "WHAT IS YOUR BASE TEN NUMBER?" ; X
40   FOR I = 10 TO 0 STEP -1
50   IF X < 16^I THEN D(I) = 0: D$(I) = "0" :
     GOTO 800
60   T = INT (16^I)
70   Z = Z + 1
80   D(I) = INT (X/16^I)
90   IF Z = 1 THEN Q = I
100  D$(I) = STR$ (D(I))
110  IF D(I) = 10 THEN D$(I) = "A"
120  IF D(I) = 11 THEN D$(I) = "B"
130  IF D(I) = 12 THEN D$(I) = "C"
140  IF D(I) = 13 THEN D$(I) = "D"
150  IF D(I) = 14 THEN D$(I) = "E"
160  IF D(I) = 15 THEN D$(I) = "F"
170  X = X - T*D(I)
800  NEXT I
900  PRINT "THE HEXADECIMAL EQUIVALENT IS"
```

```
910  FOR I = Q TO 0 STEP -1
920  PRINT D$(I);
930  NEXT I
940  PRINT
999  END
```

Run the program.

68. Refine the program above so that conversion to *any* base B (B ≤ 16) can be performed.

3.6 Memory Addressing and Data Representation in Computers

We conclude this discussion of the binary, octal, and hexadecimal systems of numeration with a brief analysis of two of their uses in computer science.

Addressing Specific Memory Locations: Bits and Bytes

At the beginning of this chapter, we referred to the fact that the binary system of numeration is best adapted for use in computers because any number is represented in the binary system as a sequence of zeros and ones. In a computer, the smallest unit of storage contains one binary digit, or **bit**.

It frequently happens, particularly in systems programming, that a programmer needs to access the information stored in a particular memory location. Since a relatively tiny 48K personal computer has almost four hundred thousand bits in primary memory, we can see that addressing each bit in a large system would be highly impractical. For this reason, bits are grouped into units called **bytes**. A typical byte might consist of 8 or 16 bits and has its own memory "address"—a number that refers to its location in the computer's main memory.

If the computer is to store an instruction or a piece of data in a specific memory location, we must communicate to the machine that particular address. Writing this address in binary notation would be most efficient from the computer's point of view, but it is very awkward for us. (Think of writing a decimal number like 49151 in base two!) Writing it in decimal notation would be easy for us; but the conversion to base two would require too much computer time. We reach a compromise by using a system like octal or hexadecimal to refer to memory addresses. These systems require less writing for us and can be rapidly converted to the binary system by the computer.

EXAMPLE 1

Suppose a programmer wants to instruct a computer to load the contents of memory location 49151 into the accumulator (or "adder") in the computer's arithmetic logic unit. Write this memory address in (a) binary, (b) octal, and (c) hexadecimal.

Solution

We will use the method of successive divisions by the new base to achieve the binary conversion. We can then convert directly from base two to the octal and hexadecimal systems.

a. $49151_{10} = ?_2$

	Remainder
2) 49151	
2) 24575	1
2) 12287	1
2) 6143	1
2) 3071	1
2) 1535	1
2) 767	1
2) 383	1
2) 191	1
2) 95	1
2) 47	1
2) 23	1
2) 11	1
2) 5	1
2) 2	1
2) 1	0
0	1

Thus, $49151_{10} = 1011111111111111_2$

b. $49151 = ?_8$

Group the bits in threes, as discussed in Section 2.5:

$$\underbrace{001}_{1}\ \underbrace{011}_{3}\ \underbrace{111}_{7}\ \underbrace{111}_{7}\ \underbrace{111}_{7}\ \underbrace{111_2}_{7_8}$$

So $49151 = 137777_8$.

c. $49151 = ?_H$

Group the binary digits in fours.

	1011	1111	1111	1111
Convert each group to a hexadecimal digit.	11	15	15	15
	B	F	F	F

$49151 = BFFF_H$

Data Representation

The binary system of numeration is used to represent numeric data in computers in a number of different ways. The simplest in concept is **straight binary coding,** in which a decimal numeral is simply stored as its binary equivalent. For example, the number whose decimal representation is 23 would be stored as 10111_2.

There are other commonly used methods of data representation in computers. We shall briefly discuss binary coded decimal representation and two of its refinements.

Binary Coded Decimal Representation of Numbers The conversion of a decimal numeral to its binary equivalent can be tedious and time consuming, even for a computer. Representing numbers in **binary coded decimal (BCD)** form greatly simplifies this conversion. As the name suggests, it involves coding each digit of the decimal numeral separately as a binary numeral. Note that any decimal digit can be written as a binary numeral with at most four digits. The BCD codes, then, generally use four bits to code each decimal digit as follows:

	bits			
Digit	8	4	2	1
0	0	0	0	0
1	0	0	0	1
2	0	0	1	0
3	0	0	1	1
4	0	1	0	0
5	0	1	0	1
6	0	1	1	0
7	0	1	1	1
8	1	0	0	0
9	1	0	0	1

Each four-bit pattern uniquely represents a decimal digit. Two examples will better illustrate how BCD data representation works.

EXAMPLE 2

Write the BCD representation of the following decimal numerals:

a. 75 **b.** 405

Solution

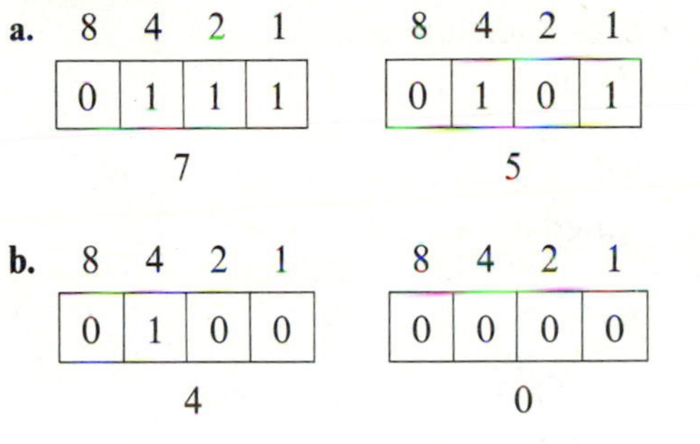

Each decimal digit requires four bits. Each bit has its own "place value."

8	4	2	1
0	1	0	1

5

EXAMPLE 3

What is the BCD representation of 9130?

Solution

8	4	2	1
1	0	0	1

9

8	4	2	1
0	0	0	1

1

8	4	2	1
0	0	1	1

3

8	4	2	1
0	0	0	0

0

Since the four-bit BCD code allows for only 16 possibilities, special characters and even the 26 letters cannot be coded. Two different eight-bit BCD codes, which allow for the coding of characters other than just digits, are discussed in the remainder of this section.

Eight-Bit Binary Coded Decimal Interchange Codes In general, an eight-bit BCD code uses two four-bit groups (one *byte*) to represent a single character. Each byte contains four numeric bits in the 8-4-2-1 place value pattern and four zone bits (the first four bits of an eight-bit code). In an eight-bit BCD code, any character (including alphabetic and special characters) can be represented uniquely by a combination of a numeric code and a particular zone-bit pattern. In fact, since we have eight bits to work with, we can represent 2^8, or 256, different characters.

One common eight-bit code is the ASCII-8 (ASCII stands for *A*merican *S*tandard *C*ode for *I*nformation *I*nterchange) code, which is used in mini- and microcomputers and in data communications applications. The EBCDIC (for *E*xtended *B*inary *C*oded *D*ecimal *I*nterchange *C*ode) code is an eight-bit code that is frequently used on large computers in business applications.

Table 3.8 below shows most common characters and their ASCII-8 representations. The abbreviated EBCDIC table provided in Table 3.9 shows the code for selected characters.

EXAMPLE 4

Write the number 375 in **(a)** binary, **(b)** ASCII-8 code, and **(c)** EBCDIC code.

TABLE 3.8

ASCII-8 codes.

	ASCII-8 Code			ASCII-8 Code	
Character	Zone	Numeric	Character	Zone	Numeric
0	0101	0000	N	1010	1110
1	0101	0001	O	1010	1111
2	0101	0010	P	1011	0000
3	0101	0011	Q	1011	0001
4	0101	0100	R	1011	0010
5	0101	0101	S	1011	0011
6	0101	0110	T	1011	0100
7	0101	0111	U	1011	0101
8	0101	1000	V	1011	0110
9	0101	1001	W	1011	0111
			X	1011	1000
A	1010	0001	Y	1011	1001
B	1010	0010	Z	1011	1010
C	1010	0011			
D	1010	0100			
E	1010	0101			
F	1010	0110			
G	1010	0111			
H	1010	1000			
I	1010	1001			
J	1010	1010			
K	1010	1011			
L	1010	1100			
M	1010	1101			

TABLE 3.9

EBCDIC codes.

EBCDIC Code			EBCDIC Code		
Character	Zone	Numeric	Character	Zone	Numeric
0	1111	0000	A	1100	0001
1	1111	0001	B	1100	0010
2	1111	0010	C	1100	0011
3	1111	0011	D	1100	0100
4	1111	0100	E	1100	0101
5	1111	0101	F	1100	0110
6	1111	0110	G	1100	0111
7	1111	0111	H	1100	1000
8	1111	1000	I	1100	1001
9	1111	1001	J	1101	0001
blank	0100	0000	K	1101	0010
.	0100	1011	L	1101	0011
<	0100	1100	M	1101	0100
(	0100	1101	N	1101	0101
+	0100	1110	O	1101	0110
&	0101	0000	P	1101	0111
$	0101	1011	Q	1101	1000
*	0101	1100	R	1101	1001
)	0101	1101	S	1110	0010
;	0101	1110	T	1110	0011
–	0110	0000	U	1110	0100
/	0110	0001	V	1110	0101
,	0110	1011	W	1110	0110
%	0110	1100	X	1110	0111
>	0110	1110	Y	1110	1000
?	0110	1111	Z	1110	1001

Solution

a.

	Remainder
2) 375	
2) 187	1
2) 93	1
2) 46	1
2) 23	0
2) 11	1
2) 5	1
2) 2	1
2) 1	0
0	1

Thus 375 = 101110111_2.

b. Each decimal digit is coded using eight bits. According to Table 3.8, we have

3	7	5
0101 0011	0101 0111	0101 0101

Thus 375 in ASCII-8 coded form is 0101 0011 0101 0111 0101 0101

c. Using the EBCDIC codes from Table 3.9, we have

3	7	5
1111 0011	1111 0111	1111 0101

So the EBCDIC form of 375 is

1111 0011 1111 0111 1111 0101

EXAMPLE 5

What message does the following EBCDIC code represent?

1100 1000 1100 1001 0100 0000 1101 0100 1101 0110 1101 0100

Solution

Using Table 3.9 we have

1100 1000	1100 1001	0100 0000	1101 0100	1101 0110	1101 0100
H	I	(blank)	M	O	M

EXERCISES FOR SECTION 3.6

In Exercises 1 through 6, write each decimal number in **(a)** binary, **(b)** octal and **(c)** hexadecimal notation (see Example 1).

1. 135 **2.** 451 **3.** 728

4. 1000 **5.** 1024 **6.** 49152

In Exercises 7 through 12, write each decimal number in its four-bit binary coded decimal (BCD) form (see Examples 2 and 3).

7. 73 **8.** 291 **9.** 451

10. 728 **11.** 1000 **12.** 1024

In Exercises 13 through 18, write each decimal number in its **(a)** ASCII-8 and **(b)** EBCDIC coded form.

13. 73 **14.** 135 **15.** 451

16. 291 **17.** 728 **18.** 1000

In Exercises 19 through 24 convert the coded message using either Table 3.8 or Table 3.9, depending on the specified code (see Example 5).

19. 1110 0011 1101 0110 1101 0101 1101 0101 1100 0100 EBCDIC

20. 1110 0011 1101 0110 0100 0000 EBCDIC
1101 0101 1101 0101 1100 0100

21. 1110 0011 1101 0110 1101 0101 EBCDIC
0100 0000 1101 0101 1100 0100

22. 1010 1101 1010 1111 1010 1101 1010 1101 1011 1001 ASCII-8

23. 1010 0100 1010 1111 0101 0110 0101 0111 ASCII-8

24. 1010 0110 1010 1001 1010 1110 ASCII-8
1010 1001 1011 0100 1010 0101

In Exercises 25 through 28, convert each message to **(a)** ASCII-8 and **(b)** EBCDIC code.

25. SANDY **26.** HB21

27. JUMP **28.** DONE

Many microcomputers use a seven-bit code, ASCII-7. The ASCII-7 codes, along with their decimal and hexadecimal representations are listed in Table 3.10. In Exercises 29 through 32, write the ASCII-7 code for each message.

29. ADD 12 **30.** JUMP!

31. $6 > 3$ **32.** $X = 7$

3.7 Chapter Review

In this chapter, we have discussed in detail the three number systems most directly associated with computer systems and applications: the binary, octal, and hexadecimal systems. In each system, we first discussed conversion of numbers between that system and the familiar decimal system. We then examined arithmetic in the new system, using our knowledge of decimal arithmetic as a basis for understanding. The relationships of the binary, octal, and hexadecimal systems to each other were also explored.

TABLE 3.10

ASCII seven-bit codes.

Decimal	Symbol	ASCII-7	Hexadecimal	Decimal	Symbol	ASCII-7	Hexadecimal
32		010 0000	20	82	R	101 0010	52
33	!	010 0001	21	83	S	101 0011	53
34	“	010 0010	22	84	T	101 0100	54
35	#	010 0011	23	85	U	101 0101	55
36	$	010 0100	24	86	V	101 0110	56
37	%	010 0101	25	87	W	101 0111	57
38	&	010 0110	26	88	X	101 1000	58
39	’	010 0111	27	89	Y	101 1001	59
40	(	010 1000	28	90	Z	101 1010	5A
41	)	010 1001	29	91	[	101 1011	5B
42	*	010 1010	2A	92	\	101 1100	5C
43	+	010 1011	2B	93	]	101 1101	5D
44	,	010 1100	2C	94	^	101 1110	5E
45	-	010 1101	2D	95	_	101 1111	5F
46	.	010 1110	2E	96	`	110 0000	60
47	/	010 1111	2F	97	a	110 0001	61
48	0	011 0000	30	98	b	110 0010	62
49	1	011 0001	31	99	c	110 0011	63
50	2	011 0010	32	100	d	110 0100	64
51	3	011 0011	33	101	e	110 0101	65
52	4	011 0100	34	102	f	110 0110	66
53	5	011 0101	35	103	g	110 0111	67
54	6	011 0110	36	104	h	110 1000	68
55	7	011 0111	37	105	i	110 1001	69
56	8	011 1000	38	106	j	110 1010	6A
57	9	011 1001	39	107	k	110 1011	6B
58	:	011 1010	3A	108	l	110 1100	6C
59	;	011 1011	3B	109	m	110 1101	6D
60	<	011 1100	3C	110	n	110 1110	6E
61	=	011 1101	3D	111	o	110 1111	6F
62	>	011 1110	3E	112	p	111 0000	70
63	?	011 1111	3F	113	q	111 0001	71
64	@	100 0000	40	114	r	111 0010	72
65	A	100 0001	41	115	s	111 0011	73
66	B	100 0010	42	116	t	111 0100	74
67	C	100 0011	43	117	u	111 0101	75
68	D	100 0100	44	118	v	111 0110	76
69	E	100 0101	45	119	w	111 0111	77
70	F	100 0110	46	120	x	111 1000	78
71	G	100 0111	47	121	y	111 1001	79
72	H	100 1000	48	122	z	111 1010	7A
73	I	100 1001	49	123	(	111 1011	7B
74	J	100 1010	4A	124	|	111 1100	7C
75	K	100 1011	4B	125	)	111 1101	7D
76	L	100 1100	4C	126	”	111 1110	7E
77	M	100 1101	4D				
78	N	100 1110	4E				
79	O	100 1111	4F				
80	P	101 0000	50				
81	Q	101 0001	51				

The chapter concluded with a discussion in Section 3.6 of memory addressing and data representation in computers. Several data encoding systems, including the ASCII and EBCDIC codes, were presented.

VOCABULARY

binary number system
base two number system
positional notation
place value
expanded form
bit
binary point
algorithm
nines complement
ones complement
twos complement
tens complement
octal number system
sevens complement
hexadecimal number system
fifteens complement
byte
memory address
binary coded decimal (BCD) representation
eight-bit code
ASCII-8 code
EBCDIC code

CHAPTER TEST

Convert the following base ten numerals to **(a)** base two, **(b)** base eight, and **(c)** base sixteen:

1. 13

2. 124

Convert each of the following numerals to base ten:

3. 10110_2

4. 206_8

5. $21B_H$

6. 1.011_2

Perform each of the following calculations in the base indicated:

7. $\begin{array}{r} 11011_2 \\ +1101_2 \\ \hline \end{array}$

8. $\begin{array}{r} 11011_2 \\ -1101_2 \\ \hline \end{array}$

9. $\begin{array}{r} 1101_2 \\ \times 101_2 \\ \hline \end{array}$

10. $11_2 \overline{)\,100001_2}$

11. $\begin{array}{r} 335_8 \\ +746_8 \\ \hline \end{array}$

12. $\begin{array}{r} 7403_8 \\ -3514_8 \\ \hline \end{array}$

13. $\begin{array}{r} 335_8 \\ \times 46_8 \\ \hline \end{array}$

14. $\begin{array}{r} ACBD_H \\ +B70A_H \\ \hline \end{array}$

15. $\begin{array}{r} BC237_H \\ -A4B3_H \\ \hline \end{array}$

16. $\begin{array}{r} 940_H \\ \times 23_H \\ \hline \end{array}$

Convert each of the following binary numbers to **(a)** octal and **(b)** hexadecimal.

17. 11001011_2

18. 100111011_2

19. 11011.01001_2

20. Write the decimal number 7325 in its four-bit binary coded decimal (BCD) form.

CHAPTER

4

The microcomputer chip and the logic board of a computer contain complex circuitry whose design is closely related to many concepts introduced in mathematical logic. In fact, the study of computer architecture begins with the study of logic and logical operators. In this chapter and the next, we introduce the basic concepts of mathematical logic and begin to describe their relationship to computer design.

Logic for Computers

Can computers think? Your answer to this question probably depends on what you mean by "thinking." To many people, computers certainly *seem* to be capable of thought. In fact, computer scientists are currently making a great deal of progress in an area called *artificial intelligence*. But the emphasis here is on the word "artificial." A computer cannot really think independently; it can do only what it has been taught (via programs) to do by some person. This chapter deals with the fundamental rules of logic and language that facilitate precise communication among people and between people and computers.

4.1 Simple and Compound Statements

The basic means of communication is language. In the English language, we call a complete thought a sentence. There are four basic types of sentences:

1. exclamatory sentences: "The mainframe is on fire!"
2. imperative sentences: "Duplicate this deck of cards."
3. interrogative sentences: "What is the baud rate of your modem?"
4. declarative sentences: "It is a 6502 microprocessor."

In the study of logic, we are concerned with *declarative* sentences that are either true or false; such sentences are called **statements**. Consider the following example:

EXAMPLE 1

Which of the following sentences are statements?

a. Mike works for IBM.
b. Is there an echo in here?
c. Forward the appropriate forms in triplicate.
d. Jennifer studies ballet and clarinet.
e. Ernie is an attorney.
f. Where's the beef?

Solution

Sentences **a**, **d** and **e** are statements since they are declarative sentences. Sentences **b** and **f** are *not* statements, since each asks a question. Sentence **c** is *not* a statement, since it is an imperative (it gives an order).

The key factor in all these cases was the ability or inability to assign a **truth value** to a given sentence—that is, to determine whether it is true or false. We do not normally think of questions or commands or exclamations as being true or false. Therefore, in logic, they are not statements.

In logic, as in grammar, there are simple and compound statements. A **simple statement** is a statement that conveys exactly one thought—for example, "Oscar is a grouch." A **compound statement** is a connection (a union or joining) of simple statements—for example, "Oscar is a grouch, *and* Ernie is an attorney." The word or words that form the connection and join the simple statements—*and* in the example above—are called **connectives.**

EXAMPLE 2

All of the following sentences are statements. Which ones are simple and which are compound?

a. I have to cash a check, and the bank is closed.
b. Yesterday was Friday the thirteenth.
c. On Friday the thirteenth, a black cat crossed my path.
d. If we use a different approach, then we can save time.
e. Jennifer studies ballet and clarinet.
f. There is an echo in here.
g. Either Gene will play golf today, or he will wash the car.

Solution

The simple statements are **b**, **c** and **f**. All the others are compound. The connective in statement **a** is *and*; in statement **d**, it is *if . . . then*; in statement **g**, it is *either . . . or*. Statement **e** bears closer examination. An English language teacher would not consider this to be a compound sentence, but, from a logic point of view, it *is* a compound statement, with *and* as the connective. To see this, we must understand that the statement

Jennifer studies ballet and clarinet.

really expresses *two* simple statements:

S_1: Jennifer studies ballet.

and

S_2: Jennifer studies clarinet.

So it has the same meaning as the compound statement S_1 *and* S_2:

Jennifer studies ballet, *and* Jennifer studies clarinet.

As we continue in this chapter, we will represent simple statements with symbols, usually the lowercase letters p, q, r, etc. Also, every statement will have (or will be assigned) a *truth value* (a determination of its truth or falsity) of T for true or F for false. This conforms with the usual conventions in mathematical logic. We may occasionally, however, in relating logical concepts to computers, use 1 for true and 0 for false. You should be aware that there is not universal agreement on this convention among all computer manufacturers; some use -1 for true. In this book, we shall conform to the $1 = \text{T}$, $0 = \text{F}$ convention.

Before concluding this introductory section, we briefly review the three basic **relational operators** in mathematics, since they occur frequently in computer applications that involve logical decisions. They are

1. $>$ (greater than)
2. $<$ (less than)
3. $=$ (equal to)

EXAMPLE 3

Indicate the *truth value* of each of the following statements:

a. $7 > 5$ **b.** $8/4 = 2$ **c.** $3 - 9 > 9 - 3$
d. $A + B < A - B$ (Assume A and B are positive integers.)

Solution

a. $7 > 5$ (7 is greater than 5) is true.
b. $8/4 = 2$ is true.
c. $3 - 9 = -6$; $9 - 3 = +6$
So $3 - 9 > 9 - 3$ ($3 - 9$ is greater than $9 - 3$) is false.
d. If A and B are positive integers, $A + B$ is always greater than $A - B$. Therefore, $A + B < A - B$ is false.

EXAMPLE 4

Is the statement $5 < x < 10$ simple or compound?

Solution

We must first translate this statement. $5 < x < 10$ means

> The value of x is both *greater than 5* ($5 < x$ is equivalent to $x > 5$) and *less than 10* ($x < 10$).

So $5 < x < 10$ is a *compound* statement; it is equivalent to

$$5 < x \qquad \textit{and} \qquad x < 10$$

Using lowercase letters to represent simple statements, we could let

p: $5 < x$

q: $x < 10$

Then $5 < x$ and $x < 10$ ($5 < x < 10$) would be represented in symbols by

p and q

In the next sections, we'll learn some shorthand notation for the common logical connectives.

EXERCISES FOR SECTION 4.1

In Exercises 1 through 10, indicate whether or not the given sentence is a statement (see Example 1).

1. Larry, take out the garbage.
2. Larry will take out the garbage.
3. Did Larry take out the garbage?
4. Columbus discovered America in 1942.
5. If I am elected president, then I will lower taxes.
6. $8 < 6$
7. Is it true that $8 < 6$?
8. Look out!
9. I will buy a new Mercedes if and only if I win the lottery.
10. Anyone who reads this book loves mathematics.

In Exercises 11 through 20, indicate whether the given statement is simple or compound (see Examples 2 and 4).

11. Candy is dandy, but I prefer ice cream.
12. Harold, the math teacher, will run for president in 1988.
13. If he is elected, he will put a computer in every household.
14. $x + 3 > 7$
15. $-7 > y > 0$
16. $2 \leq 6/3$ ($\leq$ is read "is less than or equal to")
17. Anyone who reads this book loves mathematics. (Hint: If you are a person who reads this book, then you love mathematics.)
18. Mary will go shopping at the store with the beautiful window display this afternoon.
19. If Mary goes shopping, Sue will meet her for lunch.
20. Mary will either go shopping or have lunch at home.

In Exercises 21 through 25, indicate the truth value of the given statement (see Example 3).

21. $9 + 4 = 13$

22. $9 + 4 \geq 13$ ($\geq$ is read "is greater than or equal to")

23. $16/2 \neq 8$ ($\neq$ is read "is not equal to")

24. $6 - 3 < 6 + 3$

25. $-2 > -4$

26. The sentence "This sentence is false" is *not* a statement. Why?

4.2 The NOT, AND, and OR Connectives

We mentioned in Section 4.1 that, for the sake of clarity and simplicity of expression, we shall frequently use lowercase letters to represent simple statements. As we begin our study of logical **connectives**, or **operators**, we shall find it helpful to use symbolic representations of them as well. This will enable us to represent compound statements in a convenient shorthand form.

The NOT Connective

In the English language, when we wish to change a statement to its **negation** (a statement with opposite meaning), we use the word *not* in some form. For example, suppose p stands for the statement, "Mary had a little lamb." Then the *negation* of p, symbolized by $\sim p$ (read "not p") is the statement, "Mary did *not* have a little lamb."

EXAMPLE 1

Write the negation of each of the following statements:

a. p: $x = 5$

b. p: The boy went home.

c. p: The answer to that problem is not 16.

d. p: $N < 3$

Solution

a. $\sim p$: $\sim(x = 5)$ or $x \neq 5$

b. $\sim p$: The boy did *not* go home.

c. This one is a bit tricky. Remember that $\sim p$ is a statement whose meaning is the *opposite* of p. The statement that means the opposite of "The answer to that problem is *not* 16" is the statement

$\sim p$: The answer to that problem is 16.

d. $\sim p$: $\sim(N < 3)$ or N is *not* less than 3. Note that this means the same thing as $N \geq 3$ (N is greater than or equal to 3).

Our understanding of a logical operator is usually enhanced if we examine the truth value of a statement formed by using that operator in relationship to the truth value(s) of the statement or statements it is operating on or connecting. The definition of negation suggests that the truth value of the statement $\sim p$ is the opposite of the truth value of p. Thus, if the statement p is true, then $\sim p$ is false; if p is false, then $\sim p$ is true.

In order to display truth values, it is common to list all the possibilities in a table, called a **truth table**. We use T for *true* and F for *false* and list all possible combinations of truth values for the simple statements involved. The truth table for $\sim p$ (the *negation* operator) is given in the following table:

TABLE 4.1

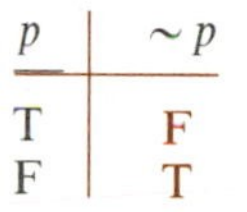

p	$\sim p$
T	F
F	T

The NOT, or negation, operator $\sim$ is very important in some computer applications. For example, if we negate the binary number 00101, we get 11010; the NOT changed each 1 (true) to 0 (false) and each 0 to 1. This is equivalent to taking the *ones complement* (see Section 3.3) of the original number, an important concept in machine and assembly language programming. Its relationship to the ones complement is one reason why negation is sometimes referred to as the *complement* operator.

The AND Connective: Conjunction

Consider the statement "I will go to the store *and* to the movies." This is a *compound* statement consisting of the simple statements

p: I will go to the store.
q: I will go to the movies.

joined by the connective *and*. In symbols, we can write: p and q or $p \wedge q$. The statement $p \wedge q$ is also called the **conjunction** of p and q.

As with negation, we need to determine under what circumstances the conjunction of two simple statements is true and when it is false. Common sense indicates that, in order for $p \wedge q$ to be true, I must do *both* things. That is, both statements ("I will go to the store" *and* "I will go to the movies") must be true for $p \wedge q$ to be true. Under any other circumstances, $p \wedge q$ is false. The truth table for $p \wedge q$ is Table 4.2.

TABLE 4.2

p	q	$p \wedge q$
T	T	T
T	F	F
F	T	F
F	F	F

This column is referred to as the *truth table* for $p \wedge q$.

Note that there are *four* rows in this truth table. This arrangement takes into account all possible combinations of truth values for p and q.

EXAMPLE 2

For each compound statement below, determine the simple statements of which it is composed, and write the compound statement symbolically.

a. Amy and her brother like to eat ice cream.
b. Jennifer studies ballet and clarinet.
c. $5 < x < 10$

Solution

a. p: Amy likes to eat ice cream.

q: Her brother likes to eat ice cream.

The compound statement is $p \wedge q$.

b. p: Jennifer studies ballet.

q: Jennifer studies clarinet.

The compound statement is $p \wedge q$.

c. $5 < x < 10$ means the value of x is "between" 5 and 10, or $x > 5$ and $x < 10$. Thus, for

$$p\text{: } x > 5$$
$$q\text{: } x < 10$$

the compound statement is $p \wedge q$.

The AND operation can also be applied to binary numbers. For example, $1 \wedge 1$ has a value of 1 and $1 \wedge 0$ has a value of 0 (remember to think of 1 as true, 0 as false).

The OR Connective: Disjunction

Consider the compound statement "I will go to the store *or* to the movies." Here, we are using the word *or* in the *inclusive* sense (we'll deal with the *exclusive or* later); so the statement is true if I perform both actions (go to the store *and* the movies). Common sense tells us that it is also true if I perform one action and not the other. Thus, the statement "I will go to the store *or* to the movies" is *false* in only *one* set of circumstances—if I don't go to the store and I don't go to the movies.

Suppose we represent the simple statements as follows:

p: I will go to the store.

q: I will go to the movies.

Our compound statement can then be expressed symbolically by p OR q or $p \vee q$. The statement $p \vee q$ is also referred to as the **disjunction** of p and q. The truth table for the disjunction, $p \vee q$, is given in Table 4.3.

TABLE 4.3

p	q	$p \vee q$
T	T	T
T	F	T
F	T	T
F	F	F

The truth table for $p \vee q$

Notice that the arrangement of this truth table (TTFF under p and TFTF under q) follows the same pattern as the truth table for conjunction ($p \wedge q$). We will follow this format whenever we set up a truth table involving two simple statements. It allows us to refer to the rightmost column (in this instance) as *the truth table for* $p \vee q$ and to compare truth tables for different compound statements.

EXAMPLE 3

If p and q are the simple statements for each part of Example 2, write the *disjunction* of p and q symbolically and in ordinary English.

Solution

a. p: Amy likes to eat ice cream.

q: Her brother likes to eat ice cream.

The *disjunction* is

$p \vee q$: Amy likes to eat ice cream, or her brother likes to eat ice cream.

This could be shortened to

Amy or her brother likes to eat ice cream.

b. p: Jennifer studies ballet.

q: Jennifer studies clarinet.

The *disjunction* is

$p \vee q$: Jennifer studies ballet, or Jennifer studies clarinet.

Or

Jennifer studies ballet or clarinet.

c. p: $x < 5$

q: $x > 10$

The *disjunction* is

$$p \vee q: x < 5 \text{ or } x > 10$$

Note that this *cannot* be written $5 > x > 10$ for two reasons:

1. The mathematical notation used implies *conjunction*, not disjunction—specifically, the conjunction of $5 > x$ ($x < 5$) with $x > 10$, rather than their disjunction.
2. The notation also implies that $5 > 10$, which is *never* true.

We conclude this section with some further examples to help clarify the $\sim$, $\wedge$ and $\vee$ connectives.

EXAMPLE 4

State whether each of the following statements is true or false:

a. $(5 > 3) \wedge (-2 < -3)$ **b.** $(5 > 3) \vee (-2 < -3)$

c. $(4 < 7) \wedge (-7 < -4)$ **d.** $(4 < 7) \vee (-7 < -4)$

e. $\sim(-4 < -7)$ **f.** $\sim(-7 < -4)$

Solution

a. $(5 > 3)$ is true; $(-2 < -3)$ is false. The conjunction $(5 > 3) \wedge (-2 < -3)$ is false.

b. Since $(5 > 3)$ is true, the disjunction $(5 > 3) \vee (-2 < -3)$ is true.

c. $(4 < 7)$ is true; $(-7 < -4)$ is true. The conjunction $(4 < 7) \wedge (-7 < -4)$ is true.

d. The disjunction $(4 < 7) \vee (-7 < -4)$ is also true.

e. $(-4 < -7)$ is false. The negation, $\sim(-4 < -7)$ is true.

f. $(-7 < -4)$ is true. The negation, $\sim(-7 < -4)$ is false.

EXAMPLE 5

For what values of x is the statement $(x > 5) \wedge (x < -3)$ true?

Solution

In order for a value of x to make the conjunction a true statement, x must make both simple statements ($x > 5$ and $x < -3$) true. That is, we must find a value of x that is both larger than 5 and smaller than -3. No such number exists; so the statement is always false.

EXAMPLE 6

Let p stand for "Sue won the contest"; let q represent "Sandy lost the contest." Translate each of the following symbolic statements into English.

a. $p \vee q$ **b.** $p \wedge q$ **c.** $p \vee (\sim q)$ **d.** $(\sim p) \wedge (\sim q)$

Solution

a. $p \vee q$: Sue won the contest or Sandy lost.

b. $p \wedge q$: Sue won the contest and Sandy lost.

c. First, we translate $\sim q$ as "Sandy did *not* lose the contest."
Then we have

$p \vee (\sim q)$: Sue won the contest or Sandy did not lose.

d. $\sim p$ is "Sue did not win the contest."
$(\sim p) \wedge (\sim q)$: Sue did not win the contest and Sandy did not lose.

Finally, an example related to programming:

EXAMPLE 7

Write a BASIC statement that prints "ERROR" if the value of N is *neither* 3 *nor* 5.

Solution

We must first translate the condition: if N is *neither* 3 *nor* 5, then N is *not* 3 *and* N is *not* 5. In symbols, this can be written, N $\neq$ 3 AND N $\neq$ 5. Since we want the computer to print a message when this condition is true, the appropriate BASIC statement is

```
IF (N<>3) AND (N<>5) THEN PRINT "ERROR"
```

[In Pascal, this would be

```
IF (N<>3) AND (N<>5) THEN WRITELN('ERROR');
```
]

EXERCISES FOR SECTION 4.2

In Exercises 1 through 5, write the negation of the given statement (see Example 1).

1. $y = 5$
2. $2 \neq 3$
3. $A \geq 1$
4. Mary is quite contrary.
5. I did not write that program.

In Exercises 6 through 10, write the *conjunction* and the *disjunction* of the given pair of statements both symbolically and in simple English (see Examples 2 and 3).

6. p: I won the lottery.
q: I will order a new Rolls-Royce tomorrow.

7. p: Today is tomorrow's yesterday.
q: Today was yesterday's tomorrow.

8. p: Mary had a little lamb.
q: Its fleece was pink with blue polka-dots.

9. p: $0 < x$
q: $x < 5$

10. p: $A > 4$
q: $A = 4$

In Exercises 11 through 20, assume that statement p is true and statement q is false. Determine the truth value of each compound statement (see Example 4).

11. $\sim p$
12. $\sim q$
13. $p \vee q$
14. $p \wedge q$
15. $(\sim p) \vee q$
16. $p \wedge (\sim q)$
17. $(\sim p) \vee (\sim q)$
18. $\sim(\sim p)$
19. $\sim(p \wedge q)$
20. $(\sim p) \wedge (\sim q)$

In Exercises 21 through 26, let p represent the statement "Mama played bass" and q represent the statement "Daddy sang tenor." Translate each symbolic statement into English (see Example 6).

21. $\sim p$ **22.** $p \vee q$ **23.** $p \wedge q$
24. $(\sim p) \vee q$ **25.** $p \wedge (\sim q)$ **26.** $(\sim p) \vee (\sim q)$

27. Write a BASIC statement that will print "HALLELUJAH!" if the value of the variable A is neither $+1$ nor -1 (see Example 7).

4.3 The Implication, Biconditional and Other Logical Operators

Implication: The IF-THEN Statement

Many statements that we use in our everyday language are **conditional**; they state that a particular conclusion will follow on the assumption that a certain condition holds. Some examples are

> If today is March 16, then tomorrow is St. Patrick's Day.
> If you are reading this book, then you must love math.
> If you don't stop playing your trumpet, I'll scream. (Here, the word "then" is omitted although it is implied in the sense of the statement.)
> If I lose the dance contest, then I'll pay you \$20.

Let's take a closer look at the last statement. We let p be "I lose the dance contest" and q be "I'll pay you \$20." Then the given statement can be written as *if p, then q*. We symbolize this as $p \rightarrow q$ and read it as "p implies q" or "if p, then q." In order to construct a truth table for $p \rightarrow q$, we need to examine its possible truth values.

Suppose Ann promises Sam, "If I lose the dance contest, then I'll pay you \$20." Under what conditions will she *break* her promise—that is, under what conditions is $p \rightarrow q$ false? The possibilities are

1. Ann loses; Ann pays.
2. Ann loses; Ann doesn't pay.
3. Ann wins; Ann pays.
4. Ann wins; Ann doesn't pay.

If Ann loses and she pays Sam the \$20, she has certainly kept her promise (if p and q are both true, then $p \rightarrow q$ is true). If she loses and fails to pay, she has *broken* her promise (if p is true and q is false, then $p \to q$ is false). On the other hand, what if Ann wins (that is, p is false)? Under these circumstances, she has made no commitment to Sam. Whether she pays Sam $20 or not, her original statement ($p \to q$) will be true. So we conclude that $p \to q$ is false *only* when p is true (Ann loses) and q is false (Ann doesn't pay). This leads to the following truth table:

TABLE 4.4

p	q	$p \to q$
T	T	T
T	F	F
F	T	T
F	F	T

Truth table for $p \to q$

Note that the only F in the truth table for $p \to q$ appears in the second row, when the truth value of p is T and the truth value of q is F.

EXAMPLE 1

Determine the truth value of each of the following conditional statements:

a. If the moon is made of green cheese, then the rain in Spain stays mainly in the plain.
b. If $1 + 1 = 2$, then every square is a rectangle.
c. If $1 + 1 = 3$, then $0 = 1$.
d. If Easter falls on a Sunday, then $2 + 2 = 5$.
e. If Easter will fall on a Wednesday next year, then Christmas is on December 25.

Solution

We begin by expressing each statement symbolically, in the form $p \to q$.

a. We have

p: The moon is made of green cheese.

q: The rain in Spain stays mainly in the plain.

Since p is false, $p \to q$ must be true whether q is true or false. (Look at the last two rows of the truth table.)

b. p: $1 + 1 = 2$

q: Every square is a rectangle.

Here, p is true and q is true, so $p \to q$ is true.

c. p: $1 + 1 = 3$
q: $0 = 1$

p is false and q is false, so $p \to q$ is true.

d. p: Easter falls on a Sunday.
q: $2 + 2 = 5$

p is true and q is false, so $p \to q$ is false.

e. p: Easter will fall on a Wednesday next year.
q: Christmas is on December 25.

p is false and q is true, so $p \to q$ is true.

In dealing with compound statements, it is often useful to look for other ways of saying the same thing—other statements involving the same simple statements that are true and false under the same conditions. We say that two statements are **logically equivalent** if they have the *same truth table*. Replacing a statement with a logically equivalent statement can sometimes be helpful in simplifying or interpreting the original statement.

Consider, for example, the statement $p \to q$. We claim that it is logically equivalent to the statement $(\sim p) \vee q$. To prove this, we construct both truth tables. Recall that $\sim p$ is true whenever p is false and false whenever p is true.

p	q	$p \to q$
T	T	T
T	F	F
F	T	T
F	F	T

p	q	$(\sim p)$	$\vee$	q
T	T	F	T	T
T	F	F	F	F
F	T	T	T	T
F	F	T	T	F

The truth tables are identical, so $p \to q$ is *logically equivalent* to $(\sim p) \vee q$.

EXAMPLE 2

Is $p \to q$ logically equivalent to $p \vee (\sim q)$?

Solution

We construct the truth tables for both statements and compare them.

p	q	$p \to q$
T	T	T
T	F	F
F	T	T
F	F	T

p	q	p	$\vee$	$(\sim q)$
T	T	T	T	F
T	F	T	T	T
F	T	F	F	F
F	F	F	T	T

This time, the truth tables are *not* identical. Therefore, $p \to q$ and $p \vee (\sim q)$ are *not* logically equivalent.

We present one final caution to the programming student. The conditional statement $p \to q$, used in the context of logic, *does not* have the same meaning or interpretation as an IF or IF-THEN statement in a programming language. We shall have more to say about this in Section 4.5.

The Biconditional Operator

A logical connective that is closely related to the conditional is the *biconditional.* If p and q are simple statements, the biconditional is written symbolically as $p \leftrightarrow q$ and read "p if and only if q." Some English language examples are

You will pass this course if and only if you study.
I will pay you $20 if and only if I lose the dance contest.
A rectangle is a square if and only if it has four equal sides.

Let's use the second example above, along with a little common sense, to determine the truth values of the biconditional statement $p \leftrightarrow q$. Suppose p and q are given by

p: I will pay you $20.
q: I lose the dance contest.

Suppose further that our friend Ann has changed her original promise to Sam. She now says, "I will pay you $20 *if and only if* I lose the dance contest." In symbols, this is the statement $p \leftrightarrow q$.

Now Ann is saying, as before, that she will pay Sam if she loses the contest. But she is also stating that if she doesn't lose (if she wins), she will *not* pay Sam the $20. Her statement will be true, then, when (and only when) p and q have the same truth value; that is, if she wins the contest and still pays Sam, she will be breaking her promise ($p \leftrightarrow q$ will be false).

The truth table for $p \leftrightarrow q$ is as follows (Table 4.5):

TABLE 4.5

p	q	$p \leftrightarrow q$
T	T	T
T	F	F
F	T	F
F	F	T

The fact that the arrow in $p \leftrightarrow q$ points in both directions might lead one to guess that $p \leftrightarrow q$ might be logically equivalent to $(p \rightarrow q) \wedge (q \rightarrow p)$. We shall examine this possibility in the next section.

EXAMPLE 3

Let p, q and r be the following statements:

p: Sandy gets a raise.

q: Harold gets a raise.

r: Larry gets a raise.

Suppose p and q are true and r is false. Write each of the following statements in English, and determine its truth value:

a. $p \leftrightarrow q$ **b.** $q \leftrightarrow p$ **c.** $q \leftrightarrow r$ **d.** $r \leftrightarrow (\sim q)$

Solution

a. $p \leftrightarrow q$ translates as

Sandy gets a raise if and only if Harold gets a raise.

Since p and q are both true, $p \leftrightarrow q$ is true.

b. $q \leftrightarrow p$ translates as

Harold gets a raise if and only if Sandy gets a raise.

Since q and p are both true, $q \leftrightarrow p$ is true. You should also note that $p \leftrightarrow q$ and $q \leftrightarrow p$ are logically equivalent. Convince yourself by constructing a truth table for $q \leftrightarrow p$ as follows:

q	p	$q \leftrightarrow p$
T	T	
F	T	
T	F	
F	F	

↑ You fill in this column.

c. $q \leftrightarrow r$ translates as

Harold gets a raise if and only if Larry gets a raise.

Since q is true and r is false, $q \leftrightarrow r$ is false.

d. $r \leftrightarrow (\sim q)$ translates as

Larry gets a raise if and only if Harold does not get a raise.

Since r is false and $\sim q$ is false (q is true), $r \leftrightarrow (\sim q)$ is true.

The NOR, NAND, and Exclusive OR Operators

We conclude this section with a discussion of two logical operators (NOR and NAND) that are important in computer design and of the exclusive OR, which is important in programming. For the first two, we begin by giving the truth tables and work from them to an interpretation.

The truth table for **NOR** (symbolized by $\downarrow$) is as follows:

TABLE 4.6

p	q	$p \downarrow q$
T	T	F
T	F	F
F	T	F
F	F	T

Suppose we compare this with the truth table for $p \vee q$:

p	q	$p \vee q$
T	T	T
T	F	T
F	T	T
F	F	F

The truth tables are *exact opposites*. Thus $p \downarrow q$ is the *negation* of $p \vee q$; that is, $p \downarrow q$ is logically equivalent to $\sim(p \vee q)$.

The truth table for **NAND** (symbolized by $\uparrow$) is

TABLE 4.7

p	q	$p \uparrow q$
T	T	F
T	F	T
F	T	T
F	F	T

Comparing this with the truth table for $p \wedge q$, we see that they are exact opposites. That is, $p \uparrow q$ is logically equivalent to $\sim(p \wedge q)$, the *negation* of $p \wedge q$.

The last logical connective we shall consider is the **exclusive OR**, symbolized by $\veebar$. Suppose p is the statement "I will buy IBM stock" and q is the statement

"I will buy Apple stock." Then $p \vee q$ symbolizes the statement "I will buy IBM stock, *or* I will buy Apple stock." This form is called the *inclusive OR*; $p \vee q$ is *true* when *both* p and q are true as well as when either p is true or q is true.

On the other hand, $p \veebar q$ stands for the statement: "I will buy IBM stock *or* I will buy Apple stock, *but not both*." Unlike the inclusive OR, this statement is *false* when *both* p and q are true; it is an *exclusive OR*.

We show the truth tables for both $p \vee q$ and $p \veebar q$ in Table 4.8.

TABLE 4.8

p	q	$p \vee q$	$p \veebar q$
T	T	T	F
T	F	T	T
F	T	T	T
F	F	F	F

EXAMPLE 4

Suppose p is false, q is true, and r is false. Find the truth value of each of the following compound statements:

a. $p \uparrow q$ **b.** $(\sim p) \uparrow q$ **c.** $q \downarrow r$
d. $(\sim q) \downarrow p$ **e.** $q \vee (\sim r)$ **f.** $q \veebar (\sim r)$

Solution

We use the truth tables.

a. p is false and q is true; so $p \uparrow q$ is true.
b. Since p is false, $\sim p$ is true; q is true. Therefore, $(\sim p) \uparrow q$ is false.
c. Since q is true and r is false, $q \downarrow r$ is false.
d. Since q is true, $(\sim q)$ is false; p is false. Therefore, $(\sim q) \downarrow p$ is true.
e. Since r is false, $\sim r$ is true; q is true. Therefore, $q \vee (\sim r)$ is true.
f. Since q is true and $\sim r$ is true, $q \veebar (\sim r)$ is false.

EXERCISES FOR SECTION 4.3

In Exercises 1 through 8, let m be the statement "Mama played bass" and d be the statement "Daddy sang tenor." Translate each statement into an English sentence (see Examples 1, 3 and 4).

1. $m \rightarrow d$ **2.** $d \rightarrow (\sim m)$ **3.** $m \leftrightarrow (\sim d)$

4. $d \leftrightarrow m$ 5. $(m \rightarrow d) \wedge (d \rightarrow m)$ 6. $m \vee d$

7. $m \veebar (\sim d)$ 8. $(\sim m) \veebar d$

For Exercises 9 through 16, suppose that m is true and d is false. Determine the truth value of each of the statements in Exercises 1 through 8.

For Exercises 17 through 22, use

p: My car starts.

q: I will be late for work.

z: I oversleep.

Translate each compound statement into symbolic form using the definitions of the connectives and Examples 1, 3 and 4.

17. If I oversleep, then I will be late for work.
18. I will be late for work if and only if my car does not start.
19. I will not be late for work if my car starts.
20. Either my car started or I was late for work, but not both.
21. I will oversleep if and only if my car does not start.
22. If I am not late for work, I did not oversleep.
23. Construct the truth table of the statement $(\sim p) \wedge (\sim q)$. Show that it is logically equivalent to $\sim(p \vee q)$ (and thus to $p \downarrow q$). The statement that $(\sim p) \wedge (\sim q)$ and $\sim(p \vee q)$ are logically equivalent, written $(\sim p) \wedge (\sim q) \Leftrightarrow \sim(p \vee q)$, is one of **DeMorgan's laws**.
24. Use truth tables to prove the second of DeMorgan's laws:

$$(\sim p) \vee (\sim q) \Leftrightarrow \sim(p \wedge q)$$

That is, $(\sim p) \vee (\sim q)$ is logically equivalent to $\sim(p \wedge q)$, and thus also to $p \uparrow q$.

4.4 Compound Statements and Truth Tables

We are now ready to deal with constructing the truth tables of more complicated compound statements. Along the way, we shall make some observations that are important in a computer programming context.

Consider the problem of constructing the truth table of the statement $(p \vee q) \wedge (\sim p)$. There are two simple statements, p and q, involved, so the truth table will have four rows. The best approach is to construct the truth table in stages. Notice that there are five entities that must be assigned truth values: p, q, $\sim p$, $(p \vee q)$, and $(p \vee q) \wedge (\sim p)$. We begin with p and number the steps for clarity.

p	q	$(p$	$\vee$	$q)$	$\wedge$	$(\sim$	$p)$
T	T	T					T
T	F	T					T
F	T	F					F
F	F	F					F
		1					1

Next, fill in the truth values for q.

p	q	$(p$	$\vee$	$q)$	$\wedge$	$(\sim$	$p)$
T	T	T		T			T
T	F	T		F			T
F	T	F		T			F
F	F	F		F			F
		1		2			1

Notice that $\wedge$ is the *major connective*; we want the *conjunction* of $(p \vee q)$ with $(\sim p)$. We fill in column 3 for $p \vee q$ and column 4 for $\sim p$.

p	q	$(p$	$\vee$	$q)$	$\wedge$	$(\sim$	$p)$
T	T	T	T	T		F	T
T	F	T	T	F		F	T
F	T	F	T	T		T	F
F	F	F	F	F		T	F
		1	3	2		4	1

The truth table is completed in column 5, by taking the conjunction of columns 3 and 4.

p	q	$(p$	$\vee$	$q)$	$\wedge$	$(\sim$	$p)$
T	T	T	T	T	F	F	T
T	F	T	T	F	F	F	T
F	T	F	T	T	T	T	F
F	F	F	F	F	F	T	F
		1	3	2	5	4	1

The two unlabelled (leftmost) columns in the truth table, the ones we start off with, are called the **reference columns**.

EXAMPLE 1

Use the method presented above to construct a truth table for the statement

$$[p \wedge (\sim q)] \vee [q \wedge (\sim p)]$$

Solution

We give the truth table below, numbering the steps.

p	q	$[p$	$\wedge$	$(\sim$	$q)]$	$\vee$	$[q$	$\wedge$	$(\sim$	$p)]$
T	T	T	F	F	T	F	T	F	F	T
T	F	T	T	T	F	T	F	F	F	T
F	T	F	F	F	T	T	T	T	T	F
F	F	F	F	T	F	F	F	F	T	F
		1	5	3	2	7	2	6	4	1

Note that $\vee$ is the main connective here; it joins columns 5 and 6. Notice also that column 7, the final result, is identical to the truth table for $p \veebar q$. So $p \veebar q$ is *logically equivalent* to $[p \wedge (\sim q)] \vee [q \wedge (\sim p)]$.

EXAMPLE 2

Construct the truth table for the statement $(p \rightarrow q) \wedge (q \rightarrow p)$. Is it logically equivalent to $p \leftrightarrow q$?

Solution

Again, we show the truth table with the steps numbered. Note that the $\wedge$ is the main connective.

p	q	$(p$	$\rightarrow$	$q)$	$\wedge$	$(q$	$\rightarrow$	$p)$
T	T	T	T	T	T	T	T	T
T	F	T	F	F	F	F	T	T
F	T	F	T	T	F	T	F	F
F	F	F	T	F	T	F	T	F
		1	3	2	5	2	4	1

Since column 5 matches the truth table for $p \leftrightarrow q$ (see Table 4.5), the given statement is logically equivalent to $p \leftrightarrow q$.

Our next example illustrates two special types of logical statements.

EXAMPLE 3

Construct truth tables for

a. $p \vee (\sim p)$ **b.** $p \wedge (\sim p)$

Solution

a. There is only *one* simple statement, p, here, so the truth table has only two rows.

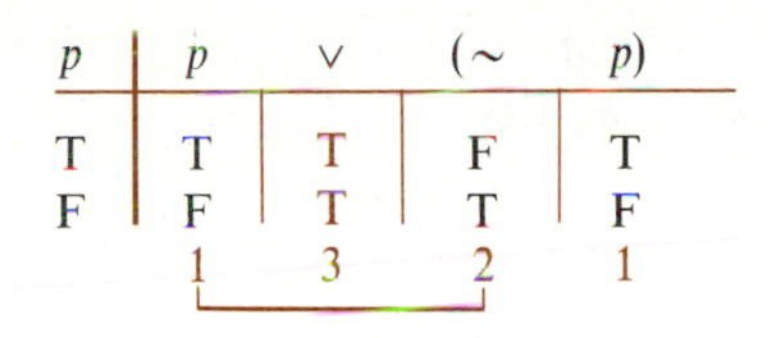

p	p	$\vee$	$(\sim$	$p)$
T	T	T	F	T
F	F	T	T	F
	1	3	2	1

The truth table for $p \vee (\sim p)$, column 3, has only Ts; that is, the statement is *always true*. Such a statement is called a **tautology**.

b.

p	p	$\wedge$	$(\sim$	$p)$
T	T	F	F	T
F	F	F	T	F
	1	3	2	1

This truth table has all Fs; the statement is *always false*. Such a statement is called a **self-contradiction**.

Let us return briefly to the first logical statement considered in this section, $(p \vee q) \wedge (\sim p)$. The parentheses in this statement make two things clear:

1. The main connective is $\wedge$.
2. In constructing the truth table, we must find the truth values for $(p \vee q)$ and $(\sim p)$ first.

Now suppose the parentheses were not there, and the statement was $p \vee q \wedge \sim p$. Is there any way to know which operator ($\vee$, $\sim$, or $\wedge$) to apply first? Luckily, the answer is "yes." Just as arithmetic operators ($\times$, /, +, −) have precedence rules, so do logical operators. The priority of logical operators is

1. ( ) Parentheses first, as in arithmetic.
2. $\sim$ (NOT)
3. $\wedge$ (AND)
4. $\vee$ (OR)

This means that, in constructing a truth table, we take care of parentheses first, followed by $\sim$, $\wedge$ and $\vee$, in that order. We apply these precedence rules in the following example:

EXAMPLE 4

Construct the truth table for the statement $p \vee q \wedge \sim p$. Is it logically equivalent to $(p \vee q) \wedge (\sim p)$—that is, do the parentheses *really* make a difference?

Solution

We must take care of $\sim$ first, then $\wedge$, then $\vee$. This gives the truth table below.

p	q	p	$\vee$	q	$\wedge$	$\sim$	p
T	T	T	T	T	F	F	T
T	F	T	T	F	F	F	T
F	T	F	T	T	T	T	F
F	F	F	F	F	F	T	F
		1	5	2	4	3	1

This statement is not logically equivalent to $(p \vee q) \wedge (\sim p)$. It is, however, logically equivalent to the statement $p \vee q$. This means that it is legitimate to substitute the simpler statement $p \vee q$ for the more complicated statement, involving more connectives, $p \vee q \wedge \sim p$. We shall find this ability quite useful in the next chapter.

The precedence rules for logical operators have a practical application to computer programming, especially in writing IF-THEN statements involving multiple conditions. This will be discussed in more detail in Section 4.5.

Up to this point, we have been constructing truth tables of compound statements that are composed of either one or two simple statements. Just for practice, let's try one with *three* simple statements.

EXAMPLE 5

Construct a truth table for $p \wedge q \vee r$.

Solution

Since three simple statements are involved, there will be $2^3 = 8$ different combinations of truth values possible. (Note that one simple statement gave $2^1 = 2$ possibilities and two simple statements gave $2^2 = 4$ possibilities.) We show the truth table below. The precedence rules require that we take care of the $\wedge$ first.

p	q	r	p	$\wedge$	q	$\vee$	r
T	T	T	T	T	T	T	T
T	T	F	T	T	T	T	F
T	F	T	T	F	F	T	T
T	F	F	T	F	F	F	F
F	T	T	F	F	T	T	T
F	T	F	F	F	T	F	F
F	F	T	F	F	F	T	T
F	F	F	F	F	F	F	F
			1	4	2	5	3

EXAMPLE 6

Construct a truth table for each of the following statements. Test each to see whether it is a tautology, a self-contradiction, or neither.

a. $(p \wedge q) \rightarrow (p \wedge r)$ **b.** $\sim(p \rightarrow q) \wedge q$ **c.** $(p \rightarrow q) \vee (q \rightarrow p)$

Solution

a. $(p \wedge q) \rightarrow (p \wedge r)$

p	q	r	$(p$	$\wedge$	$q)$	$\rightarrow$	$(p$	$\wedge$	$r)$
T	T	T	T	T	T	T	T	T	T
T	T	F	T	T	T	F	T	F	F
T	F	T	T	F	F	T	T	T	T
T	F	F	T	F	F	T	T	F	F
F	T	T	F	F	T	T	F	F	T
F	T	F	F	F	T	T	F	F	F
F	F	T	F	F	F	T	F	F	T
F	F	F	F	F	F	T	F	F	F
			1	4	2	6	1	5	3

This statement is neither a tautology nor a self-contradiction.

b. $\sim(p \rightarrow q) \wedge q$

p	q	$\sim$	$(p$	$\rightarrow$	$q)$	$\wedge$	q
T	T	F	T	T	T	F	T
T	F	T	T	F	F	F	F
F	T	F	F	T	T	F	T
F	F	F	F	T	F	F	F
		4	1	3	2	5	2

This statement is a self-contradiction.

c. $(p \to q) \vee (q \to p)$

p	q	$(p$	$\to$	$q)$	$\vee$	$(q$	$\to$	$p)$
T	T	T	T	T	T	T	T	T
T	F	T	F	F	T	F	T	T
F	T	F	T	T	T	T	F	F
F	F	F	T	F	T	F	T	F
		1	3	2	5	2	4	1

This statement is a tautology.

EXERCISES FOR SECTION 4.4

In Exercises 1 through 10, construct the truth table for the compound statement. Indicate whether it is a tautology, a self-contradiction, or neither.

1. $\sim p \to q$
2. $\sim(p \to q)$
3. $p \vee (p \to q)$
4. $(p \leftrightarrow r) \vee q$
5. $p \veebar p$
6. $(p \to q) \leftrightarrow [(\sim q) \to (\sim p)]$
7. $(p \wedge q \vee r) \wedge \sim p$
8. $(p \uparrow q) \vee r$
9. $\sim[(\sim q \vee \sim r) \to \sim(p \wedge r)]$
10. $[p \wedge (q \vee r)] \leftrightarrow [(p \wedge q) \vee (p \wedge r)]$

In Exercises 11 through 15, determine whether or not the statements in each pair are logically equivalent.

11. $\sim(p \vee q)$, $\sim p \wedge \sim q$
12. $p \wedge (q \vee r)$, $(p \wedge q) \vee r$
13. $p \leftrightarrow q$, $\sim(p \veebar q)$
14. $q \to p$, $\sim q \vee p$
15. $p \to q$, $(\sim p) \to (\sim q)$

4.5 Logical Operators and Computer Programming

In presenting applications of logic and the use of logical operators in computer programming, we shall incorporate a review of some basic concepts. Our strategy in this section is to present a series of examples with detailed explanations.

EXAMPLE 1

Solve the following inequalities:

a. $x - 4 < 6$ **b.** $xy < 0$ **c.** $xy > 0$

Solution

a. To solve this inequality, we add 4 to both sides:

$$x - 4 + 4 < 6 + 4$$

So $x < 10$ is the solution. Notice that, in general, an inequality does not have a *single value* as solution (recall the graphs of inequalities in Chapter 2). In fact, *any* value of x smaller than 10 makes $x - 4 < 6$ a true statement.

b. Here, we are looking for values of x and y that make $xy < 0$ (that is, the product of x and y is negative) a true statement. The product of two numbers is negative *if and only if* the numbers have *opposite signs*. There are two sets of possibilities:

x	y	xy
+	−	−
−	+	−

So $xy < 0$ if ($x > 0$ *and* $y < 0$) *or* ($x < 0$ *and* $y > 0$).

c. $xy > 0$ *if and only if* both values have the *same* sign; that is, both are positive or both are negative. In other words,

$$xy > 0 \quad \text{if } (x > 0 \textit{ and } y > 0) \textit{ or } (x < 0 \textit{ and } y < 0)$$

Let's examine part c more closely. Suppose, in a computer program, we had to verify that two values, x and y, had the same sign—both positive or both negative. One way would be to use a statement such as

```
IF (X > 0 AND Y > 0) OR (X < 0 AND Y < 0) THEN . . . .
```

Example 1c suggests another possible approach. We found that

$$(xy > 0) \leftrightarrow [(x > 0 \wedge y > 0) \vee (x < 0 \wedge y < 0)]$$

A biconditional statement is true on condition that both sides have the same truth value. This says that $(xy > 0)$ is true under exactly the *same conditions* as $[(x > 0 \wedge y > 0) \vee (x < 0 \wedge y < 0)]$. It would, then, be acceptable to make a substitution in the condition of the IF statement; it becomes

```
IF X * Y > 0 THEN . . .
```

A little knowledge of logic has enabled us to simplify a program statement considerably.

These considerations lead to one more digression before we continue with examples. The IF or IF-THEN statement occurs frequently in computer programs written in many languages. It is *not* the same as the logical implication ($p \rightarrow q$ or "if p, then q"). The general form of the computer IF-THEN statement is

IF (condition) THEN (execution)

Its meaning, in BASIC and other languages, is: The execution takes place *if and only if* the condition is true. (The *if and only if* is important here.)

Suppose, for example, you wrote a BASIC program that looked like this:

```
:
30 IF X > 4 THEN GO TO 60
40...
50...
60 PRINT "HI, MOM!"
70...
:
```

The "execution" part of the IF-THEN statement is GO TO 60. When line 30 is executed, the computer first compares the value of X to the value 4. If X > 4 is *true*, then execution (a branch to line 60) takes place. If X > 4 is *false* (X ≤ 4), then execution of the GO TO does not take place; instead the computer continues execution at line 40.

EXAMPLE 2

Write a BASIC statement that will print "RESPONSIBLE ADULT" if and only if a person's age is over 21.

Solution

Let AG stand for age. The statement is

```
IF AG > 21 THEN PRINT ''RESPONSIBLE ADULT''
```

This statement will cause the PRINT statement (the "execute") to be executed on condition that the value of AG is greater than 21, and *only* on that condition.

We discuss one more mathematical example to stress the importance of understanding the meaning of conjunction and disjunction before returning to strictly programming-related applications.

EXAMPLE 3

Solve each of the following inequalities:

a. $(x + 2)(x - 1) < 0$ **b.** $(x + 3)(x - 4) > 0$

Solution

We use an approach similar to that of Example 1.

a. $(x + 2)(x - 1) < 0$ if and only if $(x + 2)$ and $(x - 1)$ have opposite signs. The possibilities are

	$x + 2$	$x - 1$	$(x + 2)(x - 1)$
1	+	−	−
2	−	+	−

Possibility 1 gives us

$$x + 2 > 0 \quad \textit{and} \quad x - 1 < 0$$

Solving these inequalities gives

$$x > -2 \quad \textit{and} \quad x < 1$$

So $(x + 2)(x - 1) < 0$ whenever $x > -2$ and $x < 1$; another way of writing this is:

$$-2 < x < 1$$

Now we examine possibility 2.

$$x + 2 < 0 \quad \textit{and} \quad x - 1 > 0$$

Solving these inequalities gives

$$x < -2 \quad \textit{and} \quad x > 1$$

But this is *impossible*; no number can be *both* less than -2 *and* greater than 1. So we must *reject* possibility 2. Note the importance here of understanding the *conjunction* operator. The final solution is

all values of x between -2 and 1, excluding -2 and 1

or

$$-2 < x < 1$$

b. $(x + 3)(x - 4) > 0$ *if and only if* $(x + 3)$ and $(x - 4)$ have the *same* sign. The possibilities are

	$x + 3$	$x - 4$	$(x + 3)(x - 4)$
1	+	+	+
2	−	−	+

Possibility 1 gives us

$$x + 3 > 0 \qquad \textit{and} \qquad x - 4 > 0$$

Solving these inequalities gives

$$x > -3 \qquad \textit{and} \qquad x > 4$$

The conjunction $(x > -3) \wedge (x > 4)$ is true only for values of x that make *both* pieces true. *Any* number that is greater than 4 is also greater than -3; so any value of x that makes $x > 4$ true will also make $x > -3$ true. However, there are values of x for which $x > -3$ is true, but $x > 4$ is false. One example is $0 > -3$, but $0 < 4$. So $x > -3$ isn't enough to *guarantee* $x > 4$. The part of the solution that comes from possibility 1 then is

$$x > 4$$

Possibility 2 gives us

$$x + 3 < 0 \qquad \textit{and} \qquad x - 4 < 0$$

Solving these inequalities results in

$$x < -3 \qquad \textit{and} \qquad x < 4$$

This time $x < 4$ is not restrictive enough, since there are numbers (for example, zero) that are less than *4*, but not less than *−3*. Therefore, possibility 2 gives us

$$x < -3$$

Our final and complete solution to $(x + 3)(x - 4) > 0$ is

result of possibility 1 OR result of possibility 2

That is,

$$x > 4 \qquad \text{OR} \qquad x < -3.$$

Observe the importance of the disjunction, OR, in the final solution to part b. Could we have said $(x > 4)$ AND $(x < -3)$? That is, is there a single value of x that makes *both* $(x > 4)$ *and* $(x < -3)$ true? The answer to that question is, of course, that no such number exists.

Now we return to some programming examples.

EXAMPLE 4

Write a BASIC statement to do each of the following:

a. Print "TRUE" if the value of the variable N is strictly between 0 and 5.
b. Assign an insurance premium of $500 to a person if that person's sex is "male" and age is over 21.
c. Double the value of the variable T if T is zero or T is between 10 and 20, inclusive.

Solution

All of these cases require IF-THEN statements.

a. N is strictly between 0 and 5 means N > 0 AND N < 5.
(We could also say 0 < N AND N < 5.)
The BASIC statement is

```
IF N > 0 AND N < 5 THEN PRINT ''TRUE''
```

b. We shall use the variables AG for age and PR for premium; SX = 1 for male and SX = 2 for female. The conditions are

SX = 1 AND AG > 21

The BASIC statement is

```
IF SX = 1 AND AG > 21 THEN LET PR = 500
```

c. We want to double the value of T (LET T = 2 * T) on condition that T = 0 OR T is between 10 and 20. The requirement that T is between 10 and 20, inclusive, means

T >= 10 AND T <= 20

Combining these conditions gives

T = 0 OR (T >= 10 AND T <= 20)

The BASIC statement is

```
IF T = 0 OR (T >= 10 AND T <= 20) THEN LET T = 2 * T
```

In this statement, we included parentheses around `(T >= 10 AND T <= 20)` to make it easier to read. But are they necessary in order for the statement to work properly? In this case, they are not because the rules for precedence of logical operators will cause the AND to be done before the OR, even without parentheses. As we shall see later in this section, however, there are circumstances in which parentheses are necessary.

EXAMPLE 5

Computers can be used to monitor a continuous quantity or process such as blood pressure. Write BASIC statements that will do the following:

a. print "OK" if systolic pressure is between 100 and 140;
b. print "YOU'RE IN BIG TROUBLE" if systolic pressure is outside the acceptable range (100–140).

Solution

Let S stand for systolic pressure.

a. Systolic pressure S between 100 and 140 means

$$S > 100 \qquad \text{AND} \qquad S < 140$$

The BASIC statement is

```
IF (100 < S) AND (S < 140) THEN PRINT "OK"
```

b. S outside the acceptable range means

$$S <= 100 \qquad \text{OR} \qquad S >= 140$$

The BASIC statement is

```
IF S <= 100 OR S >= 140 THEN PRINT
"YOU'RE IN BIG TROUBLE"
```

EXAMPLE 6

A company has agreed to pay a bonus to an employee under the following conditions: The employee must be over 55 years old and have worked for the company for more than 25 years, or the employee must have made sales totaling at least $500,000.

a. Write the symbolic representation of these conditions. Use

A = age in years
E = number of years employed
D = sales in dollars

b. Write a BASIC and a Pascal statement that will print "BONUS" if the employee qualifies for a bonus.

Solution

a. $(A > 55) \wedge (E > 25) \vee (D \geq 500000)$

b. In BASIC,

```
IF (A > 55) AND (E > 25) OR (D >= 500000)
THEN PRINT "BONUS"
```

In Pascal,

```
IF ((A > 55) AND (E > 25)) OR (D >= 500000)
THEN WRITELN ('BONUS');
```

EXAMPLE 7

Rewrite the code for Example 6 to print "NO BONUS" when an employee is *not* eligible for a bonus.

Solution

An employee is *not* eligible when he fails to meet the following conditions:

$$(A > 55) \wedge (E > 25) \vee (D \geq 500000)$$

That is, if he is less than or equal to 55 years old *or* he has been employed for less than or equal to 25 years *and* in addition has sales of less than $500,000. If this is not immediately obvious to you, let's look at it another way. Suppose p is the statement $A > 55$, q is $E > 25$, and r is $D >= 500000$. The criteria for eligibility can be written as $p \wedge q \vee r$. The negation of this statement gives the criteria for ineligibility:

$$\begin{aligned} \sim(p \wedge q \vee r) &\Leftrightarrow \sim[(p \wedge q) \vee r] \\ &\Leftrightarrow \sim(p \wedge q) \wedge \sim r \\ &\Leftrightarrow (\sim p \vee \sim q) \wedge \sim r \end{aligned} \quad \text{By DeMorgan's laws}$$

But $\sim p$ is equivalent to $A \leq 55$,
$\sim q$ is equivalent to $E \leq 25$, and
$\sim r$ is equivalent to $D < 500000$.

Therefore, in order to *fail* to qualify for a bonus, a person must satisfy

$$(A \leq 55 \vee E \leq 25) \wedge (D < 500000)$$

The BASIC statement is

```
IF (A <= 55 OR E <= 25) AND D < 500000
THEN PRINT ''NO BONUS''
```

In Pascal,

```
IF ((A <= 55) OR (E <= 25)) AND (D < 500000)
THEN WRITELN ('NO BONUS');
```

The parentheses in the condition above are necessary. Suppose, for example, an employee is 50 years old ($A = 50$), has been employed for 20 years ($E = 20$), and has sales of $525,000. His sales figure clearly qualifies him for a bonus; so the BASIC statement should *not* cause "NO BONUS" to be printed. We have

$$(A <= 55 \quad \text{OR} \quad E <= 25) \qquad \text{AND} \qquad (D < 500000)$$

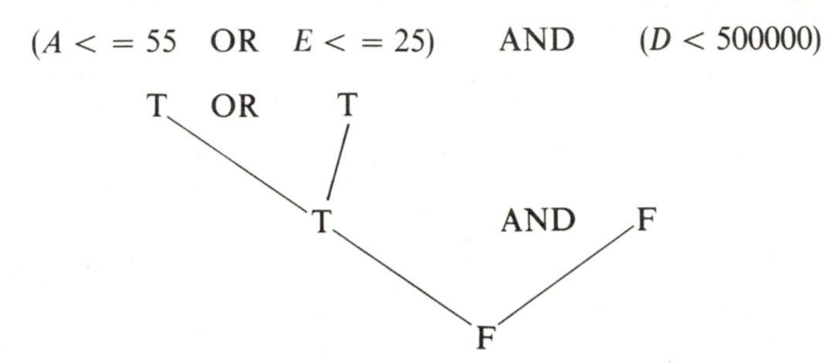

The condition is false, so execution will not take place; "NO BONUS" will *not* be printed.

Now suppose we get rid of the parentheses. The condition becomes

$$A <= 55 \quad \text{OR} \quad E <= 25 \qquad \text{AND} \qquad D < 500000$$

Because of the rules for precedence of logical operators, the AND is applied first. We have

$$A <= 55 \quad \text{OR} \quad E <= 25 \qquad \text{AND} \qquad D < 500000$$

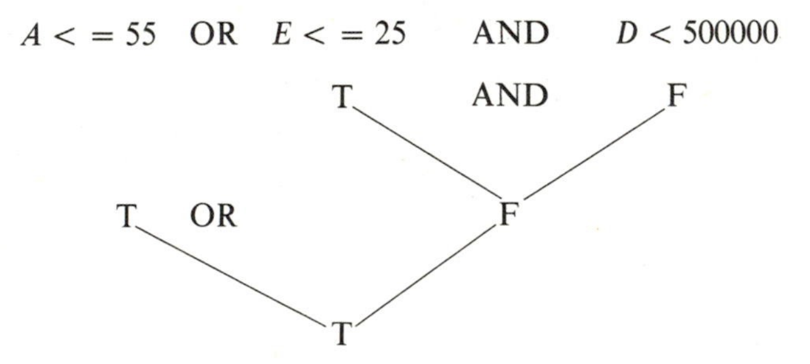

Since the condition is true, execution will take place and "NO BONUS" will be printed. This is clearly *not* what we want. The moral of this story is: Programmers must pay attention to precedence of logical operators when they write IF-THEN statements involving ANDs, ORs and NOTs.

There is much more that can be said on the relationship of logic to computer programming. Our aim in this section has been to make the reader aware of some of the possibilities.

EXERCISES FOR SECTION 4.5

In Exercises 1 through 10, solve the given inequality (see Examples 1 and 3).

1. $x - 2 < 5$
2. $x + 1 \geq -2$
3. $x + 5 > 7$
4. $x - 3 > 0$
5. $x + 2 < 0$
6. $x + 2 \geq 0$
7. $(x - 3)(x + 2) < 0$
8. $(x - 3)(x + 2) > 0$
9. $x(x - 1) < 0$
10. $x(x - 1) > 0$

In Exercises 11 through 15, first write a symbolic representation of the condition and then write an IF-THEN statement (in BASIC or Pascal) that will do what is required (see Examples 4 through 7).

11. Print "FREE ADMISSION" on a person's ticket to the roller skating rink if the person is between the ages of 16 and 25.
12. Assign a grade of B to a student whose average is at least 80 but less than 90.
13. Print "REDUCED TUITION" on a student's community college application if the student is over 65 years old or has a gross yearly income of under $12,000.
14. Deduct 20% from a student's activity fee (F) if the student is over 60 years of age and a part-time student or if the student's income is less than or equal to $10,000.
15. Reject any number between 18 and 35, except 25.
16. An employee of a certain company is eligible for a pension if (1) he or she is over 60 years of age and has been employed by the company for more than 10 years, *or* (2) he or she has worked for the company for more than 30 years.
    - **a.** Write a BASIC or Pascal statement that will cause the computer to print "ELIGIBLE" if an employee is eligible for a pension.
    - **b.** Write a statement that will cause the computer to print "NOT ELIGIBLE" if the conditions are *not* met.

    (See Examples 6 and 7.)
17. Use logical connectives to write the following program segment more efficiently:

```
 70 IF X = 7 THEN 90
 80 GO TO 110
 90 IF Y <> 4 THEN 110
100 PRINT "ACCEPT"
110 . . .
```

4.6 Sets and Set Operations

Definition of a Set

The concept of a *set* is a pervasive one in mathematics; it serves to connect a number of important fields. Many abstract concepts are better explained and understood by using set terminology and notation. Since this is a text on computer mathematics, we shall want to relate data processing concepts to the idea of a set. In order to do this, we shall digress briefly to introduce some relevant data processing concepts.

In most schools, there is a *file* cabinet of some sort containing student *records*. Each student record contains individual items of information on one particular student. Typical items, called *fields*, might be name, address, social security number, graduating class, and grade point average. A **file**, then, is a collection of related **records**—for example, the individual student records in a student file. Each item of information on the record is called a **field**. Thus, a field is a subdivision of a record. If the student records were on cards, they might look something like this:

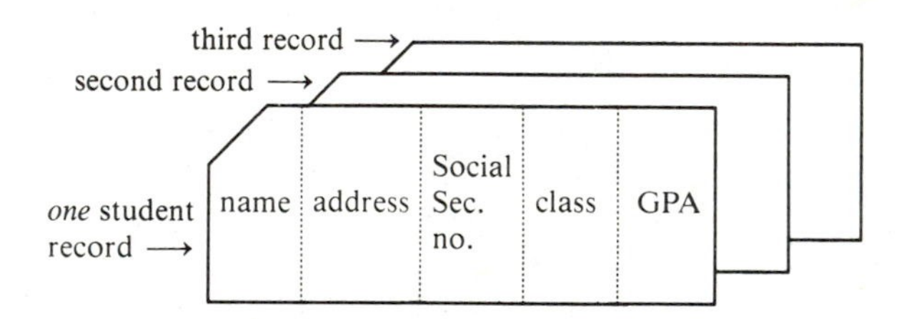

Other examples of files might be personnel files made up of employee records or inventory files containing records of items in stock or in a catalog.

Now back to mathematics. We define a **set** to be a **well-defined** collection of objects. The individual objects are called **elements** or **members** of the set. By *well defined* we mean that given any object, we must be able to determine whether or not the object is an element of the set. If we cannot determine whether or not an object is an element of a supposed set, then the collection of objects is *not well defined* and therefore is *not a set.*

Consider some examples. The collection of tall people is not a set in mathematical terms. The word *tall* is too vague; one would not be able to determine precisely whether or not a particular person belongs to the collection of tall people. In fact, a person's perception of tall is frequently related to his or her own height. On the other hand, the collection of people over six feet tall *is* a set; we have defined precisely the condition necessary for membership in the set.

A set is usually symbolized by curly braces within which we specify the members of the set. This can be done in two ways: (1) by listing all the elements or (2) by describing a property that defines what objects belong to the set (the **set-builder method**).

In the **listing method**, we list each element of the set individually. For example, if A is the set of all even numbers from 10 through 30, we have

$$A = \{10, 12, 14, 16, 18, 20, 22, 24, 26, 28, 30\}$$

We use the symbol $\in$ to mean "is a member or element of." Thus, we write $18 \in A$. We can also write $15 \notin A$; this is read "15 is *not* an element of A." We shall follow the usual convention of using a capital letter to denote a set and lowercase letters to denote elements of the set.

In the **set-builder method**, we use a property to describe the members of a set. This is especially useful when it is difficult or impossible to list the elements of the set individually. For example, let T be the set of people over six feet tall. Using the set-builder method, we write

Read "such that"

$$T = \{p \mid p \text{ is a person over 6 feet tall}\}$$

EXAMPLE 1

Which of the following are true statements?

a. $4 \in \{2, 4, 6, 8\}$
b. $3 \notin \{1, 3, 5, 10\}$
c. $3 \notin \{2, 4, 6, 8\}$
d. $10 \in \{n \mid n \text{ is an even integer}\}$
e. $10 \in \{n \mid n \text{ is an odd integer}\}$
f. George Washington $\in \{p \mid p$ is a former President of the United States$\}$

Solution

a. true **b.** false **c.** true **d.** true **e.** false **f.** true

A file can be thought of as a set of related records. The elements of the set are the individual records. Files differ from sets in that, in general, the records in a file are related in some way. They may relate, for example, to all the students in a school or to all the employees of a particular company. The elements of a set need not be related in any way; in fact, the order in which they are listed is not even important. For example, if $A = \{2, 5, 6\}$ and $B = \{5, 6, 2\}$, then A and B represent the same set.

Operations on Sets

Certain operations can be performed on sets. They are closely related to the logical operators we studied earlier in this chapter.

In general, the collection of objects under examination or discussion is called the **universal set**. It is denoted by the letter U. For example, a universal set might be the set of students in your school. This set can be further subdivided into smaller sets, such as males, females, honor students, seniors, and so on. Each of these "smaller" sets is called a **subset** of the universal set.

The *formal definition* of *subset* is: A set A is a subset of a set S, written $A \subset S$, if and only if every element of A is also an element of S. Using logical operators, we could write this definition as

$$A \subset S \Leftrightarrow a \in A \rightarrow a \in S$$

If A is *not* a subset of S, written $A \not\subset S$, then there is *at least one* element of A that does not belong to S.

EXAMPLE 2

Given the sets

$$U = \{1, 2, 3, 4, 5, 6, 7, 8\} = \text{the universal set}$$
$$A = \{1, 3, 5\}$$
$$B = \{1, 2, 3, 5, 6, 8\}$$

state whether each of the following is true or false:

a. $A \subset U$ **b.** $B \subset U$ **c.** $B \subset A$ **d.** $A \subset B$
e. $\{1, 2, 3, 4\} \subset B$ **f.** $\{1, 2, 3, 4\} \subset U$ **g.** $A \subset A$

Solution

a. $A \subset U$ is true since every element of A is an element of U.
b. $B \subset U$ is also true.
c. $B \subset A$ is false. For example, $2 \in B$ but $2 \notin A$.
d. $A \subset B$ is true.
e. $\{1, 2, 3, 4\} \subset B$ is false since $4 \in \{1, 2, 3, 4\}$ but $4 \notin B$.
f. $\{1, 2, 3, 4\} \subset U$ is true.
g. $A \subset A$ is true since every element of A is certainly an element of A.

Consider the set N of all odd integers that are evenly divisible by 2. This set is certainly well defined; we have precise criteria for membership. However, there are no integers that meet the conditions for belonging to N since *no* odd integer is evenly divisible by 2. A set like N that has *no elements* is called the **null** or **empty set**; it is denoted by the symbols $\{ \}$ or $\varnothing$.

It can be proven that the null set is a subset of any set. Let $\varnothing$ represent the null set and let S represent any other set. The definition of a subset says that $\varnothing \subset S$ if and only if $a \in \varnothing \rightarrow a \in S$ is a true statement. But $a \in \varnothing$ is always *false*. Therefore, $a \in \varnothing \rightarrow a \in S$ is *true*, whether $a \in S$ is true or false (see the truth table for $\rightarrow$, Table 4.4). Hence, $\varnothing \subset S$.

We can conclude, then, that every nonempty set has at least two subsets: the empty set and itself.

EXAMPLE 3

Let $A = \{2, 4, 6\}$. List *all* the subsets of A.

Solution

The subsets are

$$\varnothing, A, \{2\}, \{4\}, \{6\}, \{2, 4\}, \{2, 6\}, \{4, 6\}$$

The set S of female honor students constitutes a subset of the universal set of all students. It consists of all the female students who are also honor students. If F represents the set of female students and H represents the set of honor students, we can describe S by

$$S = \{x | x \in F \textit{ and } x \in H\}$$

This set is called the **intersection** of F and H, and is written $F \cap H$ (sometimes read "F intersect H").

EXAMPLE 4

Given each pair of sets A and B below, find $A \cap B$.

a. $A = \{3, 7, 9, -2\}$; $B = \{9, 7, 8, 4, 6, -2\}$

b. $A = \{2, 4, 5\}$; $B = \{1, 2, 3, 4, 5, 6\}$

c. $A = \{7, 6, 9, 2\}$; $B = \{5, 11, 3\}$

Solution

a. $A \cap B = \{7, 9, -2\}$

b. $A \cap B = \{2, 4, 5\}$

Notice here that $A \subset B$ and it turned out that $A \cap B = A$. This will be true in the general case:

$$\text{If } X \subset Y, \text{ then } X \cap Y = X.$$

c. $A \cap B = \varnothing$. That is, A and B have *no elements* in common. Two sets whose intersection is the null set are said to be **disjoint**.

It is often important in data processing to be able to select certain subsets from a file. For example, suppose we want the computer to select all female honor students (that is, all members of $F \cap S$) and print their names. Let the variable S$ represent the person's sex ("M" or "F") and A represent the student's average; the variable N$ will hold the student's name.

At this stage, we cannot properly program this problem because the set of students we wish to select is *not well defined*. What is an honor student? If we agree that honor students must have an average over 90, then we have a well-defined set and we can properly program the computer. We would use a statement such as (in BASIC)

```
IF A > 90 AND S$ = "F" THEN PRINT N$
```

Now consider the data processing problem of finding students over 65 *or* disabled. If R is the set of students who are over 65 and D is the set of disabled students, we are looking for students who are members of R or of D (or both). This set is called the **union** of R and D, and is written $R \cup D$. Thus $R \cup D$ is defined by

$$R \cup D = \{s \mid s \in R \text{ or } s \in D\}$$

EXAMPLE 5

Find the union of each of the following pairs of sets:

a. $A = \{7, -2, 9\}$; $B = \{-2, 4, 6, 11\}$
b. $A = \{1, 3, 5\}$; $B = \{1, 3, 5, 7, 9\}$

Solution

a. $A \cup B = \{7, -2, 9, 4, 6, 11\}$

Notice that, although the element -2 appears in both sets, it is listed only once in $A \cup B$. The set concept emphasizes *which* elements are being considered, not how many times they appear.

b. $A \cup B = \{1, 3, 5, 7, 9\}$

Notice here that $A \subset B$ and $A \cup B = B$. This is true in general; if $R \subset S$, then $R \cup S = S$.

EXAMPLE 6

Write in symbolic notation the data processing problem of having the computer print the names of students who are over 65 or disabled (the members of $R \cup D$).

Solution

Let S$ be the status variable, where "D" means disabled. Let A represent age and N$ represent name.

IF (S$ = "D") OR (A > 65) THEN PRINT N$

In many schools, the students can be classified as either part-time or full-time. These categories are mutually exclusive; no student belongs to both. If we denote the set of full-time students by F and the set of part-time students by P, then we may say that $F \cup P = U$ (the universal set consisting of *all* students). We can also describe the set P as the set of all members of the universal set who are *not* elements of F. In symbols, $P = \{s \mid s \notin F\}$. This set is called the **complement** of the set F, and is denoted by $\tilde{F}$. In general, we have

$$\tilde{X} = \{a \mid a \notin X\} \qquad \text{or} \qquad \tilde{X} = \{a \mid \sim(a \in X)\}$$

Note the relationship of the complement of a set to logical negation. Note also that $X \cup \tilde{X} = U$ and $X \cap \tilde{X} = \varnothing$ (*X and $\tilde{X}$* are *disjoint*).

EXAMPLE 7

Let

$$X = \{7, -1, 3, 4, 6\}$$
$$Y = \{5, 6, 2, 4\}$$
$$Z = \{11, 13, 1\}$$
$$U = \{-1, 1, 2, 3, 4, 5, 6, 7, 11, 13, 99\} = \text{the universal set}$$

Find

a. $X \cap Y$ **b.** $X \cup Y$ **c.** $X \cap Z$ **d.** $\tilde{X}$

Which pairs of sets are disjoint?

Solution

a. $X \cap Y = \{4, 6\}$

b. $X \cup Y = \{7, -1, 3, 4, 6, 5, 2\}$

c. $X \cap Z = \varnothing$

d. $\tilde{X} = \{1, 2, 5, 11, 13, 99\}$

From part c, $X \cap Z = \varnothing$, so X and Z are disjoint. Also $Y \cap Z = \varnothing$, so Y and Z are disjoint.

A reasonable question to ask is: when are two sets equal? We have been using this idea in an intuitive sense; now we shall define it precisely. Two sets are *equal* if and only if they contain exactly the same elements. Thus sets A and B are equal, $A = B$, if and only if

$$a \in A \rightarrow a \in B \qquad \textit{and} \qquad a \in B \rightarrow a \in A$$

This is another way of saying $A = B$ if and only if $A \subset B$ *and* $B \subset A$.

EXERCISES FOR SECTION 4.6

In Exercises 1 through 6, determine the truth value of the given statement (see Example 1).

1. $8 \in \{2, 4, 6\}$

2. $8 \notin \{2, 4, 6\}$

3. $1 \in \{y \mid y - 1 = 0\}$

4. $5 \in \{x \mid x \text{ is an odd integer}\}$

5. $32 \in \{n \mid n \text{ is an odd integer}\}$

6. $\varnothing \in \varnothing$

In Exercises 7 through 10, list all subsets of the given set (see Examples 2 and 3).

7. $\{a, b\}$ **8.** $\{\Delta\}$ **9.** $\varnothing$ **10.** $\{p, q, r, s\}$

In Exercises 11 through 20, let the universal set be $U = \{1, 2, 3, 4, 5, 6, 7, 8, 9\}$; $A = \{2, 3, 5, 6\}$; $B = \{3, 4, 6, 7\}$; $C = \{5, 6, 7, 8\}$; and $D = \{7\}$. Find each of the given sets (see Examples 4, 5 and 7).

11. $A \cup B$ **12.** $A \cap B$ **13.** $A \cap D$ **14.** $C \cup D$

15. $C \cap D$ **16.** $\tilde{A}$ **17.** $\widetilde{(A \cup B)}$ **18.** $\tilde{A} \cap \tilde{B}$

19. $\tilde{\varnothing}$ **20.** $B \cap C$

21. Another set operation is the *difference* of two sets, denoted by $A - B$. The difference of the sets A and B is defined as

$$A - B = A \cap \tilde{B}$$

Using A and B as in Exercises 11 through 20, find:

a. $A - B$ **b.** $B - A$ **c.** $U - A$ **d.** $B - \varnothing$

4.7 Chapter Review

In this chapter, we have discussed the fundamental concepts of mathematical logic and some related topics from the theory of sets. Whenever possible, examples of the use of these concepts in a computer programming context have been presented.

The chapter began with a discussion of simple and compound statements and of the concept of truth value. Definitions were then presented and truth tables developed for the basic logical connectives: NOT ($\sim$), AND ($\wedge$), OR ($\vee$), implication ($\rightarrow$), the biconditional operator ($\leftrightarrow$), NOR ($\downarrow$), NAND ($\uparrow$) and the exclusive OR ($\veebar$). Relationships among the various logical operators were explored in the context of logical equivalence. Section 4.4 presented examples of compound logical statements and techniques for constructing truth tables of complicated statements. In Section 4.5, we discussed examples of the use of logical operators in mathematical and programming applications.

We concluded the chapter with a section on sets and set operations, emphasizing their relationship to the logical operators and to the data processing concept of a file. This section gave definitions and examples of the concepts of set, subset, universal set, and null set, as well as the operations of union, intersection, and complement.

VOCABULARY

statement
truth value
simple statement
compound statement
connective
relational operator
logical operator
NOT connective ($\sim$)
negation
truth table
AND connective ($\wedge$)
conjunction
OR connective ($\vee$)
disjunction
inclusive OR ($\vee$)
implication ($\rightarrow$)
logical equivalence
biconditional ($\leftrightarrow$)
NOR connective ($\downarrow$)
NAND connective ($\uparrow$)
exclusive OR ($\veebar$)
reference column
tautology
self-contradiction

file
record
field
set
well-defined
element
universal set
subset
null or empty set
intersection
disjoint sets
union
complement
equality of sets

CHAPTER TEST

Let p be "It is cold," and let q be "We will have a picnic." Write each statement in Problems 1 through 5 symbolically, in terms of p and q.

1. It is cold, and we will not have a picnic.
2. We will have a picnic, or it is cold.
3. If it is cold, then we will not have a picnic.
4. We will have a picnic if and only if it is cold.
5. It is not true that we will not have a picnic.

Let p and q be as above and let r be "Someone will build a fire." Give a simple sentence that translates each of the symbolic statements in Problems 6 through 10.

6. $p \wedge r$
7. $p \wedge \sim q \vee r$
8. $(p \wedge q) \rightarrow r$
9. $(\sim p) \leftrightarrow (\sim q)$
10. $\sim(\sim r)$

Construct a truth table for each statement in Problems 11 through 15. State whether each one is a tautology, a self-contradiction, or neither.

11. $\sim(p \vee \sim q)$
12. $\sim p \leftrightarrow p$
13. $p \wedge \sim q \vee r$
14. $\sim p \rightarrow (p \wedge q)$
15. $(p \leftrightarrow q) \rightarrow (p \rightarrow q)$

Use truth tables to determine whether or not the pairs of statements in Problems 16 and 17 are logically equivalent.

16. $\sim p$, $\sim(p \vee q) \vee (\sim p \wedge q)$
17. $p \wedge \sim q$, $\sim(p \vee \sim q)$

Answer Problems 18 through 20 for the following sets:

$$\text{universal set} = U = \{a, b, c, e, i, o, q, s, u, z\}$$
$$C = \{b, c, q, s, z\}$$
$$V = \{a, e, i, o, u\}$$
$$A = \{a, b\}$$
$$B = \{q, c, z\}$$

18. List *all* subsets of B.

19. List the elements of

a. $A \cup C$ **b.** $A \cap V$ **c.** $\tilde{V}$

d. $C \cup V$ **e.** $A \cap B$

20. State whether each of the following is true or false:

a. $B \subset C$ **b.** $C \subset B$

c. $(B \cap C) \subset B$

CHAPTER

5

In the late nineteenth century mathematician George Boole published an innovative work, *An Investigation of the Laws of Thought*, in which he applied algebraic concepts to logical propositions. The mathematical theory he developed, now called **Boolean algebra**, has in modern times been applied to the study of logic circuits in electronic computers. In this chapter, we first study the fundamental concepts of Boolean algebra and then consider some of its applications to logic circuits and switching networks.

GEORGE BOOLE.

Boolean Algebra and Logic Circuits

5.1 Boolean Algebra

In the simplest form of Boole's algebra, we consider a set containing two elements. We shall denote these elements by 0 and 1. Two operations, denoted by + and ·, are defined on this set by the tables:

TABLE 5.1

+	0	1
0	0	1
1	1	1

·	0	1
0	0	0
1	0	1

Notice that these are *not* the ordinary operations of addition and multiplication although the tables are similar to the addition and multiplication tables for binary arithmetic. The 0s and 1s in the tables do *not* represent the *numbers* zero and one; they are merely symbols representing the elements of a set. The significance of the use of these particular symbols (0 and 1) will become more apparent later when we apply the concepts of Boolean algebra to circuit design and logic gates.

Before studying the properties of this Boolean algebra, let us write the tables above in a different format (Table 5.2):

TABLE 5.2

A	B	$A + B$	$A \cdot B$
1	1	1	1
1	0	1	0
0	1	1	0
0	0	0	0

Here, we have listed all possible combinations of zeros and ones, along with the results of applying each operation. Compare this with the truth table for the propositions $(p \vee q)$ and $(p \wedge q)$, where p and q are any two logical statements (Table 5.3):

TABLE 5.3

p	q	$p \vee q$	$p \wedge q$
T	T	T	T
T	F	T	F
F	T	T	F
F	F	F	F

In this table, substitute a + for $\vee$ and a $\cdot$ for $\wedge$. Now put a 0 wherever you see an F and a 1 wherever you see a T. The result is a replica of Table 5.2.

These considerations lead to one possible interpretation of the Boolean algebra we have described. The elements can be equated to

1: A given statement is true.

0: A given statement is false.

The operation + is equivalent to the logical OR; the operation $\cdot$ is equivalent to the logical AND. The Boolean algebra, then, describes the results of combining logical statements with the AND and OR operators.

EXAMPLE 1

Use Table 5.1 or Table 5.2 to evaluate the following expressions:

a. $(0 \cdot 1) \cdot 1$ **b.** $0 \cdot (1 \cdot 1)$ **c.** $1 + (0 + 1)$

d. $(1 + 0) + 1$ **e.** $1 \cdot (0 + 1)$ **f.** $(1 \cdot 0) + (1 \cdot 1)$

Solution

a. $(0 \cdot 1) \cdot 1 = 0 \cdot 1 = 0$

b. $0 \cdot (1 \cdot 1) = 0 \cdot 1 = 0$

c. $1 + (0 + 1) = 1 + 1 = 1$

d. $(1 + 0) + 1 = 1 + 1 = 1$

e. $1 \cdot (0 + 1) = 1 \cdot 1 = 1$

f. $(1 \cdot 0) + (1 \cdot 1) = 0 + 1 = 1$

Properties of a Boolean Algebra

Some of the properties of a Boolean algebra will be useful in our later attempts to design and simplify logic circuits. Here, we shall list the properties that are of interest to us, along with examples and a brief explanation of each one.

1. The operations are **commutative**. This means that the order in which they are performed does not affect the result. In other words, for all values of A and B,

$$A + B = B + A$$

and

$$A \cdot B = B \cdot A$$

This is obvious from the + and $\cdot$ tables (Table 5.1).

2. The operations are **associative**. This property is illustrated by Example 1 (parts a–d). Parts a and b show that

$$(0 \cdot 1) \cdot 1 = 0 \cdot (1 \cdot 1) = 0$$

Parts c and d show that

$$1 + (0 + 1) = (1 + 0) + 1 = 1$$

These are illustrations of the associative property: if we wish to "add" or "multiply" three elements, we may begin by operating on either the first two elements or the last two elements, and then combine the result with the remaining element.

3. Each operation is **distributive** over the other. Symbolically, this property says that

$$A \cdot (B + C) = (A \cdot B) + (A \cdot C)$$

and

$$A + (B \cdot C) = (A + B) \cdot (A + C)$$

Parts e and f of Example 1 illustrate the first of these, the distributive property of $\cdot$ over $+$. They show that

$$1 \cdot (0 + 1) = (1 \cdot 0) + (1 \cdot 1) = 1$$

EXAMPLE 2

Illustrate the distributive property of addition over multiplication by evaluating the following expressions:

a. $1 + (0 \cdot 1)$ **b.** $(1 + 0) \cdot (1 + 1)$

Solution

a. $1 + (0 \cdot 1) = 1 + 0 = 1$
b. $(1 + 0) \cdot (1 + 1) = 1 \cdot 1 = 1$

So,

$$1 + (0 \cdot 1) = (1 + 0) \cdot (1 + 1)$$

4. There is an **identity** element for each operation; that is, there is an element that, when combined with any other element by the operation, yields the original element. To find the identity element for the $+$ operation, we need an element I^+ for which the following are true:

$$0 + I^+ = 0 \qquad \text{and} \qquad 1 + I^+ = 1$$

Table 5.1 shows that 0 is such an element:

$$0 + 0 = 0 \qquad \text{and} \qquad 1 + 0 = 1$$

So 0 is the identity element for the operation $+$.

EXAMPLE 3

Find the identity element for operation $\cdot$.

Solution

Consider the $\cdot$ table. In this case, we need an element $I^{\cdot}$ that satisfies

$$0 \cdot I^{\cdot} = 0 \quad \text{and} \quad 1 \cdot I^{\cdot} = 1$$

This element is 1, since

$$0 \cdot 1 = 0 \quad \text{and} \quad 1 \cdot 1 = 1$$

Therefore, 1 is the identity element for $\cdot$.

5. Each element has a **complement**. If A is an element of a Boolean algebra, the *complement* of A, denoted A', is defined to be the element with the following properties:

$$A + A' = 1; \qquad \text{that is, } A + A' = \text{identity for } \cdot$$
$$A \cdot A' = 0; \qquad \text{that is, } A \cdot A' = \text{identity for } +$$

In our Boolean algebra, each element is the complement of the other. That is,

$$0' = 1 \quad \text{and} \quad 1' = 0$$

The reader should use Tables 5.1 and 5.2 to verify this statement. Note that the *complement* concept corresponds to the *negation* of a logical statement.

6. Each element A is **idempotent**, that is,

$$A + A = A \quad \text{and} \quad A \cdot A = A$$

Specifically,

$$0 + 0 = 0$$
$$1 + 1 = 1$$
$$0 \cdot 0 = 0$$
$$1 \cdot 1 = 1$$

EXAMPLE 4

Use the tables for + and $\cdot$ to verify the following:

a. $1 + (1 \cdot 0) = 1$ **b.** $1 \cdot (1 + 0) = 1$
c. $0 + (0 \cdot 1) = 0$ **d.** $0 \cdot (0 + 1) = 0$

Solution

a. $1 + (1 \cdot 0) = 1 + 0 = 1$ **b.** $1 \cdot (1 + 0) = 1 \cdot 1 = 1$
c. $0 + (0 \cdot 1) = 0 + 0 = 0$ **d.** $0 \cdot (0 + 1) = 0 \cdot 1 = 0$

7. Example 4 illustrates the **absorption** property: For any elements A and B of a Boolean algebra, the following are true:

$$A + (A \cdot B) = A$$

and

$$A \cdot (A + B) = A$$

EXAMPLE 5

Use the other six properties to prove that the absorption property is true.

Solution

Look first at $A + (A \cdot B)$.

$A + (A \cdot B) = (A \cdot 1) + (A \cdot B)$	1 is the identity for $\cdot$
$= A \cdot (1 + B)$	Distributive property
$= A \cdot 1$	See Table 5.1 ($1 +$ any element $= 1$)
$= A$	

Now consider $A \cdot (A + B)$.

$A \cdot (A + B) = (A \cdot A) + (A \cdot B)$	Distributive property
$= A + (A \cdot B)$	Idempotent property
$= A$	By the above $A + (A \cdot B) = A$

EXAMPLE 6

Use properties 1 through 6 above to simplify each of the following Boolean expressions:

a. $A \cdot B + A \cdot B'$ **b.** $[A' \cdot (B + C)] + (A \cdot B)$ **c.** $[A \cdot (A' + B)] \cdot C$

Solution

a.

$A \cdot B + A \cdot B' = A \cdot (B + B')$	Distributive property (3)
$= A \cdot 1$	Complement (property 5)
$= A$	1 is identity for $\cdot$ (property 4)

b.

$[A' \cdot (B + C)] + (A \cdot B) = (A' \cdot B + A' \cdot C) + A \cdot B$	Distributive property
$= (A' \cdot B + A \cdot B) + A' \cdot C$	Associative and commutative properties (1 and 2) of $+$
$= [(A' + A) \cdot B] + A' \cdot C$	Distributive property
$= 1 \cdot B + A' \cdot C$	Complement
$= B + A' \cdot C$	1 is identity for $\cdot$

c. $[A \cdot (A' + B)] \cdot C = (A \cdot A' + A \cdot B) \cdot C$ — Distributive property

$= (0 + A \cdot B) \cdot C$ — Complement

$= (A \cdot B) \cdot C$ — 0 is identity for + (property 4)

We conclude this section with a summary of the basic properties of a Boolean algebra.

PROPERTIES OF A BOOLEAN ALGEBRA

1. The operations are *commutative.*

$$A \cdot B = B \cdot A$$
$$A + B = B + A$$

2. The operations are *associative.*

$$(A + B) + C = A + (B + C)$$
$$(A \cdot B) \cdot C = A \cdot (B \cdot C)$$

3. Each operation is *distributive* over the other one.

$$A \cdot (B + C) = (A \cdot B) + (A \cdot C)$$
$$A + (B \cdot C) = (A + B) \cdot (A + C)$$

4. There is an *identity* element for each operation.

1 is the identity for $\cdot$.

0 is the identity for $+$.

5. Each element has a *complement.*

$$0' = 1$$
$$1' = 0$$

6. Each element is *idempotent.*

$$A + A = A$$
$$A \cdot A = A$$

7. The *absorption* property holds.

$$A \cdot (A + B) = A$$
$$A + (A \cdot B) = A$$

EXERCISES FOR SECTION 5.1

In Exercises 1 through 10, evaluate each expression using Table 5.1 (see Examples 1 and 4).

1. $(1 + 1) \cdot (0 + 1)$
2. $(1 \cdot 1) \cdot (1 \cdot 0)$
3. $(1 + 0) + (0 + 1)$
4. $(0 \cdot 1) + (1 + 1)$
5. $(0 + 1) \cdot (1 + 1)$
6. $(1 + 0) \cdot [1 + (0 + 1)]$
7. $(1 \cdot 0) \cdot [1 + (0 \cdot 1)]$
8. $(1 \cdot 0) + [1 \cdot (0 + 1)]$
9. $[0 + (1 \cdot 1)] + [1 + (1 \cdot 1)]$
10. $[0 + (1 \cdot 1)] \cdot [1 + (1 \cdot 1)]$

In Exercises 11 through 20, simplify each expression (see Example 6).

11. $B \cdot A + B \cdot A'$
12. $B' \cdot (A + C) + (B \cdot A)$
13. $[C \cdot (C' + A)] \cdot B$
14. $[B \cdot (B' + A)] \cdot B$
15. $[A \cdot (A' + B)] \cdot (A + B)$
16. $[A \cdot (A' + B)] \cdot (A' + B)$
17. $[(A + B) \cdot A] \cdot (A + B)$
18. $A + A \cdot (B' + C) + A \cdot (B' + C')$
19. $A \cdot (B + C) + A + A \cdot B$
20. $[(A \cdot B) \cdot C] + [(A \cdot B') \cdot C] + [A' \cdot (B' \cdot C)]$

In Exercises 21 and 22, use Examples 2 and 5 as a guide.

21. Prove DeMorgan's first law: $(A + B)' = A' \cdot B'$ by supplying reasons for each step below:

statement	*reason*
1. $(A + B) + (A' \cdot B') = (B + A) + (A' \cdot B')$	1.
2. $(B + A) + (A' \cdot B') = B + [(A + A') \cdot (A + B')]$	2.
3. $B + [(A + A') \cdot (A + B')] = B + [1 \cdot (A + B')]$	3.
4. $B + [1 \cdot (A + B')] = B + (A + B')$	4.
5. $B + (A + B') = (B + B') + A$	5.
6. $(B + B') + A = 1$	6.
7. $(A + B) + (A' \cdot B') = 1$ and thus $A + B$ and $A' \cdot B'$ are complements of each other	7. Steps 1–6 above

22. Follow the logic of Exercise 21 to prove DeMorgan's second law:

$$(A \cdot B)' = A' + B'$$

The Boolean algebra described by Table 5.1 is *one example* of a Boolean algebra. In Exercises 23 through 25, we describe three other examples of Boolean algebras.

23. Consider the collection of all eight possible three-bit "digits": 000, 001, 010, 011, 100, 101, 110, 111. If A and B are two such numbers, then define

$A + B$ to be a resulting three-bit number whose digit-by-digit sum is as in Table 5.1. Similar definitions apply for $A \cdot B$ and A'. For example,

$$010 + 011 = 011$$
$$010 \cdot 011 = 010$$
$$(010)' = 101$$

This is a Boolean algebra.

a. Evaluate $011 + 101 + 111$

b. Evaluate $010 + 001 + 110$

c. Evaluate $(011 \cdot 101) + [101 + (100)']$

d. What is the identity for $+$?

e. What is the identity for $\cdot$?

24. Consider the collection of all subsets of the set $\{x, y\}$:

$$S = \{\{x\}, \{y\}, \{x, y\}, \{\ \}\}$$

If A and B are in S then define $A + B$ as $A \cup B$ and $A \cdot B$ as $A \cap B$, the union and intersection operations. Then

$$\{x\}' = \{y\}$$
$$\{y\}' = \{x\}$$
$$\{x, y\}' = \{\ \}$$
$$\{\ \}' = \{x, y\}$$

a. Evaluate $\{x\} + [\{x\} \cdot \{x, y\}]$

b. Evaluate $\{\ \} + \{x\} \cdot [\{y\} + \{x\}]$

c. Evaluate $[\{x\} \cdot \{y\}] + [\{x, y\} \cdot \{y\}]$

d. What is the identity for $+$?

e. What is the identity for $\cdot$?

25. Consider $D = \{1, 2, 3, 4, 6, 12\}$, the collection of factors of 12. Define $+$ and $\cdot$ for two numbers, a and b, in D as

$$a + b = \text{the least common multiple of } a \text{ and } b$$
$$a \cdot b = \text{the greatest common divisor of } a \text{ and } b$$
$$a' = 12 \div a$$

For example,

$$2 + 6 = 6 \qquad 2 \cdot 6 = 2 \qquad 3' = 4$$

a. What is the identity for $+$?

b. What is the identity for $\cdot$?

c. Evaluate $[(3 + 4) \cdot 2] + [(2 \cdot 6) \cdot 3']$

5.2 Circuits

In the remaining sections of this chapter, we shall discuss some applications of the concepts of Boolean algebra and logic. We shall apply these ideas first, in this section, to the concept of a switching network or circuit.

Switches and Simple Circuits

The concepts of switches and circuits are best explained by referring to a little basic electricity. The simplest type of electrical circuit we can describe consists of a power source (for example, a battery) connected by wires to a user of power (for example, a light bulb) with an on/off switch to control the flow of electricity. A diagram of such a circuit might look like Figure 5.1.

FIGURE 5.1

A simple electrical circuit.

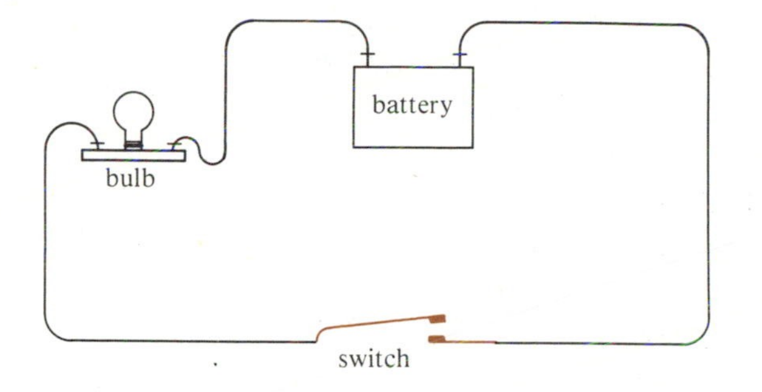

We can think of the switch as a bridge that must be *closed* in order for electric current to flow from the battery to the light bulb. The battery and the light bulb are the **terminals** of the circuit.

In this circuit, there are exactly two possibilities:

The switch is closed; or
the switch is not closed (it is open).

When the switch is closed, current flows; when the switch is open, current does not flow.

Although a circuit consisting of two terminals and one switch is easy enough to understand, it is not particularly interesting from either a theoretical or a practical point of view. Our discussion will center on **switching circuits** or **networks**, arrangements of wires and switches connecting two terminals, containing two or more switches. Since the switches are the controlling components in a network, we shall show only the switches in our circuit diagrams.

Any switching network is constructed by combining two basic types of circuits, series and parallel, in different ways.

In a **series circuit**, two (or more) switches are connected in such a way that current flows through the circuit only when *all* switches are closed. A diagram of such a circuit is:

Here, P and Q represent the two switches. A series circuit containing two switches has four possible states. They are illustrated in Table 5.4.

TABLE 5.4

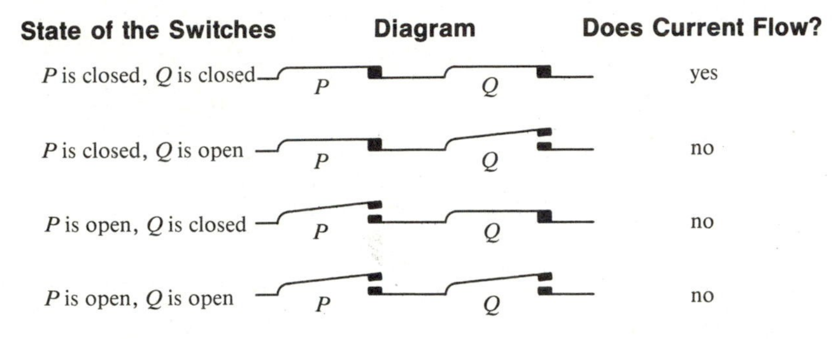

State of the Switches	Diagram	Does Current Flow?
P is closed, Q is closed	P Q	yes
P is closed, Q is open	P Q	no
P is open, Q is closed	P Q	no
P is open, Q is open	P Q	no

Now suppose that we let p represent the statement "switch P is closed" and q represent the statement "switch Q is closed." Table 5.4 can then be replaced by the following:

State of Switches			
p	q	Current Flows	
T	T	T	P closed, Q closed.
T	F	F	P closed, Q open.
F	T	F	P open, Q closed.
F	F	F	P open, Q open.

Compare this with the truth table for $p \wedge q$ (Table 4.2).

p	q	$p \wedge q$
T	T	T
T	F	F
F	T	F
F	F	F

Since the truth tables are the same, it follows that the statement "current flows" in a series circuit is logically equivalent to the statement "switch P is closed *and* switch Q is closed" ($p \wedge q$). Using this relationship, we will often find it useful to write a series circuit ——P——Q—— as the logical statement $p \wedge q$.

A **parallel circuit** is a circuit in which the switches are connected in such a way that current flows whenever *at least one* of the switches is closed. A typical diagram of a parallel circuit is

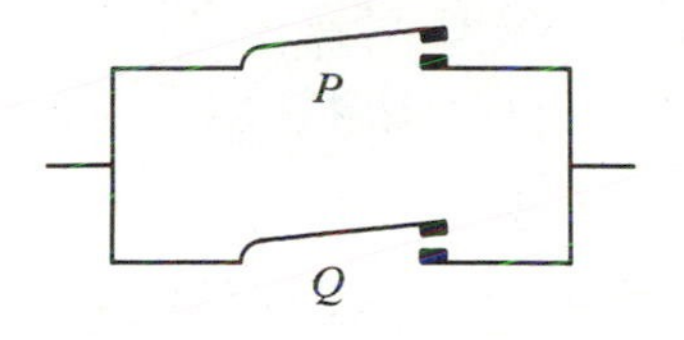

A parallel circuit containing two switches, like a series circuit with two switches, can have four possible states. They are illustrated in Table 5.5.

TABLE 5.5

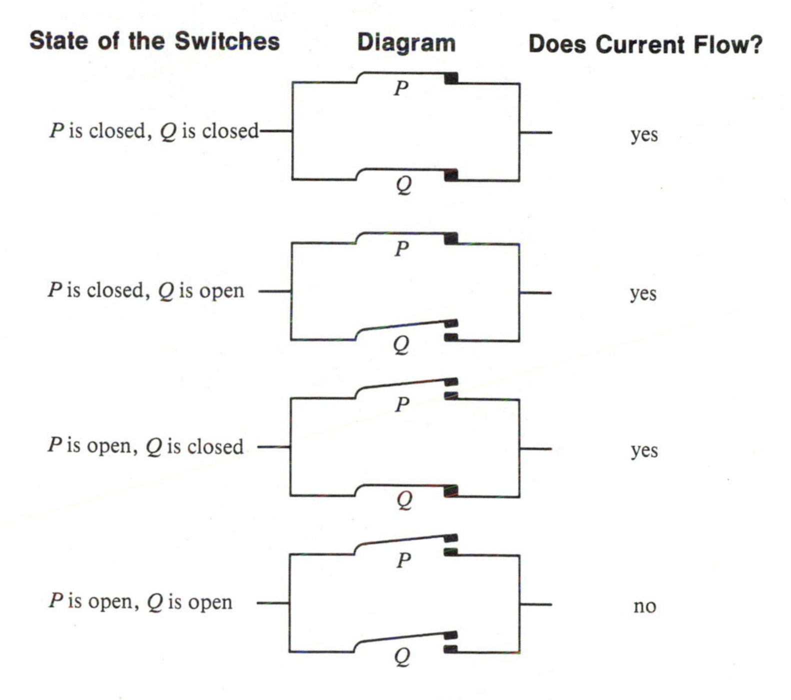

State of the Switches	Diagram	Does Current Flow?
P is closed, Q is closed		yes
P is closed, Q is open		yes
P is open, Q is closed		yes
P is open, Q is open		no

Now, as before, let p represent the statement "switch P is closed" and q represent the statement "switch Q is closed." Table 5.5 can be replaced by the following:

State of Switches p	q	Current Flows	
T	T	T	P closed, Q closed.
T	F	T	P closed, Q open.
F	T	T	P open, Q closed.
F	F	F	P open, Q open.

Compare this with the truth table for $p \vee q$ (Table 4.3).

p	q	$p \vee q$
T	T	T
T	F	T
F	T	T
F	F	F

The statement "current flows" in a parallel circuit is logically equivalent to the statement "switch P is closed *or* switch Q is closed" ($p \vee q$). With this in mind, we will frequently replace the parallel circuit

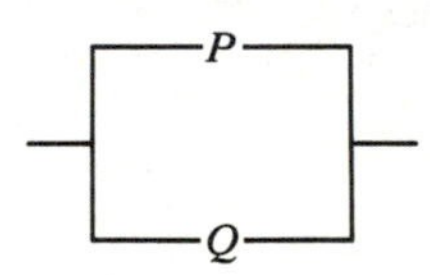

by the logical statement $p \vee q$.

More Complex Switching Networks

For practical purposes, switching networks usually contain more than two switches. We can frequently gain information about a circuit by analyzing it in terms of the equivalent logical statement.

EXAMPLE 1

Consider the switching network

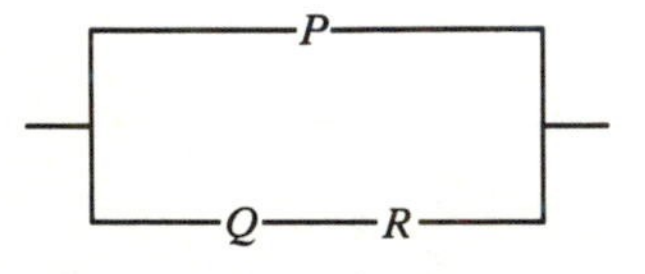

Determine the equivalent logical statement, and describe the conditions under which current flows in the circuit.

Solution

Let the following statements represent the three switches:

p: Switch P is closed.
q: Switch Q is closed.
r: Switch R is closed.

The Q,R part of the network is a series circuit. It can be represented by $q \wedge r$. This circuit is combined in parallel with P, so the statement representing the network is

$$p \vee (q \wedge r)$$

To determine the conditions under which current flows in the network, we examine the truth table of the equivalent logical statement. (Recall from Chapter 4 that there are eight possible states for a three-switch circuit, since the truth table gives eight possibilities.)

p	q	r	$p \vee (q \wedge r)$
T	T	T	T
T	T	F	T
T	F	T	T
T	F	F	T
F	T	T	T
F	T	F	F
F	F	T	F
F	F	F	F

Ts in the last column show when current flows.

Some switching networks contain **coupled switches**, that is, two switches that always operate together (both closed or both open) or always operate as opposites (whenever one is open , the other is closed). When switches are coupled to operate together, we use the same letter in the network diagram to indicate both switches. If switches are coupled to operate oppositely and we use a given letter (say Q) for one switch, we use the same letter with a prime (Q') for the other.

Let us examine opposite switches more closely. If Q is a switch and Q' is its opposite, then Q' is closed whenever Q is open and Q' is open whenever Q is closed. Suppose q is the statement "switch Q is closed" and q' is the statement "switch Q' is closed." Then the truth tables for q and q' together look like this:

	q	q'	
Q is closed.	T	F	Q' is open.
Q is open.	F	T	Q' is closed.

That is, the truth table for q' is exactly the truth table for $\sim q$. Therefore, q' and $\sim q$ are logically equivalent, and we can now use $\sim q$ to represent the statement "switch Q' is closed."

EXAMPLE 2

Write a logic statement that represents the following network:

Solution

This network consists of two parallel circuits:

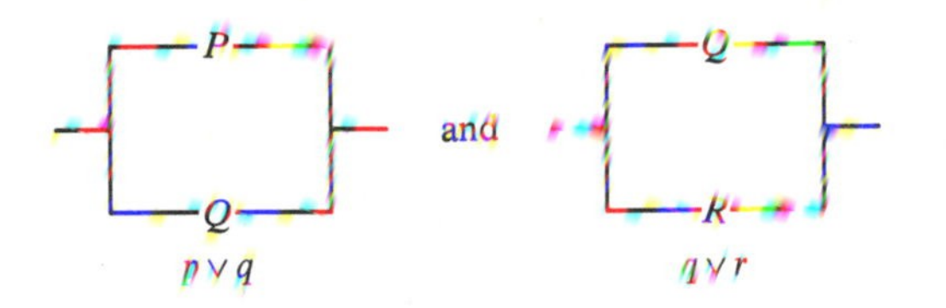

These circuits are joined in series. Recalling that $\vee$ represents parallel and $\wedge$ represents series gives us the logic statement

$$(p \vee q) \wedge (q \vee r)$$

EXAMPLE 3

Write a logic statement that represents the following network:

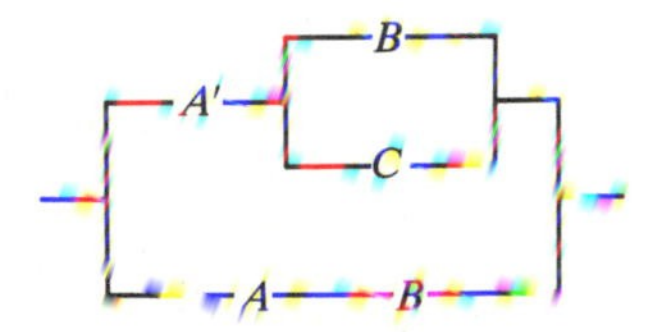

Determine the equivalent logical statement, and describe the conditions under which current flows in the circuit.

Solution

Let the following statements represent the three switches:

p: Switch P is closed.
q: Switch Q is closed.
r: Switch R is closed.

The Q,R part of the network is a series circuit. It can be represented by $q \wedge r$. This circuit is combined in parallel with P, so the statement representing the network is

$$p \vee (q \wedge r)$$

To determine the conditions under which current flows in the network, we examine the truth table of the equivalent logical statement. (Recall from Chapter 4 that there are eight possible states for a three-switch circuit, since the truth table gives eight possibilities.)

p	q	r	$p \vee (q \wedge r)$
T	T	T	T
T	T	F	T
T	F	T	T
T	F	F	T
F	T	T	T
F	T	F	F
F	F	T	F
F	F	F	F

Ts in the last column show when current flows.

Some switching networks contain **coupled switches**, that is, two switches that always operate together (both closed or both open) or always operate as opposites (whenever one is open , the other is closed). When switches are coupled to operate together, we use the same letter in the network diagram to indicate both switches. If switches are coupled to operate oppositely and we use a given letter (say Q) for one switch, we use the same letter with a prime (Q') for the other.

Let us examine opposite switches more closely. If Q is a switch and Q' is its opposite, then Q' is closed whenever Q is open and Q' is open whenever Q is closed. Suppose q is the statement "switch Q is closed" and q' is the statement "switch Q' is closed." Then the truth tables for q and q' together look like this:

	q	q'	
Q is closed.	T	F	Q' is open.
Q is open.	F	T	Q' is closed.

That is, the truth table for q' is exactly the truth table for $\sim q$. Therefore, q' and $\sim q$ are logically equivalent, and we can now use $\sim q$ to represent the statement "switch Q' is closed."

EXAMPLE 2

Write a logic statement that represents the following network:

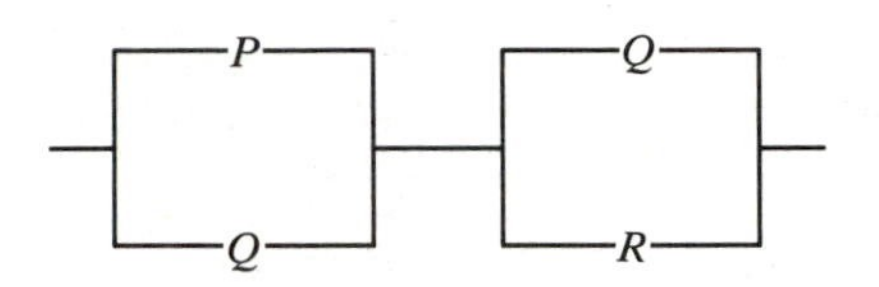

Solution

This network consists of two parallel circuits:

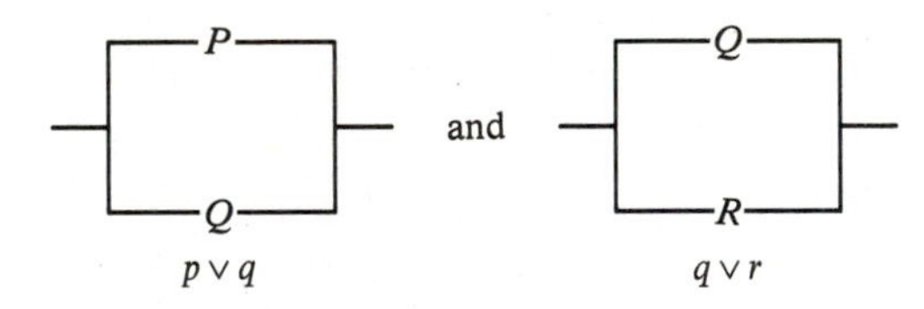

These circuits are joined in series. Recalling that $\vee$ represents parallel and $\wedge$ represents series gives us the logic statement

$$(p \vee q) \wedge (q \vee r)$$

EXAMPLE 3

Write a logic statement that represents the following network:

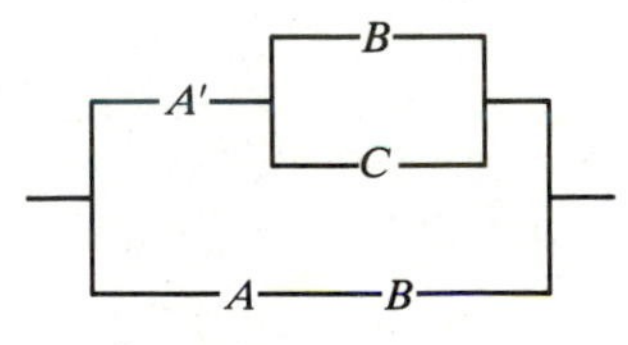

Solution

This network is more complex than the last one. We begin by breaking it down into two separate networks.

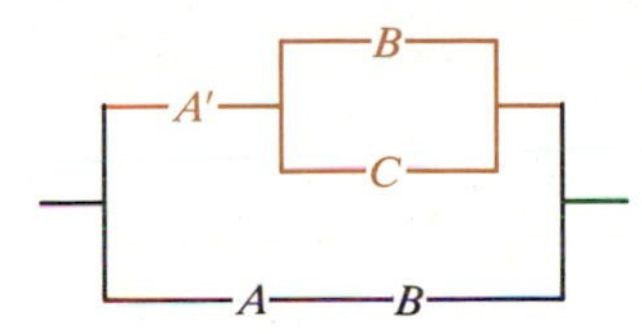

Start with the colored portion. This consists of switch A' in series with the parallel circuit

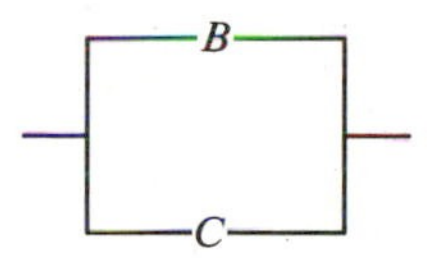

The logic statement for this part of the network is

$$(\sim a) \wedge (b \vee c)$$

Now look at the rest of the given network, the series circuit ——A——B——. This is equivalent to the logic statement $a \wedge b$. Since these two circuits are joined in parallel, the desired statement is

$$[(\sim a) \wedge (b \vee c)] \vee (a \wedge b)$$

Sometimes we may wish to work in the opposite direction—draw the switching network corresponding to a given logic statement. This approach, which is especially useful in the design of switching networks, is illustrated in the next two examples.

EXAMPLE 4

Draw a network corresponding to

$$(p \wedge q) \vee [p \wedge (\sim q \vee \sim p)]$$

Solution

Recall that $\wedge$ stands for series, $\vee$ stands for parallel, and $\sim$ becomes a prime. The $p \wedge q$ part of the statement becomes

——P——Q——

This is joined in parallel (note the $\vee$) with

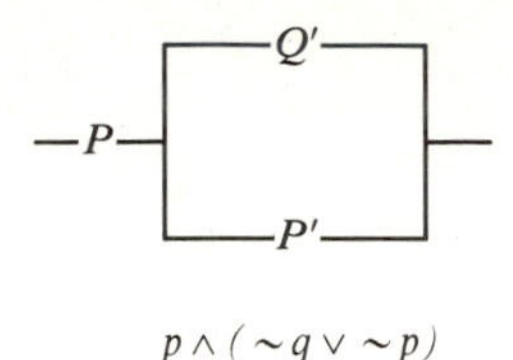

$p \wedge (\sim q \vee \sim p)$

The desired circuit is

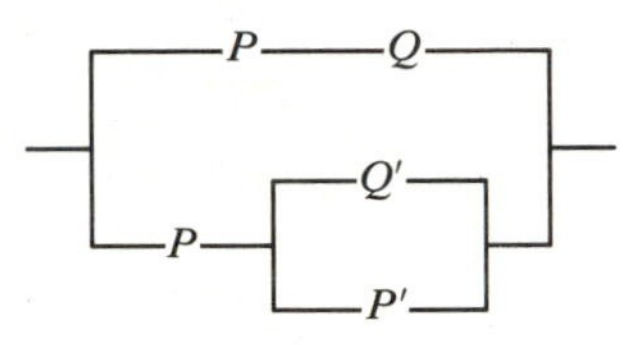

EXAMPLE 5

Zelda the electrician wishes to hook up a stairway light with two switches to operate it, one at the top of the staircase and one at the bottom. Design a network to do this.

Solution

Zelda would like to arrange the switches so that a change in the state of *exactly one* switch changes the condition of the light. Suppose she starts with both switches closed and the light off. Let p represent "switch P is closed" and q represent "switch Q is closed." The network Zelda wants can be described by the following truth table:

p	q	**Light is on**	
T	T	F	Starting condition
T	F	T	Change *just* switch Q.
F	T	T	Change *just* switch P.
F	F	F	Change *both* switches.

We must find a logical expression involving p and q that gives this truth table. One candidate is

$$(p \wedge \sim q) \vee (\sim p \wedge q)$$

(Verify that this statement has the required truth table.) Finally, a network with the desired property corresponding to the logic statement $(p \wedge \sim q) \vee (\sim p \wedge q)$ is

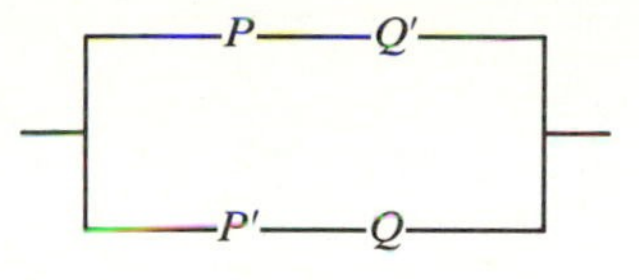

Switching Networks and Boolean Algebra

We conclude this section with a brief discussion of the relationship between Boolean algebra and switching networks. We shall examine and use it more fully in the last two sections of this chapter.

Consider a simple circuit consisting of a single switch. We shall assign the value 1 to this circuit if power is flowing through the circuit (that is, if the switch is closed). We assign the value 0 to the circuit if power is *not* flowing through the circuit (that is, if the switch is open). (Compare this with the way in which computers store information—as the presence or absence of an electrical or magnetic field.)

Now consider a series circuit with switches P and Q, as below:

——P——Q——

This circuit has "value" 1 only when switches P and Q are both closed; that is, when both P and Q have "value" 1. Otherwise, the "value" of the series circuit is 0. This can be shown in the following table (Table 5.6):

TABLE 5.6

P	Q	**Value of Series Circuit**
1	1	1
1	0	0
0	1	0
0	0	0

1 if current flows, 0 if not.

This is just the "multiplication" table for a Boolean algebra with elements denoted by P and Q:

P	Q	$P \cdot Q$
1	1	1
1	0	0
0	1	0
0	0	0

Thus, the series circuit——P——Q——may be represented by the Boolean expression $P \cdot Q$. Similarly, a parallel circuit

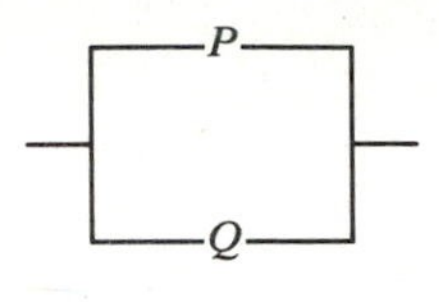

may be represented by the Boolean expression $P + Q$, as Table 5.7 illustrates:

TABLE 5.7

P	Q	Value of Parallel Circuit	$P + Q$
1	1	1	1
1	0	1	1
0	1	1	1
0	0	0	0

Finally, the *opposite* of a switch is simply represented by the Boolean *complement*.

EXAMPLE 6

Represent the networks of **(a)** Example 3 and **(b)** Example 5 by Boolean expressions.

Solution

a. (Example 3)

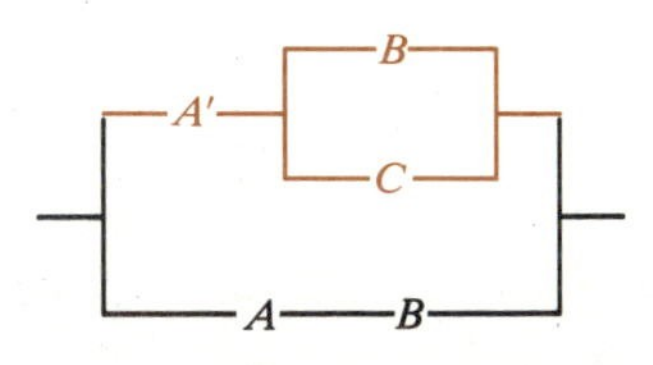

The Boolean expression for the upper (colored) part of the network is

$$A' \cdot (B + C)$$

The Boolean expression for the lower part of the network is

$$A \cdot B$$

Since these parts are joined in parallel, the desired expression is

$$[A' \cdot (B + C)] + (A \cdot B)$$

b. (Example 5)

The circuit is

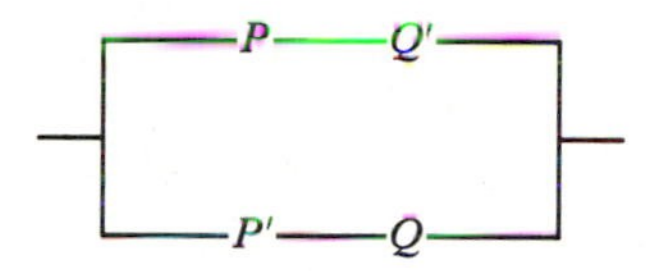

The equivalent Boolean expression is

$$(P \cdot Q') + (P' \cdot Q)$$

This ability to express complicated networks in terms of Boolean expressions, along with the properties of a Boolean algebra discussed in Section 5.1, will prove very useful in our discussion of simplifying switching networks in Section 5.3.

EXERCISES FOR SECTION 5.2

In Exercises 1 through 10, (a) determine an equivalent logical statement and (b) use truth tables to describe the conditions under which current flows (see Examples 1–3).

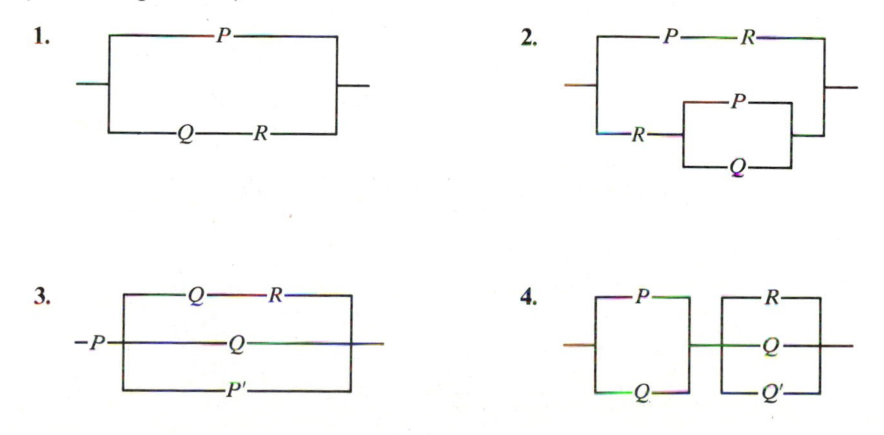

5.

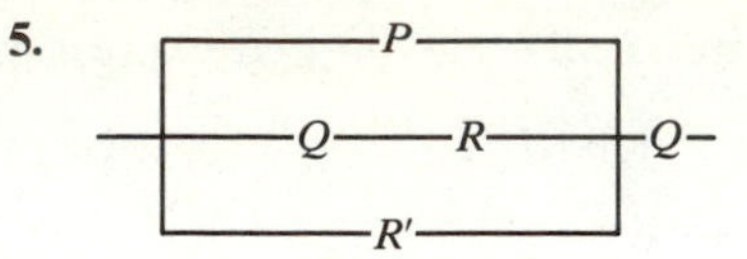

6.

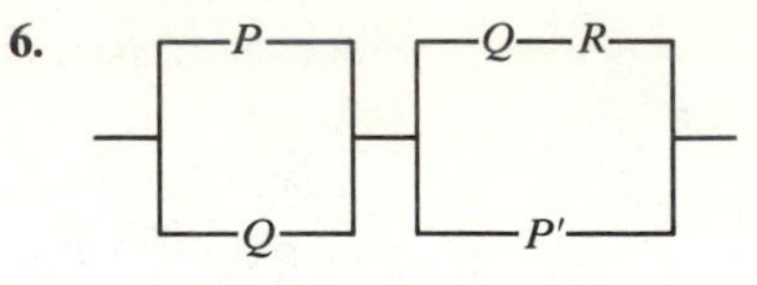

7.

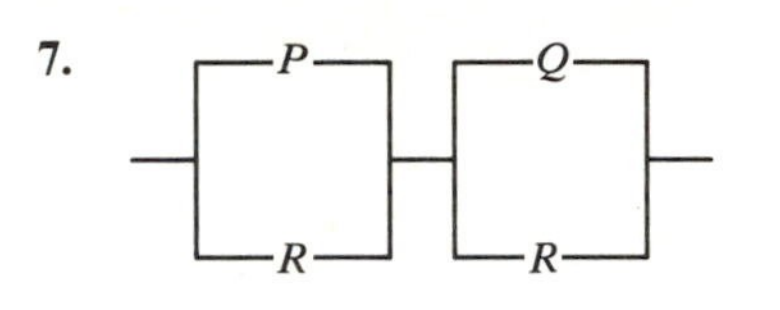

8.

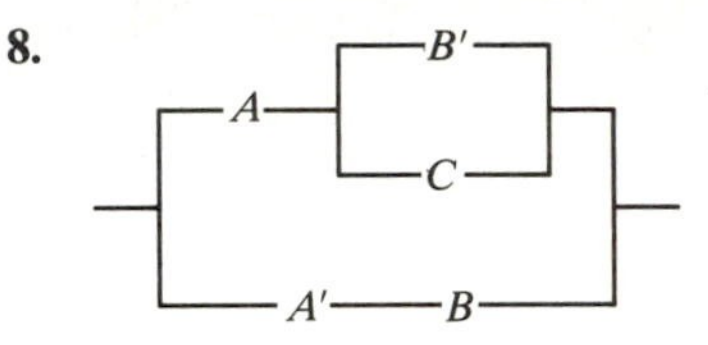

9.

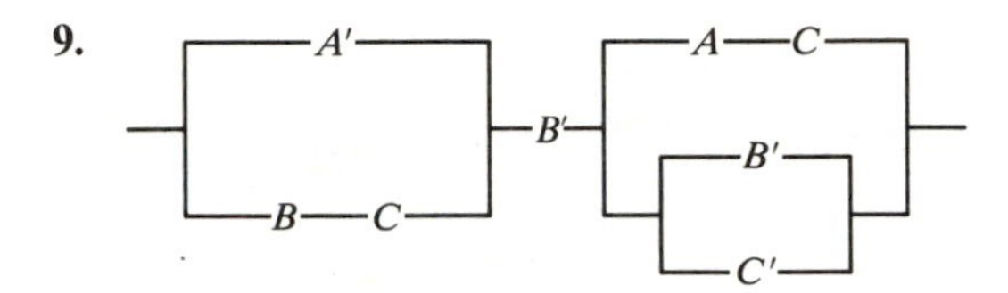

10.

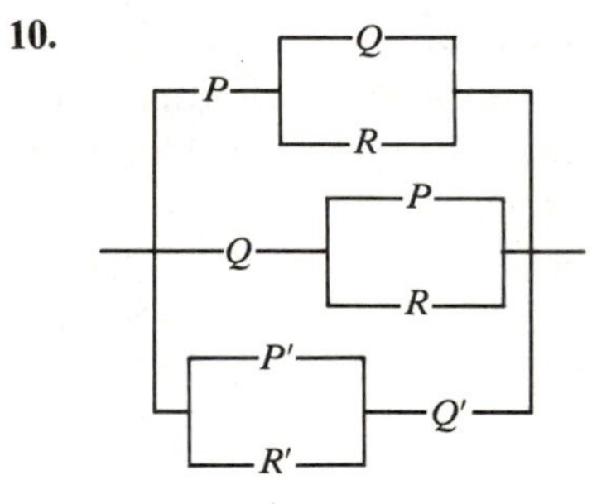

In Exercises 11 through 15, draw a network that corresponds to the given logic statement (see Example 4).

11. $(p \wedge \sim q) \vee (\sim p \wedge \sim q)$

12. $[p \vee (q \wedge r)] \wedge \sim p$

13. $[(p \vee q) \vee (q \wedge r)] \wedge (\sim p \wedge \sim q)$

14. $(p \wedge q) \vee [r \wedge (\sim r \vee \sim q)]$

15. $(p \vee q) \wedge r \wedge (p \vee q)$

16. Construct a network that corresponds to the logic statement $p \rightarrow q$. (Recall that $p \rightarrow q$ is logically equivalent to $\sim p \vee q$.)

17. **a.** Construct a network that corresponds to the logic statement

$$(p \vee q) \wedge \sim(p \wedge q)$$

(Hint: Use DeMorgan's law.)

b. Show, by means of a truth table, that $(p \vee q) \wedge \sim(p \wedge q)$ is logically equivalent to $p \veebar q$.

c. Compare the network in part a above with that of Example 5.

In Exercises 18 through 23, represent each network by a Boolean expression (see Example 6).

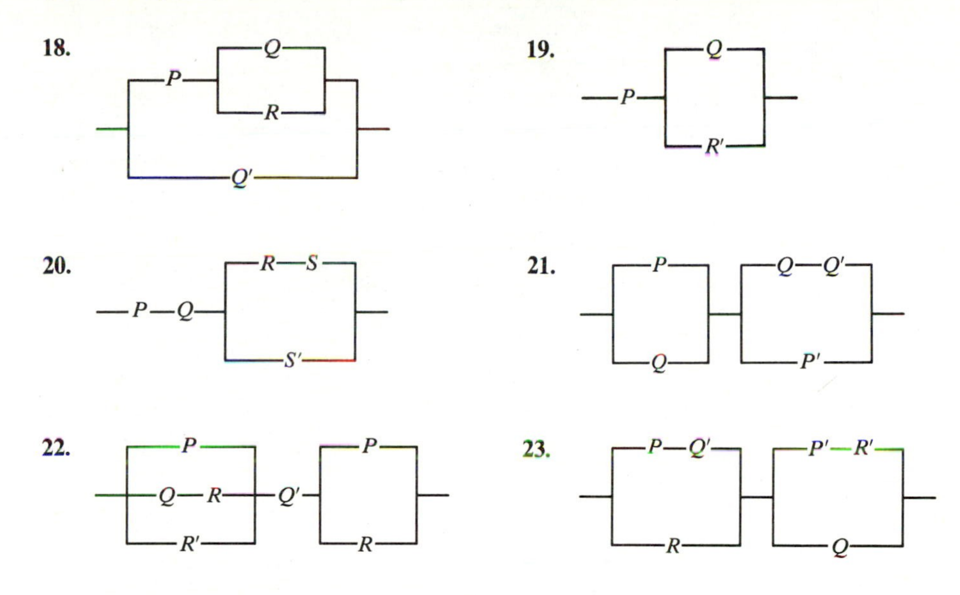

24. Recall that a tautology is a statement that is always true. Show that the network below represents a tautology:

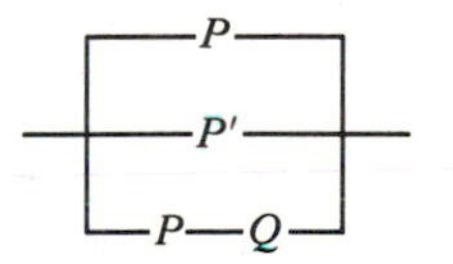

25. Recall that a self-contradiction is a statement that is never true. Show that the network below represents a self-contradiction:

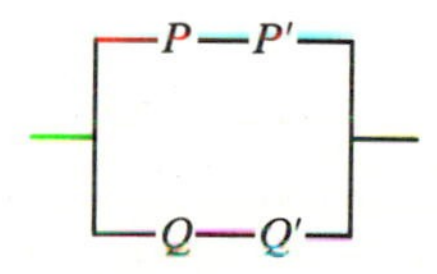

26. Construct a network that corresponds to the statement $p \leftrightarrow q$. [Hint: Recall that $p \leftrightarrow q$ is logically equivalent to $(p \wedge q) \vee (\sim p \wedge \sim q)$.]

27. Construct a network that corresponds to the statement $p \uparrow q$. (Hint: Recall $p \uparrow q$ is equivalent to $\sim p \wedge \sim q$.)

In Exercises 28 and 29, you may want to use Example 5 as a guide.

28. A society of three people wants to be able to record a majority vote secretly. Design a switching circuit that allows the flow of current when a majority votes for an issue.

29. Four people wish to record a majority vote in favor of a measure by using a switching network whereby an individual votes "yes" by closing his or her switch. In case of a tie, the outcome is to be the same as the chairperson's vote. (The chairperson is one of the four people.) Design the network.

30. **a.** Construct a logical statement that represents the network

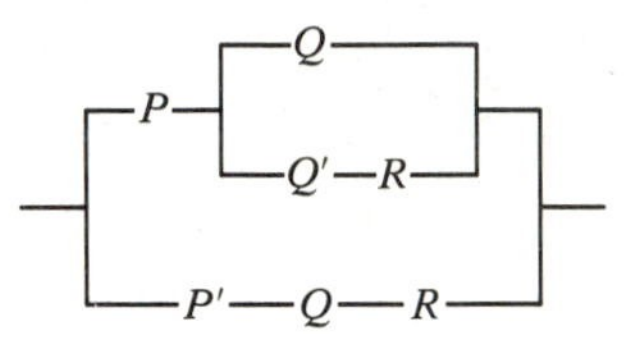

b. Construct a truth table for your answer to part a.

c. Construct a logical statement that represents the network:

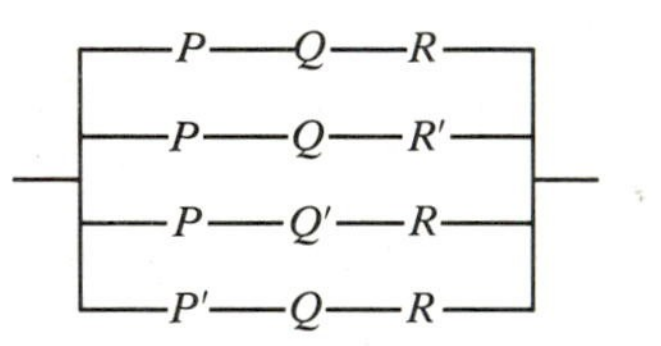

d. Construct a truth table for your answer to part c.

e. Show that the two networks are functionally the same (that is, the "light" goes on under exactly the same settings of the switches) and that each is a possible solution to Exercise 28.

5.3 Circuit Simplification

It is often useful, for economy of design as well as for financial reasons, to consider the problem of *simplifying* a given switching network. Solving such a problem involves finding an *equivalent* network—one in which current flows under the same circumstances as in the original network—that contains fewer switches and connectors.

Using Logic Statements and Truth Tables

The task of circuit simplification may be approached from two points of view. The first involves four steps:

1. Write a *logic statement* that is equivalent to the given network.
2. Construct the *truth table* for this statement.
3. Examine the truth table to determine the circumstances under which current flows in the network (that is, where Ts occur in the truth table). Use the results of this examination to find a *simpler* logic statement—one involving fewer propositions and logical connectors—that is equivalent to the original statement.
4. Draw the network that corresponds to this new logic statement. It will be equivalent to the original network.

We illustrate this first approach with some examples.

EXAMPLE 1

Simplify the following network:

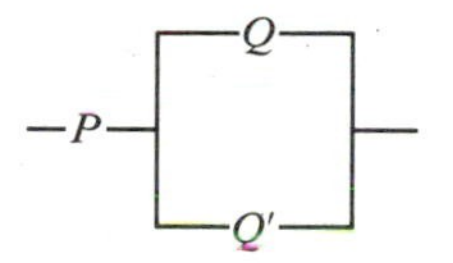

Solution

1. Let p be the statement "switch P is closed" and q be the statement "switch Q is closed." The logic statement for this network is

$$p \wedge (q \vee \sim q)$$

2. The truth table for $p \wedge (q \vee \sim q)$ is

p	q	$p \wedge (q \vee \sim q)$
T	T	T
T	F	T
F	T	F
F	F	F

Note that $q \vee \sim q$ is a *tautology*; it is always true.

3. The truth table above coincides exactly with the truth table for the statement p. Therefore, the statement p is logically equivalent to the statement $p \wedge (q \vee \sim q)$.

4. An equivalent network is

——P——

EXAMPLE 2

Simplify the following switching network.

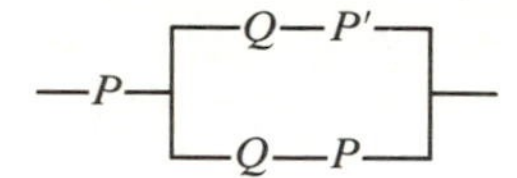

Solution

1. We define the following logic statements:

p: Switch P is closed.

q: Switch Q is closed.

The logic statement corresponding to the given network is

$$p \wedge [(q \wedge \sim p) \vee (q \wedge p)]$$

2. The desired truth table is

p	q	$p \wedge [(q \wedge \sim p) \vee (q \wedge p)]$			
T	T	T	F	T	T
T	F	F	F	F	F
F	T	F	T	T	F
F	F	F	F	F	F

Truth values are filled in only under the $\wedge$ and $\vee$ connectives to make the table easier to read.

3. This truth table is clearly identical to the truth table for $q \wedge p$. Therefore, the statement

$$q \wedge p$$

is logically equivalent to the logic statement for the original network.

4. The network corresponding to the logic statement $q \wedge p$ is

——Q——P——

This is a simplified form of the original network.

EXAMPLE 3

Simplify the switching network of Example 3 in Section 5.2.

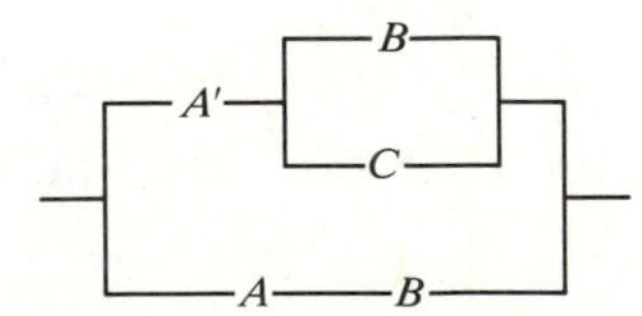

Solution

1. We define the following logic statements:

   a: Switch A is closed.

   b: Switch B is closed.

   c: Switch C is closed.

   The logic statement corresponding to the given network is

   $$[\sim a \wedge (b \vee c)] \vee (a \wedge b)$$

   (Compare Example 3 of Section 5.2.)

2.

a	b	c	$[\sim a \wedge (b \vee c)]$	$\vee$	$(a \wedge b)$	
T	T	T	F	T	T	1
T	T	F	F	T	T	2
T	F	T	F	F	F	3
T	F	F	F	F	F	4
F	T	T	T	T	F	5
F	T	F	T	T	F	6
F	F	T	T	T	F	7
F	F	F	F	F	F	8

↑ Truth table (the $\vee$ column)

3. If we consider rows 1, 2, 5, and 6 of this truth table, we see that the given disjunction is true whenever b is true. The only other T in the truth table occurs in row 7, when a is false and c is true—that is, when $\sim a$ is true and c is true. (Notice also that b is false in row 7.) This leads us to suspect that the given disjunction is equivalent to $b \vee (\sim a \wedge c)$. We verify this result by constructing the truth table for $b \vee (\sim a \wedge c)$.

a	b	c	b	$\vee$	$(\sim a \wedge c)$
T	T	T	T	T	F
T	T	F	T	T	F
T	F	T	F	F	F
T	F	F	F	F	F
F	T	T	T	T	T
F	T	F	T	T	F
F	F	T	F	T	T
F	F	F	F	F	F

   Since the truth tables are identical, the statements are logically equivalent.

4. The network corresponding to $b \vee (\sim a \wedge c)$ is

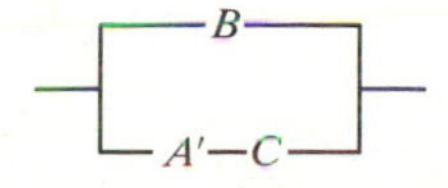

This is the desired simplification of the original network. Although there is no A switch in the simplified circuit, we keep the A *prime* notation to call attention to the relationship of the A' switch in this circuit to the A switch in the original circuit.

Using Boolean Algebra

As the networks studied become increasingly complex, this method of network simplification by considering equivalent logical statements may at times degenerate into an essentially trial-and-error approach. In such situations, another way of viewing the problem may be helpful. It involves the following steps:

1. Write a Boolean expression for the given network.
2. Simplify this Boolean expression by using the properties of a Boolean algebra discussed in Section 5.1 and summarized in the box at the end of that section.
3. Draw the circuit representing the simplified Boolean expression.

Let us apply this method to the network of Example 3.

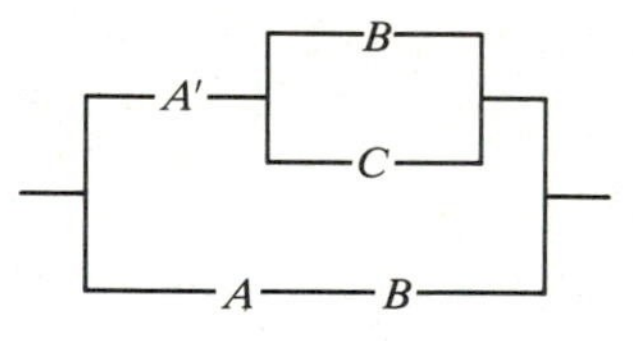

Recalling that parallel circuits correspond to the + operation and series circuits correspond to the · operation we have

1. The Boolean expression for this network is

$$[A' \cdot (B + C)] + (A \cdot B)$$

2. We simplify this expression.

$[A' \cdot (B + C)] + (A \cdot B) = (A' \cdot B + A' \cdot C) + (A \cdot B)$	Distributive property
$= A' \cdot B + (A' \cdot C + A \cdot B)$	Associative property
$= A' \cdot B + (A \cdot B + A' \cdot C)$	Commutative property
$= (A' \cdot B + A \cdot B) + A' \cdot C$	Associative property
$= (A' + A) \cdot B + A' \cdot C$	Distributive property
$= 1 \cdot B + A' \cdot C$	Definition of complement
$= B + A' \cdot C$	Identity for ·

3. The network corresponding to $B + A' \cdot C$ is

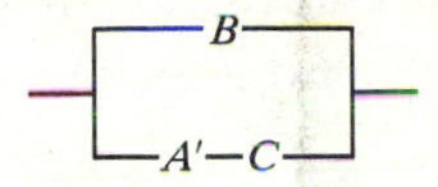

which is the same network derived in Example 3.

EXAMPLE 4

Use Boolean algebra to simplify the network of Example 4 in Section 5.2.

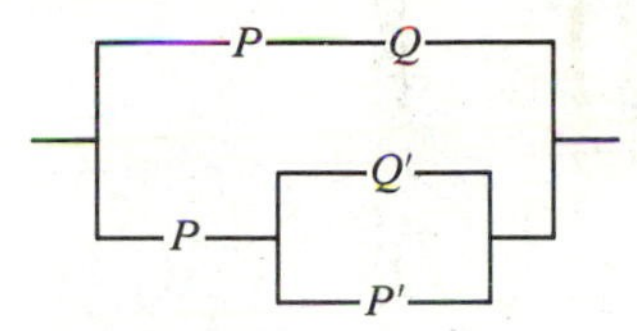

Solution

1. The Boolean expression for this network is

$$(P \cdot Q) + [P \cdot (Q' + P')]$$

2. Simplify this expression.

$(P \cdot Q) + [P \cdot (Q' + P')]$	$= P \cdot Q + (P \cdot Q' + P \cdot P')$	Distributive property
	$= P \cdot Q + P \cdot Q' + 0$	$P \cdot P' = 0$, definition of complement
	$= P \cdot Q + P \cdot Q'$	0 is identity for +
	$= P \cdot (Q + Q')$	Distributive property
	$= P \cdot 1$	Definition of complement
	$= P$	1 is identity for $\cdot$

3. The given network is equivalent to the simple circuit

——P——

EXAMPLE 5

Use Boolean algebra to show that the networks below are equivalent.

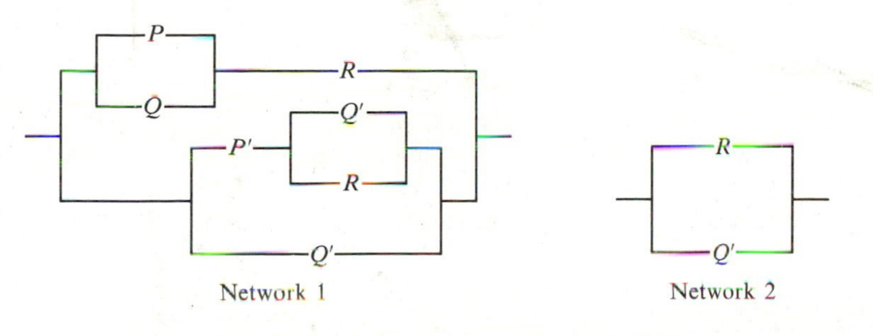

Network 1 Network 2

Solution

We first write a Boolean expression for each network and then simplify the expression for network 1, attempting to arrive at the expression for network 2. The Boolean expression for network 1 is

$$[(P + Q) \cdot R] + \{[P' \cdot (Q' + R)] + Q'\} \qquad (1)$$

The Boolean expression for network 2 is

$$R + Q' \qquad (2)$$

Expression 1 can be simplified as follows:

$$\begin{aligned}
&[(P + Q) \cdot R] + \{[P' \cdot (Q' + R)] + Q'\} \\
&\quad = (P \cdot R + Q \cdot R) + (P' \cdot Q' + P' \cdot R + Q') && \text{Distributive property} \\
&\quad = (P \cdot R + P' \cdot R) + Q \cdot R + (P' \cdot Q' + Q') && \text{Commutative and associative properties} \\
&\quad = (P + P') \cdot R + Q \cdot R + (P' + 1) \cdot Q' && \text{Distributive property} \\
&\quad = 1 \cdot R + Q \cdot R + 1 \cdot Q' && \begin{cases} P + P' = 1, \text{ complement} \\ P' + 1 = 1, \text{ Table 5.1} \end{cases} \\
&\quad = (1 + Q) \cdot R + Q' && \text{Distributive property; 1 is identity for } \cdot \\
&\quad = 1 \cdot R + Q' && 1 + Q = 1, \text{ Table 5.1} \\
&\quad = R + Q' && \text{1 is identity for } \cdot
\end{aligned}$$

Since $R + Q'$ is the Boolean expression for network 2, the proof is complete.

EXERCISES FOR SECTION 5.3

In Exercises 1 through 10, use truth tables to simplify the given network (see Examples 1–3).

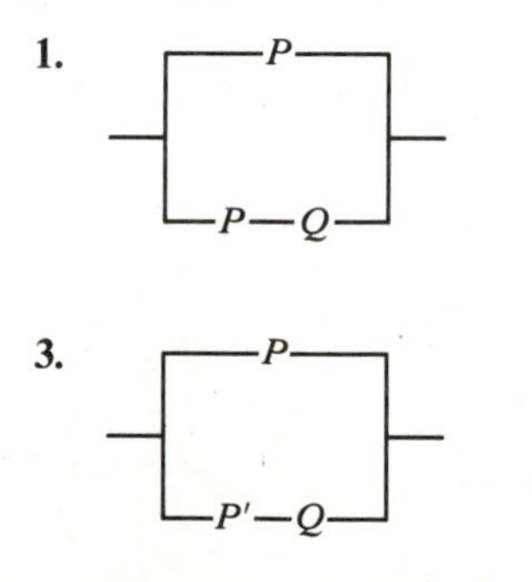

2.

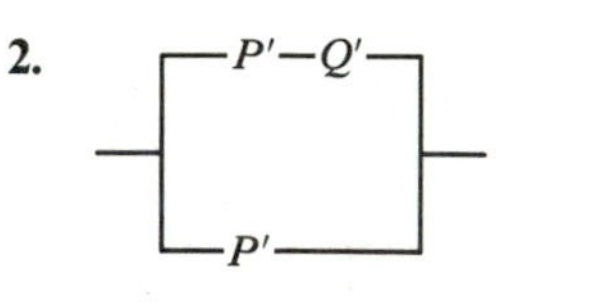

4.

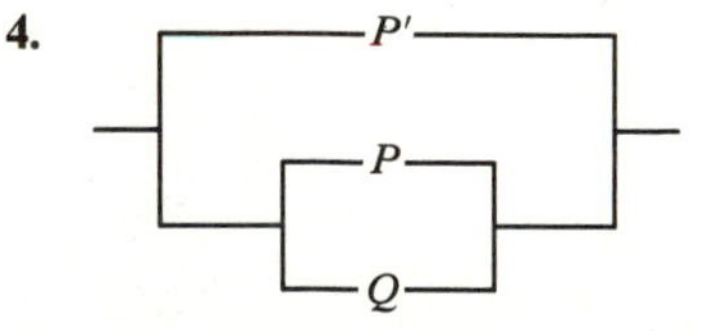

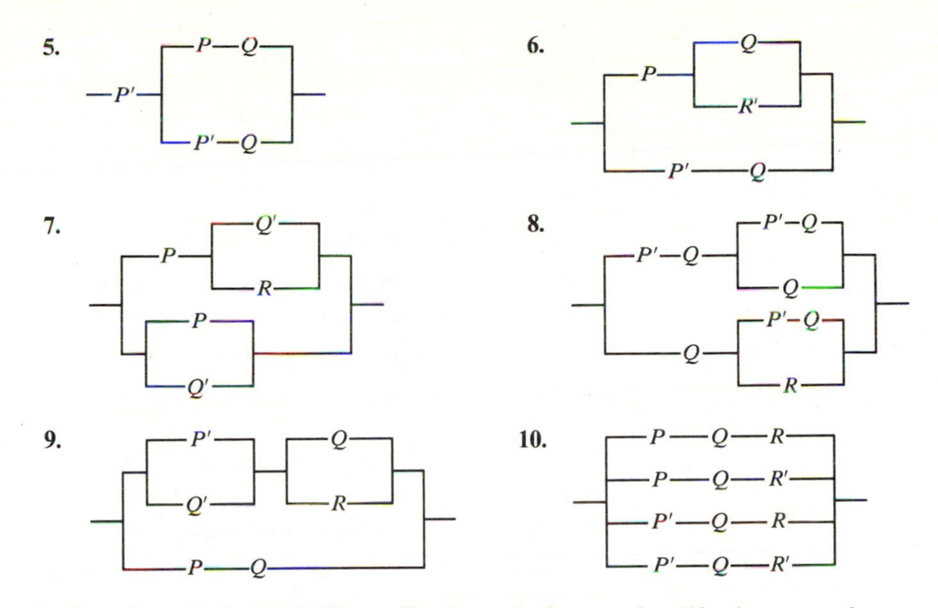

In Exercises 11 through 20, use Boolean algebra to simplify the networks for Exercises 1 through 10 (see Examples 4 and 5).

In Exercises 21 through 25, use the following network:

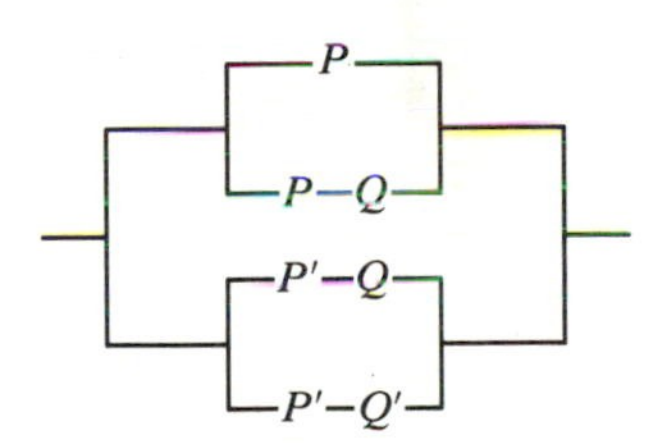

21. Construct a logic statement corresponding to the given network.

22. Construct a Boolean expression corresponding to the network.

23. Construct a truth table for your answer to Exercise 21.

24. Show that your answer to Exercise 22 is equivalent to the Boolean 1.

25. What about the current in the circuit? Under what condition(s) does it flow?

5.4 Logic Gates and Logic Circuits

The final topic of this chapter is also the topic most directly related to computer logic and computer circuitry. In fact, one way to think of a logic circuit is as a machine that accepts data from one or more input devices and produces output at exactly one output device.

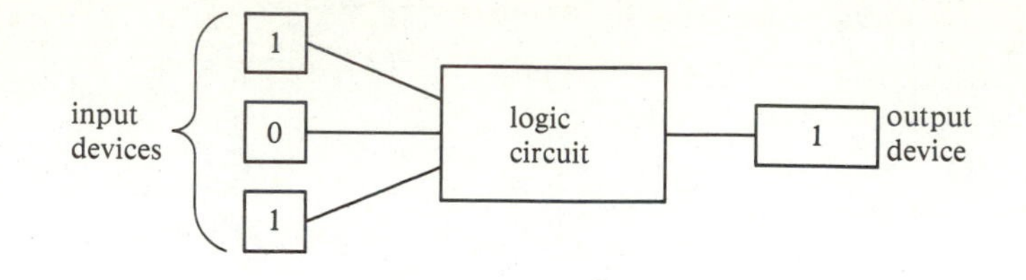

A logic circuit is capable of processing only one bit, or binary digit, at a time. Thus, at any given instant, each input device can contain at most one bit (0 or 1). Sequences of bits of equal lengths from the input devices are processed one bit at a time by the circuit to produce a sequence of bits of the same length at the output device.

Just as switching networks can be formed by combining several switching circuits, logic circuits are made up of parts called *logic gates*. We shall study the three types of logic gates that correspond to the three basic types of switching circuits we have discussed: parallel, series, and opposite coupled.

The OR Gate

The first type of logic gate we shall discuss is the **OR gate**. It is represented by the following diagram:

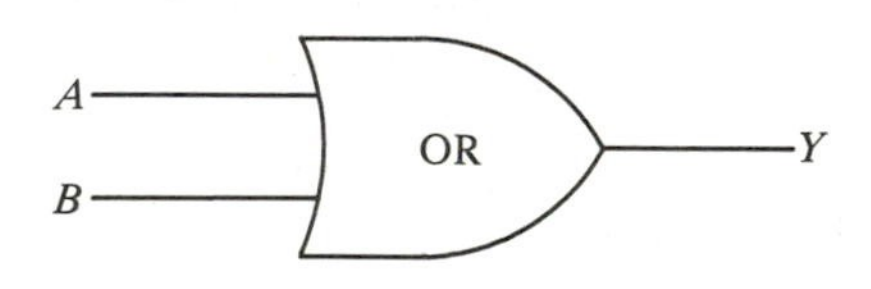

where A and B represent input values and Y represents the output. To determine how a logic gate processes information, we must know what output is produced by using all possible combinations of zeros and ones as input. An OR gate produces a zero as output if and only if it receives all zeros as input; otherwise, the output is a one. We can represent this by the following table (Table 5.8):

TABLE 5.8

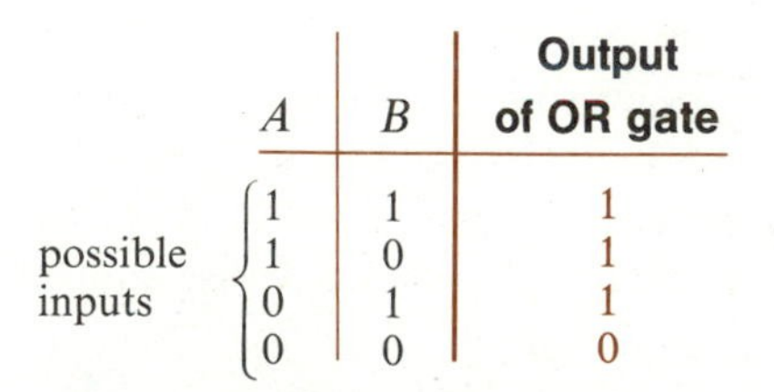

	A	B	Output of OR gate
possible inputs	1	1	1
	1	0	1
	0	1	1
	0	0	0

Compare this with the table for Boolean "addition":

A	B	$A + B$
1	1	1
1	0	1
0	1	1
0	0	0

Obviously, the tables are identical. Thus, the OR logic gate behaves just like Boolean "addition" of binary digits. Since this is true, it also follows that the OR logic gate corresponds to the parallel circuit and to the OR logical connective.

EXAMPLE 1

How would each of the following pairs of sequences of bits be processed by an OR gate?

a. 11000110
00101100

b. 00111010
01011100

c. 10001111
00111100

Solution

Recall that the output of an OR gate is a 0 if and only if all input bits are 0s, and a 1 otherwise. We process the sequences one bit at a time.

a. input $\begin{cases} 1\,1\,0\,0\,0\,1\,1\,0 \\ 0\,0\,1\,0\,1\,1\,0\,0 \end{cases}$

$\downarrow\downarrow\downarrow\downarrow\downarrow\downarrow\downarrow\downarrow$

output 1 1 1 0 1 1 1 0

b. input $\begin{cases} 0\,0\,1\,1\,1\,0\,1\,0 \\ 0\,1\,0\,1\,1\,1\,0\,0 \end{cases}$

$\downarrow\downarrow\downarrow\downarrow\downarrow\downarrow\downarrow\downarrow$

output 0 1 1 1 1 1 1 0

c. input $\begin{cases} 1\,0\,0\,0\,1\,1\,1\,1 \\ 0\,0\,1\,1\,1\,1\,0\,0 \end{cases}$

$\downarrow\downarrow\downarrow\downarrow\downarrow\downarrow\downarrow\downarrow$

output 1 0 1 1 1 1 1 1

A logic gate may receive input from more than two input devices, as the following example illustrates.

EXAMPLE 2

Suppose we have an OR gate that accepts data from three input devices:

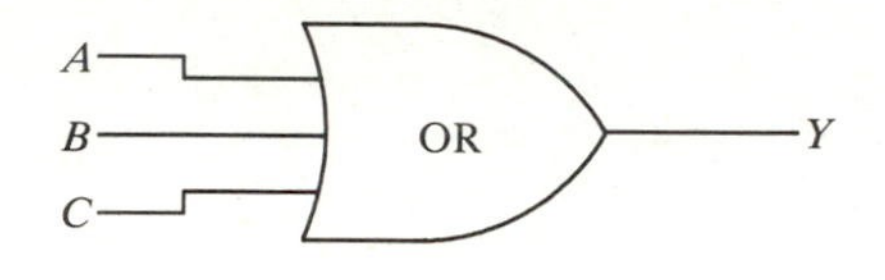

Show the output produced by the following input sequences of bits:

at A: 00001111
at B: 00110011
at C: 01010101

Solution

Again, process the sequences one bit at a time, keeping in mind that the output of an OR gate is 1 unless all input bits are 0s.

$$\text{input}\ \begin{cases} 0\ 0\ 0\ 0\ 1\ 1\ 1\ 1 \\ 0\ 0\ 1\ 1\ 0\ 0\ 1\ 1 \\ 0\ 1\ 0\ 1\ 0\ 1\ 0\ 1 \end{cases}$$

$$\downarrow\ \downarrow\ \downarrow\ \downarrow\ \downarrow\ \downarrow\ \downarrow\ \downarrow$$

$$\text{output}\quad 0\ 1\ 1\ 1\ 1\ 1\ 1\ 1$$

Notice that the three sequences of bits in Example 2 yield all possible combinations of zeros and ones in the case where there are three input devices. The output produced by using these three sequences as input (in a gate with three input devices), along with the input, is often referred to as the **truth table of the gate**.

EXAMPLE 3

Find the truth table of an OR gate with two input devices.

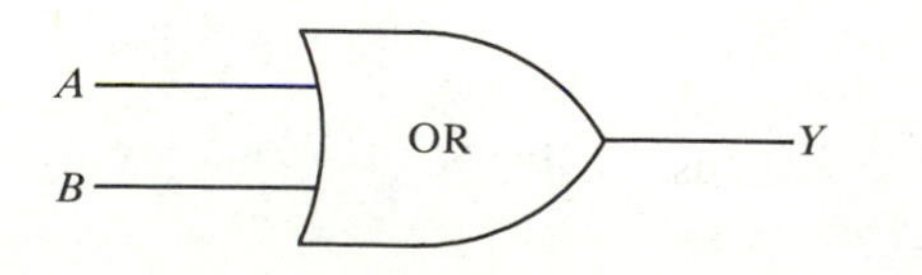

Solution

This time, the sequences 0011 and 0101 as input represent all possible combinations of 0s and 1s. Using these sequences as input gives

$$\left.\begin{array}{ll} \text{input} & \left\{\begin{array}{l} 0\,0\,1\,1 \\ 0\,1\,0\,1 \end{array}\right. \\ & \;\downarrow\downarrow\downarrow\downarrow \\ \text{output} & \;0\,1\,1\,1 \end{array}\right\} \text{ the truth table}$$

The AND Gate

Another simple logic gate is the **AND gate**, which is represented by the following diagram:

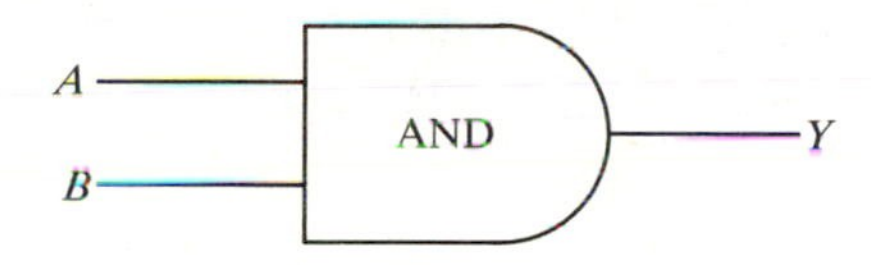

An AND gate produces a one as output if and only if it receives two ones as input; otherwise, it produces a zero. This leads to Table 5.9.

TABLE 5.9

	A	B	Output of AND Gate
possible inputs	1	1	1
	1	0	0
	0	1	0
	0	0	0

Compare this with the table for Boolean "multiplication."

A	B	$A \cdot B$
1	1	1
1	0	0
0	1	0
0	0	0

Since the tables are identical, we conclude that the AND logic gate behaves like Boolean "multiplication" of bits. Thus, it also follows that the AND logic gate will correspond to the series circuit and to the AND logical connective.

EXAMPLE 4

Show the output produced by applying an AND gate to the input sequences of Example 1.

Solution

$$\textbf{a.}\quad \begin{array}{rl} \text{input} & \begin{cases} 1\,1\,0\,0\,0\,1\,1\,0 \\ 0\,0\,1\,0\,1\,1\,0\,0 \end{cases} \\ & \;\downarrow\downarrow\downarrow\downarrow\downarrow\downarrow\downarrow\downarrow \\ \text{output} & \;0\,0\,0\,0\,0\,1\,0\,0 \end{array}$$

Note that a 1 appears on output if and only if all inputs are 1.

$$\textbf{b.}\quad \begin{array}{rl} \text{input} & \begin{cases} 0\,0\,1\,1\,1\,0\,1\,0 \\ 0\,1\,0\,1\,1\,1\,0\,0 \end{cases} \\ & \;\downarrow\downarrow\downarrow\downarrow\downarrow\downarrow\downarrow\downarrow \\ \text{output} & \;0\,0\,0\,1\,1\,0\,0\,0 \end{array}$$

$$\textbf{c.}\quad \begin{array}{rl} \text{input} & \begin{cases} 1\,0\,0\,0\,1\,1\,1\,1 \\ 0\,0\,1\,1\,1\,1\,0\,0 \end{cases} \\ & \;\downarrow\downarrow\downarrow\downarrow\downarrow\downarrow\downarrow\downarrow \\ \text{output} & \;0\,0\,0\,0\,1\,1\,0\,0 \end{array}$$

EXAMPLE 5

Construct truth tables for AND gates that have **(a)** two input devices and **(b)** three input devices.

Solution

(Compare Examples 2 and 3.)

a. The required input sequences are 0011 and 0101.

$$\left.\begin{array}{rl} \text{input} & \begin{cases} 0\,0\,1\,1 \\ 0\,1\,0\,1 \end{cases} \\ & \;\downarrow\downarrow\downarrow\downarrow \\ \text{output} & \;0\,0\,0\,1 \end{array}\right\} \text{truth table}$$

b. The input sequences in this case are: 00001111, 00110011 and 01010101.

$$\left.\begin{array}{rl} \text{input} & \begin{cases} 0\,0\,0\,0\,1\,1\,1\,1 \\ 0\,0\,1\,1\,0\,0\,1\,1 \\ 0\,1\,0\,1\,0\,1\,0\,1 \end{cases} \\ & \;\downarrow\downarrow\downarrow\downarrow\downarrow\downarrow\downarrow\downarrow \\ \text{output} & \;0\,0\,0\,0\,0\,0\,0\,1 \end{array}\right\} \text{truth table}$$

The NOT Gate

The final logic gate to be discussed is the **NOT gate**, also referred to as an **inverter gate**. This gate can have only one input device, say A, and its output device can be denoted by A'. The output of the NOT gate (think of logical negation or the Boolean complement) is zero if the input bit is one and one if the input bit is zero. This results in Table 5.9:

TABLE 5.9

A	Output of NOT gate (A')
1	0
0	1

It is clear from the table that the output of a NOT gate is the Boolean complement of the input bit. The diagram of a NOT gate is

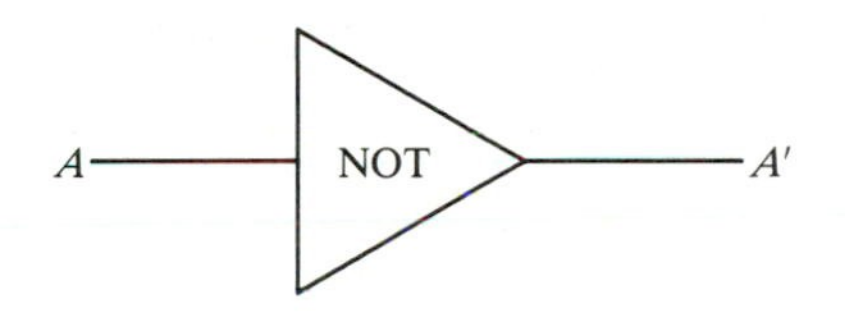

EXAMPLE 6
Show the result of applying a NOT gate to each of the following sequences of bits:

a. 11001011 **b.** 00101101

Solution

a. input 1 1 0 0 1 0 1 1
↓ ↓ ↓ ↓ ↓ ↓ ↓ ↓
output 0 0 1 1 0 1 0 0

b. input 0 0 1 0 1 1 0 1
↓ ↓ ↓ ↓ ↓ ↓ ↓ ↓
output 1 1 0 1 0 0 1 0

Before continuing, we make one final observation about the NOT gate. Suppose we have an eight-digit binary numeral. If we apply a NOT gate to this eight-bit number, the result will be the **ones complement** (see Section 3.3) of the given eight-bit number.

EXAMPLE 7

Given the eight-bit binary number 10100110, compare the output produced by applying a NOT gate to the sequence of bits in the binary numeral with the result of taking the ones complement of the number.

Solution

Putting the sequence of bits 10100110 through a NOT gate produces the output

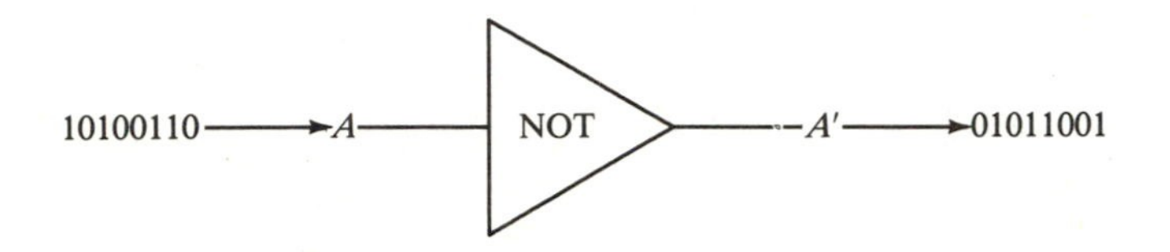

Recall that the ones complement of a binary numeral is the numeral obtained by subtracting each digit of the given binary numeral from 1. Thus, the ones complement of 10100110 is 01011001. This coincides with the result of applying a NOT gate to the sequence of bits comprising the numeral.

Logic Circuits

As we mentioned at the beginning of this section, logic gates can be combined to form logic circuits. When examining a particular logic circuit, we shall be interested in two things: the Boolean expression representing the circuit and the truth table of the circuit.

Consider the following diagram of a logic circuit, where A and B represent the input devices and Y represents the output:

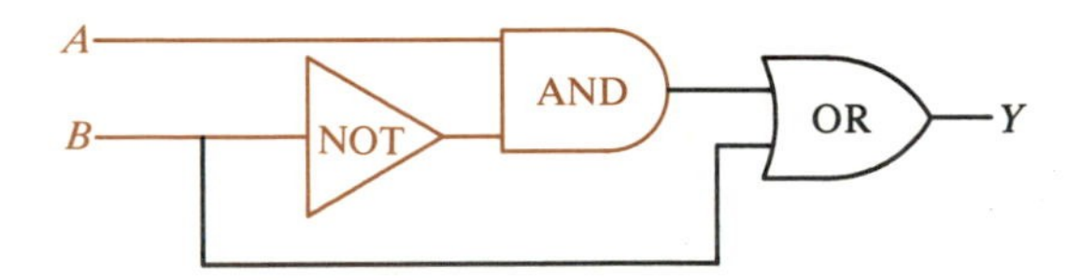

Our first task is to interpret this diagram. Inputs A and B are both connected to an AND gate; but B passes through an inverter before it reaches the AND gate. The Boolean expression for the colored portion of the diagram is (recall AND converts to Boolean ·)

$$A \cdot B'$$

This portion of the diagram is joined, along with B, to an OR gate to produce the output. Since an OR gate corresponds to Boolean +, the Boolean expression representing the entire logic circuit is

$$A \cdot B' + B$$

Now let us find the truth table for this circuit. Since there are two input devices, A and B, we need the input sequences

$$A\text{: } 0011 \qquad \text{and} \qquad B\text{: } 0101$$

First, consider the result of applying $A \cdot B'$ to these input sequences.

1. Applying an inverter to the B sequence gives

$$B'\text{: } 1010$$

2. Now $A \cdot B'$ gives:

$$\begin{array}{rl} A\text{:} & 0\,0\,1\,1 \\ B'\text{:} & 1\,0\,1\,0 \\ \cdot & \downarrow\downarrow\downarrow\downarrow \\ \text{output:} & 0\,0\,1\,0 \end{array}$$

Finally, we combine the output of $A \cdot B'$ with B to produce $A \cdot B' + B$ as follows:

$$\begin{array}{rl} A \cdot B'\text{:} & 0\,0\,1\,0 \\ B\text{:} & 0\,1\,0\,1 \\ + & \downarrow\downarrow\downarrow\downarrow \\ \text{output} & 0\,1\,1\,1 \end{array}$$

The required truth table is

$$\begin{array}{rl} \text{input} \begin{cases} A\text{:} \\ B\text{:} \end{cases} & \begin{array}{l} 0\,0\,1\,1 \\ 0\,1\,0\,1 \end{array} \\ \hline \text{output} \quad Y\text{:} & 0\,1\,1\,1 \end{array}$$

EXAMPLE 8

Consider the logic circuit represented by the following diagram:

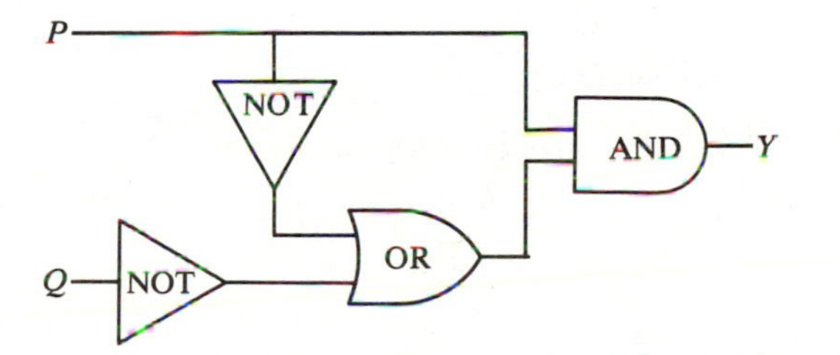

a. Find a Boolean expression representing the circuit.
b. Construct a truth table for the circuit.
c. If possible, simplify this logic circuit.

Solution

a. P and Q both pass through inverters before joining at an OR gate. This gives the Boolean expression

$$P' + Q'$$

This part of the circuit is joined, with input device P, to an AND gate. This gives the Boolean expression

$$P \cdot (P' + Q')$$

b. Since there are two input devices, we can use the input sequences

$$P\text{: } 0011 \qquad \text{and} \qquad Q\text{: } 0101$$

Then P' and Q' are given by

$$\begin{array}{r l} P'\text{:} & 1\,1\,0\,0 \\ Q'\text{:} & \underline{1\,0\,1\,0} \end{array}$$

giving

$$P' + Q'\text{: } 1\,1\,1\,0$$

Now we need $P \cdot (P' + Q')$:

$$\begin{array}{r l} P\text{:} & 0\,0\,1\,1 \\ P' + Q'\text{:} & 1\,1\,1\,0 \\ & \downarrow\downarrow\downarrow\downarrow \\ P \cdot (P' + Q')\text{:} & 0\,0\,1\,0 \end{array}$$ The final output, Y

The truth table for the given logic circuit is

$$\begin{array}{l l l} \text{input} & \begin{cases} P \\ Q \end{cases} & \begin{array}{l} 0\,0\,1\,1 \\ \underline{0\,1\,0\,1} \end{array} \\ \text{output} & Y & 0\,0\,1\,0 \end{array}$$

c. The Boolean expression for this circuit is

$$P \cdot (P' + Q')$$

The properties of a Boolean algebra allow us to simplify this expression.

$$\begin{aligned} P \cdot (P' + Q') &= P \cdot P' + P \cdot Q' && \text{Distributive property} \\ &= 0 + P \cdot Q' && \text{Definition of complement} \\ &= P \cdot Q' && \text{0 is the identity for } +. \end{aligned}$$

The original circuit is, then, equivalent to a logic circuit whose Boolean equivalent is $P \cdot Q'$. The diagram of this circuit is

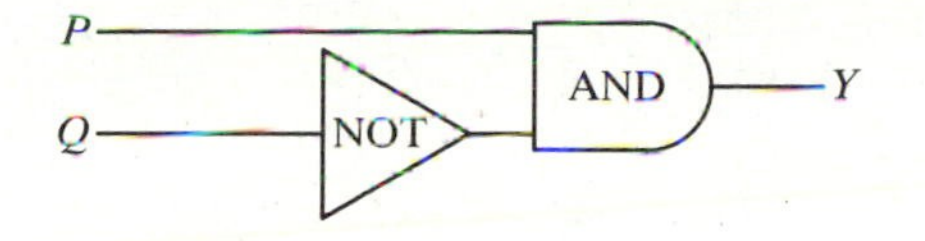

The reader should verify that this circuit has a truth table identical to the one obtained in part b of this example.

EXAMPLE 9

Consider the logic circuit represented by the following diagram:

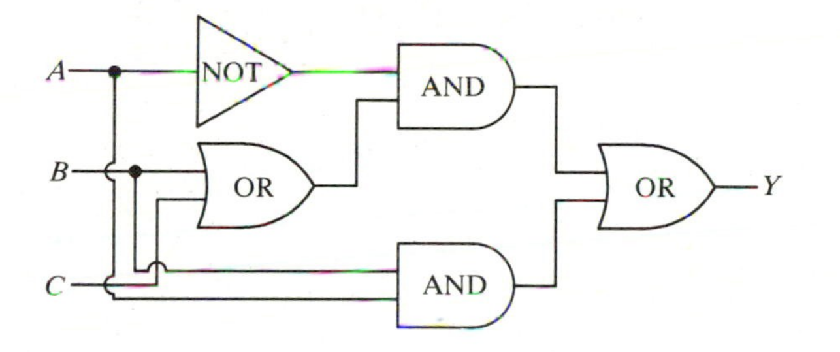

Analyze this circuit as we did the circuit of Example 8.

Solution

a. Inputs B and C pass through an OR gate

$$\text{giving } B + C$$

to join the inverted form of $A(A')$ at an AND gate

$$\text{giving } A' \cdot (B + C)$$

There is also a subcircuit of A and B joined at an AND gate

$$\text{giving } A \cdot B$$

Finally, these two circuits

$$A' \cdot (B + C) \text{ and } A \cdot B$$

meet at an OR gate. The Boolean expression for the given logic circuit is, then,

$$[A' \cdot (B + C)] + (A \cdot B)$$

b. Since there are three input devices, we use the three input sequences:

$$\begin{aligned} A&: 00001111 \\ B&: 00110011 \\ C&: 01010101 \end{aligned}$$

If A is represented by 00001111, A' is 11110000. $B + C$ is given by:

$$\begin{array}{r l} B: & 0\,0\,1\,1\,0\,0\,1\,1 \\ C: & 0\,1\,0\,1\,0\,1\,0\,1 \\ \hline B + C: & 0\,1\,1\,1\,0\,1\,1\,1 \end{array}$$

$A' \cdot (B + C)$ is given by

$$\begin{array}{r l} A': & 1\,1\,1\,1\,0\,0\,0\,0 \\ B + C: & 0\,1\,1\,1\,0\,1\,1\,1 \\ \hline A' \cdot (B + C): & 0\,1\,1\,1\,0\,0\,0\,0 \end{array}$$

$A \cdot B$ is given by

$$\begin{array}{r l} A: & 0\,0\,0\,0\,1\,1\,1\,1 \\ B: & 0\,0\,1\,1\,0\,0\,1\,1 \\ \hline A \cdot B: & 0\,0\,0\,0\,0\,0\,1\,1 \end{array}$$

Finally, $[A' \cdot (B + C)] + (A \cdot B)$ is given by

$$\begin{array}{r l} A' \cdot (B + C): & 0\,1\,1\,1\,0\,0\,0\,0 \\ A \cdot B: & 0\,0\,0\,0\,0\,0\,1\,1 \\ \hline \text{output } Y: & 0\,1\,1\,1\,0\,0\,1\,1 \end{array}$$

Therefore, the truth table is

$$\begin{array}{r r l} & A: & 0\,0\,0\,0\,1\,1\,1\,1 \\ \text{input} & B: & 0\,0\,1\,1\,0\,0\,1\,1 \\ & C: & 0\,1\,0\,1\,0\,1\,0\,1 \\ \hline \text{output} & Y: & 0\,1\,1\,1\,0\,0\,1\,1 \end{array}$$

c. For a simplified form of the Boolean expression, we can refer to Section 5.1, Example 6b. We found that

$$B + A' \cdot C$$

is equivalent to the expression

$$[A' \cdot (B + C)] + (A \cdot B)$$

A diagram of the circuit represented by $B + A' \cdot C$ appears below:

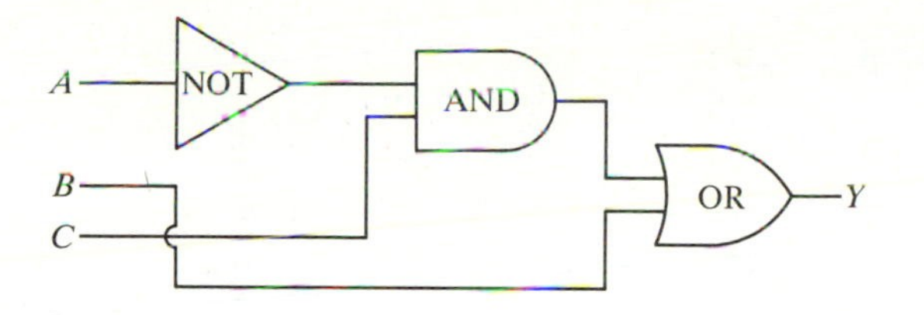

We conclude this chapter with an illustration of the use of the theory of Boolean algebra and logic gates to design a network that performs simple addition, one digit at a time.

First, we review binary addition. Suppose we wish to perform the addition $101_2 + 11_2$. The solution is as follows:

$$\begin{array}{r} {\scriptstyle 1\,1} \\ 101_2 \\ +11_2 \\ \hline 1000_2 \end{array}$$

This result was obtained by adding two bits (binary digits) at a time and recording the *result* below and any *carry* above the next column to the left. Note, for example, that

$$\begin{array}{r} 1_2 \\ +1_2 \\ \hline 10_2 \end{array}$$

So the *recorded result* in the first column is 0, and the 1 is a *carry* over to the second column. This process of adding two bits and recording the result (we'll call it the *sum*, although this term is not exactly correct) and the carry is repeated for each column. So, for example, the second column could appear as

$$\begin{array}{r} 1 \\ 0 \\ 1 \\ \hline 10 \end{array}$$

1 ← Carry from the first column.

↑ ↑— Recorded result

Carry to the third column.

We shall use these observations to illustrate the design of a **half-adder**—a network of logic gates that forms the sum of two bits, with the result being a "sum" bit and a "carry" bit. This objective is summarized in truth table form below, with P and Q representing bits.

P	Q	**Sum**	**Carry**
1	1	0	1
1	0	1	0
0	1	1	0
0	0	0	0

Note that the sum in the table refers to the recorded sum in the addition example.

We begin by observing that the carry column of the truth table corresponds to the truth table of the Boolean expression $P \cdot Q$:

P	Q	$P \cdot Q$
1	1	1
1	0	0
0	1	0
0	0	0

This tells us that, in the logic network we design, P and Q must meet at an AND gate.

The design of the sum network is not quite as obvious. We observe that we get a one in the sum column in two sets of circumstances:

1. when P is 1 and Q is 0 (compare $P \cdot Q'$); and
2. when P is 0 and Q is 1 (compare $P' \cdot Q$),

and *only* in those circumstances. So we construct the truth table of $P' \cdot Q + P \cdot Q'$.

P	Q	$P' \cdot Q$	$+$	$P \cdot Q'$
1	1	0	0	0
1	0	0	1	1
0	1	1	1	0
0	0	0	0	0

This corresponds to the sum column in our original table, so we need a logic network whose Boolean equivalent is $P' \cdot Q + P \cdot Q'$. One possible design for this network, called a **half-adder**, is given in Figure 5.2

In this brief introduction we have only scratched the surface of logic networks and Boolean algebra, and their relationship to the design of computer circuits. A more thorough treatment would be the province of a course in computer science dealing with digital circuit design.

FIGURE 5.2
A half-adder network.

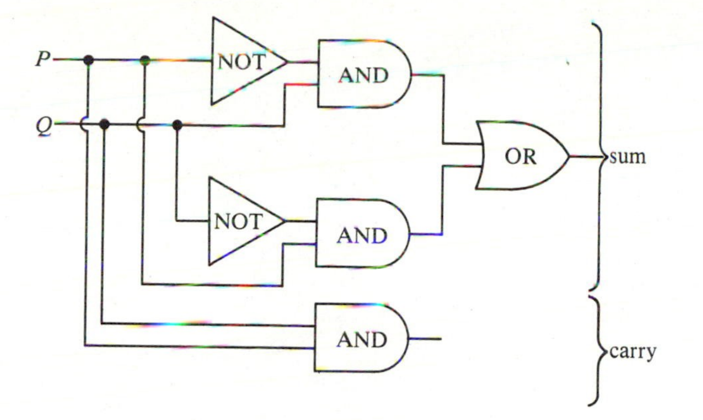

EXERCISES FOR SECTION 5.4

In Exercises 1 through 10, determine the output if the specified pairs are processed by an OR gate (see Examples 1–3).

1. 11010000
01010011

2. 10101010
11001100

3. 11011101
11010011

4. 00001010
11101011

5. 00000001
10101010

6. 10011001
01100000

7. 01101011
10000100

8. 00000000
00000000

9. 11111111
11111111

10. 11111111
01101111

In Exercises 11 through 20, determine the output if the specified pairs of Exercises 1 through 10 are processed by an AND gate (see Examples 4 and 5).

In Exercises 21 through 25, determine the output if the specified sequence of bits is processed by an inverter (see Examples 6 and 7).

21. 10010011

22. 11001101

23. 00100010

24. 10100000

25. 11100111

In Exercises 26 through 29, (a) represent the logic circuit with a Boolean expression, (b) construct a truth table for the circuit, and (c) if possible, simplify the circuit.

26.

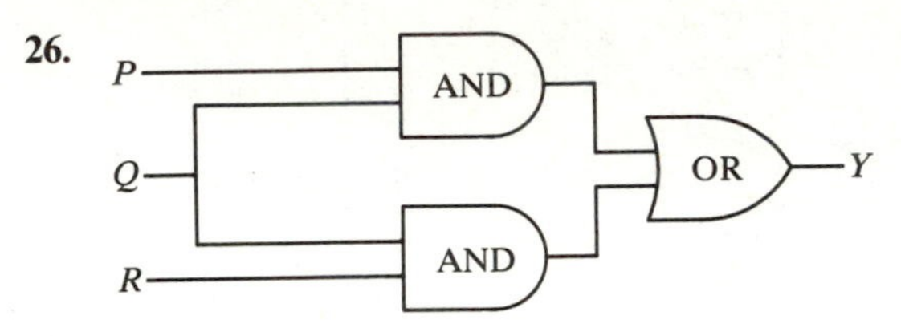

27.

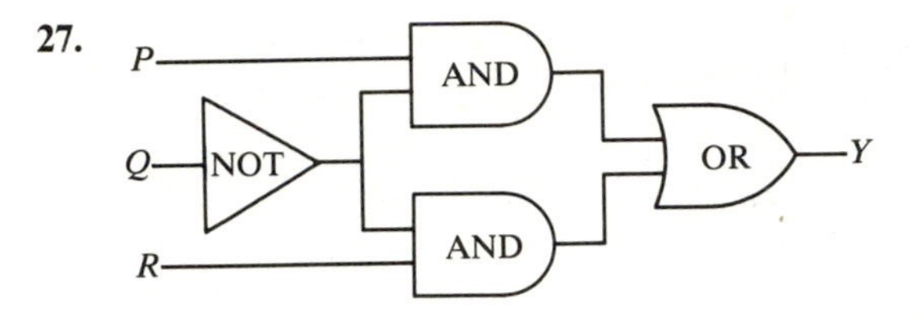

28.

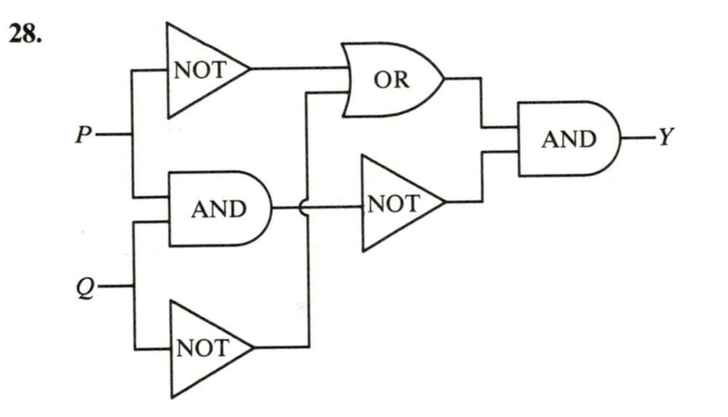

29.

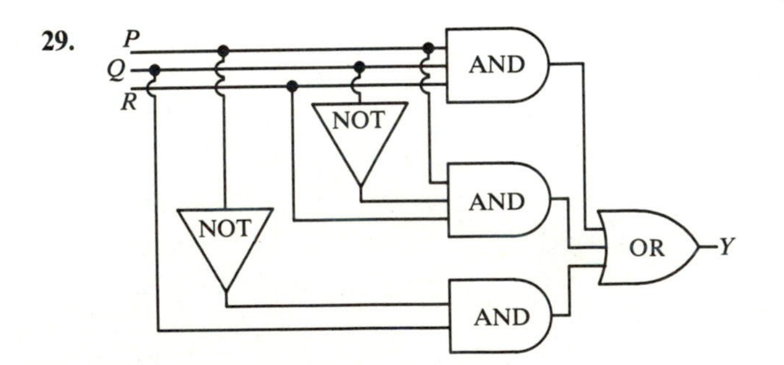

5.5 CHAPTER REVIEW

The chapter began with the definition of a Boolean algebra and its properties. The relationship between the $+$ and $\cdot$ operations on a Boolean algebra in general and the $\vee$ and $\wedge$ logical operators was shown. The properties of the logical operators and of a Boolean algebra were used in the study of switching circuits and switching networks. Section 5.2 dealt with the definitions of series and parallel circuits and of coupled switches, and with their statement in terms of logical and Boolean expressions. In Section 5.3, truth tables and the properties of a Boolean algebra were used to simplify switching networks.

The last section dealt with logic gates and circuits, an important concept in computer architecture. The OR, AND and NOT gates were studied, along with their truth tables. Finally, the Boolean algebra properties were applied to the problem of simplifying logic circuits.

VOCABULARY

Boolean algebra
commutative property
associative property
distributive property
identity
complement
idempotent property
absorption property
circuit
switching network

terminal of a circuit
series circuit
parallel circuit
coupled switches
equivalent network
logic gate
OR gate
truth table of a logic gate
AND gate
NOT (inverter) gate
half-adder

CHAPTER TEST

Simplify each of the following expressions, using the definitions of $+$, $\cdot$ and in a Boolean algebra. You may also use DeMorgan's laws: $(A + B)' = A' \cdot B'$ and $(A \cdot B)' = A' + B'$.

1. $(A + B)' + (A' \cdot B)$

2. $[B \cdot (A + B')] + (A + B)$

For each of the following networks (a) determine the equivalent logical statement; (b) represent the network by a Boolean expression; and (c) describe the conditions under which current flows.

3. P ; P' ; Q—R

4. 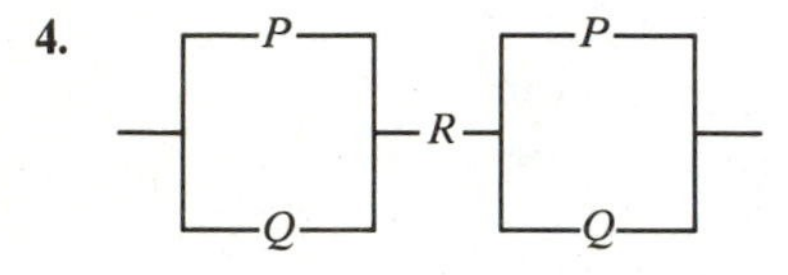

5. Draw the network that corresponds to the following logic statement:

$$(p \vee q) \wedge ((q \wedge \sim q) \vee \sim p)$$

6. Represent the network of Problem 5 by a Boolean expression.

7. Simplify the network of Problem 5 by using the truth table of the logical statement or by simplifying the Boolean expression. Draw the simplified network.

For Problems 8 through 10 use the following pair of input sequences:

A: 11010111
B: 10011001

8. Determine the output after processing by an OR gate.

9. Determine the output after processing by an AND gate.

10. Determine the output if sequence A is processed by a NOT gate.

11. a. Represent the following logic circuit with a Boolean expression.
b. Construct a truth table for the circuit.
c. If possible, simplify the circuit.

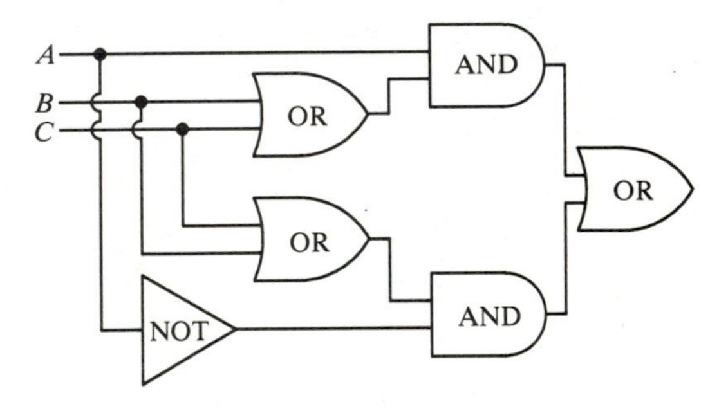

CHAPTER

6

A recipe is a set of directions for solving a problem. Following the directions brings you to the desired outcome (a tasty meatloaf, a prize-winning pie, or a nutritious casserole). In this chapter we shall study methods (or "procedures" or "directions") for solving problems; we refer to these as algorithms.

```
NEW ENGLAND POT ROAST

1 blade or bottom round pot roast-- 4 pounds
3 TBS. flour
2 tps. salt
3 TBS. bacon drippings or oil
1/2 cup freshly grated horseradish or one 4 oz. jar
1 cup whole cranberry sauce
2-3 cinnamon sticks, broken in half
6-8 whole cloves
1 cup beef broth
12 small white onions
1 bunch of carrots, cut into 2-3 inch lengths

1. Dredge the meat in the flour mixed with the salt
2. Heat the drippings or oil in a heavy skillet and brown the meat on
   all sides over high heat. Reserve the drippings.
3. In a heavy Dutch oven or casserole, mix the broth, horseradish,
   cranberry sauce, cinnamon sticks, cloves over low heat; and add
   the meat.
4. Bring the mixture to a boil, lower the heat, cover and simmer
   gently for about two hours or until the meat is tender.
5. Meanwhile, brown the carrots and onions in the drippings and
   add them to the meat broth.

Serves 6-8 people.
```

Algorithms and Problem Solving

6.1 Algorithms and Pseudocode

In this chapter we shall address ways of solving problems and ways of organizing and presenting solutions to problems. Our goal is a well-written solution that will be easy to implement on a computer and that will also allow us to detect errors readily.

The first step toward a well-written solution involves establishing a set of instructions, called an **algorithm**. An algorithm may be written in several different forms but is always *independent of a programming language* so that it can be translated, at a later stage in the problem's computerized solution, into one or several different languages. An algorithm has the following properties:

ALGORITHMS

1. An algorithm is a list of sequential instructions for solving a specific problem.
2. There are three ways of expressing an algorithm:
   a. By actually listing the steps (see Examples 1 and 2).
   b. By writing something called *pseudocode* (see Examples 3 and 4).
   c. By writing a *flowchart* (see Section 6.3).
3. The steps in an algorithm must be clear, unambiguous statements leading to as general a solution as possible.

Two examples of algorithm construction follow:

EXAMPLE 1

Suppose you are the owner of the Babot Machine Shop, an up-and-coming company that employs one hourly employee. Describe how to calculate your employee's gross pay.

Solution

1. Obtain hours worked and rate of pay.
2. Gross pay equals hours times rate.
3. Record gross pay.
4. End.

Notice that the simple algorithm describes precisely how to arrive at the gross pay. Also, the instructions are numbered for clarity and possible reference. The solution is not written in any specific computer language, but described precisely using English. The solution can now be adapted and written in any computer programming language. Thus, we emphasize that we are *programming* here, we are *not coding*.

EXAMPLE 2

It's time to face taxes! For Babot Machine Shop's one employee, state an algorithm for computing *net* pay.

Solution

1. Obtain employee's hours, rate, and number of exemptions.
2. Gross pay equals hours times rate.
3. Look up withholding using gross pay and number of exemptions. (How to look up withholding involves a separate set of instructions.)
4. Net pay equals gross pay minus withholding.
5. Record net pay.
6. End.

Notice that writing algorithms is similar to writing recipes: both involve a list of ingredients and instructions on how to combine them to get a product. In Example 2, the "ingredients" are hours, rate, and number of exemptions; the sequential set of six instructions is the algorithm.

The remaining examples in this and the next section will begin to develop the **pseudocode** variation of expressing an algorithm. Basically, pseudocode is a language-independent collection of statements that conform to certain structures (to be introduced). We mention here that the pseudocode presented, although independent of the language, has been developed with guidelines that assume **top-down** or **structured** programming.

EXAMPLE 3

The Babot Machine Shop's sole employee has been working overtime (more than 40 hours per week) and demands extra compensation. Assuming time-and-a-half for the overtime rate, develop an algorithm for computing gross pay if overtime is paid for hours worked beyond 40.

Solution

```
1.  Obtain HOURS, RATE
2.  IF HOURS > 40, then do the following
    a.  EXTRA = HOURS - 40
    b.  OVERTIME = 1.5 * RATE * EXTRA
    c.  REGULAR = 40 * RATE
    d.  GROSS = REGULAR + OVERTIME
    e.  GO TO STEP 4
3.  GROSS = HOURS * RATE
4.  Record GROSS
5.  END.
```

The algorithm above introduces three new concepts: (1) we use variables (HOURS, RATE, EXTRA, OVERTIME, REGULAR, GROSS) and symbols (>, =, −, *) for simplicity; (2) we indent for clarity; (3) a decision is made (in step 2).

The decision involves an IF statement in which the conclusion (steps 2a through 2e) is executed when the condition, or the antecedent (HOURS > 40), is true. IF statements are referred to as **conditional branching** structures. The direction in step 2e, however, instructs to skip step 3; such a branch is **unconditional**.

The algorithm in Example 3 is not, of course, a unique method of solving the problem. In fact, the unconditional transfer, **GO TO** (in step 2e), is not consistent with structured pseudocode. (There is no GO TO statement available in many versions of Pascal, for example.) An alternative is to use an IF-THEN-ELSE structure as below:

```
1. Obtain HOURS, RATE
2. IF HOURS > 40 THEN
     DO
   a. EXTRA = HOURS - 40
   b. OVERTIME = 1.5 * RATE * EXTRA
   c. REGULAR = 40 * RATE
   d. GROSS = REGULAR + OVERTIME
     END DO
3. ELSE GROSS = HOURS * RATE
4. Record GROSS
5. END.
```

EXAMPLE 4

The Babot Machine Shop has grown so that now there are many employees. If N represents the number of hourly employees, write an algorithm for calculating the gross pay of each. Ignore overtime.

Solution

```
1. Obtain N
2. Initialize counter E to 1 (* E counts employees
   as each is processed *)
3. DOWHILE (E ≤ N)
4.     Obtain HOURS, RATE for employee number E
5.     GROSS = HOURS * RATE
6.     Record GROSS
7.     E = E + 1
8. END DOWHILE
9. END.
```

The above algorithm introduces several new concepts:

Step 2 introduces **initialization**. This is particularly important when the algorithm will be used as the basis for a program run on a computer system that does not automatically set variables to certain numbers. (Think of initialization as the same as setting a gasoline pump back to zero for each new customer. Imagine what would happen if it were not!)

In step 2 we also introduce the concept of a **comment**, indicated by

(* . . . comment here . . . *)

Step 3 establishes a **loop**. A loop is used when repetition is necessary. Here steps 4 through 7 are repeated for each employee. The variable E is the loop control variable and when $E > N$, the loop is exited. The check is made in step 3 to see whether E is greater than N.

Notice that steps 4 through 6 are exactly the steps used in Example 1. Thus, if we are able to solve the problem for one employee, we need only introduce the appropriate loop to process many employees.

We conclude this section with an application of writing pseudocode to a problem that involves a mathematical formula.

EXAMPLE 5

Edi needs to do a certain number N of temperature conversions to Fahrenheit (F) from Celsius (C) using the formula

$$F = \frac{9}{5}C + 32$$

Write the pseudocode to solve Edi's task.

Solution

```
1. Obtain N (* N represents the number of
   conversions to be done *)
2. Initialize counter E to 1 (* E counts the number of
   conversions as they are done *)
3. DOWHILE (E ≤ N)
4.     Obtain C (* C is for Celsius, input variable *)
5.     F = (9/5) * C + 32 (* F is for Fahrenheit,
       output variable *)
6.     Record F
7.     E = E + 1
8. END DOWHILE
9. END.
```

EXERCISES FOR SECTION 6.1

In Exercises 1 through 5 construct an algorithm that solves the given problem (see Examples 1 and 2).

1. Joe bought a calculator in a county that has a 7% sales tax rate. Describe how to calculate the gross amount Joe had to pay.
2. Teresa purchased a guitar in a county that has a $7\frac{1}{2}\%$ sales tax rate. Describe how to calculate the gross amount Teresa had to pay.
3. Mike bowled three games last night. Describe how to calculate his average.
4. Jane drinks her coffee black. Describe how to prepare her a cup of instant coffee.
5. Suppose $Y = (A + 3B)/2$. Describe how to find Y for given values of A and B.

In Exercises 6 through 8, write an algorithm that solves the given problem using an IF statement (see Example 3).

6. Lesley purchased "something" at the store. If that something is food, then there is no sales tax; if that something is not food then the sales tax is 6%. Describe how to calculate Lesley's payment.

7. Suppose $Y = (A + 3B)/C$. Describe how to find Y for given values of A, B, and C. If C is zero, do not calculate Y.
8. Refine the algorithm in Example 3, and combine steps 2a, b, c, and d.

In Exercises 9 through 13, use a DOWHILE loop to write an algorithm that solves the given problem (see Example 4).

9. Lesley (see Exercise 6) purchased ten items. Calculate her total bill if the sales tax rate on every nonfood item is 6%.
10. Refine the algorithm in Exercise 9 to generalize the purchases to N items.
11. Rework the problem in Exercise 7 if there are
    a. Ten triplets of A, B and C for which ten values of Y must be found.
    b. N triplets of A, B and C for which N values of Y must be found.
12. Write the algorithm that calculates the *net* pay for N employees.
    a. Use E as your loop control variable.
    b. Assume the tax rate (as percent of gross) is stored in variable TR, different for each employee, instead of looking up withholding. Furthermore, assume TR is stored as a decimal (that is, 7% would be 0.07).
    c. Ignore overtime.
13. Rework Exercise 12, but this time allow for possible overtime.

6.2 Loops, Counters, and Accumulators

In Examples 4 and 5 of Section 6.1 the concepts of loops (DOWHILE) and counters (E = E + 1) were introduced. In this section, we deal more thoroughly with these concepts and also introduce the idea of an accumulator. First, let's reexamine the DOWHILE loop in general:

```
   1.  E = 1
   2.  DOWHILE (E ≤ N) (* The ''check'' to see if
       E ≤ N is made here *)
   :      :
   K.      E = E + 1 (* Automatically goes back to
           step 2 *)
 K+1.  END DOWHILE
```

Notice two features of the loop: First, in step 1, the symbol = doesn't mean "equals." Rather, it means to "evaluate the expression on the right and store it in the variable on the left." Actually, we *assign* the value of 1 to E; we could also write E ← 1. Note too that when E becomes equal to N + 1, we exit the loop (steps 2 through K) because E is no longer less than or equal to N.

The next example illustrates the use of an **accumulator**. An accumulator is a variable that is used to maintain a running total of the value of some other variable. For example,

```
7. SUM = SUM + GROSS
```

accumulates, in the variable SUM, all the values of GROSS (assuming this statement is **nested** in a loop). That is, the previous value of SUM (SUM on the right) is added to GROSS, and that new addition is stored in the variable SUM (SUM on the left). If SUM = 572 and GROSS = 100 *before* statement 7, then after the execution of statement 7, SUM = 672.

EXAMPLE 1

Let's return to the (now thriving) Babot Company's Payroll Department. Total wages paid is an important cost of doing business. Write an algorithm that totals the gross wages for all hourly employees and also computes the mean (average) gross wage.

Solution

```
 1. Obtain N        (* N is number of employees *)
 2. E = 1           (* Initialize counter *)
 3. SUM = 0         (* Initialize sum of hourly wages *)
 4. DOWHILE (E ≤ N)
 5.      Obtain HOURS, RATE
 6.      GROSS = HOURS * RATE (* GROSS is gross pay *)
 7.      SUM = SUM + GROSS
 8.      E = E + 1
 9. END DOWHILE
10. Record total gross pay for company as SUM
11. MEAN GROSS PAY = SUM/N
12. Record MEAN GROSS PAY
13. END.
```

Notice the importance of initializing the variable SUM to zero above. It is done for the same reason a cash register is set to zero before it is used to accumulate your bill.

The importance of **tracing** an algorithm and a program is crucial in problem solving. A trace is an actual pencil-and-paper "walk through" to test whether the sequence of statements (that is, the *logic*) is correct.

EXAMPLE 2

Suppose the following data applies to the Babot Corporation's (N = 5) payroll:

Employee Number	Hours	Rate
1	10	$5.00
2	20	4.00
3	25	6.00
4	40	3.50
5	40	3.00

Trace the algorithm of Example 1 by determining the values of the variables E, N, SUM, HOURS, RATE, and GROSS at each pass through the DOWHILE loop. What is the value of MEAN GROSS PAY at output?

Solution

At each pass, the values below are determined after step 7 but before step 8:

			Value of			
PASS	E	N	HOURS	RATE	GROSS	SUM
1	1	5	10	5	50	50
2	2	5	20	4	80	130
3	3	5	25	6	150	280
4	4	5	40	3.5	140	420
5	5	5	40	3	120	540

At output, MEAN GROSS PAY is SUM/N = 540/5 = 108.

The DOWHILE loop is not the only method of implementing repetition in an algorithm. The **FOR loop** is another construction that provides for repetition of an instruction or set of instructions. It should be used, however, only when the exact number of repetitions required is known in advance. The general form of the FOR loop is

```
                 FOR I = 1 TO N
Body of loop →   instructions to be repeated
                 NEXT I
```

The FOR loop construction provides for

1. initialization of a loop counter (to count the number of repetitions);
2. execution of an instruction or set of instructions (the **body** of the loop);
3. incrementation of the loop counter; and
4. testing for the loop exit. (Has the counter reached some predetermined value? If it has, exit the loop; if not, go back to step 2).

In the example above, the **loop control variable** (or loop counter) I is initialized to 1 (step 1). Then the instructions in the body of the loop are executed once (step 2). The loop control variable is then increased by 1 (step 3); and its value is compared with the value of N (step 4). If it is still less than N (that is, if not enough repetitions have taken place) steps 2, 3, and 4 are repeated. Note that, in the BASIC language, whether the value of I equals N or N + 1 when the FOR loop is exited may depend on the particular computer being used. On many microcomputer versions of BASIC, the loop control variable will reach N + 1. On some other versions, the loop is exited *before* the control variable is incremented for the last time; in this case, the value of the control variable on exit will be N. It should also be noted that different languages have slightly different implementations of the FOR loop.

EXAMPLE 3

Use a FOR loop to reconstruct the algorithm for Example 1.

Solution

```
 1. Obtain N      (* Number of employees *)
 2. SUM = 0       (* Initialize the sum *)
 3. FOR E = 1 TO N     (* E is the loop control variable *)
 4.      Obtain HOURS, RATE
 5.      GROSS PAY = HOURS * RATE
 6.      SUM = SUM + GROSS PAY
 7. NEXT E
 8. PRINT "GROSS PAY TOTAL IS" SUM
 9. MEAN GROSS PAY = SUM/N
10. PRINT (record) MEAN GROSS PAY
11. END.
```

In this example, the body of the loop consists of statements 4, 5 and 6 (note that they are indented for clarity). In step 3, the counter E (loop control variable) is initialized to 1. In the same step, the *terminal*, or *ending* value is set to N. In step 7, E is increased by 1 and compared to the value of N. The body of the loop is repeated N times.

A simple BASIC program illustrating the coding of a FOR loop follows:

```
        10  REM EXAMPLE OF FOR LOOP
        20  FOR I = 1 TO 4
        30    PRINT I
        40  NEXT I
        50  PRINT ''AT THE END, I = '' I
        60  END
   RUN
```

Depending on the version of BASIC being used, the output will be:

```
1                       or  1
2                           2
3                           3
4                           4
AT THE END, I = 4           AT THE END, I = 5
```

In order to use a FOR loop effectively, the programmer must be able to determine which instructions to include within the loop (in the body) and which to place before or after it. *Only* instructions that are to be repeated are included in the body of the loop. For example, note that statement 2 (SUM = 0) in Example 3 is *not* in the body of the loop, since we want to initialize this value only *once*. This statement is *not* repeated.

EXAMPLE 4

Suppose you are a cashier and decide to write an algorithm that will print each customer's bill and the day's total sales. Here's the algorithm:

```
 1. Obtain N               (* Number of customers *)
 2. C = 1                  (* Initialize customer counter *)
 3. TOTAL = 0              (* Total sales for all customers *)
 4. DOWHILE (C ≤ N)
 5.     Obtain NI          (* Number of items for this customer *)
 6.     I = 1              (* Initialize item counter *)
 7.     DOWHILE (I ≤ NI)
 8.         Obtain PRICE
 9.         BILL = BILL + PRICE
10.         I = I + 1
11.     END DOWHILE
12.     Print BILL
13.     TOTAL = TOTAL + BILL
14.     C = C + 1
15. END DOWHILE
16. PRINT TOTAL
17. END.
```

Trace this algorithm, finding the values of TOTAL and BILL each time statement 13 is executed for the following data (N = 2):

Customer Number	NI	Item	Price Each Item
1	2	1	$5.50
		2	7.00
2	3	1	12.00
		2	11.00
		3	10.00

Solution

Statement Number		Value of Bill	Value of Total
Customer 1	12	12.50	0
	13	12.50	12.50
Customer 2	12	45.50	12.50
	13	45.50	58.00

45.50: Should be $33.00 (Why?)

58.00: Should be $45.50. (Why?)

The value of the second customer's bill and of the day's total show that we have a "bug." The algorithm *should* initialize BILL to zero before calculations for each customer, but it doesn't. You are asked to correct this problem in Exercise 4.

We have so far studied counters (such as C or I in Example 4), accumulators (such as TOTAL in Example 4), and loops. Counters are variables to which we add 1 each time, and accumulators are more general in that they are variables used to *sum* or *accumulate* variable numerical quantities. Both counters and accumulators are set to zero (initialization) and sometimes must be reset as we saw in the previous example. Counters and accumulators can be recognized in a computer program or an algorithm by the fact that they appear on both sides of the assignment statement or equal sign (for example, SUM = SUM + PAY).

Loops, as we have seen, are used to repeat a sequence of instructions. Loops consist of initialization, process, and incrementation. In Example 4, we set C = 1 (initialization), we detailed instructions for finding the bill for a customer (process) and we incremented C each time we processed a new customer.

We pause to summarize some of the main points stressed in the examples we have presented in this section so far.

1. A problem must be clearly stated and understood before any attempt is made to solve it.

2. An *algorithm* is a finite list of instructions (steps) that solve a problem or accomplish a task. A simple example of an algorithm is a recipe.

3. Instructions are usually executed by a computer in **sequence**, that is, in the exact order written. For most algorithms, the exact order of execution is very important. However, the sequence of execution of instructions can be altered, depending on the value of a variable, by using an IF statement, or **selection** process. This allows the algorithm to branch to different instructions for different values of a particular variable.

4. Many problems involve **repetition** of a single instruction or a group of instructions. Two methods of indicating repetition are the DOWHILE loop and the FOR loop.

5. The concepts of *sequence*, *selection*, and *repetition* are central to the use of **structured programming**—a method of writing programs or algorithms that uses the sequence, selection, and repetition structures to write clear, relatively easily understood and modified sets of instructions. The longer, more complex an algorithm or program is, the more important it is to use structured programming techniques.

The examples that follow illustrate the use of structured programming techniques. We begin with an example that shows how to handle reading the last employee's record at the now blue-chip company, Babot International.

EXAMPLE 5

Assume each employee's record of hours and rate is stored on a separate card and that the number of employees (that is, cards) is unknown because of weekly changes in the company. Write an algorithm to print each employee's gross pay and the company's weekly sum of all wages.

Solution

If the cards are read by a machine, we must have a way to signal the machine that the last card has been read. This can be done by having a "trailer card" that contains impossible information in either the hours or rate *field.* (A field is one of the separate items of information in a record.) We shall use as a trailer a negative number in the hours field, and we shall write a step in the algorithm to stop reading cards and processing when this **END OF FILE (EOF)** condition is reached. The complete algorithm is as follows:

```
 1. SUM = 0     (* Initialize *)
 2. READ RATE, HOURS
 3. DOWHILE (HOURS ≥ 0)
 4.      GROSS = RATE * HOURS
 5.      PRINT GROSS
 6.      SUM = SUM + GROSS
 7.      READ RATE, HOURS
 8. END DOWHILE     (* Will go back to step 3 automatically *)
 9. PRINT SUM
10. End.
```

Notice that statements 2 and 7 are identical, but both are needed. Statement 2 reads card 1 and gives us an initial value of RATE and HOURS to test in statement 3; statement 7 reads all later cards from 2 through the trailer card.

We conclude this section with a mathematical application. Suppose we are given the lengths of the three sides of a triangle; we'll label them A, B, and C. Suppose also that we do *not* know the length of any altitude so that we cannot use the familiar formula

$$\text{Area} = \tfrac{1}{2}\text{base} \cdot \text{height}$$

The area of the triangle can still be found by first letting the *semiperimeter* $S = (A + B + C)/2$ and then applying Heron's formula:

$$\text{Area} = \sqrt{S(S - A)(S - B)(S - C)}$$

If, for example, the triangle has sides whose lengths are 3, 4, and 5 units, then $S = (3 + 4 + 5)/2 = 12/2 = 6$. Applying Heron's formula gives

$$\text{Area} = \sqrt{6(6 - 3)(6 - 4)(6 - 5)} = \sqrt{6 \cdot 3 \cdot 2 \cdot 1} = \sqrt{36} = 6$$

EXAMPLE 6

Write an algorithm that will read the lengths of the sides of four different triangles, and calculate and print the area of each one before going on to the next.

Solution

```
1. K = 1     (* Initialize the loop counter *)
2. DOWHILE (K <= 4)
3.     READ A, B, C
4.     S = (A + B + C)/2
5.     Area = √(S * (S − A) * (S − B) * (S − C))
6.     PRINT "AREA IS";AREA
7.     K = K + 1
8. END DOWHILE
9. END.
```

This algorithm *looks* good, but it may not work exactly as we expect. In fact, it presupposes that the user will be a reasonably good geometry student. Suppose, for example, that someone enters 1, 2 and 5 for the sides of the triangle. (Note that there is no "real" triangle with these sides, but our hypothetical user doesn't know that!) Then, in this case,

$$S = (1 + 2 + 5)/2 = 8/2 = 4$$

and the area of the "triangle" is given by

$$\text{Area} = \sqrt{4(4 - 1)(4 - 2)(4 - 5)} = \sqrt{4(3)(2)(-1)} = \sqrt{-24}$$

Recall (from Section 1.1) that $\sqrt{-24}$ is an *imaginary* number and that most computer languages cannot deal with imaginary numbers. They are unable to evaluate the square root of a negative number. Our algorithm would be better (software developers would say we were being "user-friendly") if we could avoid potential errors by trapping for negative values of the **radicand** (the number under the square root symbol).

EXAMPLE 7

Modify the algorithm of Example 6 to catch imaginary numbers arising from "fake" triangles.

Solution

We must trap the square root of a negative number *before* the computer is asked to evaluate it. We do this in steps 6 through 11 in the following algorithm:

```
                         1.  K = 1
                         2.  DOWHILE (K < = 4)
                      ┌  3.    READ A, B, C                            ┐
                      │  4.    S = (A + B + C)/2                       ├ Sequence
                      │  5.    X = S * (S - A) * (S - B) * (S - C)     ┘
                      │  6.    IF X > 0 THEN                             Selection
          Repetition  │  7.      BEGIN
          (loop)      │  8.        AREA = √X
                      │  9.        PRINT "AREA IS" AREA
                      │ 10.      END
                      │ 11.    ELSE PRINT "NOT A TRIANGLE"
                      └ 12.    K = K + 1
                        13.  END DOWHILE
                        14.  END.
```

This algorithm illustrates the sequence, selection, and repetition structures that are the essence of structured programming. The instructions are executed in *sequence* until step 6. At this point, *selection* occurs: if X > 0, then steps 8 and 9 are executed; otherwise, step 11 is executed. (No triangle can have an area of zero or less, so a test is performed *before* an attempt to calculate a negative square root can result in an error.) The *repetition* is controlled by the DOWHILE loop with a counter K set to 1 at step 1 and incremented in step 12.

EXERCISES FOR SECTION 6.2

Exercises 1 and 2 ask you to write an algorithm. Be sure to include a counter and loop (see Example 1).

1. Modify the algorithm in Example 1 so that it prints each employee's gross and net pay. Use TR for the variable, read as a decimal, that represents each employee's tax rate. Print out the sum of all gross pays and the sum of all net pays after the last employee is processed.
2. Salespeople at the Pingless Auto Sales Company are paid a flat rate of $200 per month plus a commission of 10% of their total sales that are less than $20,000. If their total sales equal or exceed $20,000, they are given a $600 bonus. Write an algorithm that will calculate and print each salesperson's gross pay, the total pay for all salespersons, and the company's total sales. Assume that the number of salespersons (N) and the sales for each salesperson (SALES) are read in. Write the algorithm first using a DOWHILE loop and then using a FOR . . . NEXT loop.

In Exercises 3 through 5 trace the stated algorithm with the given data (see Examples 2–4).

3. Trace the algorithm in Example 1 of this section. Use the data below (N = 4) and determine the values of the variables E, N, SUM, HOURS, RATE, GROSS at each pass through the DOWHILE loop. What is the value of MEAN GROSS PAY?

Employee Number	HOURS	RATE
1	40	\$7.00
2	25	8.00
3	30	6.50
4	20	12.00

4. a. Correct the algorithm of Example 4 by adding the new statement

BILL = 0

between old statements 4 and 5.

b. Trace the corrected algorithm using the data of Example 4.

5. a. Trace the corrected algorithm of Exercise 4a using the following data:

Customer Number	NI	Item	Price Each Item
1	2	1	10.00
		2	12.00
2	4	1	5.00
		2	6.00
		3	17.00
		4	12.50
3	3	1	5.50
		2	7.00
		3	8.00

b. What is the value of TOTAL at statement 16?

6. Consider a collection (file) of records where each record contains three fields, A, B and C, each a real number. The problem is to calculate

$$D = B^2 - 4AC$$

for each record. Assuming a trailer of $A = 0$, write an algorithm that will print out each D.

7. Repeat Exercise 6 but this time count the number of times each of the three distinct cases ($D = 0$, $D > 0$, $D < 0$) occurs.
8. Enhance the algorithm in Exercise 7 by adding the calculation

$$S = \sqrt{D}$$

only when D is nonnegative.

9. Consider the algorithm of Example 5. Would a FOR loop structure be acceptable for this problem? If so, write the pseudocode for it; if not, indicate the reason(s).
10. Rewrite the algorithm of Example 7 using a FOR loop.

Exercises 11 through 13 require you to code algorithms.

11. Code the algorithm of Example 5 in either BASIC or Pascal.
12. Code the algorithm you wrote for Exercise 8 in BASIC or Pascal.
13. Code the algorithm you wrote for Exercise 10 (or the algorithm of Example 7) in BASIC or Pascal.

6.3 Flowcharts and Algorithms

Recall from Section 6.1 that an algorithm is a sequential set of instructions that solves a specific problem. So far we have expressed algorithms in a form called pseudocode. In this section, we will begin to study a graphic form of algorithms, **flowcharts**. Flowcharts evolved when programmers and analysts found it useful to lay out solutions and picture them. Although the flowchart is still an important data processing tool, it is losing in popularity to pseudocode.

Before we begin an example, we introduce three standard flowcharting symbols:

Oval for beginning or ending.

Parallelogram for input/output—that is, READ and WRITE statements.

Rectangle for calculations or assignments—that is, processing.

EXAMPLE 1

Construct a flowchart for finding the selling price (price plus sales tax) of an item. Assume the sales tax rate is 6%.

Solution

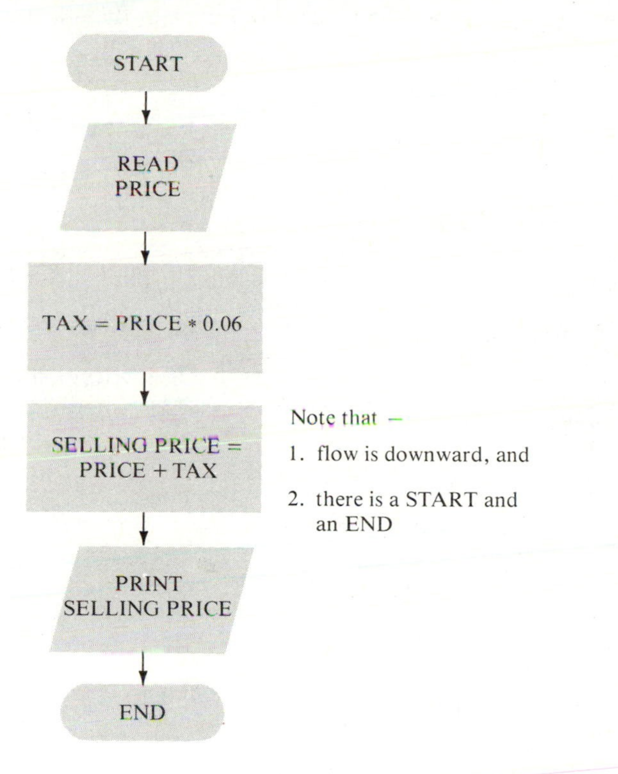

Note that –

1. flow is downward, and
2. there is a START and an END

Decision-making is a crucial part of an algorithm. When a question, such as "Is $x > 5$?," has several alternatives (but usually two: YES or NO), we use a diamond for the question:

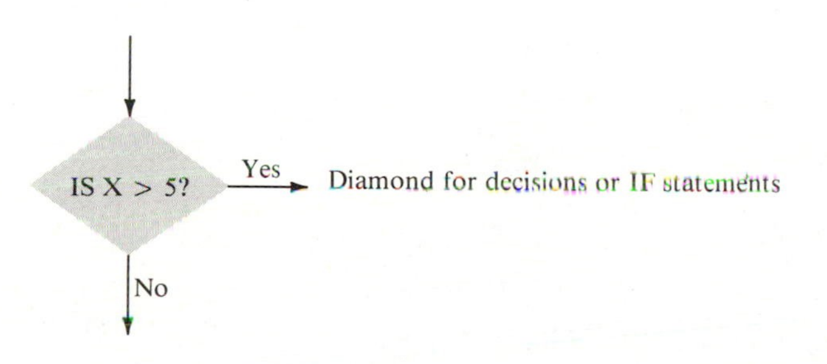

EXAMPLE 2

Refine the flowchart in Example 1 to take into account the fact that in Talahatchee County there is no sales tax for any item under $1.00.

Solution

We exhibit the flowchart, label the blocks, and explain them at the right:

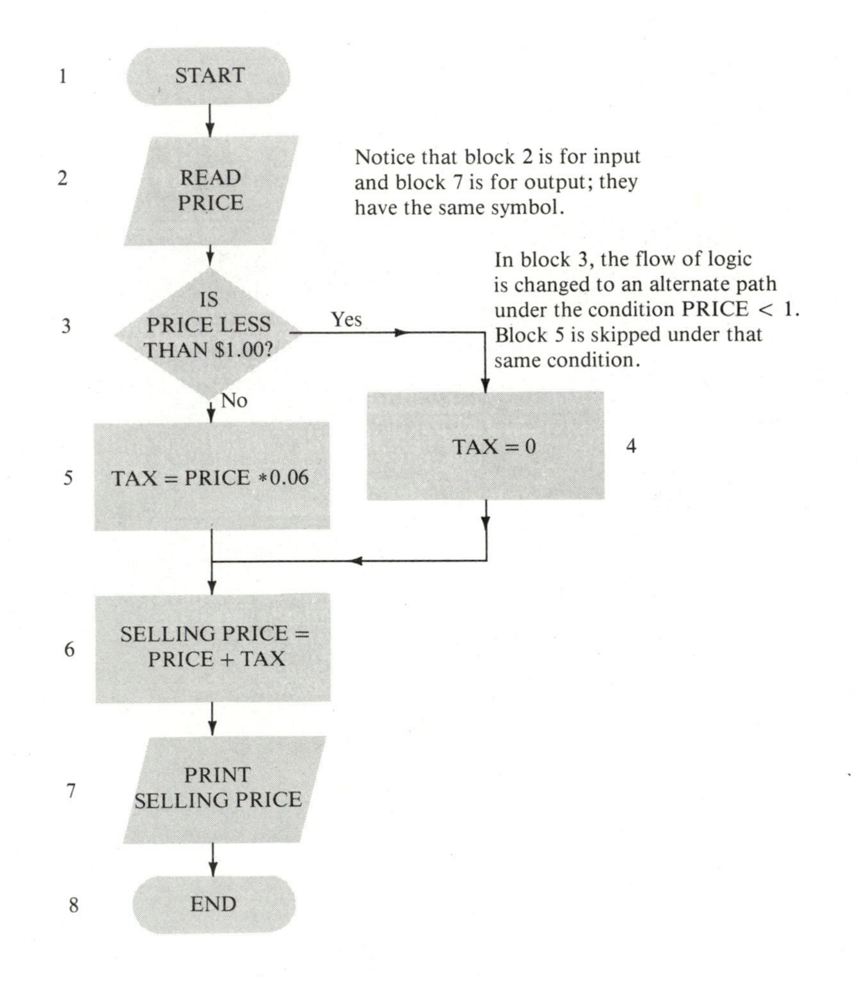

The next example makes use of the connector symbol

② Circle for indicating branching to another segment of the flowchart. Connectors are numbered or labelled in some way.

EXAMPLE 3

Write a flowchart and the pseudocode for the problem of finding the class average in a class of N students.

Solution

We will use the following variables:

S: the sum of grades
N: number of students
P: pupil number
G: grade of student
A: class average

The flowchart and pseudocode appear on page 286.

Note that in the flowchart (on page 286) for Example 3 the connector ① indicates a branch back to step 5 of the algorithm. It graphically represents the DOWHILE loop of the pseudocode version.

The real value of any algorithm, pseudocode, or flowchart is its utilization in creating a computer program. For the reader familiar with BASIC, we offer the following program using the algorithm of Example 3.

```
 10  S = 0
 20  READ N
 30  FOR P = 1 TO N
 40      READ G
 50      S = S + G
 60  NEXT P
 70  A = S/N
 80  PRINT ''CLASS AVERAGE IS'', A
 90  DATA 5, 80, 70, 90, 60, 80
100  END
```

Notice that the DATA statement appears in neither the pseudocode nor the flowchart. That's because DATA is a statement in BASIC that is really part of a READ statement and is not part of the logical solution (that is, the algorithm) of the problem.

The visual, diagrammatic form of a flowchart is apparent in the next example.

EXAMPLE 4

Construct the flowchart for the problem of printing the gross and net pay for N employees. After all employees have been processed, print the sum of all the gross pays and the sum of all the net pays (see Example 1 and Exercise 1 of Section 6.2).

FLOWCHART FOR EXAMPLE 3

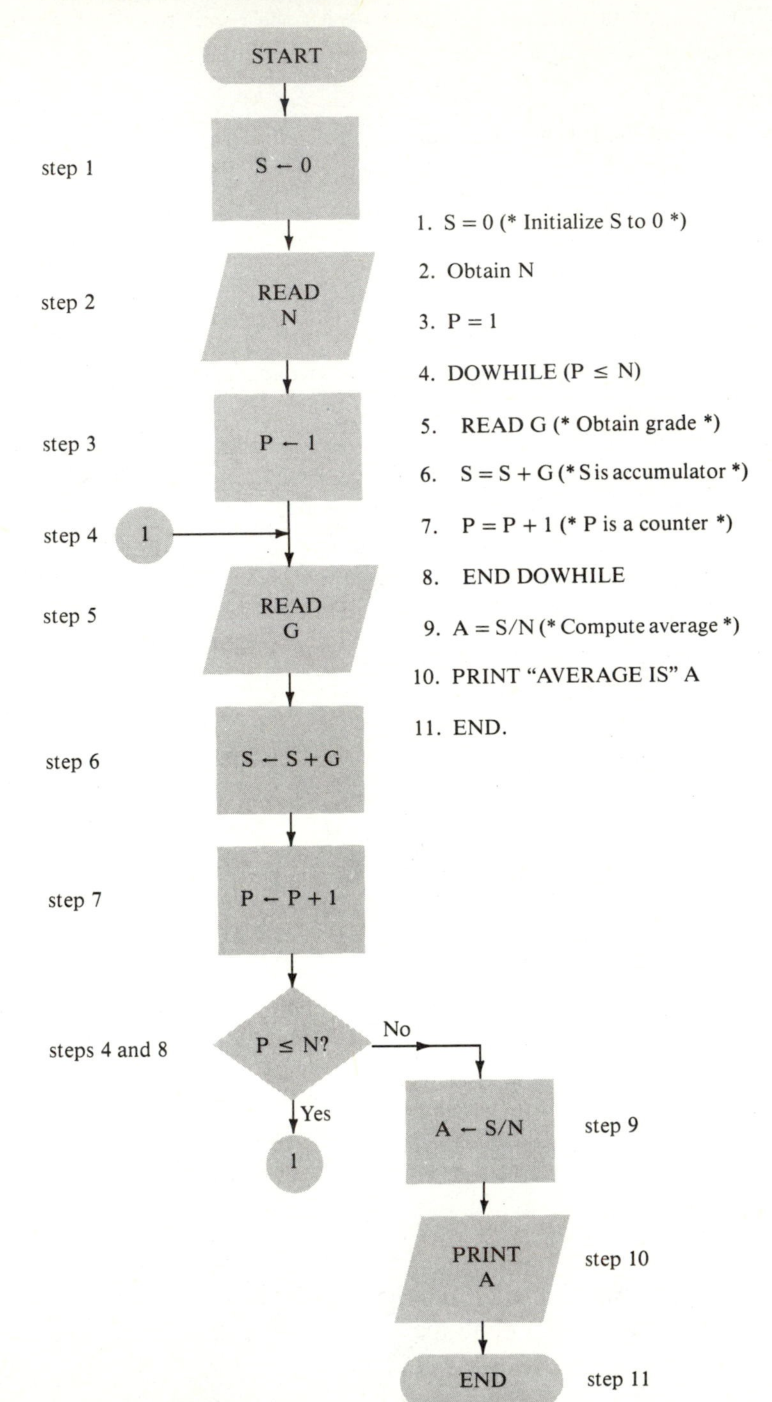

Solution (Example 4)

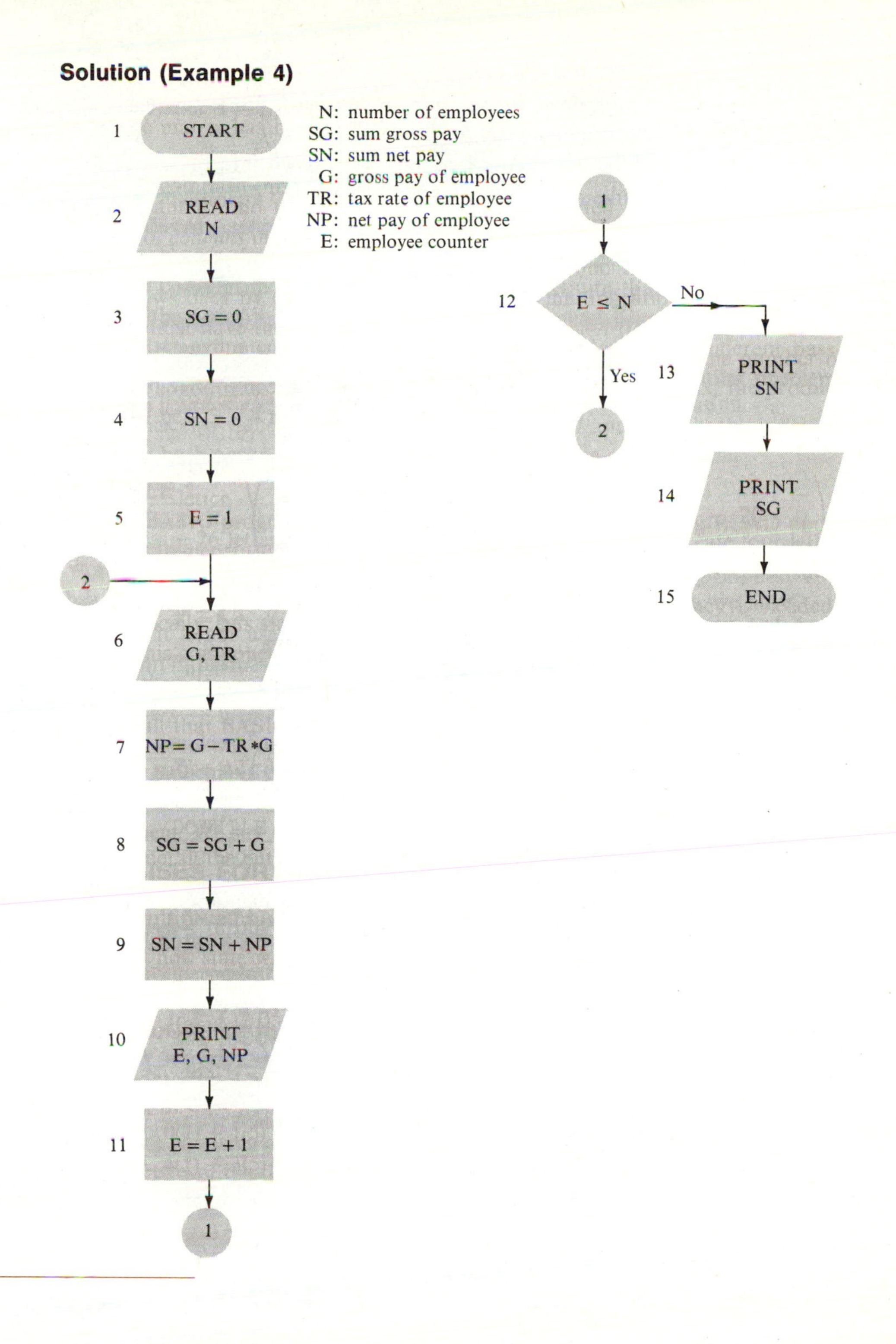

What information is conveyed by the flowchart of Example 4? Block 1 is a standard symbol for beginning a flowchart. Block 2 indicates the entering of the number of employees. A parallelogram is the symbol used for input or output (I/0). Blocks 3, 4 and 5 could be combined into one block for initialization of variables.

```
SG = 0
SN = 0
 E = 1
```

A loop begins with ②. The first instruction in the body of the loop is READ G, TR. Calculations are made in boxes 7, 8 and 9, which could have been combined as

```
NP = G - TR * G
SG = SG + G
SN = SN + NP
```

Recall that a rectangular box is used for calculations and assignments.

The body of the loop ends with block 10, and the counter is incremented in block 11 and tested in block 12 to determine whether or not to repeat the loop. Blocks 13 and 14 print the final or summary totals. Block 12, the diamond shape, is equivalent to an IF statement and is read as IF $E \leq N$:

If it is not, print SN and print SG; otherwise go back to READ G, TR (block 6).

Flowcharts are used primarily for documentation or for planning the solution to problems. Some programmers feel a visual approach is clearer than pseudocode; others disagree. The main point is that planning is advisable and often essential. Writing your solution in pseudocode or drawing flowcharts allows you to ignore the details of language syntax and particular computer requirements. You are free to concentrate on the logic and organization of the solution.

Let us consider another example.

EXAMPLE 5

Write a flowchart that illustrates the logic for finding the total sales and each customer's bill for N customers. Assume each customer buys a different number of items, NI.

Solution

First note that this is essentially the problem of Example 3 of Section 6.2.

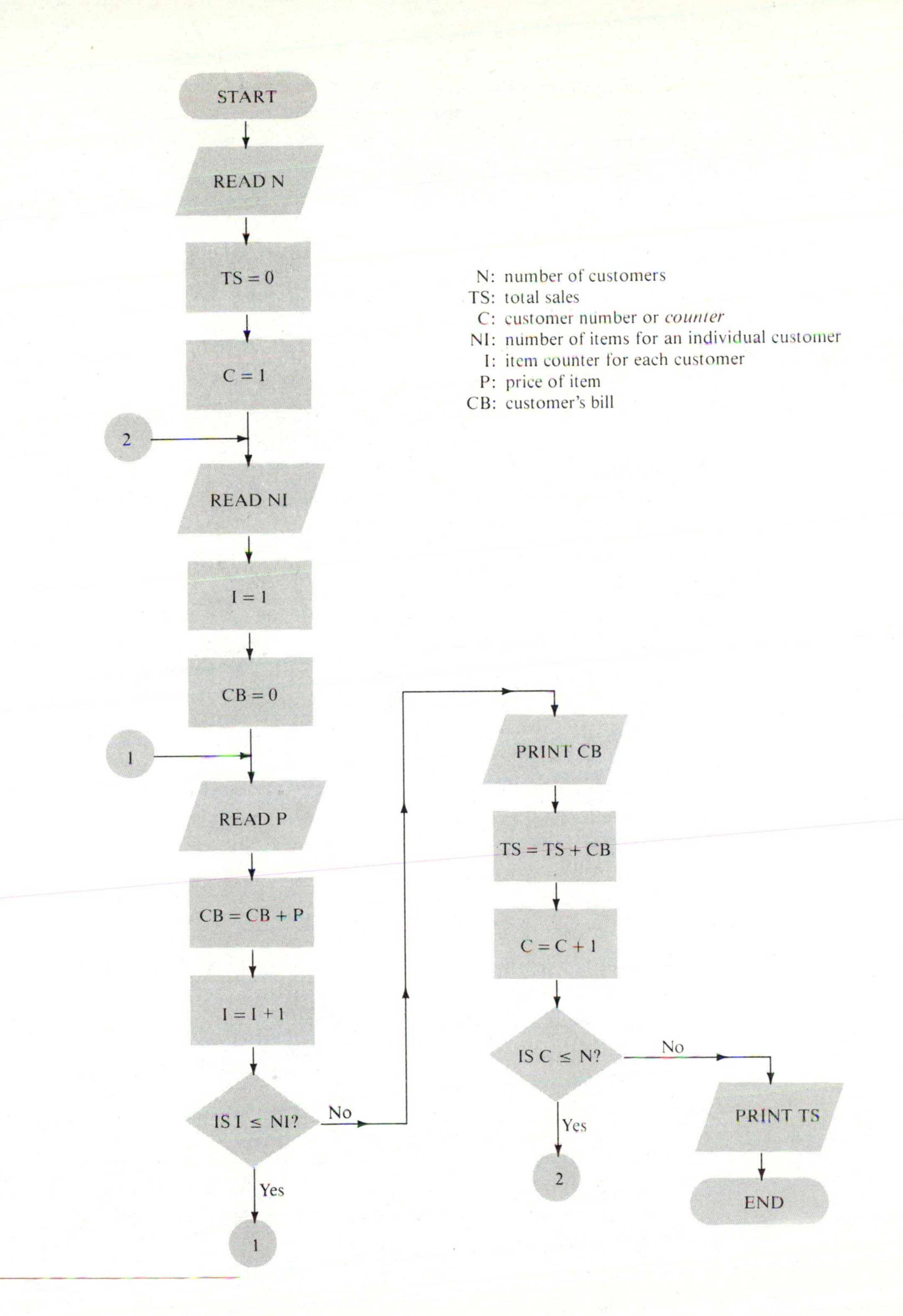
START
READ N
TS = 0
C = 1
2
READ NI
I = 1
CB = 0
1
READ P
CB = CB + P
I = I + 1
IS I ≤ NI?
No
Yes
1
PRINT CB
TS = TS + CB
C = C + 1
IS C ≤ N?
No
Yes
2
PRINT TS
END
N: number of customers
TS: total sales
C: customer number or *counter*
NI: number of items for an individual customer
I: item counter for each customer
P: price of item
CB: customer's bill

Just to emphasize the main point of this section, the flowchart is one means of conveying the essential logic of an algorithm and can be used for planning a program or documenting one. If you were asked to write a computer program to solve the problem stated in Example 3, either the flowchart or pseudocode solution should make the *coding* of the actual program relatively easy, no matter which language you choose.

EXERCISES FOR SECTION 6.3

1. Construct a flowchart to read and print the price of three items without using a loop (see Example 1).
2. Construct a flowchart to read and print the price of three items using a loop (see Example 3).
3. Modify your solution to Exercise 2 for N items instead of 3.
4. Modify the flowchart for Exercise 3 to print the total bill for the N items and the average price of the items.
5. Modify the flowchart for Example 5 to print the number of customers whose bill is over $100 as well as the total sales.
6. Suppose you want to produce a class list. Assuming that each student's name and identification number are read from DATA statements, construct a flowchart to produce a printed report consisting of a class list of students and their i.d. numbers (a) when there are 20 students and (b) when the exact number of students is unknown.
7. An honor roll is to be printed for submission to the local newspaper. Construct a flowchart to read in every student's name and average and to print a list of the names of *only* those students whose averages are above 89. Do this for the case (a) of 500 students and (b) of an unknown number of students.
8. Suppose that, as an aid in graphing the function

$$y = x^2 + \frac{5}{x}$$

you want to have a computer printout of the values of y corresponding to integer values of x, starting at $x = -10$ and ending at $x = +10$.

a. Is there any value of x between -10 and $+10$ that will cause a computer error?

b. Construct a flowchart to produce a printed list of x and y values. If there are any troublesome values in this range (part a), trap for this error in the flowchart.

9. As an aid in graphing the function

$$y = x^2 + \frac{x}{2 - x}$$

we want to produce a printout of the values of y corresponding to integer values of x from -10 to $+10$, inclusive. Construct a flowchart to do this without producing a computer error.

10. Write a flowchart for solving the following problem: N customer records are to be read in, one at a time, each consisting of customer account number, credit limit, present balance, and new charge. Update the balance by adding the new charge. Print out the customer number if the credit limit is exceeded. At the end, print out the total number of customers who exceeded their credit limits.

11. Suppose employees are categorized according to the following statements:

p: The employee is male.
q: The employee makes more than $8.00 per hour.
r: The employee has an accident-free record.

Construct a flowchart to print a list of employees who satisfy the following statement (see Chapter 4):

$$\sim p \wedge (q \vee r)$$

6.4 Chapter Review

Although it is not strictly mathematical in nature, Chapter 6 is an important one for the programming student. It treats the crucial subject of algorithm development using both pseudocode and flowcharts.

The concept of an algorithm was first presented in the context of pseudocode. The basic logical constructs for program development—including conditional and unconditional branching, initialization, loops, counters, and accumulators—were explained by the use of specific examples. The same concepts were later used in the development of flowcharts.

In this chapter the importance of algorithm development and structured programming, whether by pseudocode or flowchart, are stressed as language-independent methods of developing program logic.

VOCABULARY

algorithm
pseudocode
conditional branching
IF statement
unconditional branching
IF-THEN-ELSE
initialization
comment
loop
DOWHILE loop
counter
accumulator
trace of an algorithm
flowchart
FOR loop
loop control variable
terminal value
body of a loop
sequence
selection
repetition
radicand

CHAPTER TEST

1. Answer each of the questions below with reference to the following flowchart.
   a. Explain the purpose of each numbered box in the flowchart.
   b. Correct the flowchart by inserting the missing instruction.
   c. What is the purpose of checking X for a negative value?
   d. If the data were 8, 2, 4, 6, −2, what would be printed? Show a chart of the values of all variables at each point in the flowchart.

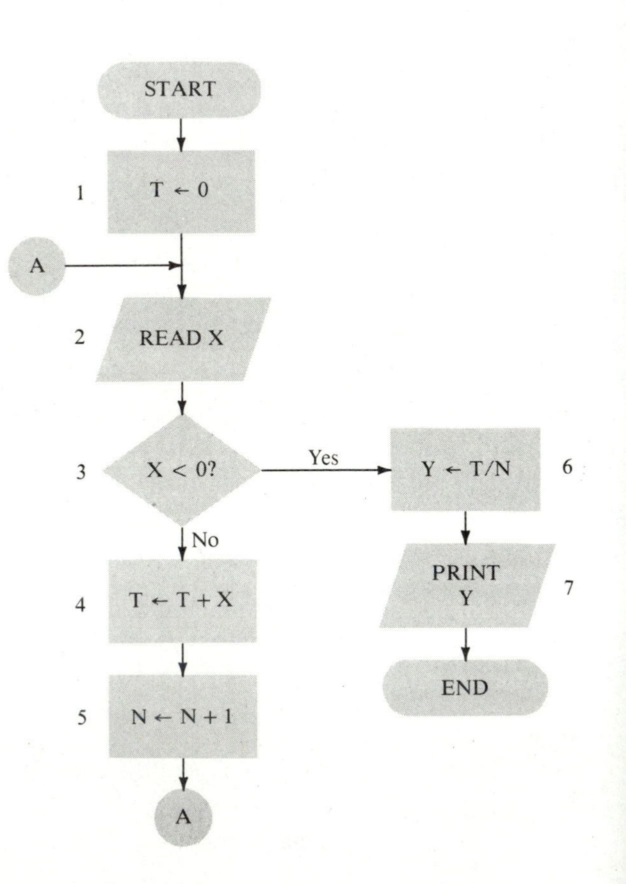

2. The formula $y = 2x + x^3/z$ is to be evaluated for a set of values. The flowchart below is a partial implementation of an algorithm to do this.

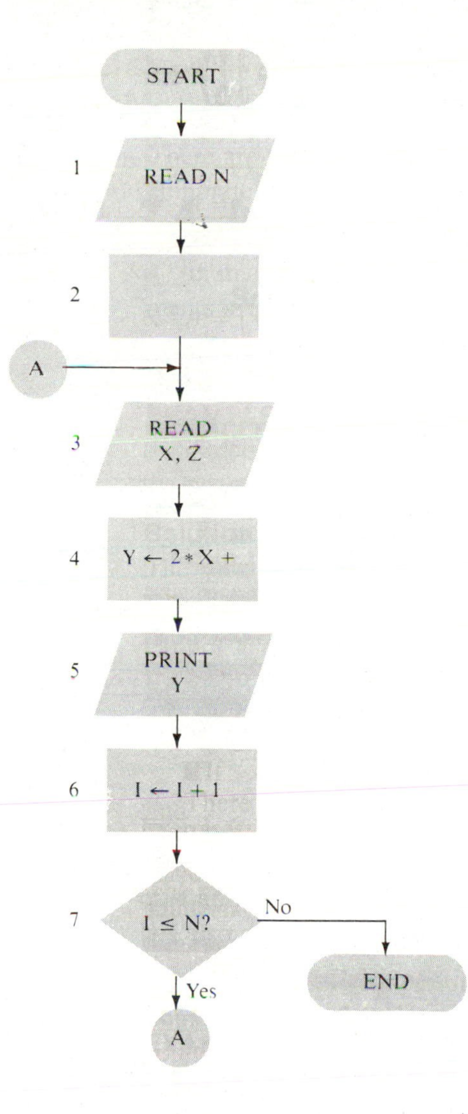

a. Explain the purpose of each numbered box.
b. Complete boxes 2 and 4.
c. Write the pseudocode for this flowchart.
d. What probably very important instruction box is omitted from the flowchart?

3. A company wishes to print the sales, salary and commission for its five salespeople. A commission chart is shown below.

Sales in Dollars	Commission as a Percentage of Sales
under 10000	0
under 20000	5%
under 40000	10%
40000 or over	15%

Using pseudocode, write an algorithm to calculate the correct commission and print the salesperson's name, sales, salary, and commission. Define all variables used. Assume the name, sales, and base salary are read in.

4. *Bonus problem.* Expand your algorithm for Problem 3 to print the final commission totals at the end of processing.

CHAPTER

7

Electronic "spreadsheets" like the VisiCalc* model offer excellent examples of how rectangular arrays of numbers, called matrices, can be applied to various business problems. The mathematics of matrices and their applications are the topic of this chapter.

IMAGINE! COMPUTER SUPPORT SERVICE
INCOME AND EXPENSE ANALYSIS

=================BUDGET TO ACTUAL =========

	JAN	FEB	MARCH	1ST QTR
*INCOME				
ESTIMATE	10500	9800	9800	30100
ACTUAL	11321	10150	10430	31901
VARIANCE	821	350.	630	1801
*EXPENSE				
ESTIMATE	2000	1800	1500	5300
ACTUAL	2318	1167	1410	4895
VARIANCE	318	-633	-90	-405
*PRE-TAX				
ESTIMATE	8500	8000	8300	24800
ACTUAL	9003	8983	9020	27006
VARIANCE	503	983	720	2206

Matrices and Subscripted Variables

* VisiCalc is a trademark of VisiCorp, Inc.

7.1 One Dimensional Arrays:
Vectors

In many data processing applications, the concept of a **matrix** (also called an **array** or a **table**) can be very useful for storing and manipulating data. In this chapter, we shall study matrices from a mathematical point of view, with frequent examples drawn from data processing applications. We begin with a discussion of one dimensional arrays, or **vectors**.

Definition of a Vector

A vector is defined as an ordered list of numbers. The numbers, or items, in the list are called the **components** of the vector. The fact that a vector is defined as an *ordered* list indicates that the order in which we list the components of a vector is important. The number of items in the list is called the **dimension** or **size** of the vector.

We shall represent a vector by a capital letter, and the components of the vector will be enclosed in parentheses. For example, the vector

$$A = (2 \quad 3 \quad 4.5 \quad -6)$$

is a vector whose dimension is 4 (there are four numbers in the list), and whose first component is 2, second component is 3, third component is 4.5, and fourth component is -6.

The components of a vector may be written either horizontally (in a row) as above, or vertically (in a column). If the components of A are listed horizontally, we call A a **row vector**. If the components of A are listed vertically,

$$A = \begin{pmatrix} 2 \\ 3 \\ 4.5 \\ -6 \end{pmatrix}$$

we call A a **column vector**. At this point in our discussion, row vectors and column vectors are essentially the same. It will, however, be necessary to distinguish between them when we discuss matrices of more than one dimension.

EXAMPLE 1

Let the vector V be given by

$$V = (12.6 \quad -5 \quad -723.8 \quad 17 \quad 42)$$

a. What is the dimension of V?
b. List the components of V.

Solution

a. The dimension of V is 5.
b.

first component	12.6
second component	−5
third component	−723.8
fourth component	17
fifth component	42

Equality of Vectors

EQUALITY OF VECTORS
Two vectors are **equal** if they have the same dimension and corresponding components are equal.

Suppose, for example, we have two vectors A and B of dimension 5. If

$$A = (a \quad b \quad c \quad d \quad e) \qquad \text{and} \qquad B = (v \quad w \quad x \quad y \quad z)$$

then

$$A = B \qquad \text{if} \quad a = v, b = w, c = x, d = y, \text{ and } e = z$$

To simplify our notation and avoid confusion, we generally use subscripted variables to refer to vector components. In the vector A, then, we would refer to the first component as a_1, the second component as a_2, the third component as a_3, and so on. The example above would then become

If

$$A = (a_1 \quad a_2 \quad a_3 \quad a_4 \quad a_5) \qquad \text{and} \qquad B = (b_1 \quad b_2 \quad b_3 \quad b_4 \quad b_5)$$

then

$$A = B \qquad \text{if} \quad a_1 = b_1, a_2 = b_2, a_3 = b_3, a_4 = b_4, \text{ and } a_5 = b_5$$

EXAMPLE 2

Let $A = (1 \quad 2 \quad 3 \quad -5)$
$B = (1 \quad 2 \quad 3)$
$C = (1 \quad 2 \quad 3 \quad 5)$
$D = (1 \quad 2 \quad 3 \quad 10/2)$

State whether each of the following is true or false:

a. $A = C$ **b.** $B = C$ **c.** $C = D$

Solution

a. $A = C$ is false. The vectors have the same dimension, but the fourth component of A (-5) is not equal to the fourth component of C $(+5)$.

b. $B = C$ is false since the two vectors have different dimensions.

c. $C = D$ is true. The vectors have the same dimension, and corresponding components are equal.

Vector Addition

Before discussing vector arithmetic, we shall present a simplified real-life application in which vectors would be useful for storing and manipulating data. We shall frequently refer to this example to help motivate and clarify explanations of vector arithmetic.

Our friend Sam, owner of Micro Edge Products, is a distributor of computer supplies. He ships five different products to the two stores, store A and store B, in a small (but growing) chain of computer outlets called Computer Heaven. In preparation for computerizing his own operation, Sam is studying ways of organizing his shipping and billing information.

Suppose the five items shipped by Micro Edge Products to the two stores are

1. $5\frac{1}{4}$-inch diskettes (box of 10)
2. 8-inch diskettes (box of 10)
3. black-and-white monitors
4. color monitors
5. boxes of $8\frac{1}{2} \times 11$-inch paper

If Sam wants to keep track of the number of units of each item shipped to each store, he can do so by using a vector for store A (denoted by A) and a vector for store B (denoted by B). The first component of each vector is the number of units of item 1 shipped, the second component is the number of units of item 2 shipped, and so on.

Suppose Sam makes the following shipments:

to store A: 10 boxes of $5\frac{1}{4}$-inch diskettes,
5 boxes of 8-inch diskettes,
4 black-and-white monitors,
1 color monitor, and
12 boxes of paper;

to store B: 15 boxes of $5\frac{1}{4}$-inch diskettes,
0 boxes of 8-inch diskettes,
2 black-and-white monitors,
6 color monitors, and
8 boxes of paper.

The vectors representing these shipments are

$$A = (10 \quad 5 \quad 4 \quad 1 \quad 12)$$
$$B = (15 \quad 0 \quad 2 \quad 6 \quad 8)$$

Note that if a store received no units of a given item, we indicate this by putting a zero in the appropriate place in the vector.

Now Sam wants to calculate the total number of units of each item shipped to all Computer Heaven stores. He can accomplish this by adding the corresponding components of vector A and vector B, creating a new vector, the *sum* of A and B, denoted by $A + B$.

Thus,

$$\begin{aligned} A + B &= (10 \quad 5 \quad 4 \quad 1 \quad 12) + (15 \quad 0 \quad 2 \quad 6 \quad 8) \\ &= (10 + 15 \quad 5 + 0 \quad 4 + 2 \quad 1 + 6 \quad 12 + 8) \end{aligned}$$

So

$$A + B = (25 \quad 5 \quad 6 \quad 7 \quad 20)$$

ADDITION OF VECTORS
The *sum* of two row (or column) vectors, A and B, *of the same dimension,* is the vector formed by adding the corresponding components of A and B.

Note that the sum of two vectors can be found only if the vectors are of the same dimension.

EXAMPLE 3

Let $R = (5 \quad 6 \quad -11 \quad 4.1)$ $\quad S = (0 \quad 0 \quad 0 \quad 9)$ $\quad T = (1 \quad 3 \quad 5 \quad 6)$

$$U = \begin{pmatrix} 9 \\ 3 \\ 0.5 \\ 8.6 \end{pmatrix} \qquad V = \begin{pmatrix} 5 \\ 7 \\ 9 \\ 11 \end{pmatrix} \qquad W = \begin{pmatrix} 1 \\ 2 \\ 3 \end{pmatrix}$$

Find (if possible):

a. $R + S$ **b.** $R - T$ **c.** $R + W$ **d.** $U + V$
e. $V + W$ **f.** $T - R$ **g.** $V - U$ **h.** $U - V$

Solution

a.
$$\begin{aligned} R + S &= (5 \quad 6 \quad -11 \quad 4.1) + (0 \quad 0 \quad 0 \quad 9) \\ &= (5+0 \quad 6+0 \quad -11+0 \quad 4.1+9) \\ &= (5 \quad 6 \quad -11 \quad 13.1) \end{aligned}$$

b. To subtract one vector from another, we simply subtract corresponding components.

$$\begin{aligned} R - T &= (5 \quad 6 \quad -11 \quad 4.1) - (1 \quad 3 \quad 5 \quad 6) \\ &= (5-1 \quad 6-3 \quad -11-5 \quad 4.1-6) \\ &= (4 \quad 3 \quad -16 \quad -1.9) \end{aligned}$$

c. $R + W$ has no meaning, because vectors R and W have different dimensions. Also, we usually add row vectors to row vectors or column vectors to column vectors. This helps to avoid confusion when we discuss matrices of more than one dimension.

d.
$$U + V = \begin{pmatrix} 9 \\ 3 \\ 0.5 \\ 8.6 \end{pmatrix} + \begin{pmatrix} 5 \\ 7 \\ 9 \\ 11 \end{pmatrix} = \begin{pmatrix} 14 \\ 10 \\ 9.5 \\ 19.6 \end{pmatrix}$$

e. $V + W$ has no meaning because V and W have different dimensions.

f.
$$\begin{aligned} T - R &= (1 \quad 3 \quad 5 \quad 6) - (5 \quad 6 \quad -11 \quad 4.1) \\ &= (1-5 \quad 3-6 \quad 5-(-11) \quad 6-4.1) \\ &= (-4 \quad -3 \quad 16 \quad 1.9) \end{aligned}$$

g.
$$V - U = \begin{pmatrix} 5 \\ 7 \\ 9 \\ 11 \end{pmatrix} - \begin{pmatrix} 9 \\ 3 \\ 0.5 \\ 8.6 \end{pmatrix} = \begin{pmatrix} -4 \\ 4 \\ 8.5 \\ 2.4 \end{pmatrix}$$

h.
$$U - V = \begin{pmatrix} 9 \\ 3 \\ 0.5 \\ 8.6 \end{pmatrix} - \begin{pmatrix} 5 \\ 7 \\ 9 \\ 11 \end{pmatrix} = \begin{pmatrix} 4 \\ -4 \\ -8.5 \\ -2.4 \end{pmatrix}$$

Multiplication of a Vector by a Scalar

In matrix and vector terminology, a number is frequently referred to as a **scalar**. When we talk about multiplying a vector by a scalar, then, we are simply referring to multiplying the vector by some number.

Let's get back to Sam of Micro Edge Products. If Sam wants to send bills to the stores in the Computer Heaven chain, he needs a list of the prices of the various items he ships. For convenience, he puts these prices into a vector which we'll call C, with the price of item 1 in position 1, the price of item 2 in position 2, and so on. The prices in the C vector will be in the same relative positions as the quantities ordered in the A and B vectors.

Suppose Sam's price list is as follows:

item 1	$5\frac{1}{4}$-inch diskettes	\$20.00 per box
item 2	8-inch diskettes	\$35.00 per box
item 3	black-and-white monitors	\$100.00 each
item 4	color monitors	\$250.00 each
item 5	$8\frac{1}{2} \times 11$-inch paper	\$25.00 per box

Then the vector containing the prices is

$$C = (20 \quad 35 \quad 100 \quad 250 \quad 25)$$

Now, Sam has decided that he wants to encourage his customers to order more color monitors. He offers a 5% discount on all items purchased to any individual store that orders more than five color monitors in one shipment. He may find it useful to build a vector D to hold the discounts. Each component of D will be obtained by multiplying the corresponding component of C by the scalar (number) 0.05. We write

$$D = 0.05 \cdot C \text{ or } D = 0.05C$$

The vector D is

$$\begin{aligned} D &= 0.05(20 \quad 35 \quad 100 \quad 250 \quad 25) \\ &= (0.05 \cdot 20 \quad 0.05 \cdot 35 \quad 0.05 \cdot 100 \quad 0.05 \cdot 250 \quad 0.05 \cdot 25) \end{aligned}$$

So

$$D = (1 \quad 1.75 \quad 5 \quad 12.50 \quad 1.25)$$

The price vector, $C1$, for a store that qualifies for a discount, will then be $C - D$. So

$$\begin{aligned} C1 &= (20 \quad 35 \quad 100 \quad 250 \quad 25) - (1 \quad 1.75 \quad 5 \quad 12.50 \quad 1.25) \\ &= (19 \quad 33.25 \quad 95 \quad 237.50 \quad 23.75) \end{aligned}$$

In general, the product of a scalar s times a vector V is defined to be the vector $s \cdot V$ obtained by multiplying each component of V by the number s.

EXAMPLE 4

Let $V = (5 \quad 6 \quad 7 \quad 8)$

Find:

a. $2 \cdot V$ **b.** $0.5 \cdot V$ **c.** $-1 \cdot V$ **d.** $-3 \cdot V$

Solution

a. $2 \cdot (5 \quad 6 \quad 7 \quad 8) = (2 \cdot 5 \quad 2 \cdot 6 \quad 2 \cdot 7 \quad 2 \cdot 8)$
$= (10 \quad 12 \quad 14 \quad 16)$

b. $0.5 \cdot (5 \quad 6 \quad 7 \quad 8) = (0.5 \cdot 5 \quad 0.5 \cdot 6 \quad 0.5 \cdot 7 \quad 0.5 \cdot 8)$
$= (2.5 \quad 3 \quad 3.5 \quad 4)$

c. $-1 \cdot (5 \quad 6 \quad 7 \quad 8) = (-1 \cdot 5 \quad -1 \cdot 6 \quad -1 \cdot 7 \quad -1 \cdot 8)$
$= (-5 \quad -6 \quad -7 \quad -8)$

d. $-3 \cdot (5 \quad 6 \quad 7 \quad 8) = (-3 \cdot 5 \quad -3 \cdot 6 \quad -3 \cdot 7 \quad -3 \cdot 8)$
$= (-15 \quad -18 \quad -21 \quad -24)$

Vector Multiplication

Finally, Sam has to prepare an invoice to send to store A with its shipment. This invoice must, of course, include the total cost of the items shipped. The information needed to calculate this amount is contained in the vector A, which contains the number of units of each item shipped to store A, and the vector C, which contains the prices of the items. Recall that A and C are

$$A = (10 \quad 5 \quad 4 \quad 1 \quad 12)$$
$$C = (20 \quad 35 \quad 100 \quad 250 \quad 25)$$

To determine the invoice total, we must first calculate the total cost of each individual item (number of units $\cdot$ price per unit) and then add these amounts. This number will turn out to be the *product* of the vectors A and C denoted by $A \cdot C$. To be consistent with our later discussion of matrices, we will first write

C as a column vector:

$$C = \begin{pmatrix} 20 \\ 35 \\ 100 \\ 250 \\ 25 \end{pmatrix}$$

Also, convention requires that we write the row vector first. Then, the invoice total is given by

$$\begin{aligned} A \cdot C &= (10 \quad 5 \quad 4 \quad 1 \quad 12) \cdot \begin{pmatrix} 20 \\ 35 \\ 100 \\ 250 \\ 25 \end{pmatrix} \\ &= (10 \cdot 20) + (5 \cdot 35) + (4 \cdot 100) + (1 \cdot 250) + (12 \cdot 25) \\ &= 200 + 175 + 400 + 250 + 300 \\ &= 1325 \end{aligned}$$

Thus, the invoice total for store A, before shipping charges, is \$1325.

PRODUCT OF VECTORS

In general, the product of two vectors A and B, of the *same dimension*, denoted by $A \cdot B$, is the product of the first components plus the product of the second components plus the product of the third components, and so on.

Observe that the result of vector multiplication is a number (scalar), *not* a vector.

EXAMPLE 5

Find the total invoice amount, before shipping charges, for store B above.

Solution

The vector containing the amounts shipped to store B is

$$B = (15 \quad 0 \quad 2 \quad 6 \quad 8)$$

Since store B ordered six color monitors, it qualifies for a discount. The price vector is

$$C1 = \begin{pmatrix} 19 \\ 33.25 \\ 95 \\ 237.50 \\ 23.75 \end{pmatrix}$$

The total invoice amount is

$$B \cdot C1 = (15 \quad 0 \quad 2 \quad 6 \quad 8) \cdot \begin{pmatrix} 19 \\ 33.25 \\ 95 \\ 237.50 \\ 23.75 \end{pmatrix}$$

$$\begin{aligned} &= (15 \cdot 19) + (0 \cdot 33.25) + (2 \cdot 95) \\ &\quad + (6 \cdot 237.50) + (8 \cdot 23.75) \\ &= 285 + 0 + 190 + 1425 + 190 \\ &= 2090 \end{aligned}$$

The total invoice amount for store B, before shipping charges, is $2090.

EXAMPLE 6

Let

$$A = (1 \quad 5 \quad -3 \quad 7) \qquad B = \begin{pmatrix} 2 \\ 4 \\ 6 \\ 8 \end{pmatrix} \qquad C = \begin{pmatrix} 1 \\ 3 \\ 5 \\ 7 \end{pmatrix} \qquad D = \begin{pmatrix} 2 \\ 6 \\ 5 \end{pmatrix}$$

Find (if possible):

a. $A \cdot B$ **b.** $A \cdot C$ **c.** $A \cdot D$

Solution

a. $A \cdot B = (1 \quad 5 \quad -3 \quad 7) \cdot \begin{pmatrix} 2 \\ 4 \\ 6 \\ 8 \end{pmatrix}$

$$\begin{aligned} &= (1 \cdot 2) + (5 \cdot 4) + (-3 \cdot 6) + (7 \cdot 8) \\ &= 2 + 20 + (-18) + 56 = 60 \end{aligned}$$

b. $A \cdot C = (1 \quad 5 \quad -3 \quad 7) \cdot \begin{pmatrix} 1 \\ 3 \\ 5 \\ 7 \end{pmatrix} = (1 \cdot 1) + (5 \cdot 3) + (-3 \cdot 5) + (7 \cdot 7)$

$$= 1 + 15 + (-15) + 49 = 50$$

c. $A \cdot D$ has no meaning because vector multiplication is defined *only* on a row vector and a column vector of the same dimension.

TABLE 7.1

Programming statements that establish arrays.

Language	Statement	Explanation
BASIC	`10 DIM V(100)`	Establishes an array of 100 real numbers, V(1) through V(100)
	`20 DIM X%(50)`	Establishes an array of 50 integers, X%(1) through X%(50)
	`30 DIM A$(20)`	Establishes 20 string variables, A$(1) through A$(20)*
FORTRAN	`DIMENSION COST(5)`	Establishes an array of 5 real numbers, COST(1) through COST(5)
	`DIMENSION I(6), K(10)`	Establishes two integer-variable arrays, I(1) through I(6) and K(1) through K(10)
Pascal	`VAR VECTOR: ARRAY[1 . . 20] OF INTEGER`	Establishes an array called VECTOR indexed by the integers 1 through 20. Each component in VECTOR is an integer
	`VAR GRADE: ARRAY['A' . . 'F'] OF REAL`	Establishes an array variable called GRADE indexed by the letters A, B, C, D, E, F, where each component is a real number
COBOL	`02 NAME OCCURS 100 TIMES`	The variable NAME is indexed by the integers 1 through 100.

* The use of the % to define an integer variable is not available in all versions of BASIC. Also, string variables are handled quite differently in different versions of BASIC, so DIM A$(20) may create *one* string of length 20 characters. Finally, most versions of BASIC will actually create 101 storage locations with DIM V(100): V(0) through V(100)!

The array data type is a very important tool in programming. By indexing a variable (usually with integers), we can use the index notation to manipulate the variables. In BASIC, for example, early in a program we can use a DIM statement to establish the size or length of an array

```
10  DIM X(20)
```

This statement creates twenty variables, $X_1, X_2, X_3, \ldots, X_{20}$, which would appear in the program as X(1), X(2), . . . , X(20).

The utility of the subscripts is shown in the abridged BASIC program below where the vector multiplication of the Micro Edge Corporation's billing is done:

```
 10  DIM A(5), C(5)
  :
 90  TOTAL = 0
100  FOR I = 1 TO 5
110  LET M = A(I) * C(I)
120  TOTAL = TOTAL + M
130  NEXT I
  :
200  PRINT "TOTAL INVOICE = "; TOTAL
```

The distinction between row and column vectors is not made in the program above, but it will be addressed later in this chapter. See also Exercise 36 in this section.

Table 7.1 provides sample statements from BASIC, FORTRAN, COBOL, and Pascal, each establishing various arrays.

EXERCISES FOR SECTION 7.1

In Exercises 1 through 6 state the dimension of each vector (see Example 1).

1. $A = (5 \quad -7 \quad 3)$

2. $B = (1 \quad 2 \quad 3 \quad 4 \quad 6)$

3. $C = \begin{pmatrix} 5 \\ -7 \\ 18 \\ 0 \\ 0 \end{pmatrix}$

4. $D = \begin{pmatrix} -1 \\ -2 \\ -3 \\ 0 \\ -5 \end{pmatrix}$

5. $X = (-20 \quad 5 \quad 7 \quad \frac{1}{2} \quad 0.8)$

6. $Y = (1 \quad 5 \quad 10 \quad 31 \quad 87 \quad 61 \quad 121)$

In Exercises 7 through 12 use vectors R, S, T, U and V below to answer "true" or "false" to each statement (see Example 2).

$$R = (8 \quad 6 \quad 10 \quad 12) \qquad S = (6 \quad 8 \quad 10 \quad 12)$$

$$T = (0 \quad 6 \quad 8 \quad 10 \quad 12) \qquad U = \begin{pmatrix} 0 \\ 6 \\ 8 \\ 10 \\ 12 \end{pmatrix} \qquad V = \begin{pmatrix} 6 \\ 8 \\ 10 \\ 12 \end{pmatrix}$$

7. $R = S$ **8.** $S = T$ **9.** $T = U$
10. $U = V$ **11.** $S = V$ **12.** $\frac{1}{2}(S + S) = S$

In Exercises 13 through 30 use vectors A through F below to find, where possible, the result of the indicated operation (see Examples 3–6).

$$A = (1 \quad 2 \quad 3 \quad 4 \quad 5) \qquad B = (0 \quad 5 \quad 10 \quad 12 \quad 14)$$

$$C = \begin{pmatrix} 8 \\ 5 \\ -10 \\ 6 \\ 0 \end{pmatrix} \qquad D = \begin{pmatrix} 1 \\ 2 \\ 3 \\ 4 \end{pmatrix}$$

$$E = \begin{pmatrix} 0 \\ 0 \\ 0 \\ 3 \end{pmatrix} \qquad F = (0.1 \quad 0.2 \quad 0.3 \quad 0.4)$$

13. $A + B$ **14.** $A - B$ **15.** $B - A$
16. $A + C$ **17.** $D + E$ **18.** $\frac{1}{2}B$
19. $\frac{1}{2}B + A$ **20.** $\frac{1}{2}(B + A)$ **21.** $6(D + E)$
22. $A \cdot C$ **23.** $F \cdot D$ **24.** $F \cdot E$
25. $B \cdot C$ **26.** $C \cdot A$ **27.** $(A + B) \cdot C$
28. $A \cdot C + B \cdot C$ **29.** $(4A) \cdot C$ **30.** $(A + B) \cdot (2C)$

31. Nelson was in a diving competition and chose six dives, each with a degree of difficulty as described below:

Dive	Degree of Difficulty
1. Forward dive, layout position	1.2
2. Backward dive, tuck position	1.3
3. Reverse dive, two somersaults, pike position	2.6
4. Inward dive, layout position	1.6
5. Inward $1\frac{1}{2}$ somersaults, tuck position	2.1
6. Backward dive with half twist, pike position	1.9

The judges awarded him the following scores:

Dive	Score
1	28
2	26
3	30
4	30
5	42
6	28

a. Represent the degree of difficulty information by a row vector X.

b. Represent the scores information by a column vector Y.

c. Calculate Nelson's total score, $X \cdot Y$.

32. Phil is on a diet and is restricted to nine foods. The chart below depicts the food, its calorie content, and number of servings Phil had last Friday:

Food	Calories per Serving	Friday's Food Consumption
apple	70	2
celery	5	1
lima beans	260	3
steak	240	4
whole wheat bread	55	2
grapefruit	55	5
liver	60	0
cabbage	25	0
water	0	10

Using a row vector for calories and a column vector for number of servings, find Phil's total calorie intake for Friday by using vector multiplication.

33. Last semester Roseann received two As, one B, one C, and one F. If an A is worth four points, a B is worth three points, a C is worth two points, a D is worth 1 point and an F is worth zero points, find Roseann's total points by vector multiplication.

34. A stockbroker is about to purchase some stock as follows:

Stock	Number of Shares	Price per Share
General Motors	10	$78.50
United Park Mines	100	3.75
SFN Industries	200	38.25
Bally Manufacturing	100	23.50

a. Find the total cost of stock by using a row vector A for the number of shares, a column vector B for price per share, and computing $A \cdot B$.
b. The stockbroker predicts a 25% increase in all stock. Find the projected worth of stock, $A \cdot (1.25B)$.

35. The Adams family made the following purchases last week:

Item	Quantity	Cost per Item
color TV	1	$375.00
shirts	5	15.00
grass seed	2	3.00
gasoline	50	1.50
baseballs	2	2.00

a. Using a row vector for quantity Q, a column vector for cost C, find their subtotal $Q \cdot C$.
b. Assuming a 7% sales tax rate, find the total spent, $Q \cdot (1.07C)$.
c. Show that $1.07(Q \cdot C) = Q \cdot (1.07C)$.

36. The BASIC program at the end of Section 7.1 was somewhat abbreviated because it did not read in any variables' values. Run the following variation of it to do the calculation for Micro Edge developed in this section.

```
 10  DIM A(5),C(5)
 20  REM                  . . . STATEMENTS 30–50 READ IN THE
 21  REM                        5 VALUES OF A
 22  REM                          FROM DATA STATEMENT 990
 30  FOR I = 1 TO 5
 40    READ A(I)
 50  NEXT I
 60  REM                  . . . STATEMENTS 70–90 READ IN
 61  REM                        5 VALUES OF C
 65  REM                          FROM DATA STATEMENT 991
 70  FOR I = 1 TO 5
 80    READ C(I)
 90  NEXT I
 95  LET SUM = 0
100  FOR I = 1 TO 5
110    LET M = A(I) * C(I)
120    LET SUM = SUM + M
130  NEXT I
```

```
 200  PRINT "TOTAL INVOICE = "; SUM
 990  DATA    10,5,4,1,12
 991  DATA    20,35,100,250,25
9999  END
```

37. Refine the BASIC program in Exercise 36 to accept INPUT values of A and C.
38. Generalize the BASIC program of Exercise 36 to multiply vectors of dimension n where $2 \le n \le 50$.

7.2 Two Dimensional Arrays: Matrices

As we have seen, vectors can be very useful tools for storing data in an organized way. When we have to store large amounts of information, however, the vector (or list) concept may not be adequate.

We can return to our friend Sam for an example of such a situation. Sam's business is expanding rapidly, in large part due to the rapid expansion of his favorite customer, the Computer Heaven chain of stores. The chain has grown from the original two stores, A and B, to a chain of 50 stores, one in every state. Sam could, if he chose to, record the items ordered by these stores in 50 separate vectors, but it would be much more convenient if he recorded this information in a table like the one below.

Store Number	Item Number 1	2	3	4	5
1	10	2	0	1	5
2	8	3	2	6	4
3	0	7	10	3	0
4					
⋮	⋮	⋮	⋮	⋮	⋮
50					

In each row of the table he records the quantities of the individual items ordered by the store whose number corresponds to the number of that row. The quantities ordered by store 1 are in row 1; the quantities ordered by store 2 are in row 2; and so on. For example, the table indicates that store 3 ordered 10 units of item 3 and store 2 ordered 6 units of item 4. Also, Sam can find the total number of units of any particular item he must ship by adding the entries in the column for that item.

Mathematicians refer to such a table as a **matrix**. That is, a matrix is defined to be a rectangular arrangement (array, or table) of numbers. The entries in a matrix, referred to as **components**, are arranged in rows and columns. In computer programming applications, this allows for rapid access to information; any entry in a matrix can be referenced—pointed to—by identifying the row and column in which it resides. As with vectors, the entries in a matrix are enclosed in parentheses. For example, consider the following matrix:

$$M = \begin{pmatrix} 2 & 3 & 5 & 7 \\ -10 & 12 & 1 & 0 \\ 5 & -5 & 6 & -7.2 \end{pmatrix}$$

Notice that M contains 12 elements (entries, components) arranged in three rows and four columns. We say that M is a 3×4 (read "3 by 4") matrix, and 3×4 are called the dimensions of the matrix M. In general, the dimensions of a matrix are given by (number of rows)×(number of columns). The matrix Sam uses to store his order information for the Computer Heaven chain has 50 rows and 5 columns, so it is a 50×5 matrix.

EXAMPLE 1

What are the dimensions of the following matrices?

$$B = \begin{pmatrix} 1 & 0 \\ 0 & 0 \end{pmatrix} \quad C = \begin{pmatrix} 5 & -4 & 5 \\ -11 & 0 & -1 \\ -17 & 0 & -1 \end{pmatrix} \quad D = \begin{pmatrix} 1 & 9 & 0 & 0 & 6 \end{pmatrix}$$

$$E = \begin{pmatrix} 5 \\ 9 \\ -11 \\ 1.6 \\ 1.7 \end{pmatrix} \quad F = \begin{pmatrix} 2 & 2 & 2 \\ 1 & 1 & 1 \end{pmatrix}$$

Solution

B is a 2×2 matrix
C is a 3×3 matrix
D is a 1×5 matrix
E is a 5×1 matrix
F is a 2×3 matrix

Matrices D and E in Example 1 should look familiar; we saw matrices like them in Section 7.1, where they were described as vectors. In fact, a matrix (like

D in Example 1) that has exactly one row is a *row vector*; and a matrix like E in Example 1 with exactly one column is a *column vector*. Row and column vectors are, then, special types of matrices.

Matrices like B and C in Example 1 with the same number of rows as columns, are called **square matrices**. They will have a special role to play in our discussion of matrix arithmetic.

EQUALITY OF MATRICES

Two matrices are said to be *equal* if they have the same dimensions and the corresponding components are equal.

We noted earlier that computer programming applications frequently make use of matrices because of the ease with which data in a matrix can be accessed. We can find a given matrix component if we know its row number and its column number. Recall that in discussing vectors, we mentioned that the components of a vector can be listed in terms of subscripted variables; the subscript of each component identifies its position in the vector. Thus if $A = (a_1\ a_2\ a_3\ a_4\ a_5 \cdots a_n)$, where n is some positive integer, then the dimension of A (a *one dimensional array*) is n; and the first component of A is represented by a_1, the second by a_2, the third by a_3, and so on. Also, if $B = (b_1\ b_2\ b_3\ b_4 \cdots b_n)$, the sum $A + B$ can be represented by

$$A + B = (a_1 + b_1 \quad a_2 + b_2 \quad a_3 + b_3 \quad \cdots \quad a_n + b_n)$$

We are simply extending the concepts we discussed in Section 7.1 to vectors of any dimension.

A similar notation can be used to represent components of matrices with several rows and columns. We said that we could identify a component of a matrix by its *row* and *column* position. This means that to pinpoint a specific element of a matrix we will need *two* subscripts, one to identify the row and one to identify the column. For the sake of uniformity, the subscript identifying the row will always be listed first. This notation allows us to represent an $m \times n$ matrix A (a matrix with m rows and n columns) as follows:

$$A = \begin{pmatrix} a_{11} & a_{12} & a_{13} & a_{14} & \cdots & a_{1n} \\ a_{21} & a_{22} & a_{23} & a_{24} & \cdots & a_{2n} \\ a_{31} & a_{32} & a_{33} & a_{34} & \cdots & a_{3n} \\ \vdots & \vdots & \vdots & \vdots & & \vdots \\ a_{m1} & a_{m2} & a_{m3} & a_{m4} & \cdots & a_{mn} \end{pmatrix}$$

EXAMPLE 2

Let $B = \begin{pmatrix} 2 & 4 & 0 & 6 & -1 \\ -5 & 2 & 1 & 0 & 3 \\ 9 & 17 & 4.6 & -7 & -8 \end{pmatrix}$

Identify the following components of B:

a. b_{13} **b.** b_{31} **c.** b_{25} **d.** b_{33}

Solution

a. b_{13} is in row 1, column 3; $b_{13} = 0$
b. b_{31} is in row 3, column 1; $b_{31} = 9$
c. b_{25} is in row 2, column 5; $b_{25} = 3$
d. b_{33} is in row 3, column 3; $b_{33} = 4.6$

Although mathematicians use lowercase letters with double subscripts to represent components of a matrix, computer programmers use a slightly different notation because subscripts can't be typed as easily as ordinary letters and numbers on a computer keyboard. A programmer would identify the element of the matrix B above in row 1, column 3 as B(1, 3), instead of b_{13}. Similarly, she would refer to the element of row 2, column 5 as B(2, 5). This convention is simply writing the mathematician's subscripts in parentheses next to the name of the matrix; the convention of always listing the row number first still holds. In fact,

TABLE 7.2
Statements that establish matrices.

Language	Statement	Explanation
BASIC	`10 DIM X(5,6), Y(2,3)`	Establishes two matrices, X having 30 real number components (5 rows, 6 columns) and Y having 6 components (2 rows, 3 columns)
FORTRAN	`DIMENSION TABLE(5,20)`	Establishes a 100-component matrix called TABLE that has 5 rows and 20 columns
Pascal	`VAR CHART: ARRAY[1 . . 5, 1 . . 10] OF INTEGER`	Establishes an array (matrix) variable called CHART with 50 components, 5 rows and 10 columns; each component must be an integer.

to *establish* a 3×5 matrix B as in Example 2 in BASIC we would incorporate the following DIM statement

```
10  DIM B(3,5)
```

If we write a calculation statement, however, such as

```
100  LET Y = 6 * B(3,5)
```

we are multiplying 6 by the *value of the component* in the third row, fifth column.

We show in Table 7.2 selected languages' ways of establishing matrices (two dimensional arrays) similar to the table we presented for one dimensional arrays, vectors, in Section 7.1 (Table 7.1).

EXERCISES FOR SECTION 7.2

In Exercises 1 through 6, determine the dimensions of each matrix (see Example 1).

1. $P = \begin{pmatrix} 4 & -1 & 2 \\ 6 & 0 & 1 \end{pmatrix}$

2. $Q = \begin{pmatrix} 8 & 1 & 0 & -6 & 3 \\ 2 & 4 & 6 & -1 & 5 \\ -11 & 0 & 0 & 3 & 6 \end{pmatrix}$

3. $R = (1 \quad 5 \quad 7)$

4. $S = \begin{pmatrix} 1 & 0 & 0 & 0 \\ 0 & 1 & 0 & 0 \\ 0 & 0 & 1 & 0 \\ 0 & 0 & 0 & 1 \end{pmatrix}$

5. $T = \begin{pmatrix} -8 \\ 7 \\ 12 \\ 5.6 \\ 0 \end{pmatrix}$

6. $U = \begin{pmatrix} 1 & 2 \\ 3 & 4 \\ 5 & 6 \\ 7 & 8 \end{pmatrix}$

In Exercises 7 through 25, use matrices A and B below to find the indicated components (see Example 2).

$$A = \begin{pmatrix} -1 & 3 & 5 & 0 \\ 2 & 5 & 6 & 8 \end{pmatrix} \qquad B = \begin{pmatrix} 3 & 4 & 7 \\ 1 & 2 & 8 \\ 3 & 4 & 6 \\ 0 & 0 & 1 \end{pmatrix}$$

7. a_{13} **8.** a_{24} **9.** a_{12} **10.** a_{21}
11. b_{12} **12.** b_{21} **13.** b_{32} **14.** a_{14}
15. b_{41} **16.** b_{43} **17.** b_{23} **18.** b_{33}
19. $A(2, 2)$ **20.** $A(1, 4)$ **21.** $A(2, 3)$ **22.** $B(2, 2)$
23. $B(1, 1)$ **24.** $B(4, 2)$ **25.** $B(1, 3)$

7.3 Matrix Arithmetic

In this section, we extend our knowledge of the arithmetic of vectors, introduced in Section 7.1, by operating on the two dimensional matrices studied in Section 7.2.

Addition of Matrices

In Section 7.1, we added and subtracted vectors. We add matrices similarly by using the following rules:

ADDITION OF MATRICES

1. The matrices must have the same dimensions.
2. Corresponding components are added (or subtracted) to form the sum (or difference).

For example, if $A = \begin{pmatrix} 2 & 3 & 4 \\ 0 & 1 & -2 \end{pmatrix}$ and $B = \begin{pmatrix} -4 & 2 & 0 \\ 1 & 3 & -5 \end{pmatrix}$

then

$$A + B = \begin{pmatrix} 2 + (-4) & 3 + 2 & 4 + 0 \\ 0 + 1 & 1 + 3 & -2 + (-5) \end{pmatrix} = \begin{pmatrix} -2 & 5 & 4 \\ 1 & 4 & -7 \end{pmatrix}$$

and

$$A - B = \begin{pmatrix} 2-(-4) & 3-2 & 4-0 \\ 0-1 & 1-3 & -2-(-5) \end{pmatrix} = \begin{pmatrix} 6 & 1 & 4 \\ -1 & -2 & 3 \end{pmatrix}$$

In the general case, if A and B are $m \times n$ matrices,

$$A = \begin{pmatrix} a_{11} & a_{12} & a_{13} & \cdots & a_{1n} \\ a_{21} & a_{22} & a_{23} & \cdots & a_{2n} \\ a_{31} & a_{32} & a_{33} & \cdots & a_{3n} \\ \vdots & \vdots & \vdots & & \vdots \\ a_{m1} & a_{m2} & a_{m3} & \cdots & a_{mn} \end{pmatrix}$$

and

$$B = \begin{pmatrix} b_{11} & b_{12} & b_{13} & \cdots & b_{1n} \\ b_{21} & b_{22} & b_{23} & \cdots & b_{2n} \\ b_{31} & b_{32} & b_{33} & \cdots & b_{3n} \\ \vdots & \vdots & \vdots & & \vdots \\ b_{m1} & b_{m2} & b_{m3} & \cdots & b_{mn} \end{pmatrix}$$

then

$$A + B = \begin{pmatrix} a_{11}+b_{11} & a_{12}+b_{12} & a_{13}+b_{13} & \cdots & a_{1n}+b_{1n} \\ a_{21}+b_{21} & a_{22}+b_{22} & a_{23}+b_{23} & \cdots & a_{2n}+b_{2n} \\ \vdots & \vdots & \vdots & & \vdots \\ a_{m1}+b_{m1} & a_{m2}+b_{m2} & a_{m3}+b_{m3} & \cdots & a_{mn}+b_{mn} \end{pmatrix}$$

EXAMPLE 1

Let $X = \begin{pmatrix} 1 & 2 & 3 & 4 \\ 9 & 8 & 7 & 6 \\ 0 & 5 & -2 & -4 \end{pmatrix}$ and $Y = \begin{pmatrix} 4 & 3 & -2 & 1 \\ 0 & 6 & 9 & -7 \\ 3 & 4 & 1 & 5 \end{pmatrix}$.

Find

a. $X + Y$ **b.** $X - Y$

Solution

a. $X + Y = \begin{pmatrix} 1+4 & 2+3 & 3+(-2) & 4+1 \\ 9+0 & 8+6 & 7+9 & 6+(-7) \\ 0+3 & 5+4 & -2+1 & -4+5 \end{pmatrix}$

$$= \begin{pmatrix} 5 & 5 & 1 & 5 \\ 9 & 14 & 16 & -1 \\ 3 & 9 & -1 & 1 \end{pmatrix}$$

$$\text{b. } X - Y = \begin{pmatrix} 1-4 & 2-3 & 3-(-2) & 4-1 \\ 9-0 & 8-6 & 7-9 & 6-(-7) \\ 0-3 & 5-4 & -2-1 & -4-5 \end{pmatrix}$$

$$= \begin{pmatrix} -3 & -1 & 5 & 3 \\ 9 & 2 & -2 & 13 \\ -3 & 1 & -3 & -9 \end{pmatrix}$$

Note that when two $m \times n$ matrices are added or subtracted, the result is an $m \times n$ matrix.

EXAMPLE 2

Let us revisit our old friend Sam of Micro Edge Products. To make our illustrations manageable, we'll shrink the Computer Heaven chain to three stores for this and future examples.

Suppose Sam's sales figures to the Computer Heaven chain were as follows for the first three months of 1985:

January

Store	Item Number				
	1	2	3	4	5
1	10	5	4	1	12
2	15	0	2	6	8
3	6	10	7	3	5

February

Store	Item Number				
	1	2	3	4	5
1	12	2	7	5	0
2	0	6	8	1	4
3	11	1	0	7	9

March	Store	Item Number				
		1	2	3	4	5
	1	0	6	3	2	6
	2	7	5	13	0	12
	3	14	6	5	6	7

Represent the data for each month in matrix form and, using these matrices, construct a matrix that will contain the total sales figures by store and product for the quarter.

Solution

Matrix for January:

$$J = \begin{pmatrix} 10 & 5 & 4 & 1 & 12 \\ 15 & 0 & 2 & 6 & 8 \\ 6 & 10 & 7 & 3 & 5 \end{pmatrix}$$

Matrix for February:

$$F = \begin{pmatrix} 12 & 2 & 7 & 5 & 0 \\ 0 & 6 & 8 & 1 & 4 \\ 11 & 1 & 0 & 7 & 9 \end{pmatrix}$$

Matrix for March:

$$M = \begin{pmatrix} 0 & 6 & 3 & 2 & 6 \\ 7 & 5 & 13 & 0 & 12 \\ 14 & 6 & 5 & 6 & 7 \end{pmatrix}$$

To obtain the total quarterly sales for a given product to a given store, we add the entries corresponding to that product and store in each of the three tables. For example, the total amount sold of item 2 in Store 3 is $10 + 1 + 6 = 17$. To construct a *matrix* Q for quarterly sales, then, we add matrices J, F and M.

$$Q = J + F + M$$

$$= \begin{pmatrix} 10+12+0 & 5+2+6 & 4+7+3 & 1+5+2 & 12+0+6 \\ 15+0+7 & 0+6+5 & 2+8+13 & 6+1+0 & 8+4+12 \\ 6+11+14 & 10+1+6 & 7+0+5 & 3+7+6 & 5+9+7 \end{pmatrix}$$

$$= \begin{pmatrix} 22 & 13 & 14 & 8 & 18 \\ 22 & 11 & 23 & 7 & 24 \\ 31 & 17 & 12 & 16 & 21 \end{pmatrix}$$

Multiplication of a Matrix by a Scalar

A matrix can be multiplied by a number just as vectors were in Section 7.1. We simply multiply each component by the number.

If

$$A = \begin{pmatrix} 3 & 6 & -1 \\ 9 & -2 & -9 \end{pmatrix}$$

then

$$(\tfrac{1}{3})A = \begin{pmatrix} 1 & 2 & -\frac{1}{3} \\ 3 & -\frac{2}{3} & -3 \end{pmatrix}$$

and

$$5A = \begin{pmatrix} 15 & 30 & -5 \\ 45 & -10 & -45 \end{pmatrix}$$

EXAMPLE 3

Sam wants to project what his sales figures will be in the second quarter of 1983 if he manages to double his first quarter sales. Write the matrix that shows this projection.

Solution

The first quarter sales matrix (see Example 2) is

$$Q = \begin{pmatrix} 22 & 13 & 14 & 8 & 18 \\ 22 & 11 & 23 & 7 & 24 \\ 31 & 17 & 12 & 16 & 21 \end{pmatrix}$$

The second quarter projection is

$$2Q = \begin{pmatrix} 2(22) & 2(13) & 2(14) & 2(8) & 2(18) \\ 2(22) & 2(11) & 2(23) & 2(7) & 2(24) \\ 2(31) & 2(17) & 2(12) & 2(16) & 2(21) \end{pmatrix}$$

$$= \begin{pmatrix} 44 & 26 & 28 & 16 & 36 \\ 44 & 22 & 46 & 14 & 48 \\ 62 & 34 & 24 & 32 & 42 \end{pmatrix}$$

Matrix Multiplication

The multiplication of two matrices is a more complex task. It involves following specific rules. Before stating these rules, we shall illustrate with an example.

$$\text{Let } A = \begin{pmatrix} -1 & 2 \\ 0 & 1 \\ 4 & 3 \end{pmatrix} \text{ and } B = \begin{pmatrix} 1 & 2 & 10 & 4 \\ 5 & -2 & 20 & 10 \end{pmatrix}.$$

We want to find the product of A with B, denoted $A \cdot B$. Notice that the number of *columns* in matrix A is equal to the number of *rows* in matrix B. This is necessary in matrix multiplication since we will find the product by multiplying rows of A by columns of B using vector multiplication. In order to do this, we must have the number of entries in a row of A (that is, the number of *columns* of A) equal to the number of entries in a column of B (the number of *rows* of B). Also, since A is a 3×2 matrix and B is a 2×4 matrix, the product $A \cdot B$ will be a 3×4 matrix:

$$A \cdot B = \begin{pmatrix} -1 & 2 \\ 0 & 1 \\ 4 & 3 \end{pmatrix} \cdot \begin{pmatrix} 1 & 2 & 10 & 4 \\ 5 & -2 & 20 & 10 \end{pmatrix} = \begin{pmatrix} \_ & \_ & \_ & \_ \\ \_ & \_ & 20 & \_ \\ \_ & \_ & \_ & \_ \end{pmatrix}$$

Specifically, let's start by finding the component in row 2, column 3 of $A \cdot B$. We get this component by multiplying row 2 of A by column 3 of B. This component, then, is

$$(0 \quad 1) \cdot \begin{pmatrix} 10 \\ 20 \end{pmatrix} = 0(10) + 1(20) = 20$$

In general, the entry in row r, column c of $A \cdot B$ is the product of row r of A with column c of B. Completing the multiplication process gives us, after twelve vector multiplications

$$\begin{aligned} A \cdot B &= \begin{pmatrix} -1 & 2 \\ 0 & 1 \\ 4 & 3 \end{pmatrix} \cdot \begin{pmatrix} 1 & 2 & 10 & 4 \\ 5 & -2 & 20 & 10 \end{pmatrix} \\ &= \begin{pmatrix} -1(1) + 2(5) & -1(2) + 2(-2) & -1(10) + 2(20) & -1(4) + 2(10) \\ 0(1) + 1(5) & 0(2) + 1(-2) & 0(10) + 1(20) & 0(4) + 1(10) \\ 4(1) + 3(5) & 4(2) + 3(-2) & 4(10) + 3(20) & 4(4) + 3(10) \end{pmatrix} \\ &= \begin{pmatrix} 9 & -6 & 30 & 16 \\ 5 & -2 & 20 & 10 \\ 19 & 2 & 100 & 46 \end{pmatrix} \end{aligned}$$

MATRIX MULTIPLICATION

To obtain the product $A \cdot B$ of two matrices A and B, the following rules must be followed:

1. The number of columns in matrix A must equal the number of rows in matrix B.
2. If A is an $a \times m$ matrix and B is an $m \times b$ matrix, then $A \cdot B$ is an $a \times b$ matrix.
3. To calculate the component in row r and column c, of $A \cdot B$, multiply row r of matrix A by column c of matrix B using vector multiplication.

EXAMPLE 4

Let $A = \begin{pmatrix} 2 & 0 & -1 \\ 7 & 6 & 5 \end{pmatrix}$, $B = \begin{pmatrix} 4 & 3 & 2 \\ -1 & -3 & -5 \\ 0 & 4 & 4 \end{pmatrix}$ and $C = \begin{pmatrix} 7 & -6 \\ 5 & 4 \end{pmatrix}$.

Find the following products:

a. $A \cdot B$ **b.** $A \cdot C$ **c.** $C \cdot A$

Solution

a. $A \cdot B$ is a 2×3 matrix, since A has 2 rows and B has 3 columns.

$$A \cdot B = \begin{pmatrix} 2 & 0 & -1 \\ 7 & 6 & 5 \end{pmatrix} \cdot \begin{pmatrix} 4 & 3 & 2 \\ -1 & -3 & -5 \\ 0 & 4 & 4 \end{pmatrix}$$

$$= \begin{pmatrix} 2(4)+0(-1)+(-1)(0) & 2(3)+0(-3)+(-1)(4) & 2(2)+0(-5)+(-1)(4) \\ 7(4)+6(-1)+5(0) & 7(3)+6(-3)+5(4) & 7(2)+6(-5)+5(4) \end{pmatrix}$$

$$= \begin{pmatrix} 8 & 2 & 0 \\ 22 & 23 & 4 \end{pmatrix}$$

b. The product $A \cdot C$ cannot be found, since A has 3 columns and C has 2 rows.

c. $C \cdot A$ is a 2×3 matrix.

$$C \cdot A = \begin{pmatrix} 7 & -6 \\ 5 & 4 \end{pmatrix} \cdot \begin{pmatrix} 2 & 0 & -1 \\ 7 & 6 & 5 \end{pmatrix}$$

$$= \begin{pmatrix} 7(2)+(-6)(7) & 7(0)+(-6)(6) & 7(-1)+(-6)(5) \\ 5(2)+4(7) & 5(0)+4(6) & (5)(-1)+4(5) \end{pmatrix}$$

$$= \begin{pmatrix} -28 & -36 & -37 \\ 38 & 24 & 15 \end{pmatrix}$$

EXAMPLE 5

Let $C = \begin{pmatrix} 7 & -6 \\ 5 & 4 \end{pmatrix}$, $D = \begin{pmatrix} 1 & 0 \\ 0 & 1 \end{pmatrix}$ and $E = \begin{pmatrix} -1 & 2 \\ 1 & 0 \end{pmatrix}$.

Find the following:

a. $C \cdot E$ **b.** $E \cdot C$ **c.** $D \cdot C$

Solution

a. $C \cdot E$ is a 2×2 matrix.

$$\begin{aligned} C \cdot E &= \begin{pmatrix} 7 & -6 \\ 5 & 4 \end{pmatrix} \cdot \begin{pmatrix} -1 & 2 \\ 1 & 0 \end{pmatrix} \\ &= \begin{pmatrix} 7(-1) + (-6)(1) & 7(2) + (-6)(0) \\ 5(-1) + 4(1) & 5(2) + 4(0) \end{pmatrix} \\ &= \begin{pmatrix} -13 & 14 \\ -1 & 10 \end{pmatrix} \end{aligned}$$

b. $$E \cdot C = \begin{pmatrix} -1 & 2 \\ 1 & 0 \end{pmatrix} \cdot \begin{pmatrix} 7 & -6 \\ 5 & 4 \end{pmatrix} = \begin{pmatrix} (-1)(7) + 2(5) & (-1)(-6) + 2(4) \\ 1(7) + 0(5) & (1)(-6) + 0(4) \end{pmatrix}$$

$$= \begin{pmatrix} 3 & 14 \\ 7 & -6 \end{pmatrix}$$

c. $$D \cdot C = \begin{pmatrix} 1 & 0 \\ 0 & 1 \end{pmatrix} \cdot \begin{pmatrix} 7 & -6 \\ 5 & 4 \end{pmatrix} = \begin{pmatrix} 1(7) + 0(5) & 1(-6) + 0(4) \\ 0(7) + 1(5) & 0(-6) + 1(4) \end{pmatrix}$$

$$= \begin{pmatrix} 7 & -6 \\ 5 & 4 \end{pmatrix}$$

Note that some basic rules pertaining to the multiplication of real numbers do *not* apply to matrix multiplication. We can see in Examples 4 and 5 that $A \cdot C$ is not equal to $C \cdot A$; and $C \cdot E$ is not equal to $E \cdot C$. That is, matrix multiplication is *not commutative*.

EXAMPLE 6

Sam wants to calculate the total amount of his bill (before shipping charges) to each store of the Computer Heaven chain for the month of March. Use matrix multiplication to obtain these amounts.

Solution

The sales matrix for March (see Example 2) is

$$M = \begin{pmatrix} 0 & 6 & 3 & 2 & 6 \\ 7 & 5 & 13 & 0 & 12 \\ 14 & 6 & 5 & 6 & 7 \end{pmatrix}$$

The amount billed to each store can be obtained by multiplying each item's price by the quantity sold to that store and adding the resulting products.

Let's look at the price vector we used for Sam's business in Section 7.1. We'll write it as a column vector,

$$C = \begin{pmatrix} 20 \\ 35 \\ 100 \\ 250 \\ 25 \end{pmatrix}$$

Since M is a $\mathbf{3} \times 5$ matrix and C is a $5 \times \mathbf{1}$ matrix, we can form the product $M \cdot C$; it will be a 3×1 matrix. Each entry in $M \cdot C$ is the billing amount for one store in the chain.

$$M \cdot C = \overbrace{\begin{pmatrix} 0 & 6 & 3 & 2 & 6 \\ 7 & 5 & 13 & 0 & 12 \\ 14 & 6 & 5 & 6 & 7 \end{pmatrix}}^{\textit{quantities of items}} \cdot \left.\begin{pmatrix} 20 \\ 35 \\ 100 \\ 250 \\ 25 \end{pmatrix}\right\} \textit{prices of items}$$

$$= \begin{pmatrix} 0(20) + 6(35) + 3(100) + 2(250) + 6(25) \\ 7(20) + 5(35) + 13(100) + 0(250) + 12(25) \\ 14(20) + 6(35) + 5(100) + 6(250) + 7(25) \end{pmatrix} \quad \text{sums of quantities times prices}$$

$$= \begin{pmatrix} 1160 \\ 1915 \\ 2665 \end{pmatrix} \quad \begin{matrix} \text{Amount billed to store 1} \\ \text{Amount billed to store 2} \\ \text{Amount billed to store 3} \end{matrix}$$

This example provides some justification for our definition of matrix multiplication. It illustrates that our rules for multiplication of matrices produce meaningful results in a real-world application. This type of application, in which data are stored and manipulated in matrices, occurs frequently in business data processing.

EXERCISES FOR SECTION 7.3

In Exercises 1 through 12, perform the indicated matrix operation, if possible. Use A, B, X, Y, and Z as below (see Examples 1–3).

$$A = \begin{pmatrix} 5 & 6 & -2 \\ 0 & 3 & 10 \end{pmatrix} \qquad B = \begin{pmatrix} 8 & 3 & 6 \\ 2 & 10 & 20 \end{pmatrix} \qquad X = \begin{pmatrix} 5 & 0 \\ 1 & 3 \\ 6 & 5 \end{pmatrix}$$

$$Y = \begin{pmatrix} 1 & 10 \\ 2 & 20 \\ 3 & 30 \end{pmatrix} \qquad Z = \begin{pmatrix} 1 & 5 & 6 \\ 3 & 5 & 7 \\ 1 & 2 & 3 \end{pmatrix}$$

1. $A + B$	**2.** $A - B$	**3.** $4A$	**4.** $4A + B$
5. $A + X$	**6.** $X - Y$	**7.** $3X + 2Y$	**8.** $3X - 2Y$
9. $X + Z$	**10.** $7A - 3B$	**11.** $X + \frac{1}{10}Y$	**12.** $Z + 3Z$

In Exercises 13 through 33, find the indicated matrix product, if possible. Use A, B, X, Y, and Z as given in Exercises 1–12 as well as C, D and E below (see Examples 4 and 5).

$$C = \begin{pmatrix} 1 & 2 \\ 5 & 3 \end{pmatrix} \qquad D = \begin{pmatrix} 1 & 0 & 0 \\ 0 & 1 & 0 \\ 0 & 0 & 1 \end{pmatrix} \qquad E = \begin{pmatrix} 4 & 5 & -2 \\ 3 & 1 & 6 \\ 0 & 0 & 8 \end{pmatrix}$$

13. $A \cdot X$	**14.** $X \cdot A$	**15.** $A \cdot B$
16. $C \cdot A$	**17.** $A \cdot C$	**18.** $B \cdot E$
19. $E \cdot B$	**20.** $D \cdot E$	**21.** $E \cdot D$
22. $Y \cdot A$	**23.** $Y \cdot B$	**24.** $Y \cdot C$
25. $E \cdot X$	**26.** $E \cdot Y$	**27.** $E \cdot Z$
28. $(A + B) \cdot E$	**29.** $Z \cdot Y$	**30.** $(2A + 3B) \cdot D$
31. $C \cdot C$	**32.** $D \cdot D$	**33.** $E \cdot E$

In Exercises 34 through 36, solve the applied problem using matrices (see Example 6).

34. Suppose the Threadfine Machine Company produces three kinds of bolts: 3-inch, 5-inch, and 8-inch. It has two factories. The Belmont, CA factory produces 400, 500, and 200, respectively, of each type bolt; the Waterbury, CT factory produces 200, 300, 300, respectively, of each type of bolt. Also, it is known that each 3-inch bolt costs \$.02 to produce and \$.01 to ship; each 5-inch bolt costs \$.03 to produce and \$.02 to ship; and each 8-inch bolt costs \$.05 to produce and \$.03 to ship. Let A be the 2×3 matrix representing factory output and let B be the 3×2 matrix showing costs for each bolt as below:

$$\begin{array}{l} \text{Belmont factory} \\ \text{Waterbury factory} \end{array} \begin{array}{c} \begin{array}{ccc} \text{3-in} & \text{5-in} & \text{8-in} \end{array} \\ \begin{pmatrix} 400 & 500 & 200 \\ 200 & 300 & 300 \end{pmatrix} \end{array} = A$$

$$\begin{array}{l} \\ \\ \text{3-in} \\ \text{5-in} \\ \text{8-in} \end{array} \begin{array}{c} \begin{array}{cc} \text{Production} & \text{Shipping} \\ \text{costs} & \text{costs} \end{array} \\ \begin{pmatrix} 0.02 & 0.01 \\ 0.03 & 0.02 \\ 0.05 & 0.03 \end{pmatrix} \end{array} = B$$

Find the costs per factory, $A \cdot B$.

35. Interpret each component of $A \cdot B$ in Exercise 34.

36. A textbook publisher incurs the following costs: typesetting, payroll, advertising, and material. For a certain biology book, they are 40%, 15%, 5%, and 5% of the wholesale price, respectively. For a literature book they are 20%, 20%, 5% and 10%, respectively. For a mathematics text they are 40%, 15%, 10% and 5%, respectively.

a. Depict the cost percentages in a 4×3 matrix, X.

b. The wholesale price of the biology book is \$21.50, of the literature text \$14.00, and of the math text \$23.50. Using appropriate dimension, construct a wholesale price matrix, Y, and find the publisher's costs by category by multiplying X and Y in the correct order.

37. Multiply the 4×1 matrix of Exercise 36b by the vector

$$Z = (1 \quad 1 \quad 1 \quad 1)$$

What does this result represent?

If A is a square matrix, that is, an $n \times n$ matrix, then we can write $A \cdot A = A^2$, $A \cdot A^2 = A^3$, etc., as we would ordinarily use exponents. In Exercises 38 through 43 let

$$A = \begin{pmatrix} 1 & 2 \\ 0 & 3 \end{pmatrix} \quad \text{and} \quad B = \begin{pmatrix} 0.2 & 0.8 \\ 0.1 & 0.9 \end{pmatrix}$$

Find

38. A^2 **39.** A^3 **40.** B^2

41. B^3 **42.** A^2B^2 **43.** $(AB)^2$

The *transpose* of an $m \times n$ matrix A, denoted A^T, is an $n \times m$ matrix found by writing the rows of A as columns. For example, if

$$A = \begin{pmatrix} 1 & 2 & 3 \\ 4 & 5 & 6 \end{pmatrix} \quad \text{then} \quad A^T = \begin{pmatrix} 1 & 4 \\ 2 & 5 \\ 3 & 6 \end{pmatrix}$$

In Exercises 44 through 48, find the transpose of the given matrix.

44. $\begin{pmatrix} 1 & -2 & 3 \\ 5 & 0 & 6 \end{pmatrix}$

45. $\begin{pmatrix} 1 & 2 \\ 3 & 4 \\ 5 & 6 \end{pmatrix}$

46. $\begin{pmatrix} 1 & 0 & 0 \\ 0 & 1 & 0 \\ 0 & 0 & 1 \end{pmatrix}$

47. $\begin{pmatrix} 1 & 5 \\ 0 & -7 \end{pmatrix}$

48. $\begin{pmatrix} 0.1 & 0.2 & 0.7 \\ 0 & 0.5 & 0.5 \\ 0.1 & 0.1 & 0.8 \end{pmatrix}$

7.4 Invertible Matrices

Square Matrices

In this section and in Section 7.5, we shall confine our discussion to matrices with the same number of rows as columns. Such a matrix is called a **square matrix**. For example,

$$A = \begin{pmatrix} 1 & 2 & 3 \\ 4 & 5 & 6 \\ 7 & 8 & 9 \end{pmatrix}$$

is a 3×3 matrix. It is a square matrix because its dimensions (its number of rows and number of columns) are equal. The individual components of A whose row number and column number are equal form the **main diagonal** of A. These components lie in a straight line going from the upper left-hand corner of A to the lower right-hand corner.

If

$$A = \begin{pmatrix} 1 & 2 & 3 \\ 4 & 5 & 6 \\ 7 & 8 & 9 \end{pmatrix}$$

the main diagonal of A consists of the components 1, 5 and 9. In general, if A is an $n \times n$ matrix,

$$A = \begin{pmatrix} a_{11} & a_{12} & a_{13} & \cdots & a_{1n} \\ a_{21} & a_{22} & a_{23} & \cdots & a_{2n} \\ \vdots & \vdots & \vdots & & \vdots \\ a_{n1} & a_{n2} & a_{n3} & \cdots & a_{nn} \end{pmatrix}$$

the main diagonal of A consists of the components $a_{11}, a_{22}, a_{33}, \cdots, a_{nn}$. Since the two dimensions of a square matrix are equal, an $n \times n$ matrix is frequently referred to as a **matrix of order n**. The matrix A above is a matrix of order 3 since it is a 3×3 matrix.

EXAMPLE 1

$$X = \begin{pmatrix} 1 & 2 & 4 & 0 \\ 5 & 6 & 3 & 1 \\ -2 & 4 & -6 & 8 \end{pmatrix} \qquad B = \begin{pmatrix} 1 & 2 \\ 0 & 1 \end{pmatrix} \qquad C = \begin{pmatrix} 1 & 2 \\ 3 & 4 \\ 5 & 6 \end{pmatrix}$$

$$D = \begin{pmatrix} 1 & 2 & 3 & 4 \\ 5 & 6 & 7 & 8 \\ 9 & 0 & 0 & 0 \\ 8 & 7 & 6 & 5 \end{pmatrix} \qquad I = \begin{pmatrix} 1 & 0 & 0 \\ 0 & 1 & 0 \\ 0 & 0 & 1 \end{pmatrix}$$

State whether or not each matrix is a square matrix. If the matrix is square, find its order.

Solution

X is not a square matrix; it is a 3×4 matrix.
B is a square matrix of order 2.
C is not a square matrix; it is a 3×2 matrix.
D is a square matrix of order 4.
I is a square matrix of order 3.

Identity Matrix

Consider matrix I of Example 1. It is a square matrix of order 3 with ones on its main diagonal and zeros elsewhere.

Let

$$A = \begin{pmatrix} 1 & 2 & 3 \\ 4 & 5 & 6 \\ 7 & 8 & 9 \end{pmatrix}$$

Note that A is also a matrix of order 3. Now consider the product $A \cdot I$.

$$A \cdot I = \begin{pmatrix} 1 & 2 & 3 \\ 4 & 5 & 6 \\ 7 & 8 & 9 \end{pmatrix} \cdot \begin{pmatrix} 1 & 0 & 0 \\ 0 & 1 & 0 \\ 0 & 0 & 1 \end{pmatrix}$$

$$= \begin{pmatrix} 1(1) + 2(0) + 3(0) & 1(0) + 2(1) + 3(0) & 1(0) + 2(0) + 3(1) \\ 4(1) + 5(0) + 6(0) & 4(0) + 5(1) + 6(0) & 4(0) + 5(0) + 6(1) \\ 7(1) + 8(0) + 9(0) & 7(0) + 8(1) + 9(0) & 7(0) + 8(0) + 9(1) \end{pmatrix}$$

$$= \begin{pmatrix} 1 & 2 & 3 \\ 4 & 5 & 6 \\ 7 & 8 & 9 \end{pmatrix}$$

That is, $A \cdot I = A$. Now suppose B is *any* 3×3 matrix, and consider the product $B \cdot I$. If

$$B = \begin{pmatrix} b_{11} & b_{12} & b_{13} \\ b_{21} & b_{22} & b_{23} \\ b_{31} & b_{32} & b_{33} \end{pmatrix}$$

then

$$B \cdot I = \begin{pmatrix} b_{11} & b_{12} & b_{13} \\ b_{21} & b_{22} & b_{23} \\ b_{31} & b_{32} & b_{33} \end{pmatrix} \cdot \begin{pmatrix} 1 & 0 & 0 \\ 0 & 1 & 0 \\ 0 & 0 & 1 \end{pmatrix}$$

$$= \begin{pmatrix} b_{11}(1)+b_{12}(0)+b_{13}(0) & b_{11}(0)+b_{12}(1)+b_{13}(0) & b_{11}(0)+b_{12}(0)+b_{13}(1) \\ b_{21}(1)+b_{22}(0)+b_{23}(0) & b_{21}(0)+b_{22}(1)+b_{23}(0) & b_{21}(0)+b_{22}(0)+b_{23}(1) \\ b_{31}(1)+b_{32}(0)+b_{33}(0) & b_{31}(0)+b_{32}(1)+b_{33}(0) & b_{31}(0)+b_{32}(0)+b_{33}(1) \end{pmatrix}$$

$$= \begin{pmatrix} b_{11} & b_{12} & b_{13} \\ b_{21} & b_{22} & b_{23} \\ b_{31} & b_{32} & b_{33} \end{pmatrix} = B$$

If we did the multiplication $I \cdot B$, we would also get $I \cdot B = B$.

We say that I is the **identity matrix** for matrices of order 3, since the product of any matrix of order 3 with the matrix I is always the original matrix. There is an identity matrix I for the set of square matrices of any order. It is simply the matrix of that order formed by putting ones on the main diagonal and zeros elsewhere.

EXAMPLE 2

a. Find the identity matrix for matrices of order 2. Justify your result by multiplying a specific matrix of order 2 by the identity matrix.

b. Do the same for matrices of order 4.

Solution

a. The identity matrix of order 2 is

$$I = \begin{pmatrix} 1 & 0 \\ 0 & 1 \end{pmatrix}$$

Let

$$A = \begin{pmatrix} -2 & 3 \\ 7 & -6 \end{pmatrix}$$

Then

$$A \cdot I = \begin{pmatrix} -2 & 3 \\ 7 & -6 \end{pmatrix} \cdot \begin{pmatrix} 1 & 0 \\ 0 & 1 \end{pmatrix}$$

$$= \begin{pmatrix} -2(1) + 3(0) & -2(0) + 3(1) \\ 7(1) + (-6)(0) & 7(0) + (-6)(1) \end{pmatrix}$$

$$= \begin{pmatrix} -2 & 3 \\ 7 & -6 \end{pmatrix} = A$$

Also

$$I \cdot A = \begin{pmatrix} 1 & 0 \\ 0 & 1 \end{pmatrix} \cdot \begin{pmatrix} -2 & 3 \\ 7 & -6 \end{pmatrix}$$

$$= \begin{pmatrix} -2+0 & 3+0 \\ 0+7 & 0+(-6) \end{pmatrix} = \begin{pmatrix} -2 & 3 \\ 7 & -6 \end{pmatrix} = A$$

b. The identity matrix of order 4 is

$$I = \begin{pmatrix} 1 & 0 & 0 & 0 \\ 0 & 1 & 0 & 0 \\ 0 & 0 & 1 & 0 \\ 0 & 0 & 0 & 1 \end{pmatrix}$$

For a specific matrix of order 4 we shall use C:

$$C = \begin{pmatrix} 1 & 2 & 3 & 4 \\ -4 & -3 & -2 & -1 \\ 5 & 6 & 0 & 8 \\ 0 & 8 & 7 & 6 \end{pmatrix}$$

$$C \cdot I = \begin{pmatrix} 1 & 2 & 3 & 4 \\ -4 & -3 & -2 & -1 \\ 5 & 6 & 0 & 8 \\ 0 & 8 & 7 & 6 \end{pmatrix} \cdot \begin{pmatrix} 1 & 0 & 0 & 0 \\ 0 & 1 & 0 & 0 \\ 0 & 0 & 1 & 0 \\ 0 & 0 & 0 & 1 \end{pmatrix}$$

$$= \begin{pmatrix} 1 & 2 & 3 & 4 \\ -4 & -3 & -2 & -1 \\ 5 & 6 & 0 & 8 \\ 0 & 8 & 7 & 6 \end{pmatrix} = C$$

The reader is urged to verify this result and the fact that $I \cdot C = C$.

EXAMPLE 3

Let $X = \begin{pmatrix} -5 & 2 \\ 3 & -1 \end{pmatrix}$ and $Y = \begin{pmatrix} 1 & 2 \\ 3 & 5 \end{pmatrix}$.

Find $X \cdot Y$ and $Y \cdot X$.

Solution

$$X \cdot Y = \begin{pmatrix} -5 & 2 \\ 3 & -1 \end{pmatrix} \cdot \begin{pmatrix} 1 & 2 \\ 3 & 5 \end{pmatrix} = \begin{pmatrix} -5+6 & -10+10 \\ 3-3 & 6-5 \end{pmatrix} = \begin{pmatrix} 1 & 0 \\ 0 & 1 \end{pmatrix}$$

$$Y \cdot X = \begin{pmatrix} 1 & 2 \\ 3 & 5 \end{pmatrix} \cdot \begin{pmatrix} -5 & 2 \\ 3 & -1 \end{pmatrix} = \begin{pmatrix} -5+6 & 2-2 \\ -15+15 & 6-5 \end{pmatrix} = \begin{pmatrix} 1 & 0 \\ 0 & 1 \end{pmatrix}$$

EXAMPLE 4

Find $P \cdot Q$ and $Q \cdot P$ for the following matrices:

$$P = \begin{pmatrix} 12 & -9 & 12 \\ -2 & 3 & -4 \\ -6 & 6 & -6 \end{pmatrix} \qquad Q = \begin{pmatrix} \frac{1}{6} & \frac{1}{2} & 0 \\ \frac{1}{3} & 0 & \frac{2}{3} \\ \frac{1}{6} & -\frac{1}{2} & \frac{1}{2} \end{pmatrix}$$

Solution

$$P \cdot Q = \begin{pmatrix} 12 & -9 & 12 \\ -2 & 3 & -4 \\ -6 & 6 & -6 \end{pmatrix} \cdot \begin{pmatrix} \frac{1}{6} & \frac{1}{2} & 0 \\ \frac{1}{3} & 0 & \frac{2}{3} \\ \frac{1}{6} & -\frac{1}{2} & \frac{1}{2} \end{pmatrix} = \begin{pmatrix} 1 & 0 & 0 \\ 0 & 1 & 0 \\ 0 & 0 & 1 \end{pmatrix}$$

$$Q \cdot P = \begin{pmatrix} \frac{1}{6} & \frac{1}{2} & 0 \\ \frac{1}{3} & 0 & \frac{2}{3} \\ \frac{1}{6} & -\frac{1}{2} & \frac{1}{2} \end{pmatrix} \cdot \begin{pmatrix} 12 & -9 & 12 \\ -2 & 3 & -4 \\ -6 & 6 & -6 \end{pmatrix} = \begin{pmatrix} 1 & 0 & 0 \\ 0 & 1 & 0 \\ 0 & 0 & 1 \end{pmatrix}$$

Notice that, in Examples 3 and 4, the product of the two given matrices yielded the identity matrix. Also observe that, although in general the product of two matrices is *not* commutative, when the product is the identity matrix the two matrices *can* be multiplied in either order.

The Inverse of a Matrix

THE INVERSE OF A MATRIX

When a matrix A can be multiplied by a matrix B to yield the identity matrix, we say that B is the **inverse** of A (or that A is the **inverse** of B, or that A and B are inverses).

If A is a square matrix, it follows that B must be a square matrix of the same order, since two square matrices cannot be multiplied unless they are of the same order (why?). If the matrix A has an inverse, we denote the inverse of A by A^{-1}.

Consider the matrices A and B:

$$A = \begin{pmatrix} 4 & 3 \\ 2 & 1 \end{pmatrix} \qquad B = \begin{pmatrix} -\frac{1}{2} & \frac{3}{2} \\ 1 & -2 \end{pmatrix}$$

$$A \cdot B = \begin{pmatrix} 4 & 3 \\ 2 & 1 \end{pmatrix} \cdot \begin{pmatrix} -\frac{1}{2} & \frac{3}{2} \\ 1 & -2 \end{pmatrix} = \begin{pmatrix} 4(-\frac{1}{2}) + 3(1) & 4(\frac{3}{2}) + 3(-2) \\ 2(-\frac{1}{2}) + 1(1) & 2(\frac{3}{2}) + 1(-2) \end{pmatrix}$$

$$= \begin{pmatrix} -2+3 & 6+(-6) \\ -1+1 & 3+(-2) \end{pmatrix} = \begin{pmatrix} 1 & 0 \\ 0 & 1 \end{pmatrix} = I$$

Therefore, B is the inverse of A, that is, $B = A^{-1}$.

Given two matrices of the same order, we can determine whether or not they are inverses of each other by finding their product. If the product is the identity matrix, then the two matrices are inverses.

EXAMPLE 5

Show that $B = A^{-1}$ for

$$A = \begin{pmatrix} 1 & 2 & 0 \\ -1 & 1 & 2 \\ 1 & 0 & -1 \end{pmatrix} \quad \text{and} \quad B = \begin{pmatrix} -1 & 2 & 4 \\ 1 & -1 & -2 \\ -1 & 2 & 3 \end{pmatrix}$$

Solution

To prove that $B = A^{-1}$, we need to show that $A \cdot B = I$.

$$A \cdot B = \begin{pmatrix} 1 & 2 & 0 \\ -1 & 1 & 2 \\ 1 & 0 & -1 \end{pmatrix} \cdot \begin{pmatrix} -1 & 2 & 4 \\ 1 & -1 & -2 \\ -1 & 2 & 3 \end{pmatrix}$$

$$= \begin{pmatrix} 1(-1) + 2(1) + 0(-1) & 1(2) + 2(-1) + 0(2) & 1(4) + 2(-2) + 0(3) \\ (-1)(-1) + 1(1) + 2(-1) & -1(2) + 1(-1) + 2(2) & -1(4) + 1(-2) + 2(3) \\ 1(-1) + 0(1) + (-1)(-1) & 1(2) + 0(-1) + (-1)(2) & 1(4) + 0(-2) + (-1)(3) \end{pmatrix}$$

$$= \begin{pmatrix} -1+2+0 & 2+(-2)+0 & 4+(-4)+0 \\ 1+1+(-2) & -2+(-1)+4 & -4+(-2)+6 \\ -1+0+1 & 2+0+(-2) & 4+0+(-3) \end{pmatrix}$$

$$= \begin{pmatrix} 1 & 0 & 0 \\ 0 & 1 & 0 \\ 0 & 0 & 1 \end{pmatrix} = I$$

Therefore, $B = A^{-1}$.

Finding the Inverse of a Square Matrix

As we have seen, determining whether or not two matrices are inverses of each other is relatively simple; we look at their product. The problem of *finding the inverse* of a square matrix is more complex because there is no simple formula we can apply. The method we shall use involves performing special operations, called *row transformations*, on the given matrix until it has been converted to the identity matrix. At the same time, we will apply the same row transformations to the identity matrix. The matrix derived in this way from the identity matrix will be the inverse of the original matrix.

An example will clarify this process. We wish to find M^{-1}, the inverse of matrix M.

$$M = \begin{pmatrix} 1 & 3 \\ 4 & 6 \end{pmatrix}$$

ROW TRANSFORMATIONS

The row transformations we may use are

1. interchange of any two rows of M;
2. multiplication of any row of M by a nonzero constant; and
3. replacement of any row of M, say row r, by the result of multiplying any other row by a nonzero constant and adding it (using addition of row vectors) to row r.

Our object will be convert M to the identity matrix of order 2,

$$I = \begin{pmatrix} 1 & 0 \\ 0 & 1 \end{pmatrix}$$

Whenever we apply a row transformation to M, we will apply the same row transformation to I. It will be useful to write M and I in the same parentheses, separated by a vertical line, so that we can easily see the results of both transformations.

The process goes as follows:

$$\begin{array}{c} \quad M \qquad\qquad I \\ \left(\begin{array}{cc|cc} 1 & 3 & 1 & 0 \\ 4 & 6 & 0 & 1 \end{array}\right) \end{array}$$

$\downarrow$ Multiply row 1 by -4 and add to row 2, putting result in row 2.

$$\left(\begin{array}{cc|cc} 1 & 3 & 1 & 0 \\ 0 & -6 & -4 & 1 \end{array}\right)$$

↓ Multiply row 2 by $-\frac{1}{6}$.

$$\left(\begin{array}{cc|cc} 1 & 3 & 1 & 0 \\ 0 & 1 & \frac{2}{3} & -\frac{1}{6} \end{array}\right)$$

↓ Multiply row 2 by -3 and add to row 1, putting result in row 1.

$$\left(\begin{array}{cc|cc} 1 & 0 & -1 & \frac{1}{2} \\ 0 & 1 & \frac{2}{3} & -\frac{1}{6} \end{array}\right)$$

New $M = I$ New $I = M^{-1}$

Since M has been transformed to the identity matrix, the process is complete. The transformed version of the original identity matrix is the inverse of M.

$$M^{-1} = \begin{pmatrix} -1 & \frac{1}{2} \\ \frac{2}{3} & -\frac{1}{6} \end{pmatrix}$$

The reader should verify this by finding the product $M \cdot M^{-1}$. (It should be I.)

EXAMPLE 6

Let A be the matrix of Example 5,

$$A = \begin{pmatrix} 1 & 2 & 0 \\ -1 & 1 & 2 \\ 1 & 0 & -1 \end{pmatrix}$$

Use row transformations to find A^{-1}.

Solution

$$\begin{array}{c} \qquad A \qquad\qquad\qquad I \\ \left(\begin{array}{ccc|ccc} 1 & 2 & 0 & 1 & 0 & 0 \\ -1 & 1 & 2 & 0 & 1 & 0 \\ 1 & 0 & -1 & 0 & 0 & 1 \end{array}\right) \end{array}$$

↓ Add row 1 to row 2, putting result in row 2.

$$\left(\begin{array}{ccc|ccc} 1 & 2 & 0 & 1 & 0 & 0 \\ 0 & 3 & 2 & 1 & 1 & 0 \\ 1 & 0 & -1 & 0 & 0 & 1 \end{array}\right)$$

↓ Multiply row 1 by -1 and add to row 3, putting result in row 3.

$$\left(\begin{array}{rrr|rrr} 1 & 2 & 0 & 1 & 0 & 0 \\ 0 & 3 & 2 & 1 & 1 & 0 \\ 0 & -2 & -1 & -1 & 0 & 1 \end{array}\right)$$

$\downarrow$ Multiply row 2 by $\frac{1}{3}$.

$$\left(\begin{array}{rrr|rrr} 1 & 2 & 0 & 1 & 0 & 0 \\ 0 & 1 & \frac{2}{3} & \frac{1}{3} & \frac{1}{3} & 0 \\ 0 & -2 & -1 & -1 & 0 & 1 \end{array}\right)$$

$\downarrow$ Multiply row 2 by -2 and add to row 1, putting result in row 1.

$$\left(\begin{array}{rrr|rrr} 1 & 0 & -\frac{4}{3} & \frac{1}{3} & -\frac{2}{3} & 0 \\ 0 & 1 & \frac{2}{3} & \frac{1}{3} & \frac{1}{3} & 0 \\ 0 & -2 & -1 & -1 & 0 & 1 \end{array}\right)$$

$\downarrow$ Multiply row 2 by 2 and add to row 3, putting the result in row 3.

$$\left(\begin{array}{rrr|rrr} 1 & 0 & -\frac{4}{3} & \frac{1}{3} & -\frac{2}{3} & 0 \\ 0 & 1 & \frac{2}{3} & \frac{1}{3} & \frac{1}{3} & 0 \\ 0 & 0 & \frac{1}{3} & -\frac{1}{3} & \frac{2}{3} & 1 \end{array}\right)$$

$\downarrow$ Multiply row 3 by 3.

$$\left(\begin{array}{rrr|rrr} 1 & 0 & -\frac{4}{3} & \frac{1}{3} & -\frac{2}{3} & 0 \\ 0 & 1 & \frac{2}{3} & \frac{1}{3} & \frac{1}{3} & 0 \\ 0 & 0 & 1 & -1 & 2 & 3 \end{array}\right)$$

$\downarrow$ Multiply row 3 by $\frac{4}{3}$ and add to row 1, putting result in row 1.

$$\left(\begin{array}{rrr|rrr} 1 & 0 & 0 & -1 & 2 & 4 \\ 0 & 1 & \frac{2}{3} & \frac{1}{3} & \frac{1}{3} & 0 \\ 0 & 0 & 1 & -1 & 2 & 3 \end{array}\right)$$

$\downarrow$ Multiply row 3 by $-\frac{2}{3}$ and add to row 2, putting result in row 2.

$$\left(\begin{array}{rrr|rrr} 1 & 0 & 0 & -1 & 2 & 4 \\ 0 & 1 & 0 & 1 & -1 & -2 \\ 0 & 0 & 1 & -1 & 2 & 3 \end{array}\right)$$

New $A = I$ New $I = A^{-1}$

At last,

$$A^{-1} = \begin{pmatrix} -1 & 2 & 4 \\ 1 & -1 & -2 \\ -1 & 2 & 3 \end{pmatrix}$$

Although this process of using row transformations to transform a given matrix to the identity matrix may seem cumbersome, it will be very important in solving systems of linear equations in Chapter 8. At that point, we shall also consider some practical applications of this technique.

Singular and Nonsingular Matrices

Our algorithm for finding the inverse of a square matrix depends on being able to transform that matrix to the identity matrix. Are there any situations in which this is impossible?

Consider the following matrix:

$$D = \begin{pmatrix} 2 & 4 & 6 \\ 1 & 2 & 0 \\ -3 & -6 & 0 \end{pmatrix}$$

Let's use row transformation to try to find D^{-1}.

$$\left(\begin{array}{ccc|ccc} 2 & 4 & 6 & 1 & 0 & 0 \\ 1 & 2 & 0 & 0 & 1 & 0 \\ -3 & -6 & 0 & 0 & 0 & 1 \end{array}\right)$$

$\downarrow$ Multiply row 2 by -2 and add to row 1.

$$\left(\begin{array}{ccc|ccc} 0 & 0 & 6 & 1 & -2 & 0 \\ 1 & 2 & 0 & 0 & 1 & 0 \\ -3 & -6 & 0 & 0 & 0 & 1 \end{array}\right)$$

$\downarrow$ Multiply row 1 by $\frac{1}{6}$.

$$\left(\begin{array}{ccc|ccc} 0 & 0 & 1 & \frac{1}{6} & -\frac{1}{3} & 0 \\ 1 & 2 & 0 & 0 & 1 & 0 \\ -3 & -6 & 0 & 0 & 0 & 1 \end{array}\right)$$

$\downarrow$ Multiply row 2 by 3 and add to row 3.

$$\left(\begin{array}{ccc|ccc} 0 & 0 & 1 & \frac{1}{6} & -\frac{1}{3} & 0 \\ 1 & 2 & 0 & 0 & 1 & 0 \\ 0 & 0 & 0 & 0 & 3 & 1 \end{array}\right)$$

In the matrix on the left, none of the row transformations will allow us to get a zero as the first component of the second row, along with a one as the second component (convince yourself of this). Therefore, it is impossible to convert the matrix D to the identity matrix by using row transformations. It follows that matrix D has *no inverse*. Such a matrix is called a **singular** matrix; a matrix that has an inverse is **nonsingular**.

The reason we could not transform the matrix D to the identity matrix is that, during the transformation process, we arrived at a matrix containing a row of zeros. Whenever this occurs, we can conclude that the matrix with which we are working is singular.

In Chapter 8, we shall see the notion of the matrix inverse applied to several areas. Exercise 26 also involves an application of the matrix inverse. For additional applications, consult Chapter 5 of *Finite Mathematics with Applications, 2nd Edition* by L. Gilligan and R. Nenno (Glenview, IL: Scott, Foresman, 1979).

EXERCISES FOR SECTION 7.4

In Exercises 1 through 5, determine whether or not the given matrix is square. If it is square, find its order (see Example 1).

1. $\begin{pmatrix} 1 & 3 & 5 \\ 6 & 7 & 8 \end{pmatrix}$

2. $\begin{pmatrix} 1 & 4 \\ -3 & 6 \end{pmatrix}$

3. $\begin{pmatrix} 0 & 0 \\ 0 & 7 \end{pmatrix}$

4. $\begin{pmatrix} 1 & 0 & -6 \\ 10 & 15 & -3 \\ -2 & 10 & 45 \end{pmatrix}$

5. $\begin{pmatrix} 1 & 0 & 0 & 0 \\ 0 & 1 & 0 & 0 \\ 0 & 0 & 1 & 0 \\ 0 & 0 & 0 & 1 \end{pmatrix}$

6. **a.** What is the identity matrix of order 5?

b. Verify your answer to part a by multiplying your answer by matrix F below (see Example 2).

$$F = \begin{pmatrix} 2 & 3 & -5 & 6 & 0 \\ 1 & 2 & 7 & -5 & 3 \\ 0 & 0 & 4 & 1 & -2 \\ 1 & 1 & 1 & 3 & 1 \\ 0 & 2 & -3 & 0 & -1 \end{pmatrix}$$

In Exercises 7 through 15, determine whether or not $B = A^{-1}$. That is, determine whether or not the given matrices are inverses of each other (see Examples 3–5).

7. $A = \begin{pmatrix} 4 & 1 \\ 7 & 2 \end{pmatrix} \quad B = \begin{pmatrix} 2 & -1 \\ -7 & 4 \end{pmatrix}$

8. $A = \begin{pmatrix} 8 & 2 \\ 14 & 4 \end{pmatrix} \qquad B = \begin{pmatrix} 1 & -\frac{1}{2} \\ \frac{7}{2} & 2 \end{pmatrix}$

9. $A = \begin{pmatrix} 2 & 1 \\ 3 & 2 \end{pmatrix} \qquad B = \begin{pmatrix} 2 & -1 \\ -3 & 2 \end{pmatrix}$

10. $A = \begin{pmatrix} 4 & 2 \\ 6 & 4 \end{pmatrix} \qquad B = \begin{pmatrix} 1 & -\frac{1}{2} \\ -\frac{3}{2} & 1 \end{pmatrix}$

11. $A = \begin{pmatrix} 1 & 2 \\ 3 & 4 \end{pmatrix} \qquad B = \begin{pmatrix} -2 & 1 \\ \frac{3}{2} & -\frac{1}{2} \end{pmatrix}$

12. $A = \begin{pmatrix} -4 & 2 \\ 3 & -1 \end{pmatrix} \qquad B = \begin{pmatrix} 0.5 & 1 \\ 1.5 & 1 \end{pmatrix}$

13. $A = \begin{pmatrix} 1 & 0 & 8 \\ 5 & 10 & -20 \\ 0 & 0 & 2 \end{pmatrix} \qquad B = \begin{pmatrix} 1 & 0 & 4 \\ \frac{1}{2} & \frac{1}{10} & 3 \\ 0 & 0 & \frac{1}{2} \end{pmatrix}$

14. $A = \begin{pmatrix} 1 & 3 & 0 \\ 2 & 0 & 4 \\ 1 & -3 & 3 \end{pmatrix} \qquad B = \begin{pmatrix} 2 & \frac{3}{2} & 2 \\ \frac{1}{3} & \frac{1}{2} & -\frac{2}{3} \\ -1 & 1 & 1 \end{pmatrix}$

15. $A = \begin{pmatrix} 2 & 3 & -4 \\ 1 & 1 & 1 \\ -1 & 7 & -10 \end{pmatrix} \qquad B = \begin{pmatrix} \frac{17}{39} & -\frac{2}{39} & -\frac{7}{39} \\ -\frac{9}{39} & \frac{24}{39} & \frac{6}{39} \\ -\frac{8}{39} & \frac{17}{39} & \frac{1}{39} \end{pmatrix}$

In Exercises 16 through 25, use row transformations to find the inverse of each matrix, if possible. If not possible, write "singular." Check your results by multiplying your inverse by the original matrix to obtain I (see Example 6).

16. $\begin{pmatrix} 2 & -1 \\ -3 & 2 \end{pmatrix}$

17. $\begin{pmatrix} 1 & 2 \\ 2 & 3 \end{pmatrix}$

18. $\begin{pmatrix} 7 & 4 \\ 3 & 2 \end{pmatrix}$

19. $\begin{pmatrix} -3 & 7 \\ -2 & 5 \end{pmatrix}$

20. $\begin{pmatrix} 2 & 7 \\ 4 & 14 \end{pmatrix}$

21. $\begin{pmatrix} -6 & 3 \\ -4 & 2 \end{pmatrix}$

22. $\begin{pmatrix} 2 & 0 & 0 \\ 0 & 2 & 0 \\ 0 & 0 & 2 \end{pmatrix}$

23. $\begin{pmatrix} 0 & 0 & 4 \\ 0 & 4 & 0 \\ 4 & 0 & 0 \end{pmatrix}$

24. $\begin{pmatrix} -5 & 2 & -1 \\ 2 & -2 & 2 \\ 1 & 0 & -1 \end{pmatrix}$

25. $\begin{pmatrix} 2 & 7 & -1 \\ 1 & 3 & 4 \\ -1 & -4 & 5 \end{pmatrix}$

26. One application of the matrix inverse is to the *Leontief input-output analysis* in economics. There, if we are given a matrix

$$A = \begin{pmatrix} 0.4 & 0.2 & 0.1 \\ 0.3 & 0.1 & 0.2 \\ 0.6 & 0.2 & 0.1 \end{pmatrix}$$

it is necessary (because of the theory behind the model) to find $(I - A)^{-1}$.

a. For A above, find $I - A$.

b. Find $(I - A)^{-1}$.

In Exercises 27 through 30, consult Exercise 26 and find $(I - A)^{-1}$ for each given matrix, A.

27. $A = \begin{pmatrix} 0.5 & 0.2 \\ 0.1 & 0.3 \end{pmatrix}$

28. $A = \begin{pmatrix} 0.1 & 0.2 \\ 0.3 & 0.5 \end{pmatrix}$

29. $A = \begin{pmatrix} 0.1 & 0.3 & 0.2 \\ 0.5 & 0 & 0.1 \\ 0.3 & 0.4 & 0 \end{pmatrix}$

30. $A = \begin{pmatrix} 0 & 0 & 0.3 \\ 0.5 & 0.2 & 0.1 \\ 0.6 & 0.2 & 0.1 \end{pmatrix}$

7.5 Determinants

With every square matrix is associated a numerical value called a **determinant**. As we shall see in Section 8.4, the notion of a determinant is very useful in solving systems of linear equations. For now, we examine how to calculate this numerical value.

Determinant of a 2 × 2 Matrix

Let A be a general 2×2 matrix, that is,

$$A = \begin{pmatrix} a_{11} & a_{12} \\ a_{21} & a_{22} \end{pmatrix}$$

The **determinant** of A is denoted

$$|A| \qquad \text{or} \qquad \begin{vmatrix} a_{11} & a_{12} \\ a_{21} & a_{22} \end{vmatrix}$$

Notice that the components of A are enclosed by straight lines in the determinant. The numerical value of the determinant is defined as follows:

$$\begin{vmatrix} a_{11} & a_{12} \\ a_{21} & a_{22} \end{vmatrix} = a_{11} \cdot a_{22} - a_{12} \cdot a_{21}$$

The determinant is the difference of the diagonal products. The product of the components on the diagonal going from upper left to lower right is preceded by a plus sign; the product of the components on the diagonal going from upper right to lower left is preceded by a minus sign.

EXAMPLE 1

Find the determinants of the following matrices:

a. $A = \begin{pmatrix} 1 & 2 \\ 3 & 4 \end{pmatrix}$ **b.** $B = \begin{pmatrix} 7 & 5 \\ -2 & 0 \end{pmatrix}$ **c.** $C = \begin{pmatrix} 4 & \frac{1}{2} \\ 8 & 1 \end{pmatrix}$

Solution

a. $\begin{vmatrix} 1 & 2 \\ 3 & 4 \end{vmatrix} = (1 \cdot 4) - (2 \cdot 3) = 4 - 6 = -2$

b. $\begin{vmatrix} 7 & 5 \\ -2 & 0 \end{vmatrix} = (7 \cdot 0) - [5 \cdot (-2)] = 0 - (-10) = 10$

c. $\begin{vmatrix} 4 & \frac{1}{2} \\ 8 & 1 \end{vmatrix} = (4 \cdot 1) - (\frac{1}{2} \cdot 8) = 4 - 4 = 0$

Notice that the determinant of matrix C above is zero. Whenever the determinant of a square matrix is zero, it turns out that that matrix is a *singular* matrix—it has no inverse (see Section 7.4). If the determinant of a matrix is not zero, the matrix is *nonsingular*.

In addition to its potential in solving systems of linear equations, the determinant offers us an immediate bonus. It can be used in a more systematic (but not necessarily shorter) algorithm for finding the inverse of a nonsingular matrix. This new algorithm is also adaptable for use in computer programming.

We illustrate by finding the inverse of matrix A of Example 1. There are four steps in the procedure:

1. Find the determinant of A

$$\begin{vmatrix} 1 & 2 \\ 3 & 4 \end{vmatrix} = -2 \qquad \text{(See Example 1)}$$

   A has an inverse, since $|A| \neq 0$.

2. Interchange the components on the main diagonal of A. This gives the matrix

$$\begin{pmatrix} 4 & 2 \\ 3 & 1 \end{pmatrix}$$

3. Change the signs of a_{12} and a_{21} (the two remaining components of A)—that is, multiply a_{12} and a_{21} by -1. This gives the matrix

$$\begin{pmatrix} 4 & -2 \\ -3 & 1 \end{pmatrix}$$

4. Finally, multiply this matrix by the multiplicative inverse (reciprocal) of the determinant of A. The result is the matrix A^{-1}. The reciprocal of -2 is $-\frac{1}{2}$, so

$$A^{-1} = -\tfrac{1}{2}\begin{pmatrix} 4 & -2 \\ -3 & 1 \end{pmatrix} = \begin{pmatrix} -2 & 1 \\ \frac{3}{2} & -\frac{1}{2} \end{pmatrix}$$

The reader should verify that this matrix really is the inverse of A by checking the product $A \cdot A^{-1}$ (and $A^{-1} \cdot A$).

EXAMPLE 2

Find the inverse of each of the following matrices:

a. $B = \begin{pmatrix} 7 & 5 \\ -2 & 0 \end{pmatrix}$ **b.** $X = \begin{pmatrix} -3 & 5 \\ 6 & -10 \end{pmatrix}$

Solution

a. 1. $\begin{vmatrix} 7 & 5 \\ -2 & 0 \end{vmatrix} = 10$ The determinant of B (see Example 1).

2. $\begin{pmatrix} 0 & 5 \\ -2 & 7 \end{pmatrix}$ Interchanging the components on the main diagonal.

3. $\begin{pmatrix} 0 & -5 \\ 2 & 7 \end{pmatrix}$ Multiplying the other components by -1.

4. $B^{-1} = \frac{1}{10}\begin{pmatrix} 0 & -5 \\ 2 & 7 \end{pmatrix} = \begin{pmatrix} 0 & -\frac{1}{2} \\ \frac{2}{10} & \frac{7}{10} \end{pmatrix} = \begin{pmatrix} 0 & -\frac{1}{2} \\ \frac{1}{5} & \frac{7}{10} \end{pmatrix}$ $\frac{1}{10}$ is the reciprocal of 10.

Check

$$B \cdot B^{-1} = \begin{pmatrix} 7 & 5 \\ -2 & 0 \end{pmatrix} \cdot \begin{pmatrix} 0 & -\frac{1}{2} \\ \frac{1}{5} & \frac{7}{10} \end{pmatrix}$$

$$= \begin{pmatrix} 7(0) + 5(\frac{1}{5}) & 7(-\frac{1}{2}) + 5(\frac{7}{10}) \\ -2(0) + 0(\frac{1}{5}) & -2(-\frac{1}{2}) + 0(\frac{7}{10}) \end{pmatrix} = \begin{pmatrix} 1 & 0 \\ 0 & 1 \end{pmatrix} = I$$

b. 1. $|X| = \begin{vmatrix} -3 & 5 \\ 6 & -10 \end{vmatrix} = (-3) \cdot (-10) - (5 \cdot 6)$

$$= 30 - 30 = 0$$

Since $|X| = 0$, X has no inverse; X is a *singular* matrix.

Determinant of a 3 × 3 Matrix

One method of finding the determinant of a 3 × 3 matrix is to extend the process, used for 2 × 2 matrices, of multiplying diagonal components. We will not, however, be able to extend this process beyond a 3 × 3 matrix.

If

$$A = \begin{pmatrix} a_{11} & a_{12} & a_{13} \\ a_{21} & a_{22} & a_{23} \\ a_{31} & a_{32} & a_{33} \end{pmatrix}$$

Then

$$|A| = \begin{vmatrix} a_{11} & a_{12} & a_{13} \\ a_{21} & a_{22} & a_{23} \\ a_{31} & a_{32} & a_{33} \end{vmatrix}$$

$$= a_{11}a_{22}a_{33} + a_{12}a_{23}a_{31} + a_{13}a_{21}a_{32}$$
$$- a_{13}a_{22}a_{31} - a_{11}a_{23}a_{32} - a_{12}a_{21}a_{33}$$

Luckily, it will not be necessary to memorize this formula. We illustrate with an example.

EXAMPLE 3

Find the determinants of the following matrices:

a. $\begin{pmatrix} 1 & 2 & 3 \\ 4 & 5 & 6 \\ 7 & 8 & 9 \end{pmatrix}$ **b.** $\begin{pmatrix} 1 & 2 & 0 \\ -1 & 1 & 2 \\ 1 & 0 & -1 \end{pmatrix}$

Solution

a. To find

$$\begin{vmatrix} 1 & 2 & 3 \\ 4 & 5 & 6 \\ 7 & 8 & 9 \end{vmatrix}$$

we first reproduce the first two columns of the matrix to the right of the determinant:

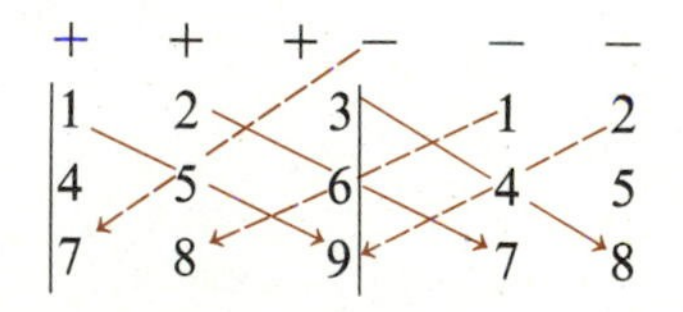

Now we draw all three-element diagonals. To calculate the determinant, we will *add* the products of the components on diagonals that go from upper left to lower right (solid arrows) and *subtract* the products of the components on diagonals that go from upper right to lower left (dashed arrows). So

$$\begin{vmatrix} 1 & 2 & 3 \\ 4 & 5 & 6 \\ 7 & 8 & 9 \end{vmatrix} = \overbrace{(1 \cdot 5 \cdot 9) + (2 \cdot 6 \cdot 7) + (3 \cdot 4 \cdot 8)}^{(a_{11}a_{22}a_{33}) + (a_{12}a_{23}a_{31}) + (a_{13}a_{21}a_{32})}$$

$$\underbrace{-(3 \cdot 5 \cdot 7) - (1 \cdot 6 \cdot 8) - (2 \cdot 4 \cdot 9)}_{-(a_{13}a_{22}a_{31}) - (a_{11}a_{23}a_{32}) - (a_{12}a_{21}a_{33})}$$

$$= 45 + 84 + 96 - 105 - 48 - 72$$

$$= 0 \qquad \text{We have a } \textit{singular} \text{ matrix.}$$

Note that our results coincide with the formula above for the determinant of any 3×3 matrix.

b. We write

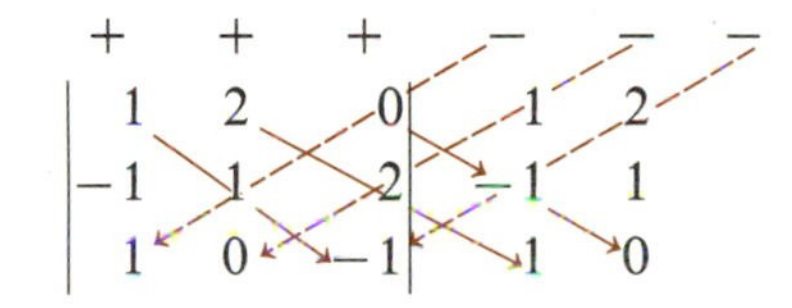

So

$$\begin{vmatrix} 1 & 2 & 0 \\ -1 & 1 & 2 \\ 1 & 0 & -1 \end{vmatrix} = (1 \cdot 1 \cdot (-1)) + (2 \cdot 2 \cdot 1) + (0 \cdot (-1) \cdot 0) - (0 \cdot 1 \cdot 1) - (1 \cdot 2 \cdot 0) - (2 \cdot (-1) \cdot (-1))$$

$$= -1 + 4 + 0 - 0 - 0 - 2 = 1$$

Using Cofactors to Evaluate Determinants

Another method of evaluating determinants of 3×3 matrices, which can be easily extended to matrices of order greater than 3, involves the use of *cofactors*. We shall confine our examples to 3×3 matrices; the concepts illustrated, however, apply to matrices of any order.

We first define the *minor* of an element, say m_{rc}, of a matrix M. The **minor** of the component m_{rc} is the value obtained by deleting row r and column c

from the matrix M and then calculating the determinant of the resulting matrix. Some examples will illustrate this concept.

EXAMPLE 4

Find the minors of the following elements of M:

$$M = \begin{pmatrix} 4 & 5 & 0 \\ 2 & 3 & -1 \\ -7 & 0 & 1 \end{pmatrix}$$

a. m_{11} **b.** m_{23} **c.** m_{31} **d.** m_{32}

Solution

a. Deleting row 1 and column 1 gives

$$\text{minor of } m_{11} = \begin{vmatrix} 3 & -1 \\ 0 & 1 \end{vmatrix} = (3 \cdot 1) - (-1 \cdot 0) = 3$$

b. Deleting row 2 and column 3 gives

$$\text{minor of } m_{23} = \begin{vmatrix} 4 & 5 \\ -7 & 0 \end{vmatrix} = (4 \cdot 0) - (5 \cdot (-7)) = 35$$

c. Deleting row 3 and column 1 gives

$$\text{minor of } m_{31} = \begin{vmatrix} 5 & 0 \\ 3 & -1 \end{vmatrix} = 5 \cdot (-1) - (0 \cdot 3) = -5$$

d. Deleting row 3 and column 2 gives

$$\text{minor of } m_{32} = \begin{vmatrix} 4 & 0 \\ 2 & -1 \end{vmatrix} = 4 \cdot (-1) - (0 \cdot 2) = -4$$

COFACTOR OF A MATRIX

The *cofactor* of the element m_{rc} of a matrix M is given by

1. $+1$ times the minor of m_{rc} if the sum $r + c$ is even, or
2. -1 times the minor of m_{rc} if the sum $r + c$ is odd.

Alternatively, the cofactor of $m_{rc} = (-1)^{r+c} \cdot$ minor of m_{rc}.

EXAMPLE 5

Let M be the matrix of Example 4,

$$M = \begin{pmatrix} 4 & 5 & 0 \\ 2 & 3 & -1 \\ -7 & 0 & 1 \end{pmatrix}$$

Find the cofactors of the following elements of M:

a. m_{11} **b.** m_{23} **c.** m_{31} **d.** m_{32}

Solution

We'll use the results of Example 4.

a. $r + c = 1 + 1 = 2$, which is even. So the cofactor of $m_{11} = +1 \cdot \text{minor of } m_{11} = +1 \cdot 3 = 3$

b. $2 + 3 = 5$, which is odd. So the cofactor of $m_{23} = -1 \cdot 35 = -35$.

c. $3 + 1 = 4$, which is even. So the cofactor of $m_{31} = +1 \cdot (-5) = -5$.

d. $3 + 2 = 5$, which is odd. So the cofactor of $m_{32} = -1 \cdot (-4) = 4$.

We will now explain how we can use cofactors to evaluate the determinant of a matrix. For a 3×3 matrix, this will *not* be shorter or easier than the method we have already discussed. It has the advantage, however, of making it possible to reduce the problem of finding the determinant of *any* square matrix to one of finding determinants of 2×2 matrices.

We illustrate by evaluating the determinant of the matrix in part a of Example 3.

$$\begin{vmatrix} 1 & 2 & 3 \\ 4 & 5 & 6 \\ 7 & 8 & 9 \end{vmatrix}$$

We first choose any row of this matrix (say row 1); we shall find the determinant by "expanding" on this row. We find the cofactor of each component in the chosen row, and multiply each cofactor by the corresponding component. The determinant is the sum of these products.

$$\begin{aligned} \begin{vmatrix} 1 & 2 & 3 \\ 4 & 5 & 6 \\ 7 & 8 & 9 \end{vmatrix} &= 1 \cdot \text{cofactor of } a_{11} + 2 \cdot \text{cofactor of } a_{12} + 3 \cdot \text{cofactor of } a_{13} \\ &= 1 \cdot (+1) \cdot \begin{vmatrix} 5 & 6 \\ 8 & 9 \end{vmatrix} + 2 \cdot (-1) \cdot \begin{vmatrix} 4 & 6 \\ 7 & 9 \end{vmatrix} + 3 \cdot (+1) \cdot \begin{vmatrix} 4 & 5 \\ 7 & 8 \end{vmatrix} \\ &= 1 \cdot (45 - 48) - 2 \cdot (36 - 42) + 3 \cdot (32 - 35) \\ &= 1 \cdot (-3) - 2 \cdot (-6) + 3 \cdot (-3) \\ &= -3 + 12 - 9 = 0 \end{aligned}$$

Note that this coincides with the result of Example 3.

EXAMPLE 6

Find the determinants of the following matrices using cofactors:

a. $A = \begin{pmatrix} 1 & 2 & 0 \\ -1 & 1 & 2 \\ 1 & 0 & -1 \end{pmatrix}$ **b.** $M = \begin{pmatrix} 4 & 5 & 0 \\ 2 & 3 & -1 \\ -7 & 0 & 1 \end{pmatrix}$

Solution

a. This time, we shall expand on row 3.

$$\begin{vmatrix} 1 & 2 & 0 \\ -1 & 1 & 2 \\ 1 & 0 & -1 \end{vmatrix} = \underset{a_{31}}{\underset{\uparrow}{+1}} \cdot \underbrace{(+1) \cdot \begin{vmatrix} 2 & 0 \\ 1 & 2 \end{vmatrix}}_{\text{Cofactor of } a_{31}} + \underset{a_{32}}{\underset{\uparrow}{0}} \cdot \underbrace{(-1) \cdot \begin{vmatrix} 1 & 0 \\ -1 & 2 \end{vmatrix}}_{\text{Cofactor of } a_{32}}$$

$$+ \underset{a_{33}}{\underset{\uparrow}{(-1)}} \cdot \underbrace{(+1) \cdot \begin{vmatrix} 1 & 2 \\ -1 & 1 \end{vmatrix}}_{\text{Cofactor of } a_{33}}$$

$$= 1 \cdot (4 - 0) + 0 + (-1) \cdot [1 - (-2)]$$

$$= 4 + 0 - 3 = 1$$

b. Expanding on row 1 gives:

$$\begin{vmatrix} 4 & 5 & 0 \\ 2 & 3 & -1 \\ -7 & 0 & 1 \end{vmatrix} = \underset{m_{11}}{\underset{\uparrow}{4}} \cdot \underbrace{(+1) \cdot \begin{vmatrix} 3 & -1 \\ 0 & 1 \end{vmatrix}}_{\text{Cofactor of } m_{11}} + \underset{m_{12}}{\underset{\uparrow}{5}} \cdot \underbrace{(-1) \cdot \begin{vmatrix} 2 & -1 \\ -7 & 1 \end{vmatrix}}_{\text{Cofactor of } m_{12}}$$

$$+ \underset{m_{13}}{\underset{\uparrow}{0}} \cdot \underbrace{(+1) \cdot \begin{vmatrix} 2 & 3 \\ -7 & 0 \end{vmatrix}}_{\text{Cofactor of } m_{13}}$$

$$= 4 \cdot (3 - 0) + 5 \cdot [-(2 - 7)] + 0$$

$$= (4 \cdot 3) + (5 \cdot 5)$$

$$= 12 + 25 = 37$$

Advantages of this method of evaluating determinants include:

1. It is adaptable to square matrices of any order.
2. For 3×3 matrices, it usually involves fewer calculations with smaller numbers than does the "diagonal product" method.
3. Since we can expand on any row, we can choose the easiest one (preferably one with a zero component!)

Using Cofactors to Find the Inverse of a Matrix

Cofactors can also be used to find the inverse of a given square matrix, if it exists. Suppose, for example, we wish to find the inverse of the matrix.

$$A = \begin{pmatrix} 1 & 2 & 0 \\ -1 & 1 & 2 \\ 1 & 0 & -1 \end{pmatrix}$$

of Example 6. The process involves four steps:

1. Find the determinant of A.

$$\begin{vmatrix} 1 & 2 & 0 \\ -1 & 1 & 2 \\ 1 & 0 & -1 \end{vmatrix} = 1$$

   If the determinant of A were zero, we would be finished since A would be a singular matrix and therefore have no inverse.

2. Replace each component of A with its cofactor, forming a new matrix. This requires calculating the cofactor of each component of A. A table helps organize the information.

	Component	Cofactor	
a_{11}	1	-1	$(+1)\cdot\begin{vmatrix} 1 & 2 \\ 0 & -1 \end{vmatrix} = -1$
a_{12}	2	1	$(-1)\cdot\begin{vmatrix} -1 & 2 \\ 1 & -1 \end{vmatrix} = -(1-2)$
a_{13}	0	-1	$(+1)\cdot\begin{vmatrix} -1 & 1 \\ 1 & 0 \end{vmatrix} = 0 - 1$
a_{21}	-1	2	$(-1)\cdot\begin{vmatrix} 2 & 0 \\ 0 & -1 \end{vmatrix} = (-1)\cdot(-2)$
a_{22}	1	-1	$(+1)\cdot\begin{vmatrix} 1 & 0 \\ 1 & -1 \end{vmatrix} = -1 - 0$
a_{23}	2	2	$(-1)\cdot\begin{vmatrix} 1 & 2 \\ 1 & 0 \end{vmatrix} = -(0-2)$

	Component	Cofactor	
a_{31}	1	4	$(+1)\cdot\begin{vmatrix} 2 & 0 \\ 1 & 2 \end{vmatrix} = 4 - 0$
a_{32}	0	-2	$(-1)\cdot\begin{vmatrix} 1 & 0 \\ -1 & 2 \end{vmatrix} = -(2 - 0)$
a_{33}	-1	3	$(+1)\cdot\begin{vmatrix} 1 & 2 \\ -1 & 1 \end{vmatrix} = 1 - (-2)$

The new matrix of the cofactors is

$$\begin{pmatrix} -1 & 1 & -1 \\ 2 & -1 & 2 \\ 4 & -2 & 3 \end{pmatrix}$$

3. Form the *transpose* of the new matrix. The transpose of a matrix is the matrix formed by interchanging the rows and columns of the original matrix:

   row 1 becomes column 1,
   row 2 becomes column 2,
   row 3 becomes column 3, etc.

   In this case, the transpose is

$$\begin{pmatrix} -1 & 2 & 4 \\ 1 & -1 & -2 \\ -1 & 2 & 3 \end{pmatrix}$$

4. This matrix multiplied by the multiplicative inverse (reciprocal) of the determinant of A, is the inverse of A. Since the determinant of A is 1, its reciprocal is also 1:

$$A^{-1} = 1\cdot\begin{pmatrix} -1 & 2 & 4 \\ 1 & -1 & -2 \\ -1 & 2 & 3 \end{pmatrix} = \begin{pmatrix} -1 & 2 & 4 \\ 1 & -1 & -2 \\ -1 & 2 & 3 \end{pmatrix}$$

The reader should find the product $A \cdot A^{-1}$ to verify that these calculations are correct.

EXAMPLE 7

Find the inverse, if possible, of each of the following matrices:

a. $X = \begin{pmatrix} -3 & 4 & -3 \\ 2 & 3 & 2 \\ 1 & -5 & 1 \end{pmatrix}$ **b.** $Y = \begin{pmatrix} 3 & -1 & 0 \\ 5 & 1 & 2 \\ 2 & 1 & 0 \end{pmatrix}$

Solution

a. 1. Find the determinant of X. We expand on row 2:

$$\begin{aligned} |X| &= \begin{vmatrix} -3 & 4 & -3 \\ 2 & 3 & 2 \\ 1 & -5 & 1 \end{vmatrix} \\ &= 2 \cdot (-1) \begin{vmatrix} 4 & -3 \\ -5 & 1 \end{vmatrix} + 3 \cdot (+1) \begin{vmatrix} -3 & -3 \\ 1 & 1 \end{vmatrix} \\ &\quad + 2 \cdot (-1) \begin{vmatrix} -3 & 4 \\ 1 & -5 \end{vmatrix} \\ &= -2(4 - 15) + 3(-3 + 3) - 2(15 - 4) \\ &= -2(-11) + 3(0) - 2(11) \\ &= 22 + 0 - 22 = 0 \end{aligned}$$

Since $|X| = 0$, X is a singular matrix. No inverse exists, so we're done.

b. 1. Find $|Y|$.
We expand on row 1:

$$\begin{aligned} |Y| &= \begin{vmatrix} 3 & -1 & 0 \\ 5 & 1 & 2 \\ 2 & 1 & 0 \end{vmatrix} \\ &= 3 \cdot (1) \begin{vmatrix} 1 & 2 \\ 1 & 0 \end{vmatrix} + (-1) \cdot (-1) \begin{vmatrix} 5 & 2 \\ 2 & 0 \end{vmatrix} + 0 \cdot (1) \begin{vmatrix} 5 & 1 \\ 2 & 1 \end{vmatrix} \\ &= 3(0 - 2) + 1(0 - 4) + 0 \\ &= -6 - 4 = -10 \end{aligned}$$

2. Replace each component of Y with its cofactor.

	Component	Cofactor	
y_{11}	3	-2	$+1 \cdot \begin{vmatrix} 1 & 2 \\ 1 & 0 \end{vmatrix} = 0 - 2$
y_{12}	-1	4	$-1 \cdot \begin{vmatrix} 5 & 2 \\ 2 & 0 \end{vmatrix} = -(0 - 4)$
y_{13}	0	3	$+1 \cdot \begin{vmatrix} 5 & 1 \\ 2 & 1 \end{vmatrix} = 5 - 2$
y_{21}	5	0	$-1 \cdot \begin{vmatrix} -1 & 0 \\ 1 & 0 \end{vmatrix} = -(0 - 0) = 0$
y_{22}	1	0	$+1 \cdot \begin{vmatrix} 3 & 0 \\ 2 & 0 \end{vmatrix} = 0 - 0 = 0$
y_{23}	2	-5	$-1 \cdot \begin{vmatrix} 3 & -1 \\ 2 & 1 \end{vmatrix} = -(3 + 2)$
y_{31}	2	-2	$+1 \cdot \begin{vmatrix} -1 & 0 \\ 1 & 2 \end{vmatrix} = -2 - 0$
y_{32}	1	-6	$-1 \cdot \begin{vmatrix} 3 & 0 \\ 5 & 2 \end{vmatrix} = -(6 - 0)$
y_{33}	0	8	$+1 \cdot \begin{vmatrix} 3 & -1 \\ 5 & 1 \end{vmatrix} = 3 + 5$

Replacing each component by its cofactor gives the new matrix:

$$\begin{pmatrix} -2 & 4 & 3 \\ 0 & 0 & -5 \\ -2 & -6 & 8 \end{pmatrix}$$

3. Transpose the new matrix:

$$\begin{pmatrix} -2 & 0 & -2 \\ 4 & 0 & -6 \\ 3 & -5 & 8 \end{pmatrix}$$

4. Multiply by the reciprocal of -10:

$$Y^{-1} = -\tfrac{1}{10} \cdot \begin{pmatrix} -2 & 0 & -2 \\ 4 & 0 & -6 \\ 3 & -5 & 8 \end{pmatrix}$$

$$Y^{-1} = \begin{pmatrix} \frac{1}{5} & 0 & \frac{1}{5} \\ -\frac{2}{5} & 0 & \frac{3}{5} \\ -\frac{3}{10} & \frac{1}{2} & -\frac{4}{5} \end{pmatrix}$$

The reader should verify that $Y \cdot Y^{-1}$ is the identity.

EXERCISES FOR SECTION 7.5

In Exercises 1 through 10, find the determinant for each 2×2 matrix (see Example 1).

1. $\begin{pmatrix} 3 & 6 \\ 1 & 5 \end{pmatrix}$ **2.** $\begin{pmatrix} 4 & 10 \\ 1 & 6 \end{pmatrix}$ **3.** $\begin{pmatrix} 5 & 25 \\ 2 & 8 \end{pmatrix}$

4. $\begin{pmatrix} 10 & 6 \\ 15 & 4 \end{pmatrix}$ **5.** $\begin{pmatrix} 1 & 3 \\ 2 & -2 \end{pmatrix}$ **6.** $\begin{pmatrix} 0 & 3 \\ 5 & 2 \end{pmatrix}$

7. $\begin{pmatrix} 0 & 3 \\ 5 & -2 \end{pmatrix}$ **8.** $\begin{pmatrix} 0 & 3 \\ -5 & -2 \end{pmatrix}$ **9.** $\begin{pmatrix} 1 & -8 \\ 2 & 5 \end{pmatrix}$

10. $\begin{pmatrix} -1 & -7 \\ 5 & -3 \end{pmatrix}$

In Exercises 11 through 20, find the inverse of each matrix by using determinants (see Example 2).

11. $\begin{pmatrix} 3 & 1 \\ 11 & 4 \end{pmatrix}$ **12.** $\begin{pmatrix} 7 & 2 \\ 4 & 1 \end{pmatrix}$ **13.** $\begin{pmatrix} 4 & 1 \\ 7 & 2 \end{pmatrix}$

14. $\begin{pmatrix} -2 & 1 \\ -7 & 4 \end{pmatrix}$ **15.** $\begin{pmatrix} 2 & 1 \\ 3 & 2 \end{pmatrix}$ **16.** $\begin{pmatrix} 5 & 4 \\ 4 & 2 \end{pmatrix}$

17. $\begin{pmatrix} 1 & 2 \\ 3 & 1 \end{pmatrix}$ **18.** $\begin{pmatrix} 1 & 2 \\ 2 & 5 \end{pmatrix}$ **19.** $\begin{pmatrix} 3 & 4 \\ -2 & 6 \end{pmatrix}$

20. $\begin{pmatrix} \frac{1}{2} & \frac{1}{2} \\ 1 & 2 \end{pmatrix}$

In Exercises 21 through 26, find (a) the minor of the given component; (b) the cofactor of the given component; and (c) the determinant of the 3 × 3 matrix (see Examples 4–6).

21. a_{12} of $A = \begin{pmatrix} 4 & 5 & 0 \\ 2 & 3 & -1 \\ -7 & 0 & 1 \end{pmatrix}$

22. a_{13} of $A = \begin{pmatrix} 4 & 5 & 0 \\ 2 & 3 & -1 \\ -7 & 0 & 1 \end{pmatrix}$

23. a_{31} of $A = \begin{pmatrix} 1 & 0 & 8 \\ 5 & 10 & -20 \\ 0 & 0 & 2 \end{pmatrix}$

24. a_{12} of $A = \begin{pmatrix} 1 & 0 & 8 \\ 5 & 10 & -20 \\ 0 & 0 & 2 \end{pmatrix}$

25. a_{22} of $A = \begin{pmatrix} 2 & 3 & -4 \\ 1 & 1 & 1 \\ -1 & 7 & -10 \end{pmatrix}$

26. a_{23} of $A = \begin{pmatrix} 2 & 3 & -4 \\ 1 & 1 & 1 \\ -1 & 7 & -10 \end{pmatrix}$

In Exercises 27 through 30, use determinants to find the inverse of each matrix (see Example 7).

27. Use the matrix of Exercise 21.

28. Use the matrix of Exercise 23.

29. Use the matrix of Exercise 25.

30. $\begin{pmatrix} 1 & 5 & 3 \\ -2 & 6 & 0 \\ 5 & -4 & 1 \end{pmatrix}$

7.6 Matrices and BASIC (Optional)

We can combine the concepts of algorithms and flowcharts from Chapter 6 with the ideas of matrix operations to see how we might program a computer in BASIC to perform matrix arithmetic.

We know that the DIM statement creates a subscripted variable. For example

```
10  DIM A(10,10), B(10,10)
```

establishes 100 variables in the A array and 100 subscripted B variables. For the remainder of this section, we shall assume that no matrix need be larger than a 10 × 10 matrix.

Data and BASIC

Once the matrix has been established with a DIM statement, data for its components must be put into the storage locations. In BASIC there are three ways we can accomplish this:

1. have the data stored in the program by using READ and DATA statements.
2. input the data from the keyboard during program execution using an INPUT statement.
3. input the data from a file where data is stored on a disk, tape, or other external medium.

We shall concentrate on the first two options only.

EXAMPLE 1

Write a BASIC program to create the following two arrays using READ and DATA statements:

$$A = \begin{pmatrix} 1 & 4 & 16 \\ 6 & -7 & 39 \end{pmatrix} \quad \text{and} \quad B = \begin{pmatrix} 5 & 0 \\ 46 & 3 \end{pmatrix}$$

Solution

We use the variables R and C as subscripts in the programs below.
Also recall that BASIC's FOR/NEXT statements create loops of instructions.

The algorithm	*The FOR/NEXT loop*
R = 1	
DOWHILE R <= 2	100 FOR R = 1 TO 2
C = 1	
DOWHILE C <= 3	110 FOR C = 1 TO 3
READ A(R, C)	120 READ A(R, C)
C = C + 1	
END C DOWHILE	130 NEXT C
R = R + 1	
END R DOWHILE	140 NEXT R

The data for A and B are placed in DATA statements very carefully *according to rows* in statement 990 below:

```
 10  DIM A(2, 3), B(2, 2)
 20  REM A IS A 2-BY-3 MATRIX AND B IS 2-BY-2
 75  REM MATRIX A IS READ IN IN STATEMENTS 100-140
 85  REM R STANDS FOR ROW, C FOR COLUMN
100  FOR R = 1 TO 2
```

```
110     FOR C = 1 TO 3
120         READ A(R,C)
130     NEXT C
140   NEXT R
150   REM MATRIX B IS READ IN IN STATEMENTS 200-240
200   FOR R = 1 TO 2
210     FOR C = 1 TO 2
220       READ B(R, C)
230     NEXT C
240   NEXT R
990   DATA 1, 4, 16, 6, -7, 39, 5, 0, 46, 3
999   END
```

EXAMPLE 2

How would the program of Example 1 be changed to allow data to be INPUT from the keyboard?

Solution

Statements 120 and 220 could be replaced as follows:

Replace statement 120 with

```
120   PRINT "A'S COMPONENT IN ROW"; R;
      ", COLUMN"; C; "IS";
125   INPUT A(R,C)
```

Replace statement 220 with

```
220   PRINT "B'S COMPONENT IN ROW"; R;
      ", COLUMN"; C; "IS";
225   INPUT B(R, C)
```

The DATA statement 990 would not be necessary in this case; the final, modified program is as follows:

```
 10   DIM A(2,3), B(2,2)
100   FOR R = 1 TO 2
110     FOR C = 1 TO 3
120       PRINT "A'S COMPONENT IN ROW"; R;
          ", COLUMN"; C; "IS";
125         INPUT A(R,C)
130     NEXT C
140   NEXT R
200   FOR R = 1 TO 2
```

```
210     FOR C = 1 TO 2
220       PRINT "B'S COMPONENT IN ROW"; R;
          ", COLUMN"; C; "IS";
225         INPUT B(R,C)
230     NEXT C
240   NEXT R
999   END
```

The program of Example 2 is more versatile than the program of Example 1 because the data is not part of the program. It is still, however, fairly restrictive because it limits A to be 2×3 matrix and B to be a 2×2 matrix. The program in the next example is more general; it inputs components for two matrices of *any* dimension less than 10.

EXAMPLE 3

Write a BASIC program that will input components for two matrices, A and B, of any size.

Solution

```
  5   REM STATEMENT 10 PLACES A
  6   REM REASONABLE RESTRICTION
  7   REM ON MATRICES' DIMENSIONS.
 10   DIM A(10,10), B(10,10)
 20   PRINT "HOW MANY ROWS IN MATRIX A";
 30   INPUT RA
 40   PRINT "HOW MANY COLUMNS IN MATRIX A";
 50   INPUT CA
100   FOR R = 1 TO RA
110     FOR C = 1 TO CA
120       PRINT "MATRIX A ROW"; R;
          "COL"; C; ": ";
125       INPUT A(R,C)
130     NEXT C
140   NEXT R
160   PRINT "HOW MANY ROWS IN MATRIX B";
170   INPUT RB
180   PRINT "HOW MANY COLUMNS IN MATRIX B";
190   INPUT CB
200   FOR R = 1 TO RB
210     FOR C = 1 TO CB
```

```
220       PRINT "MATRIX B ROW"; R;
          "COL"; C; ": ";
225       INPUT B(R,C)
230     NEXT C
240   NEXT R
999   END
```

The next three examples illustrate how to perform arithmetic on matrices.

EXAMPLE 4

Assume matrices A and B have the same dimensions and that their components have already been input. Write a program *segment* that will compute $A + B$.

Solution

We assume that matrix X will equal $A + B$ and that NR represents the number of rows of each matrix and NC represents the number of columns of each matrix. Lines 700–740 represent the computation; lines 800–860 print out the results:

```
 11  DIM X(10,10)
  :
700  FOR R = 1 TO NR
710    FOR C = 1 TO NC
720      X(R,C) = A(R,C) + B(R, C)
730    NEXT C
740  NEXT R
800  PRINT "THE SUM A + B = "
810  FOR R = 1 TO NR
820    FOR C = 1 TO NC
830      PRINT X(R, C); "    ";
840    NEXT C
850    PRINT
860  NEXT R
999  END
```

EXAMPLE 5

Construct a flowchart for solving the problem of multiplying two 3×3 matrices, A and B. Assume the components have already been read in and that the product is stored in matrix X.

Solution

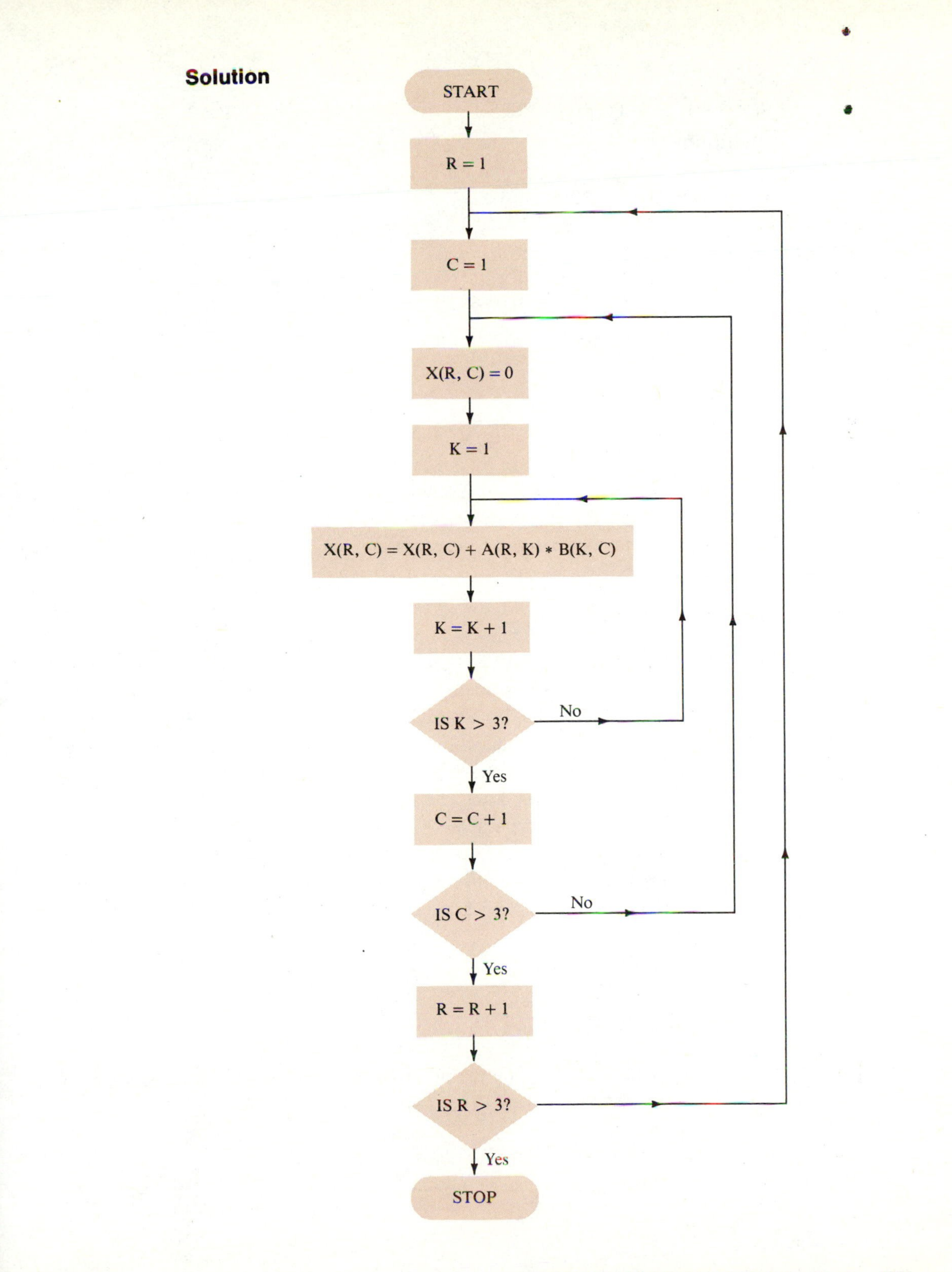

START
R = 1
C = 1
X(R, C) = 0
K = 1
X(R, C) = X(R, C) + A(R, K) * B(K, C)
K = K + 1
IS K > 3?
No
Yes
C = C + 1
IS C > 3?
No
Yes
R = R + 1
IS R > 3?
Yes
STOP

EXAMPLE 6

Code a BASIC program segment to the flowchart of Example 5.

Solution

```
      ⋮
1000  FOR R = 1 TO 3
1020    FOR C = 1 TO 3
1025      X(R, C) = 0
1030      FOR K = 1 TO 3
1040        X(R, C) = X(R, C) + A(R, K) * B(K, C)
1050    NEXT K
1060    NEXT C
1070  NEXT R
      ⋮
```

Note that in the program of Example 6 we have assumed that A, B and X are pre-dimensioned 3×3 matrices. Note also that both the flowchart and the program segment can easily be generalized to products of square matrices of *any* order.

The VisiCalc Program

We introduced this chapter with a picture of a VisiCalc spreadsheet. That and other versions of electronic spreadsheet/accounting programs utilize a matrix-like component system. In the VisiCalc model, for example, its 254 rows are labeled from 1 to 254 and its 63 columns are labeled A through Z, AA through AZ, and BA through BK. Thus, the entire spreadsheet is a 254×63 matrix capable of holding 16,002 entries. Each entry can be a value, a label, even a formula based on other entries. Thus, much capability for design is built in; the user can create "forms" for payroll, cash flow, inventory, tax reports or many other operations. The possibilities seem endless. Such electronic spreadsheets are just one example of the widespread application of the concept of a rectangular array.

EXERCISES FOR SECTION 7.6

In Exercises 1 through 5, refine the program of Example 1 to accommodate the given matrices and then run it.

1. $A = \begin{pmatrix} 1 & 5 & 0 \\ 7 & -3 & 4 \end{pmatrix}$ $\qquad B = \begin{pmatrix} 1 & 6 \\ 7 & 9 \end{pmatrix}$

2. $A = \begin{pmatrix} 1 & 5 & 7 & 3 \\ 2 & 0 & -4 & 6 \\ 1 & 0 & 3 & 9 \end{pmatrix}$ $\quad B = \begin{pmatrix} 8 & 4 & 6 \\ 3 & 9 & -5 \end{pmatrix}$

3. $A = \begin{pmatrix} 1 & 2 & 1 \\ 5 & 0 & 0 \\ 7 & -4 & 3 \\ 3 & 6 & 9 \end{pmatrix}$ $\quad B = \begin{pmatrix} 8 & 3 \\ 4 & 9 \\ 6 & -5 \end{pmatrix}$

4. $A = \begin{pmatrix} 5 & 6 & 3 \\ 1 & 4 & 9 \end{pmatrix}$ $\quad B = (3 \quad 1 \quad 5)$

5. $A = \begin{pmatrix} 1 & 0 \\ 0 & 1 \\ 5 & 7 \end{pmatrix}$ $\quad B = (8 \quad 9)$

In Exercises 6 through 10, use the matrices of Exercises 1 through 5 to refine the program of Example 2. Then run the program.

11. Write an algorithm for solving the problem of multiplying a given $n \times n$ matrix A by a given scalar k.
12. Write a BASIC program for the algorithm of Exercise 11.
13. Assume that matrices A and B have the same dimensions and that their components have already been input. Write a program segment that will compute $k_1 A + k_2 B$ where k_1 and k_2 are INPUT by the user (see Example 4).

Exercises 14 through 17 refer to the following problem: Two matrices A and B are to be multiplied to obtain $A \cdot B$, which is stored as matrix X. The components of A and B have already been read into the appropriate variables and the user has also supplied A's dimensions (RA, CA) and B's dimensions (RB, CB).

14. What will the dimensions of X be?
15. Construct a flowchart to solve the problem. Be sure to test that the product is *possible* (see Example 5).
16. Code a BASIC program to the flowchart of Exercise 15 (see Example 6).
17. Test your program in Exercise 16 with the following:

    a. $A = \begin{pmatrix} 4 & 7 & 3 \\ -2 & 0 & 2 \end{pmatrix}$ $\quad B = \begin{pmatrix} 1 & 2 & 1 \\ 5 & 7 & 1 \\ 8 & 6 & 0 \end{pmatrix}$

b. $A = \begin{pmatrix} 6 & 1 \\ 8 & 0 \\ 0 & 9 \end{pmatrix}$ $\qquad$ $B = \begin{pmatrix} 1 & 1 & 2 & 3 \\ 0 & 1 & 4 & 6 \\ 8 & 0 & -2 & 1 \end{pmatrix}$

c. $A = (8 \quad 0 \quad 9 \quad 3)$ $\qquad$ $B = \begin{pmatrix} 1 \\ 0 \\ 0 \\ 9 \end{pmatrix}$

18. Some versions of BASIC have a MAT READ statement that reads in all components of the matrix. In effect, an *implied nested loop* reads them in just as in Example 1. The program equivalent to the one in Example 1 would then be

```
 10  DIM A(2,3), B(2,2)
125  MAT READ A
135  MAT READ B
990  DATA 1, 4, 16, 6, −7, 39, 5, 0, 46, 3
999  END
```

a. If your version of BASIC has such a statement, run the program above.

b. A MAT PRINT statement, available on some versions of BASIC, automatically prints out the matrix. If the statement

```
200  MAT PRINT A;
```

is added to the program above, the following output occurs:

```
1    4  16
6   −7  39
```

Adapt the program above to print out A and B after they are read in.

19. Use the MAT PRINT statement to modify lines 810–860 of the program in Example 4.

20. Matrix operations are built into some versions of BASIC. For example, the statement

```
400  MAT X = A * B
```

automatically calculates $A * B$ and stores the result in matrix X (provided X has been suitably dimensioned). Use this as well as its analogous statement MAT Y = A + B to write a program that (a) has a menu that asks users whether they want to add matrices or multiply them; (b) inputs elements of the two given matrices, A and B; and (c) performs the desired operation and prints out the results.

7.7 Chapter Review

In Chapter 7, we studied matrices, their arithmetic, and some of their properties. We began by defining a vector (a one dimensional array). We considered some practical applications of vector addition and multiplication.

We then went on to the study of two dimensional arrays—matrices. Section 7.3 presented the addition and multiplication of matrices. Section 7.4 concentrated on some of the properties of square matrices. The concept of an identity matrix was defined, and we learned how to find the inverse of a square matrix, if one exists. In Section 7.5, the cofactor of an element of a matrix was defined, and we studied the role of cofactors in evaluating a determinant and in finding the inverse of a nonsingular matrix.

The chapter concluded with an optional section on matrix manipulation in the BASIC programming language.

VOCABULARY

vector
component
dimension
row vector
column vector
equality of vectors
scalar
DIM statement
matrix
dimensions of a matrix
equality of matrices
square matrix
main diagonal of a square matrix
matrix of order n
identity matrix
inverse of a matrix
row transformations
singular matrix
nonsingular matrix
determinant
minor of an element
cofactor of an element
transpose of a matrix

CHAPTER TEST

Let $A = (1 \quad -3 \quad 6)$, $B = (0 \quad 2 \quad 6)$, and $C = \begin{pmatrix} 2 \\ 1 \\ 3 \end{pmatrix}$. Find the result of performing each of the following operations:

1. $A + B$
2. $B \cdot C$

3. $B - 2A$

4. $\frac{1}{2}(B - 2A)$

Given the matrix

$$M = \begin{pmatrix} 3 & -2 & 4 \\ 1 & 0 & 5 \\ 6 & -6 & -3 \\ -4 & 2 & -7 \end{pmatrix}$$

5. What are the dimensions of M? Is M a square matrix?

6. Identify the following elements of M:

a. m_{23} **b.** m_{32} **c.** m_{13} **d.** m_{41}

7. Find the matrix $-3M$.

8. Find the matrix $(1 \quad 0 \quad 0 \quad 1) \cdot M$.

9. What is the identity matrix of order 3? Define singular matrix and nonsingular matrix.

Let $A = \begin{pmatrix} 1 & 2 \\ 3 & 4 \end{pmatrix}$ $B = \begin{pmatrix} 2 & 0 \\ 0 & 2 \end{pmatrix}$

$C = \begin{pmatrix} -1 & 2 \\ 4 & -8 \end{pmatrix}$

10. Find $A + \frac{1}{2}B$.

11. Find $A \cdot \frac{1}{2}B$.

12. Find the determinants of matrices A and C.

13. Find the inverses of matrices A and C, if they exist.

14. Let $R = \begin{pmatrix} 1 & 3 & 0 \\ 2 & 0 & 4 \\ 1 & -3 & 3 \end{pmatrix}$

a. Find the determinant of R.
b. Find R^{-1} by using row transformations.
c. Find R^{-1} by using cofactors.

CHAPTER

8

Equations and systems of equations play an important role in design of all types, including manufacturing, construction, and architecture. The complex structure of a highway bridge, the stresses on a door hinge, the fuel efficiency of an automobile—all may require the careful analysis of situations involving many interrelated variable factors. In this chapter, you will learn several methods for solving systems of equations, groups of equations dealing with the same variables.

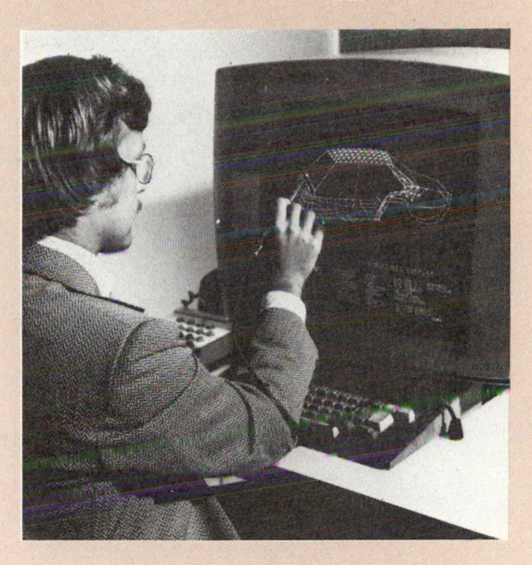

Linear Algebra

8.1 Introduction to Systems of Equations: Two Linear Equations in Two Variables

In Chapter 2, we discussed methods of solving first degree equations in one variable—equations like $3x - 1 = 11$, for example. We also examined first degree equations in two variables (like $3x - 2y = 7$) and found that there is not just one solution (one pair of x and y values) that solves the equation; rather, there are infinitely many pairs of values of x and y that satisfy the equation; Those pairs of values are represented as points in the plane and the graph of such an equation is a straight line. For this reason, equations such as $3x - 2y = 7$ are called **linear equations in two variables** (or unknowns). We say that

$$ax + by = c$$

is the *general form* of a linear equation in two unknowns where a, b, and c represent real numbers.

In this chapter, we shall be interested in finding, when possible, solutions for **systems** of linear equations. The object will be to find a set of values of the variables that satisfies *two* or *more* linear equations in the system. We begin, in this section, by studying systems of two linear equations in two unknowns.

Solution by Graphing

Consider the following system of linear equations in two variables:

$$\begin{aligned} x + 2y &= 5 \\ 3x - 2y &= 7 \end{aligned}$$

To solve the system, we try to find a *single* pair of values, one for x and one for y, that can be substituted in *both* equations to make them true statements.

The graph of each of these equations is a straight line. Suppose we graph both equations on the same rectangular coordinate system. Recall that the graph of a linear equation represents all the pairs of x and y values that make the equation a true statement. It seems reasonable to conclude that the point where the two lines (graphs) intersect represents the solution of the system since the x and y coordinates of that point satisfy both equations.

We illustrate this by graphing the equations (Figure 8.1).

FIGURE 8.1

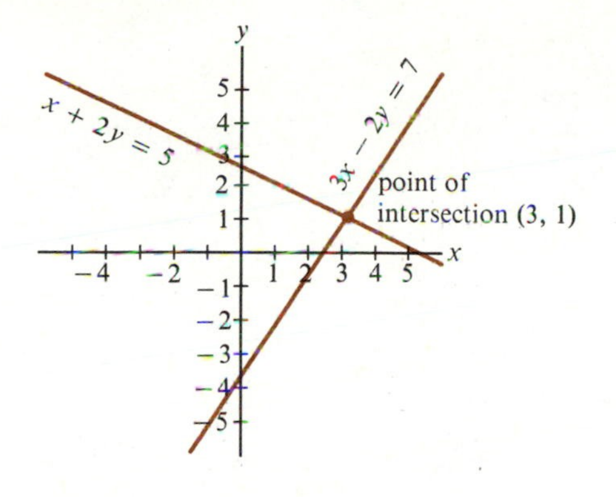

The point of intersection of the two lines is the point (3, 1). Thus $x = 3$, $y = 1$ is the solution of the given system of linear equations.

To verify that $x = 3$, $y = 1$ is indeed a solution of the system

$$\begin{aligned} x + 2y &= 5 \\ 3x - 2y &= 7 \end{aligned}$$

substitute these values for x and y in both equations.

$x + 2y = 5$	$3x - 2y = 7$
$3 + 2(1) = 5$	$3(3) - 2(1) = 7$
$3 + 2 = 5$	$9 - 2 = 7$
$5 = 5$ ✓	$7 = 7$ ✓

EXAMPLE 1

Solve each of the following systems of linear equations in two variables by graphing both equations in each system on the same rectangular coordinate system:

a. $2y - 5x = 0$; $y = x + 3$ **b.** $x - 3 = 0$; $2x + y = 4$

Solution

a. Graph the equations on the same rectangular coordinate system (Figure 8.2).

FIGURE 8.2

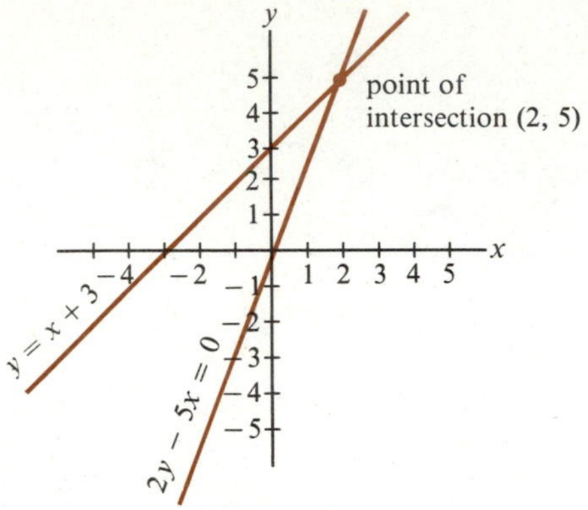

The point of intersection is the point (2, 5). Therefore, the solution of the system is $x = 2$, $y = 5$.

b. We graph the two equations in Figure 8.3.

FIGURE 8.3

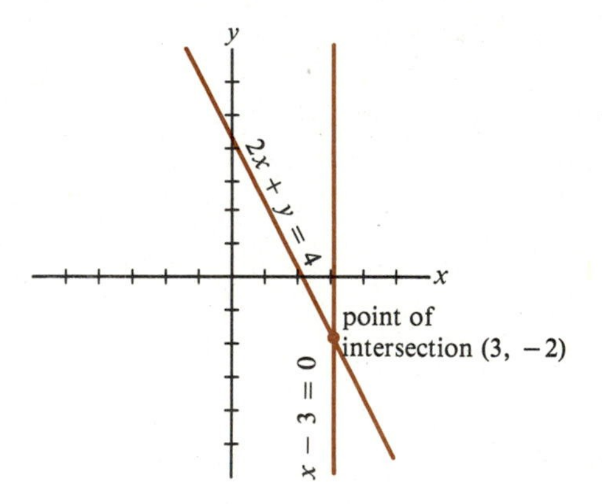

The point of intersection of the graphs is $(3, -2)$; so the solution of the system of equations is $x = 3$, $y = -2$.

There are some circumstances in which the graphical method of solving a system of two linear equations in two variables will not be adequate. These circumstances fall into two general categories:

1. there is no unique solution; or
2. there is a unique solution, but graphing does not enable us to determine it with sufficient accuracy.

Systems of Equations with No Unique Solution

Among systems with no unique solutions, there are two possibilities: either there may be no common solution to the two equations; or there may be infinitely many common solutions—that is, a solution exists but is not unique.

EXAMPLE 2

If possible, solve each of the following systems of equations graphically.

a. $12x - 9y = 36$
$3y - 4x - 24 = 0$

b. $10x + 25y = 20$
$\frac{5}{6}y + \frac{1}{3}x = \frac{2}{3}$

Solution

a. The two equations are graphed in Figure 8.4.

FIGURE 8.4

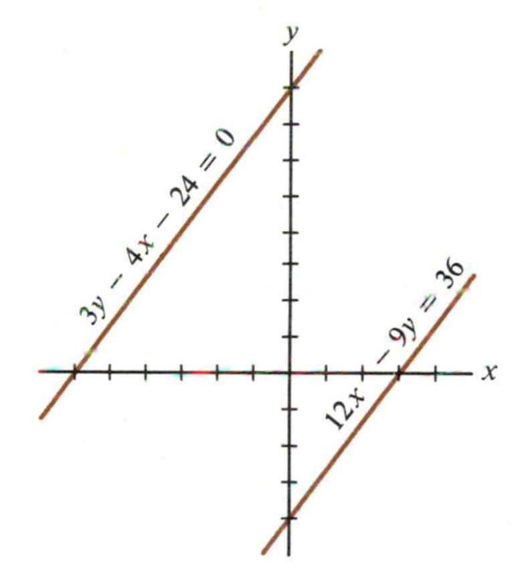

The graphs of these two equations do *not* intersect since the lines are parallel. This indicates that the given equations have *no common solution.* Such a system of equations is said to be **inconsistent**.

b.

FIGURE 8.5

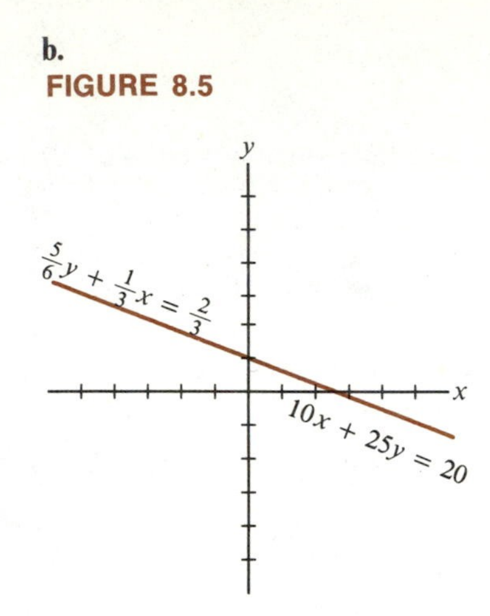

This time, the graphs of the two equations *coincide* (Figure 8.5)—they are the same line. This means that any (x, y) pair that is a solution of one equation is also a solution of the other. Thus, there are infinitely many common solutions; there is no unique solution to the system of equations. These equations are said to be **dependent**.

When Is a Graphic Solution Inadequate?

When a unique solution exists (that is, the straight-line graphs intersect in exactly one point), reading the point of intersection from the graph may be difficult and subject to human error. The next example illustrates this problem.

EXAMPLE 3

Try to solve the following system of linear equations:

$$2x - 3y = 5$$
$$3x + 4y = 7$$

Solution

Graph the equations on the same set of axes (Figure 8.6):

FIGURE 8.6

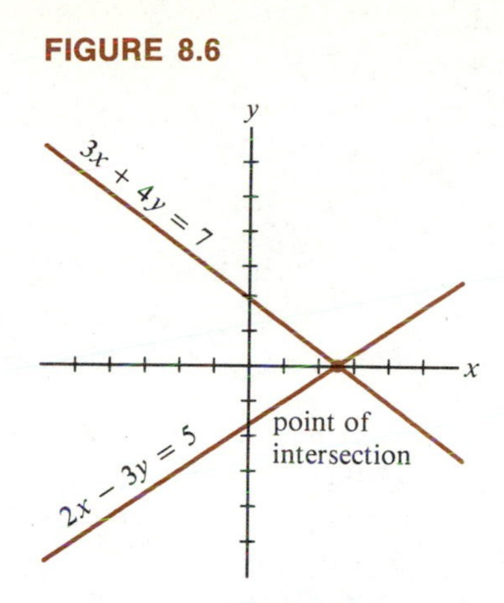

Note that the x and y coordinates of the point of intersection are *not* whole numbers. It would be very difficult to determine their exact values from the graph. We will see in Section 8.2 that the solution is actually $(41/17, -1/17)$!

In such a situation, we need a more accurate method of solution. We shall discuss several such methods in this chapter. These methods will also resolve another shortcoming of the method of graphic solution—the fact that it is not useful for solving systems of linear equations in more than two unknowns.

EXERCISES FOR SECTION 8.1

In Exercises 1 through 15, graph the system and determine its unique solution, if one exists. If not, state whether the system is inconsistent or dependent (see Examples 1 and 2).

1. $x + 2y = 10$
$3x - y = 2$

2. $2x + y = 10$
$3x + y = 11$

3. $3x - y = 1$
$2x + y = 4$

4. $3x + y = 13$
$4x - y = 1$

5. $-2x + y = 3$
$-x + y = 5$

6. $-4x + y = 4$
$3x - 2y = -13$

7. $5x + y = 4$
$2x - 3y = -12$

8. $5x - y = -4$
$2x + 3y = 12$

9. $6x - 3y = 15$
$8x - 4y = 10$

10. $6x - 3y = 15$
$8x - 4y = 20$

11. $x + 2y = 6$
$-2x - 4y = -12$

12. $x + 2y = 6$
$-2x - 4y = 10$

13. $-x + y = 2$
$3x - 5y = -2$

14. $-x + y = 2$
$-3x + 3y = 5$

15. $x + 2y = 9$
$2x - 4y = 12$

The variables don't always have to be x and y! In Exercises 16 through 20 graph the system, using x_2 as the vertical axis and x_1 as the horizontal axis. State the solution, if one exists; if not, state whether the system is inconsistent or dependent.

16. $x_1 - 3x_2 = 8$
$3x_1 + x_2 = 4$

17. $3x_1 + x_2 = 4$
$-x_1 + 2x_2 = 8$

18. $3x_1 + x_2 = 4$
$-x_1 + 2x_2 = 0$

19. $10x_1 + 5x_2 = 0$
$-2x_1 - x_2 = 1$

20. $4x_1 - 7x_2 = 0$
$-5x_1 + 9x_2 = 0$

8.2 Algebraic Solution of Systems of Linear Equations in Two Variables

In Example 3 of Section 8.1 we encountered a system of two linear equations in two variables for which a graphical method of solution was inadequate because of its lack of accuracy. We also noted that the graphical method of solution would not work for systems of equations involving three or more variables. In this section and the next we present an algebraic method for solving such systems. We begin here with systems of two equations in two unknowns.

Method of Substitution

There are several possible algebraic approaches to solving systems of linear equations. In some cases, the **method of substitution** will do the job.

EXAMPLE 1

As a winner in a recent state lottery, Mona Moneybags received an unexpected windfall of \$2,400.00. She consulted her stockbroker and the local psychic for advice on how best to invest this money. Her broker said she should purchase two stocks: Robert's Robots at \$10.00 a share and Ye Olde Colonial Computer Shoppe at \$15.00 a share. The psychic said she should buy twice as many shares of computer stock as of robot stock. If she wishes to follow the advice of both the broker and the psychic, how many shares of each stock should she buy with her \$2,400.00?

Solution

Let r stand for the number of shares of robot stock at $10.00 a share and c stand for the number of shares of computer stock at $15.00 a share. Mona will spend

$10 \cdot r$ on robot stock and
$15 \cdot c$ on computer stock

Since she has $2,400.00 to spend, this yields the equation:

$$10r + 15c = 2400$$

According to the psychic, the number of shares of computer stock should be twice the number of shares of robot stock,

$$c = 2r$$

To find the number of shares of each stock, we need a solution to the following system of equations:

$$10r + 15c = 2400 \qquad (1)$$
$$c = 2r \qquad (2)$$

Since c is expressed in terms of r in equation 2, we can *substitute* for c in equation 1 and solve the resulting equation for r

$$10r + 15(2r) = 2400$$
$$10r + 30r = 2400$$
$$40r = 2400$$
$$r = 60$$

From equation 2,

$$c = 2r = 120$$

Therefore, Mona should buy 60 shares of robot stock and 120 shares of computer stock.

The method of substitution can be used in solving systems of linear equations whenever one variable can be readily expressed in terms of another.

EXAMPLE 2

Solve the following system of linear equations:

$$15x + 3y = 12 \qquad (1)$$
$$5x - 4y = 9 \qquad (2)$$

Solution

Solve equation 1 for y:

$$15x + 3y = 12$$

$$3y = 12 - 15x \qquad \text{Subtract } 15x \text{ from both sides.}$$

$$y = 4 - 5x \qquad \text{Multiply both sides by } \tfrac{1}{3}.$$

Now substitute for y in equation 2 and solve the resulting equation for x. (Note: It is important to use equation 2 at this stage since we used equation 1 to solve for y.)

$$5x - 4(4 - 5x) = 9$$

$$5x - 16 + 20x = 9$$

$$25x - 16 = 9$$

$$25x = 9 + 16$$

$$25x = 25$$

$$x = 1$$

Finally, substitute this value ($x = 1$) for x in either of the *original* equations.

$$15x + 3y = 12$$

$$15(1) + 3y = 12$$

$$3y = -3$$

$$y = -1$$

The solution is $x = 1$, $y = -1$.

The reader should verify this by substituting these values for x and y in the two original equations.

Another Algebraic Method: Gaussian Elimination

In some cases, the method of substitution is not so easily applied. Consider, for example, the system of Example 3 in Section 8.1:

$$2x - 3y = 5 \qquad (1)$$

$$3x + 4y = 7 \qquad (2)$$

Neither equation can easily be solved for one variable in terms of the other without introducing fractions. Instead, our aim will be to *transform* this system into one that looks like:

$$1x + 0y = ?$$

$$0x + 1y = !$$

or

$$x = ?$$
$$y = !$$

This transformation is accomplished by using one or several of the following operations:

1. Multiply any equation in the system by a nonzero constant.
2. Add any nonzero multiple of one equation to another equation.
3. Interchange two equations.

Applying these operations to a system of linear equations results in an *equivalent* system—one with the same solution as the original system.

Let us get back to the problem of solving the system

$$2x - 3y = 5 \qquad (1)$$
$$3x + 4y = 7 \qquad (2)$$

The first objective will be to multiply one (or both) of the equations by a constant (or constants) to produce a new system in which the coefficient of x in equation 1 is the negative of the coefficient of x in equation 2. (We could have chosen to apply this process to the variable y instead.) We can accomplish this by multiplying equation 1 by -3 and equation 2 by 2. (Notice that we are multiplying each equation by $+$ or $-$ the coefficient of x in the other equation. This will *always* work, but shortcuts are often available.) This yields the equivalent system

$$-6x + 9y = -15 \qquad (1)$$
$$6x + 8y = 14 \qquad (2)$$

Now add equation 1 to equation 2, giving a new equation 2 in which the coefficient of x is zero

$$-6x + 9y = -15 \qquad (1)$$
$$0x + 17y = -1 \qquad (2)$$

Multiply equation 2 by $\frac{1}{17}$.

$$-6x + 9y = -15 \qquad (1)$$
$$0x + 1y = -\frac{1}{17} \qquad (2)$$
$$y = -\frac{1}{17} \qquad \text{We have solved for } y.$$

Now multiply equation 2 by -9 and add it to equation 1, giving a new equation 1 in which the coefficient of y is zero.

$$-6x + 0y = -15 + \frac{9}{17} = -\frac{246}{17} \qquad (1)$$

$$0x + 1y = -\tfrac{1}{17} \qquad (2)$$

Finally, multiply equation 1 by $-\frac{1}{6}$.

$$1x + 0y = \frac{246}{102} = \frac{41}{17} \qquad (1)$$

$$0x + 1y = -\tfrac{1}{17} \qquad (2)$$

The solution of the original system is

$$x = \tfrac{41}{17}$$
$$y = -\tfrac{1}{17}$$

(Verify that this solution is correct by substituting it in the original equations.)

Because this process of *elimination* becomes more complex with larger (3×3, 4×4, etc.) systems, it will be beneficial for us to refine the procedure for obtaining the

$$\begin{matrix} 1x & 0y \\ 0x & 1y \end{matrix}$$

configuration. The reader should examine the next example carefully to realize that the following objectives are motivating each step:

1. Transform the coefficient of x_1 in the first equation to one by multiplying the equation by its reciprocal.
2. Transform the coefficient of x_1 in the second equation to zero by adding an appropriate multiple of the first equation to the second equation.
3. Transform the coefficient of x_2 in the second equation to one by multiplying the equation by its reciprocal.
4. Transform the coefficient of x_2 in the first equation to zero by adding an appropriate multiple of the second equation to the first equation.

EXAMPLE 3

Solve the following system of linear equations:

$$5x_1 - 2x_2 = 6 \qquad (1)$$

$$3x_1 + 4x_2 = 10 \qquad (2)$$

Solution

First, we transform the 5 to a 1 in equation 1 by multiplying by $\frac{1}{5}$ (the reciprocal of 5).

$$\begin{array}{ll} 1x_1 - \frac{2}{5}x_2 = \frac{6}{5} & (1) \\ 3x_1 + 4x_2 = 10 & (2) \end{array}$$

Now we add -3 times the first equation to the second equation:

$$\begin{array}{ll} 1x_1 - \frac{2}{5}x_2 = \frac{6}{5} & (1) \\ 0x_1 + \frac{26}{5}x_2 = \frac{32}{5} & (2) \end{array}$$

$$\left[\begin{array}{c} -3x_1 + \frac{6}{5}x_2 = -\frac{18}{5} \\ 3x_1 + 4x_2 = 10 \\ \hline 0x_1 + \frac{26}{5}x_2 = \frac{32}{5} \end{array}\right] \quad (2)$$

Multiplying (2) by $\frac{5}{26}$ (the reciprocal of $\frac{26}{5}$) gives

$$\begin{array}{ll} 1x_1 - \frac{2}{5}x_2 = \frac{6}{5} & (1) \\ 0x_1 + 1x_2 = \frac{16}{13} & (2) \end{array}$$

To transform $-\frac{2}{5}$ to 0, we add $\frac{2}{5}$ times (2) to (1):

$$\begin{array}{ll} 1x_1 + 0x_2 = \frac{22}{13} & (1) \\ 0x_1 + 1x_2 = \frac{16}{13} & (2) \end{array}$$

$$\left[\begin{array}{c} 1x_1 - \frac{2}{5}x_2 = \frac{6}{5} \\ 0x_1 + \frac{2}{5}x_2 = \frac{32}{65} \\ \hline 1x_1 + 0x_2 = \frac{22}{13} \end{array}\right] \quad (1)$$

Finally, the solution is $x_1 = \frac{22}{13}$, $x_2 = \frac{16}{13}$.

Using Matrices to Solve Systems of Equations

If we examine what we have been doing, it becomes apparent that the kinds of operations we have been performing on systems of equations are very similar to operations we performed on matrices in Chapter 7. In fact, we could save a good deal of writing if we worked with the matrix formed by the coefficients of the variables in a system of equations, along with the constants, rather than with the equations themselves. Consider for example, the system of Example 3:

$$\begin{array}{ll} 5x_1 - 2x_2 = 6 & (1) \\ 3x_1 + 4x_2 = 10 & (2) \end{array}$$

We form the matrix

$$\left(\begin{array}{cc|c} 5 & -2 & 6 \\ 3 & 4 & 10 \end{array}\right)$$

drawing a vertical line to separate the coefficients from the constants. This type of matrix is called an **augmented matrix**. We shall transform this matrix to

$$\left(\begin{array}{cc|c} 1 & 0 & \square \\ 0 & 1 & \triangle \end{array}\right)$$

by using row transformations. We can then read the solution from this matrix.

$$\left.\begin{array}{l} 1x_1 + 0x_2 = \square \\ 0x_1 + 1x_2 = \triangle \end{array}\right\} \quad \text{or} \quad \left\{\begin{array}{l} x_1 = \square \\ x_2 = \triangle \end{array}\right.$$

The operations we used on the equations correspond to the familiar row transformations on matrices studied in Chapter 7:

1. Multiply a row by a nonzero constant;
2. add a nonzero multiple of one row to another row; and
3. interchange two rows.

This is illustrated below by repeating Example 3, this time using the augmented matrix.

$$\left(\begin{array}{cc|c} 5 & -2 & 6 \\ 3 & 4 & 10 \end{array}\right)$$

EXAMPLE 4

Solve the system of Example 3 using matrices.

Solution

Multiply row 1 by $\frac{1}{5}$:

$$\left(\begin{array}{cc|c} 1 & -\frac{2}{5} & \frac{6}{5} \\ 3 & 4 & 10 \end{array}\right)$$

↓ Add -3 times the top row to the bottom row.

$$\left(\begin{array}{cc|c} 1 & -\frac{2}{5} & \frac{6}{5} \\ 0 & \frac{26}{5} & \frac{32}{5} \end{array}\right)$$

↓ Multiply the bottom row by $\frac{5}{26}$.

$$\left(\begin{array}{cc|c} 1 & -\frac{2}{5} & \frac{6}{5} \\ 0 & 1 & \frac{16}{13} \end{array}\right)$$

↓ Add $\frac{2}{5}$ times the bottom row to the top row.

$$\left(\begin{array}{cc|c} 1 & 0 & \frac{22}{13} \\ 0 & 1 & \frac{16}{13} \end{array}\right)$$

The solution can now be read from the matrix:

$$x_1 = \tfrac{22}{13}$$
$$x_2 = \tfrac{16}{13}$$

We conclude this section with a summary of the method of Example 4. In general, the algorithm for solving a system using the row transformations is:

USING MATRICES TO SOLVE SYSTEMS OF EQUATIONS

1. Establish the augmented matrix that represents the system. In the general 2×2 case, this step looks like

$$\begin{aligned} a_{11}x + a_{12}y &= b_1 \\ a_{21}x + a_{22}y &= b_2 \end{aligned} \Rightarrow \left(\begin{array}{cc|c} a_{11} & a_{12} & b_1 \\ a_{21} & a_{22} & b_2 \end{array}\right)$$

2. If necessary, interchange two rows so that the component in the first row, first column, a_{11}, is nonzero.
3. Multiply each component in row 1 by $1/a_{11}$. Thus, a_{11} becomes 1.
4. Convert all other components in the first column to zero by adding an appropriate multiple of the first row to all other rows.
5. Repeat steps 2 through 4 for each diagonal component. (That is, convert a_{kk} to one and all other components in column k to zero for $k = 1, 2, \ldots, n$.)

EXERCISES FOR SECTION 8.2

In Exercises 1 through 6, use the substitution method to solve each system (see Examples 1 and 2).

1. $\begin{aligned} 2x - 3y &= 7 \\ x &= y + 4 \end{aligned}$

2. $\begin{aligned} 3x - y &= 4 \\ x &= y \end{aligned}$

3. $\begin{aligned} 5x - 2y &= 0 \\ x &= y \end{aligned}$

4. $\begin{aligned} 5x - 2y &= 0 \\ y &= x + 6 \end{aligned}$

5. $\begin{aligned} 5x + 2y &= 0 \\ x &= y - 14 \end{aligned}$

6. $\begin{aligned} 2x - 3y &= 2 \\ 2x &= -3y \end{aligned}$

In Exercises 7 through 12, use the method of Example 3 to solve the system.

7. $\begin{aligned} 2x - 3y &= 7 \\ x - y &= 4 \end{aligned}$

8. $\begin{aligned} 3x - y &= 4 \\ x - y &= 0 \end{aligned}$

9. $\begin{aligned} 5x - 2y &= 0 \\ x - y &= 0 \end{aligned}$

10. $\begin{aligned} 5x - 2y &= 0 \\ x - y &= -6 \end{aligned}$

11. $\begin{aligned} 5x + 2y &= 0 \\ 4x + 3y &= 14 \end{aligned}$

12. $\begin{aligned} 2x - 3y &= 2 \\ 6x - 3y &= 4 \end{aligned}$

In Exercises 13 through 20, use matrices and row transformations to solve each system (see Example 4).

13. $2x + 3y = 17$
$5x - y = 0$

14. $2x + 3y = 14$
$5x - y = 1$

15. $3x + 4y = 6$
$2x + y = -1$

16. $3x - y = -9$
$4x + 5y = 7$

17. $5x + 10y = 350$
$3x - 2y = -30$

18. $2x - y = 1$
$\frac{1}{2}x + \frac{1}{2}y = 1$

19. $3x - 6y = 24$
$5x + 2y = -8$

20. $3x - 4y = 9$
$-2x + 7y = 0$

21. The sum of two numbers is 19; their difference is 5. Find the two numbers.

22. If $3600 is going to be used to purchase two stocks, *A* and *B*, and twice as much is going to be used to purchase *A* as *B*, how much is going to be spent on each?

23. When two adults and three children enter a movie theater the cost of tickets is $19.00. When one adult and four children enter the same performance the cost is $17.00. Find the price of one adult's ticket and one child's ticket.

24. The Ajax Construction Company has two major cost concerns: materials (measured in tons) and labor (measured in work-hours). In June 3 tons of material and 100 work-hours cost $2500. In July, 4 tons of material and 150 work-hours cost $3500. Find Ajax's cost of one ton of material and the cost of one work-hour.

8.3 Systems of Linear Equations in More Than Two Variables

The matrix method outlined in Section 8.2 is especially useful for solving systems of linear equations in three or more variables. Before proceeding with the examples that follow, the reader should carefully study the five-step procedure at the close of Section 8.2, for these are precisely the steps we use below.

EXAMPLE 1

Solve the following system of linear equations:

$$x - y + z = 8$$
$$2x + 3y - z = 3$$
$$3x + 6y - 4z = -1$$

Solution

The augmented matrix is

$$\left(\begin{array}{rrr|r} 1 & -1 & 1 & 8 \\ 2 & 3 & -1 & 3 \\ 3 & 6 & -4 & -1 \end{array}\right)$$

Our goal is to use row transformations to get it in the form

$$\left(\begin{array}{rrr|r} 1 & 0 & 0 & a \\ 0 & 1 & 0 & b \\ 0 & 0 & 1 & c \end{array}\right)$$

Adding -2 times row 1 to row 2 of the augmented matrix gives

$$\left(\begin{array}{rrr|r} 1 & -1 & 1 & 8 \\ 0 & 5 & -3 & -13 \\ 3 & 6 & -4 & -1 \end{array}\right)$$

Add -3 times row 1 to row 3.

$$\left(\begin{array}{rrr|r} 1 & -1 & 1 & 8 \\ 0 & 5 & -3 & -13 \\ 0 & 9 & -7 & -25 \end{array}\right)$$

Multiply row 2 by $\frac{1}{5}$ to get a 1 in the second column.

$$\left(\begin{array}{rrr|r} 1 & -1 & 1 & 8 \\ 0 & 1 & -\frac{3}{5} & -\frac{13}{5} \\ 0 & 9 & -7 & -25 \end{array}\right)$$

Add row 2 to row 1.

$$\left(\begin{array}{rrr|r} 1 & 0 & \frac{2}{5} & \frac{27}{5} \\ 0 & 1 & -\frac{3}{5} & -\frac{13}{5} \\ 0 & 9 & -7 & -25 \end{array}\right)$$

Add -9 times row 2 to row 3.

$$\left(\begin{array}{rrr|r} 1 & 0 & \frac{2}{5} & \frac{27}{5} \\ 0 & 1 & -\frac{3}{5} & -\frac{13}{5} \\ 0 & 0 & -\frac{8}{5} & -\frac{8}{5} \end{array}\right)$$

Multiply row 3 by $-\frac{5}{8}$ to get a 1 in the third column.

$$\left(\begin{array}{ccc|c} 1 & 0 & \frac{2}{5} & \frac{27}{5} \\ 0 & 1 & -\frac{3}{5} & -\frac{13}{5} \\ 0 & 0 & 1 & 1 \end{array}\right)$$

Add $-\frac{2}{5}$ times row 3 to row 1.

$$\left(\begin{array}{ccc|c} 1 & 0 & 0 & 5 \\ 0 & 1 & -\frac{3}{5} & -\frac{13}{5} \\ 0 & 0 & 1 & 1 \end{array}\right)$$

Add $\frac{3}{5}$ times row 3 to row 2.

$$\left(\begin{array}{ccc|c} 1 & 0 & 0 & 5 \\ 0 & 1 & 0 & -2 \\ 0 & 0 & 1 & 1 \end{array}\right)$$

This is the final form of the matrix.

The solution is $x = 5$, $y = -2$, $z = 1$.

The reader should verify that this solution is correct by substituting in the original equations.

EXAMPLE 2

Solve the following system:

$$\begin{aligned} x_1 - x_2 + 5x_3 &= 9 \\ 2x_1 + x_2 &= -4 \\ 2x_1 - 5x_2 - 10x_3 &= 28 \end{aligned}$$

Solution

The augmented matrix is

$$\left(\begin{array}{ccc|c} 1 & -1 & 5 & 9 \\ 2 & 1 & 0 & -4 \\ 2 & -5 & -10 & 28 \end{array}\right)$$

Note that the second equation is equivalent to $2x_1 + 1x_2 + 0x_3 = -4$

Add -2 times row 1 to row 2 and add -2 times row 1 to row 3.

$$\left(\begin{array}{ccc|c} 1 & -1 & 5 & 9 \\ 0 & 3 & -10 & -22 \\ 0 & -3 & -20 & 10 \end{array}\right)$$

Multiply row 2 by $\frac{1}{3}$.

$$\begin{pmatrix} 1 & -1 & 5 & | & 9 \\ 0 & 1 & -\frac{10}{3} & | & -\frac{22}{3} \\ 0 & -3 & -20 & | & 10 \end{pmatrix}$$

Add 3 times row 2 to row 3 and add row 2 to row 1.

$$\begin{pmatrix} 1 & 0 & \frac{5}{3} & | & \frac{5}{3} \\ 0 & 1 & -\frac{10}{3} & | & -\frac{22}{3} \\ 0 & 0 & -30 & | & -12 \end{pmatrix}.$$

Multiply row 3 by $-\frac{1}{30}$.

$$\begin{pmatrix} 1 & 0 & \frac{5}{3} & | & \frac{5}{3} \\ 0 & 1 & -\frac{10}{3} & | & -\frac{22}{3} \\ 0 & 0 & 1 & | & \frac{2}{5} \end{pmatrix}$$

Add $\frac{10}{3}$ times row 3 to row 2 and $-\frac{5}{3}$ times row 3 to row 1.

$$\begin{pmatrix} 1 & 0 & 0 & | & 1 \\ 0 & 1 & 0 & | & -6 \\ 0 & 0 & 1 & | & \frac{2}{5} \end{pmatrix}$$

The solution is $x_1 = 1$, $x_2 = -6$, and $x_3 = \frac{2}{5}$.

This method can be used to solve any system of linear equations for which a unique solution exists. If the system has n equations in n variables, the augmented matrix will have n rows (one for each equation) and $n + 1$ columns (the extra column holding the constants). The objective will be to transform it, if possible, to a matrix of the form

$$\begin{pmatrix} 1 & 0 & 0 & \cdots & 0 & | & c_1 \\ 0 & 1 & 0 & \cdots & 0 & | & c_2 \\ 0 & 0 & 1 & \cdots & 0 & | & c_3 \\ \vdots & \vdots & \vdots & & \vdots & & \vdots \\ 0 & 0 & 0 & \cdots & 1 & | & c_n \end{pmatrix}$$

What if, however, such a transformation is *not* possible? This will occur if the system of equations is either inconsistent (no solution) or dependent (infinitely many solutions).

EXAMPLE 3

If possible, solve the following system of linear equations:

$$\begin{aligned} a + b - c &= 6 \\ 3a + b + 2c &= 12 \\ 2a + 3c &= 1 \end{aligned}$$

Solution

The augmented matrix is

$$\left(\begin{array}{ccc|c} 1 & 1 & -1 & 6 \\ 3 & 1 & 2 & 12 \\ 2 & 0 & 3 & 1 \end{array}\right)$$

A zero is added to row 3 because $2a + 3c = 1$ is really equivalent to $2a + 0b + 3c = 1$.

$\downarrow$ Add -3 times row 1 to row 2 and add -2 times row 1 to row 3.

$$\left(\begin{array}{ccc|c} 1 & 1 & -1 & 6 \\ 0 & -2 & 5 & -6 \\ 0 & -2 & 5 & -11 \end{array}\right)$$

$\downarrow$ Multiply row 2 by $-\frac{1}{2}$; then, add 2 times row 2 to row 3.

$$\left(\begin{array}{ccc|c} 1 & 1 & -1 & 6 \\ 0 & 1 & -\frac{5}{2} & 3 \\ 0 & 0 & 0 & -5 \end{array}\right)$$

Notice that row 3 in this matrix represents an impossible situation, for it corresponds to the equation

$$0a + 0b + 0c = -5$$

or

$$0 = -5$$

Since this can never be true, the given system of equations has no solution—it is *inconsistent*. In general, whenever we get a row containing all zeros to the left of the vertical line and a nonzero number to the right, we conclude that the associated system of equations is inconsistent.

EXAMPLE 4

If possible, solve the following system of linear equations:

$$\begin{aligned} 3x + y - z &= 2 \\ 3x - 2y + z &= 0 \\ -9x - 3y + 3z &= -6 \end{aligned}$$

Solution

The augmented matrix for this system is

$$\left(\begin{array}{ccc|c} 3 & 1 & -1 & 2 \\ 3 & -2 & 1 & 0 \\ -9 & -3 & 3 & -6 \end{array}\right)$$

To save some space, we divide row 1 by 3, add -3 times that row to row 2 and add 9 times row 1 to row 3 in a single step.

$$\left(\begin{array}{ccc|c} 1 & \frac{1}{3} & -\frac{1}{3} & \frac{2}{3} \\ 0 & -3 & 2 & -2 \\ 0 & 0 & 0 & 0 \end{array}\right)$$

Note that row 3 of this matrix consists entirely of zeros. This row can be eliminated from the matrix. Its presence indicates that the system of equations is *dependent*; it will have infinitely many solutions.

Let's continue working with the matrix obtained by eliminating row 3:

$$\left(\begin{array}{ccc|c} 1 & \frac{1}{3} & -\frac{1}{3} & \frac{2}{3} \\ 0 & -3 & 2 & -2 \end{array}\right)$$

$\downarrow$ Multiply row 2 by $-\frac{1}{3}$.

$$\left(\begin{array}{ccc|c} 1 & \frac{1}{3} & -\frac{1}{3} & \frac{2}{3} \\ 0 & 1 & -\frac{2}{3} & \frac{2}{3} \end{array}\right)$$

$\downarrow$ Add $-\frac{1}{3}$ times row 2 to row 1.

$$\left(\begin{array}{ccc|c} 1 & 0 & -\frac{1}{9} & \frac{4}{9} \\ 0 & 1 & -\frac{2}{3} & \frac{2}{3} \end{array}\right)$$

This is as far as we can go.

This matrix may be interpreted as

$$\begin{aligned} x - \tfrac{1}{9}z &= \tfrac{4}{9} \\ y - \tfrac{2}{3}z &= \tfrac{2}{3} \end{aligned} \qquad \text{or} \qquad \begin{aligned} x &= \tfrac{1}{9}z + \tfrac{4}{9} \\ y &= \tfrac{2}{3}z + \tfrac{2}{3} \end{aligned}$$

Therefore, the given system of equations has infinitely many solutions, which can be obtained by substituting values for z in the two equations above and then solving for x and y. For example, when $z = 1$, $x = \frac{5}{9}$ and $y = \frac{4}{3}$, and when $z = 5$, $x = 1$ and $y = 4$.

One final note: the decision to solve for x and y in terms of z above was arbitrary. We could have solved for x and z in terms of y, or for y and z in terms of x. For example, return to the point in Example 4 at which we dropped the row of zeros, yielding the 2×4 matrix:

$$\left(\begin{array}{ccc|c} 1 & \frac{1}{3} & -\frac{1}{3} & \frac{2}{3} \\ 0 & -3 & 2 & -2 \end{array}\right)$$

Suppose we multiply row 2 by $\frac{1}{6}$ and add it to row 1:

$$\left(\begin{array}{ccc|c} 1 & -\frac{1}{6} & 0 & \frac{1}{3} \\ 0 & -3 & 2 & -2 \end{array}\right)$$

Now we multiply row 2 by $\frac{1}{2}$:

$$\left(\begin{array}{ccc|c} 1 & -\frac{1}{6} & 0 & \frac{1}{3} \\ 0 & -\frac{3}{2} & 1 & -1 \end{array}\right)$$

This matrix represents the equations:

$$\begin{aligned} x - \tfrac{1}{6}y &= \tfrac{1}{3} \\ -\tfrac{3}{2}y + z &= -1 \end{aligned} \qquad \text{or} \qquad \begin{aligned} x &= \tfrac{1}{3} + \tfrac{1}{6}y \\ z &= -1 + \tfrac{3}{2}y \end{aligned}$$

We have thus expressed x and z in terms of y, again showing that the original system of equations is *dependent*.

We conclude this section with another matrixlike method of solving a system. In general, if there are n linear equations in n variables $(x_1, x_2, \ldots, x_n)$, we can write the system as

$$\begin{array}{l} a_{11}x_1 + a_{12}x_2 + \cdots + a_{1n}x_n = b_1 \\ a_{21}x_1 + a_{22}x_2 + \cdots + a_{2n}x_n = b_2 \\ \vdots \qquad\quad \vdots \qquad\qquad\qquad\quad \vdots \\ a_{n1}x_1 + a_{n2}x_2 + \cdots + a_{nn}x_n = b_n \end{array}$$

Furthermore, the matrix representation is

$$A \cdot X = B$$

where

$$A = \begin{pmatrix} a_{11} & a_{12} & \cdots & a_{1n} \\ a_{21} & a_{22} & \cdots & a_{2n} \\ \vdots & & & \\ a_{n1} & a_{n2} & \cdots & a_{nn} \end{pmatrix}, \quad X = \begin{pmatrix} x_1 \\ x_2 \\ \vdots \\ x_n \end{pmatrix} \quad \text{and} \quad B = \begin{pmatrix} b_1 \\ b_2 \\ \vdots \\ b_n \end{pmatrix}$$

Now, we can solve

$$A \cdot X = B$$

by "left-multiplying" both sides by A^{-1} (if it exists). Note that since matrix multiplication is not commutative, we must multiply on the *left* on *both* sides of the equation.

$$\begin{aligned} (A^{-1} \cdot A) \cdot X &= A^{-1} \cdot B \\ I \cdot X &= A^{-1} \cdot B \\ X &= A^{-1} \cdot B \end{aligned}$$

So X, the vector of variables, can be obtained by multiplying A^{-1} times B. The next example uses this approach.

EXAMPLE 5

Solve the 3×3 system below:

$$\begin{aligned} 2x + 3y - 4z &= 44 \\ x + y + z &= 14 \\ -x + 7y - 10z &= 75 \end{aligned}$$

Solution

$$A = \begin{pmatrix} 2 & 3 & -4 \\ 1 & 1 & 1 \\ -1 & 7 & -10 \end{pmatrix} \qquad X = \begin{pmatrix} x \\ y \\ z \end{pmatrix} \qquad B = \begin{pmatrix} 44 \\ 14 \\ 75 \end{pmatrix}$$

We leave it to the reader (in Exercise 16) to determine that

$$A^{-1} = \begin{pmatrix} \frac{17}{39} & -\frac{2}{39} & -\frac{7}{39} \\ -\frac{9}{39} & \frac{24}{39} & \frac{6}{39} \\ -\frac{8}{39} & \frac{17}{39} & \frac{1}{39} \end{pmatrix}$$

Now $X = A^{-1} \cdot B$

$$X = \begin{pmatrix} \frac{17}{39} & -\frac{2}{39} & -\frac{7}{39} \\ -\frac{9}{39} & \frac{24}{39} & \frac{6}{39} \\ -\frac{8}{39} & \frac{17}{39} & \frac{1}{39} \end{pmatrix} \cdot \begin{pmatrix} 44 \\ 14 \\ 75 \end{pmatrix} = \begin{pmatrix} 5 \\ 10 \\ -1 \end{pmatrix}$$

Thus, $x = 5$, $y = 10$, and $z = -1$.

EXERCISES FOR SECTION 8.3

In Exercises 1 through 12, solve the given system, if it has a unique solution. If it does not, state whether the system is inconsistent or dependent (see Examples 1–4).

1. $$\begin{aligned} x + 2y + z &= 2 \\ -3x + 6y - z &= 10 \\ 6x - y + 3z &= 1 \end{aligned}$$

2. $$\begin{aligned} x - y + z &= 10 \\ 3x - y + 4z &= 16 \\ -x + 3y - z &= 32 \end{aligned}$$

3. $$\begin{aligned} x_1 + x_2 - x_3 &= 4 \\ 3x_1 - x_2 + 5x_3 &= -4 \\ 8x_1 - x_2 - x_3 &= 7 \end{aligned}$$

4. $$\begin{aligned} 3x_1 - x_2 - x_3 &= 15 \\ 2x_1 + 4x_2 - x_3 &= 10 \\ -x_1 - 7x_2 &= -5 \end{aligned}$$

5. $$\begin{aligned} x + 5y + z &= 2 \\ x + 3y + 2z &= 1 \\ 3x + 7y + 10z &= 5 \end{aligned}$$

6. $$\begin{aligned} 4x + y + z &= 6 \\ 2x - 3y + 5z &= 4 \\ x - y + 8z &= 8 \end{aligned}$$

7. $$\begin{aligned} x + 5y + z &= 2 \\ x + 3y + 2z &= 1 \\ 3x + 13y + 4z &= 5 \end{aligned}$$

8. $$\begin{aligned} x + 5y + z &= 2 \\ x + 3y + 2z &= 1 \\ 3x + 13y + 4z &= 6 \end{aligned}$$

9. $$\begin{aligned} x_1 + 2x_2 + x_3 &= 2 \\ -3x_1 + 6x_2 - x_3 &= 10 \\ 12x_2 + 2x_3 &= 16 \end{aligned}$$

10. $$\begin{aligned} x_1 + 2x_2 + x_3 &= 2 \\ -3x_1 + 6x_2 - x_3 &= 10 \\ 12x_2 + 2x_3 &= 15 \end{aligned}$$

11. $$\begin{aligned} x_1 + 5x_2 &= 3 \\ x_2 + 3x_3 &= -2 \\ x_1 - 4x_2 &= -3 \end{aligned}$$

12. $$\begin{aligned} x_1 + 5x_2 &= 3 \\ x_2 + 3x_3 &= -2 \\ x_1 - 3x_2 - 4x_3 &= 8 \end{aligned}$$

In Exercises 13 through 15 use the method of the matrix inverse to find the solution, if it exists, to each system (see Example 5).

13. $$\begin{aligned} x_1 + x_2 - x_3 &= -17 \\ x_1 - 2x_2 + x_3 &= 3 \\ 7x_2 + x_3 &= 40 \end{aligned}$$

14. $$\begin{aligned} x + 2y &= 8 \\ 2x - y &= 1 \end{aligned}$$

15. $$\begin{aligned} x_1 + x_2 - x_3 &= 5 \\ x_1 - 2x_2 + x_3 &= 0 \\ 7x_2 + x_3 &= 40 \end{aligned}$$

16. Verify that A^{-1} is, in fact, the inverse of the matrix in Example 5.
17. Refer to Exercise 24, Section 8.2. Suppose July's figures are 6 tons of material and 200 work-hours cost \$5000 and June's are 3 tons of material plus 100 work-hours cost \$2500. Can you find the cost of one ton of material? Explain.

8.4 Cramer's Rule

Using Determinants to Solve Systems of Equations

As our final method for solving systems of linear equations, we shall discuss **Cramer's rule**. Developed by Gabriel Cramer, an eighteenth-century Swiss mathematician, it is a strictly numerical method that involves no matrix transformations.

Consider the system of two equations in two variables, x and y:

$$\begin{aligned} a_1x + b_1y &= c_1 \\ a_2x + b_2y &= c_2 \end{aligned}$$

It can be easily shown (see Exercise 20) that the solution to the system is given by

$$x = \frac{b_2c_1 - b_1c_2}{a_1b_2 - a_2b_1} \qquad \text{and} \qquad y = \frac{a_1c_2 - a_2c_1}{a_1b_2 - a_2b_1}$$

Notice that each numerator and each denominator above is the evaluation of one of these three determinants:

$$Dx = \begin{vmatrix} c_1 & b_1 \\ c_2 & b_2 \end{vmatrix} = b_2c_1 - b_1c_2 \qquad Dy = \begin{vmatrix} a_1 & c_1 \\ a_2 & c_2 \end{vmatrix} = a_1c_2 - a_2c_1$$

$$D = \begin{vmatrix} a_1 & b_1 \\ a_2 & b_2 \end{vmatrix} = a_1b_2 - a_2b_1$$

Consider the following system:

$$3x - 2y = 5 \qquad (1)$$
$$x = 7 - 2y \qquad (2)$$

The first step is to write both equations in the form $ax + by = c$. This results in the system

$$3x - 2y = 5 \qquad (1)$$
$$x + 2y = 7 \qquad (2)$$

Next, we form the **coefficient matrix** of the system. This is the matrix consisting of the numerical coefficients of the variables as they appear in the equations. We shall want to calculate the determinant of this matrix, which we shall call D. Thus,

$$D = \begin{vmatrix} 3 & -2 \\ 1 & 2 \end{vmatrix} = (3 \cdot 2) - (-2 \cdot 1) = 6 - (-2) = 8$$

Now, we evaluate the determinant, denoted Dx, of the matrix formed by replacing the coefficients of the variable x in the coefficient matrix with the constants (5 and 7) in the linear system.

$$Dx = \begin{vmatrix} 5 & -2 \\ 7 & 2 \end{vmatrix} = (5 \cdot 2) - (-2 \cdot 7) = 10 - (-14) = 24$$

Similarly, evaluate the determinant Dy obtained by substituting the constants for the coefficients of y:

$$Dy = \begin{vmatrix} 3 & 5 \\ 1 & 7 \end{vmatrix} = (3 \cdot 7) - (5 \cdot 1) = 21 - 5 = 16$$

We can now use these determinants to solve the system:

$$x = \frac{Dx}{D} \qquad y = \frac{Dy}{D}$$

So, for this system,

$$x = \tfrac{24}{8} \quad \text{or} \quad x = 3$$
$$y = \tfrac{16}{8} \quad \text{or} \quad y = 2$$

We verify this solution by substituting in the original equations:

(1) $3x - 2y = 5$	(2) $x = 7 - 2y$
$3(3) - 2(2) = 5$	$3 = 7 - 2(2)$
$9 - 4 = 5$	$3 = 7 - 4$
$5 = 5$ ✓	$3 = 3$ ✓

EXAMPLE 1

Use Cramer's rule to solve the following system of linear equations:

$$y = 2x + 2 \qquad (1)$$
$$x + 2y = 9 \qquad (2)$$

Solution

First, we write both equations in the form $ax + by = c$. This yields

$$-2x + y = 2 \qquad (1)$$
$$x + 2y = 9 \qquad (2)$$

We evaluate D, Dx, and Dy:

$$D = \begin{vmatrix} -2 & 1 \\ 1 & 2 \end{vmatrix} = (-4) - 1 = -5$$

$$Dx = \begin{vmatrix} 2 & 1 \\ 9 & 2 \end{vmatrix} = 4 - 9 = -5$$

$$Dy = \begin{vmatrix} -2 & 2 \\ 1 & 9 \end{vmatrix} = -18 - 2 = -20$$

Solve for x and y:

$$x = \frac{Dx}{D} = \frac{-5}{-5}$$

$$y = \frac{Dy}{D} = \frac{-20}{-5}$$

The solution is $x = 1$, $y = 4$.

The reader should verify this by substituting in the original equations.

Cramer's rule can be applied to systems of equations with more than two variables. We shall need the methods of Section 7.5, specifically expanding on a row of the matrix using cofactors, to evaluate the determinants involved.

Recall that, expanding on row 1,

$$\begin{vmatrix} 1 & 2 & 3 \\ 4 & 5 & 6 \\ 7 & 8 & 9 \end{vmatrix} = 1 \cdot (\text{cofactor of } 1) + 2 \cdot (\text{cofactor of } 2) + 3 \cdot (\text{cofactor of } 3)$$

where the cofactor of an element is obtained by taking the determinant of the matrix found by deleting the row and column containing that element from the original matrix and then multiplying that determinant by $+1$ or -1 (depending on whether the sum of the element's row number and column number is even or odd).

Thus,

$$\begin{vmatrix} 1 & 2 & 3 \\ 4 & 5 & 6 \\ 7 & 8 & 9 \end{vmatrix} = 1 \cdot (+1) \cdot \begin{vmatrix} 5 & 6 \\ 8 & 9 \end{vmatrix} + 2 \cdot (-1) \cdot \begin{vmatrix} 4 & 6 \\ 7 & 9 \end{vmatrix} + 3 \cdot (+1) \cdot \begin{vmatrix} 4 & 5 \\ 7 & 8 \end{vmatrix}$$

$$= (45 - 48) - 2 \cdot (36 - 42) + 3 \cdot (32 - 35)$$

$$= -3 - 2(-6) + 3(-3) = -3 + 12 - 9 = 0$$

EXAMPLE 2

Solve the following system of linear equations using Cramer's rule:

$$x - y + 10z = 13$$

$$2x + y = 0$$

$$2x - 5y - 5z = 34$$

Solution

$$D = \begin{vmatrix} 1 & -1 & 10 \\ 2 & 1 & 0 \\ 2 & -5 & -5 \end{vmatrix}$$

If a variable does not appear in an equation, its coefficient in that equation is zero.

So

$$D = 1(+1)\begin{vmatrix} 1 & 0 \\ -5 & -5 \end{vmatrix} + (-1)(-1)\begin{vmatrix} 2 & 0 \\ 2 & -5 \end{vmatrix} + 10(+1)\begin{vmatrix} 2 & 1 \\ 2 & -5 \end{vmatrix}$$

$$= (-5 - 0) + (-10 - 0) + 10(-10 - 2)$$

$$= -5 - 10 - 120 = -135$$

$$Dx = \begin{vmatrix} 13 & -1 & 10 \\ 0 & 1 & 0 \\ 34 & -5 & -5 \end{vmatrix}$$

$$= 13(+1)\begin{vmatrix} 1 & 0 \\ -5 & -5 \end{vmatrix} + (-1)(-1)\begin{vmatrix} 0 & 0 \\ 34 & -5 \end{vmatrix} + 10(+1)\begin{vmatrix} 0 & 1 \\ 34 & -5 \end{vmatrix}$$

$$= 13 \cdot (-5 - 0) + 1 \cdot (0) + 10 \cdot (0 - 34)$$

$$= -65 - 340 = -405$$

$$Dy = \begin{vmatrix} 1 & 13 & 10 \\ 2 & 0 & 0 \\ 2 & 34 & -5 \end{vmatrix}$$

$$= 1(+1)\begin{vmatrix} 0 & 0 \\ 34 & -5 \end{vmatrix} + 13(-1)\begin{vmatrix} 2 & 0 \\ 2 & -5 \end{vmatrix} + 10(+1)\begin{vmatrix} 2 & 0 \\ 2 & 34 \end{vmatrix}$$

$$= 0 - 13 \cdot (-10) + 10 \cdot (68) = 130 + 680 = 810$$

$$Dz = \begin{vmatrix} 1 & -1 & 13 \\ 2 & 1 & 0 \\ 2 & -5 & 34 \end{vmatrix}$$

$$= 1(+1)\begin{vmatrix} 1 & 0 \\ -5 & 34 \end{vmatrix} + (-1)(-1)\begin{vmatrix} 2 & 0 \\ 2 & 34 \end{vmatrix} + 13(+1)\begin{vmatrix} 2 & 1 \\ 2 & -5 \end{vmatrix}$$

$$= 34 + 68 + 13 \cdot (-10 - 2) = 102 - 156 = -54$$

Therefore,

$$x = \frac{Dx}{D} = \frac{-405}{-135}$$

$$y = \frac{Dy}{D} = \frac{810}{-135}$$

$$z = \frac{Dz}{D} = \frac{-54}{-135}$$

The solution is $x = 3$, $y = -6$, $z = \frac{2}{5}$.

EXAMPLE 3

Solve the following system of linear equations using Cramer's rule:

$$x_1 + 2x_2 = 0$$
$$x_1 + 2x_2 - x_3 = -2$$
$$x_1 + x_2 + x_3 = 3$$

Solution

$$D = \begin{vmatrix} 1 & 2 & 0 \\ 1 & 2 & -1 \\ 1 & 1 & 1 \end{vmatrix} = -1$$

The reader should verify the calculation of the determinants.

$$Dx_1 = \begin{vmatrix} 0 & 2 & 0 \\ -2 & 2 & -1 \\ 3 & 1 & 1 \end{vmatrix} = -2$$

$$Dx_2 = \begin{vmatrix} 1 & 0 & 0 \\ 1 & -2 & -1 \\ 1 & 3 & 1 \end{vmatrix} = 1$$

$$Dx_3 = \begin{vmatrix} 1 & 2 & 0 \\ 1 & 2 & -2 \\ 1 & 1 & 3 \end{vmatrix} = -2$$

So,

$$x_1 = \frac{Dx_1}{D} = \frac{-2}{-1} = 2$$

$$x_2 = \frac{Dx_2}{D} = \frac{1}{-1} = -1$$

$$x_3 = \frac{Dx_3}{D} = \frac{-2}{-1} = 2$$

There are two final situations to consider. Suppose the given system of linear equations is inconsistent or dependent. What will happen if we attempt to solve it using Cramer's rule? We should expect to be unable to obtain a unique solution.

EXAMPLE 4

Use Cramer's rule to attempt to solve the systems of linear equations of Examples 3 and 4 in Section 8.3.

a.
$$\begin{aligned} a + b - c &= 6 \\ 3a + b + 2c &= 12 \\ 2a + 3c &= 1 \end{aligned}$$

b.
$$\begin{aligned} 3x + y - z &= 2 \\ 3x - 2y + z &= 0 \\ -9x - 3y + 3z &= -6 \end{aligned}$$

Solution

a. $D = \begin{vmatrix} 1 & 1 & -1 \\ 3 & 1 & 2 \\ 2 & 0 & 3 \end{vmatrix}$

$= 1\begin{vmatrix} 1 & 2 \\ 0 & 3 \end{vmatrix} - 1\begin{vmatrix} 3 & 2 \\ 2 & 3 \end{vmatrix} - 1\begin{vmatrix} 3 & 1 \\ 2 & 0 \end{vmatrix}$

$= 3 - (9 - 4) - (-2) = 3 - 5 + 2 = 0$

Note that, to obtain values of a, b and c using Cramer's rule, we must divide the determinants Da, Db and Dc by D. Since $D = 0$, this is impossible. Thus, Cramer's rule will *not* provide us with a solution. This is as it should be since we determined in Example 3 that the system was inconsistent.

b. $D = \begin{vmatrix} 3 & 1 & -1 \\ 3 & -2 & 1 \\ -9 & -3 & 3 \end{vmatrix}$

$= 3\begin{vmatrix} -2 & 1 \\ -3 & 3 \end{vmatrix} - 1\begin{vmatrix} 3 & 1 \\ -9 & 3 \end{vmatrix} - 1\begin{vmatrix} 3 & -2 \\ -9 & -3 \end{vmatrix}$

$= 3 \cdot (-6 + 3) - (9 + 9) - (-9 - 18)$

$= -9 - 18 + 27 = 0$

Since $D = 0$, Cramer's rule does not yield a solution for this system. In Example 4, we determined that the system was *dependent*.

In general, if a system of linear equations is either inconsistent or dependent, the determinant of the matrix of coefficients will be zero and no unique solution can be found using Cramer's rule. In fact, no unique solution exists! (See Exercise 22 for a further explanation.)

EXERCISES FOR SECTION 8.4

In Exercises 1 through 5, use Cramer's rule to solve the given 2×2 linear system, if possible (see Example 1).

1. $2x + 3y = 5$
$3x + 5y = 9$

2. $3x - 2y = -6$
$4x + 3y = 9$

3. $2x + y = 5$
$4x - 3y = -10$

4. $6x + 9y = 3$
$2x - 3y = -\frac{1}{3}$

5. $2x + 5y = 7$
$-3x + 7y = 9$

In Exercises 6 through 19, use Cramer's rule to solve the given 3×3 linear system, if possible (see Examples 2–4).

6. $$\begin{aligned} x + 2y + z &= 12 \\ 3x - y + 2z &= 5 \\ x + y &= 5 \end{aligned}$$

7. $$\begin{aligned} 3x + z &= 6 \\ y - 2z &= -2 \\ x + y &= 5 \end{aligned}$$

8. $$\begin{aligned} 7x + y + 2z &= -5 \\ 3x - y + 4z &= 2 \\ x - 2z &= 1 \end{aligned}$$

9. $$\begin{aligned} x + y + z &= 0 \\ x - y &= 0 \\ 6x - 9y + 3z &= -9 \end{aligned}$$

10. $$\begin{aligned} 2x + y + z &= 8 \\ 3x + 10y - z &= 93 \\ -x + y + z &= 20 \end{aligned}$$

11. $$\begin{aligned} 4x - y &= 3 \\ x - y + 8z &= 5 \\ 2x - 7z &= -8 \end{aligned}$$

12. $$\begin{aligned} x + y + z &= 11 \\ 3x + 5y - 2z &= 11 \\ x + y - z &= -1 \end{aligned}$$

13. $$\begin{aligned} x_1 + x_2 + x_3 &= 6 \\ x_1 + x_2 - x_3 &= 0 \\ 8x_1 - x_2 + 3x_3 &= 15 \end{aligned}$$

14. $$\begin{aligned} x - 2y - 2z &= -15 \\ -x - 6y + z &= -1 \\ 3x - 10y + z &= 7 \end{aligned}$$

15. $$\begin{aligned} 2x + 2y - z &= 16 \\ x - 6y - z &= -6 \\ x - 10y + 3z &= -14 \end{aligned}$$

16. $$\begin{aligned} x_1 + 7x_2 - 8x_3 &= 4 \\ 2x_1 - x_2 - 10x_3 &= 7 \\ 3x_1 + 5x_2 + 7x_3 &= 0 \end{aligned}$$

17. $$\begin{aligned} x + y + z &= 4 \\ 2x - y + z &= 0 \\ 4x + y + 3z &= 8 \end{aligned}$$

18. $$\begin{aligned} x + y + z &= 4 \\ 2x - y + z &= 0 \\ 4x + y + 3z &= 9 \end{aligned}$$

19. $$\begin{aligned} x + y + z &= 6 \\ -2x + 3y - 4z &= 2 \\ 7x - 8y + 13z &= 0 \end{aligned}$$

20. Use the algebraic elimination method to find the solution for x to

$$a_1x + b_1y = c_1 \qquad (1)$$
$$a_2x + b_2y = c_2 \qquad (2)$$

by

(1) multiplying equation 1 by $-b_2$;
(2) multiplying equation 2 by b_1; and
(3) adding the two resulting equations and solving for x.

21. Repeat Exercise 20, but this time solve for y. (You will have to use different multipliers.)

22. The determinant of the coefficient matrix is zero for each system in Exercises 17, 18, and 19; and therefore, no unique solution exists. Notice that in Exercise 18, at least one of the determinants Dx, Dy and Dz is not zero. Notice further that in Exercise 17, Dx, Dy and Dz are all zero. For this reason, the system in Exercise 17 is dependent; the system in Exercise 18 is inconsistent. What about the system of Exercise 19?

8.5 Linear Programming

A type of problem frequently encountered in computer programming applications is the problem of determining how to maximize some quantity (for example, profit) or minimize some quantity (for example, cost) under a particular set of conditions. The study of problems of this type is referred to as **linear programming**.

Graphical Solution

There are several possible approaches to solving linear programming problems. When there are only two quantities that are subject to change (variables) involved in the equation that defines the **objective** (profit or cost, for example) and in the restrictions, or **constraints**, imposed, a solution can frequently be found by using graphs.

Suppose, for example, that GTB Industries manufactures two products, widgets and filberts, in its Connecticut plant. The company's profit on these items is \$5.00 per widget and \$7.00 per filbert. Thus if w is the number of widgets produced in a day and f is the number of filberts produced, the potential profit P, for that day can be expressed by the equation

$$P = 7f + 5w$$

The objective of GTB Industries is, of course, to maximize potential profit—that is, to find the values of f and w that make P, called the **objective function**, as large as possible.

Suppose, also, that GTB's market researchers have determined that the company should produce at most ten more widgets than filberts, and that at least five widgets should be produced each day. Finally, the physical limitations of the plant dictate that no more than 26 units (widgets and filberts combined) can be produced in a day.

We would like to determine the maximum possible value of P and the number of filberts and widgets required to achieve that maximum, given the constraints on w and f listed. We must first express the constraints in terms of linear inequalities. The problem becomes

Maximize $P = 7f + 5w$ subject to the following constraints:

(C1)	$w \leq f + 10$	At most 10 more widgets than filberts.
(C2)	$w \geq 5$	At least 5 widgets must be produced.
(C3)	$w + f \leq 26$	No more than 26 units can be produced.

(C4)	$f \geq 0$	A common sense constraint—GTB *can't* produce a negative number of filberts.

We concentrate first on finding the set of possible values for w and f that satisfies all four constraints. That is, any possible solution to the problem of maximizing profits must satisfy

C1 *and* C2 *and* C3 *and* C4

We can find all such points by graphing all four linear inequalities on the same rectangular coordinate system and finding the region of intersection of the four graphs. The maximization problem will then be reduced to finding the point in that region whose f and w coordinates make the value of P greatest.

That point will always be a vertex, or corner, of the region created by graphing the constraints. (The justification of this statement is beyond the scope of this text. It is, however, the basic principle in the theory of linear programming, and the interested reader can consult a text on the subject for more detail.)

We begin by graphing C1 and C2, and noting the region of intersection of the graphs (Figure 8.7).

FIGURE 8.7

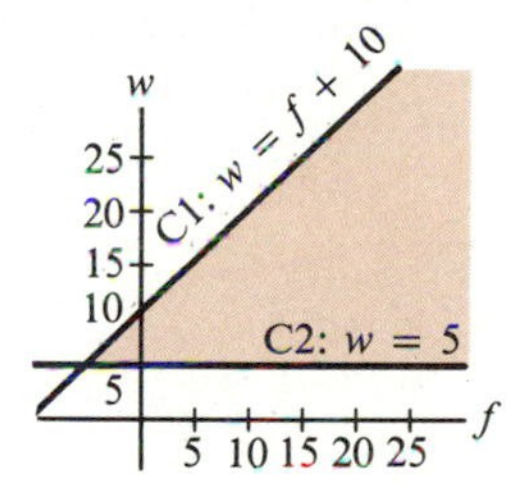

Next, we add C3 (Figure 8.8).

FIGURE 8.8

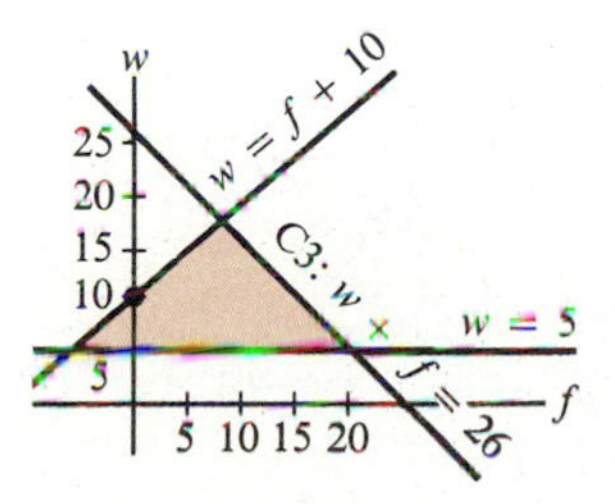

Finally, we add C4 (Figure 8.9).

FIGURE 8.9

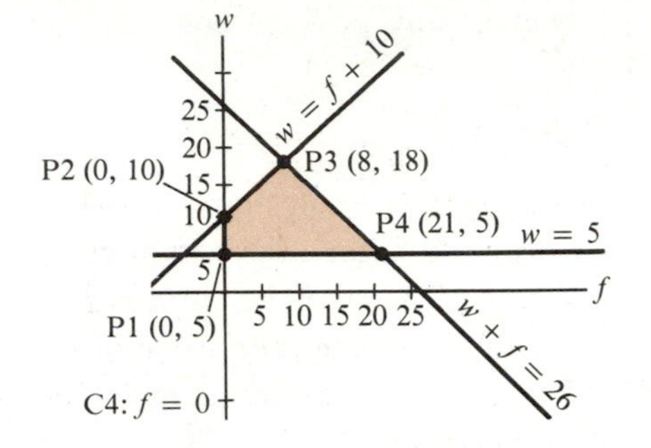

We have now located the region of intersection of the graphs of all four constraints. The points of interest to us are the corners or vertices of this region, labelled P1, P2, P3, and P4 on the final graph, Figure 8.9. The values of f and w for *one* of these points are the values that maximize profit.

The next step, then, is to determine the values of f and w for each of the four vertices of the region. These are called the **feasible solutions** of the problem.

P1 is the point of intersection of the graphs of the equations $w = 5$ and $f = 0$. So

$$\text{P1} = (0, 5)$$

P2 is at the intersection of $f = 0$ and $w = f + 10$. So

$$\text{P2} = (0, 10)$$

P3 is at the intersection of $w = f + 10$ and $w + f = 26$. So

$$\text{P3} = (8, 18)$$ **(Solve the equations simultaneously.)**

P4 is at the intersection of $w = 5$ and $w + f = 26$. So

$$\text{P4} = (21, 5)$$

The final step is to substitute the f and w coordinates of each of the above points in the objective equation:

$$\text{P} = 7f + 5w$$

P1 = (0, 5)	$P = 7(0) + 5(5) = 25$
P2 = (0, 10)	$P = 7(0) + 5(10) = 50$
P3 = (8, 18)	$P = 7(8) + 5(18) = 56 + 90 = 146$
P4 = (21, 5)	$P = 7(21) + 5(5) = 147 + 25 = 172$

This shows that P is greatest (172), that is, profit is maximized, when $f = 21$ and $w = 5$ (at P4). So a maximum profit of $172.00 will be realized when the plant produces 21 filberts and 5 widgets per day.

EXAMPLE 1

Floppy Tech, Inc. operates two factories that produce the following accessories for microcomputer diskettes: diskette envelopes, diskette labels, and write-protect tabs. The daily output of each factory is given by the following table.

Factory	Diskette Envelopes	Labels	Write-Protect Tabs
Factory 1	100	400	400
Factory 2	100	900	200

The cost of operating factory 1 is $10,000 per day; the cost of operating factory 2 is $20,000 per day. To meet its customers' demands, Floppy Tech must produce its products in the following daily quantities:

at least 5000 envelopes
at least 30,000 labels, and
at least 16,000 tabs.

In addition, the union contract stipulates that the two factories can be operated for a combined total of at most 100 days per quarter.

Determine how many days per quarter each factory should operate in order to minimize cost while also meeting the other constraints.

Solution

If factory 1 operates for x days and factory 2 operates for y days, the total cost for the quarter is given by the objective equation:

$$C = 10000x + 20000y$$

The constraints are:

(C1)	$100x + 100y \geq 5000$	Produce at least 5,000 envelopes.
(C2)	$400x + 900y \geq 30000$	Produce at least 30,000 labels.
(C3)	$400x + 200y \geq 16000$	Produce at least 16,000 tabs.
(C4)	$x + y \leq 100$	Operate for at most 100 days in total.
(C5)	$x \geq 0$	Common sense
(C6)	$y \geq 0$	

We first simplify the first three constraints, giving

(C1)	$x + y \geq 50$	Divide both sides by 100.
(C2)	$4x + 9y \geq 300$	Divide both sides by 100.
(C3)	$2x + y \geq 80$	Divide both sides by 200.
(C4)	$x + y \leq 100$	
(C5)	$x \geq 0$	
(C6)	$y \geq 0$	

Next, graph the inequalities for all six constraints on the same rectangular coordinate system (Figure 8.10).

FIGURE 8.10

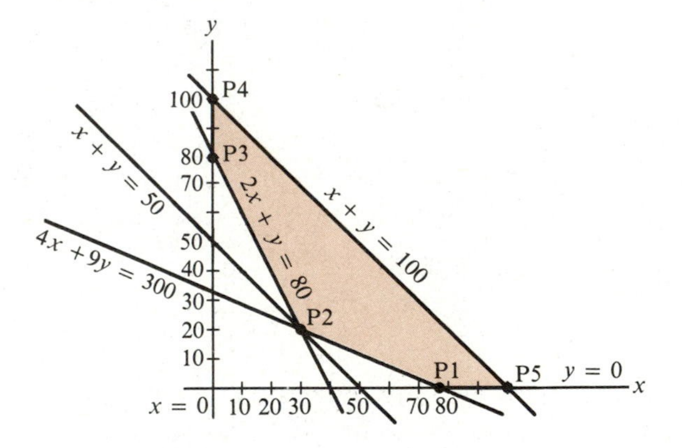

The feasible solutions are the sets of coordinates of the corners of the shaded region in Figure 8.10: P1, P2, P3, P4 and P5.

P1 is the intersection of

$$4x + 9y = 300 \text{ and}$$
$$y = 0$$

So $\qquad P1 = (75, 0)$

P2 is the intersection of three equations:

$$4x + 9y = 300$$
$$x + y = 50$$
$$2x + y = 80$$

So $\qquad P2 = (30, 20)$

P3 is the intersection of

$$x = 0 \quad \text{and}$$
$$2x + y = 80$$

So $\quad P3 = (0, 80)$

P4 is the intersection of

$$x = 0 \quad \text{and}$$
$$x + y = 100$$

So $\quad P4 = (0, 100)$

P5 is the intersection of

$$y = 0 \quad \text{and}$$
$$x + y = 100$$

So $\quad P5 = (100, 0)$

Finally, we make a table showing the value of C for each of the feasible solutions:

Point	x	y	$C = 10{,}000x + 20{,}000y$
P1 (75, 0)	75	0	\$750,000
P2 (30, 20)	30	20	\$700,000
P3 (0, 80)	0	80	\$1,600,000
P4 (0, 100)	0	100	\$2,000,000
P5 (100, 0)	100	0	\$1,000,000

The table shows that the cost C is least, \$700,000, when $x = 30$ and $y = 20$. Therefore, Floppy Tech will minimize its costs at \$700,000 if it operates factory 1 for 30 days and factory 2 for 20 days per quarter.

Algebraic Solution

This type of linear programming problem, as well as problems involving more than two variables, can also be solved algebraically. We illustrate by referring to the maximization problem for GTB Industries discussed at the beginning of this section.

The problem was to maximize

$$P = 7f + 5w$$

subject to four constraints:

$$\begin{aligned} &(C1) \quad & w &\leq f + 10 \\ &(C2) \quad & w &\geq 5 \\ &(C3) \quad & w + f &\leq 26 \\ &(C4) \quad & f &\geq 0 \end{aligned}$$

When we solved this problem graphically, we found that each feasible solution was the simultaneous solution to two of the following equations, each of which corresponds to one of the constraints:

$$\begin{aligned} &(E1) \quad & w &= f + 10 \\ &(E2) \quad & w &= 5 \\ &(E3) \quad & w + f &= 26 \\ &(E4) \quad & f &= 0 \end{aligned}$$

To solve this problem algebraically, we first pair the four equations in all possible combinations and find the simultaneous solution of each pair:

Pair of Equations	**Simultaneous Solution**
E1, E2	$f = -5, w = 5$
E1, E3	$f = 8, w = 18$
E1, E4	$f = 0, w = 10$
E2, E3	$f = 21, w = 5$
E2, E4	$f = 0, w = 5$
E3, E4	$f = 0, w = 26$

The next step is to test each simultaneous solution for feasibility—that is, verify that it does not violate any of the constraints. Finally, we evaluate P for each of the feasible solutions.

Equations	**Simultaneous Solution**	**Feasible?**	**Value of P**
E1, E2	$(-5, 5)$	no (violates C4)	—
E1, E3	$(8, 18)$	yes	146
E1, E4	$(0, 10)$	yes	50
E2, E3	$(21, 5)$	yes	172
E2, E4	$(0, 5)$	yes	25
E3, E4	$(0, 26)$	no (violates C1)	—

The table shows that the maximum profit (largest value of P) occurs when $f = 21$ and $w = 5$—that is, when GTB produces 21 filberts and 5 widgets.

Although this algebraic method of solving linear programming problems can be useful, it can also become quite lengthy, as the following example illustrates.

EXAMPLE 2

Maximize $P = 2a + b - c$ subject to

$$\begin{aligned} &(C1) & 3a - b + 2c &\leq 24 \\ &(C2) & a + b + c &\leq 22 \\ &(C3) & a + 2b + 3c &\leq 36 \\ &(C4) & a &\geq 0 \\ &(C5) & b &\geq 0 \\ &(C6) & c &\geq 0 \end{aligned}$$

Solution

The equations associated with the constraints are

$$\begin{aligned} &(E1) & 3a - b + 2c &= 24 \\ &(E2) & a + b + c &= 22 \\ &(E3) & a + 2b + 3c &= 36 \\ &(E4) & a &= 0 \\ &(E5) & b &= 0 \\ &(E6) & c &= 0 \end{aligned}$$

Since there are three variables in this problem, we must group the equations in sets of three to find simultaneous solutions. We then check each simultaneous solution to see whether it is feasible and evaluate P for each of the feasible solutions.

Possibility	Equations	Simultaneous Solution	Feasible?	Value of P
1	E1, E2, E3	(10, 10, 2)	yes	28
2	E1, E2, E4	$(0, 6\frac{2}{3}, 15\frac{1}{3})$	no (violates C3)	—
3	E1, E2, E5	$(-20, 0, 42)$	no (violates C4)	—
4	E1, E2, E6	$(11\frac{1}{2}, 10\frac{1}{2}, 0)$	yes	$33\frac{1}{2}$
5	E1, E3, E4	(0, 0, 12)	yes	-12
6	E1, E3, E5	(0, 0, 12)	yes	-12
7	E1, E3, E6	(12, 12, 0)	no (violates C2)	—
8	E1, E4, E5	(0, 0, 12)	yes	-12
9	E1, E4, E6	$(0, -24, 0)$	no (violates C5)	—
10	E1, E5, E6	(8, 0, 0)	yes	16
11	E2, E3, E4	$(0, 30, -8)$	no (violates C6)	—
12	E2, E3, E5	(15, 0, 7)	no (violates C1)	—
13	E2, E3, E6	(8, 14, 0)	yes	30
14	E2, E4, E5	(0, 0, 22)	no (violates C1)	—
15	E2, E4, E6	(0, 22, 0)	no (violates C3)	—
16	E2, E5, E6	(22, 0, 0)	no (violates C1)	—
17	E3, E4, E5	(0, 0, 12)	yes	-12
18	E3, E4, E6	(0, 18, 0)	yes	18
19	E3, E5, E6	(36, 0, 0)	no (violates C1)	—
20	E4, E5, E6	(0, 0, 0)	yes	0

The feasible solution that gives the largest value of P is possibility 4, when

$$a = 11\tfrac{1}{2}, \qquad b = 10\tfrac{1}{2}, \qquad c = 0 \qquad \text{and} \qquad P = 33\tfrac{1}{2}$$

This method can obviously get out of hand rather quickly. In the next section, we shall discuss the *simplex* method, a more systematic approach to solving linear programming problems.

EXERCISES FOR SECTION 8.5

In Exercises 1 through 10, solve each linear programming problem graphically (see Example 1).

1. Maximize $P = 300x + 200y$ subject to
$$\begin{aligned} x + y &\le 10 \\ 2x + y &\le 12 \\ 4x + 3y &\le 30 \\ x &\ge 0 \\ y &\ge 0 \end{aligned}$$

2. Maximize $P = 600x + 200y$ subject to
$$\begin{aligned} x + y &\le 10 \\ 2x + y &\le 12 \\ 4x + 3y &\le 30 \\ x &\ge 0 \\ y &\ge 0 \end{aligned}$$

3. Maximize $P = 200x + 250y$ subject to
$$\begin{aligned} x + y &\le 10 \\ 2x + y &\le 12 \\ 4x + 3y &\le 30 \\ x &\ge 0 \\ y &\ge 0 \end{aligned}$$

4. Maximize $P = 10x + 100y$ subject to
$$\begin{aligned} x + y &\ge 6 \\ 5x + y &\le 60 \\ x + 5y &\le 60 \\ x &\ge 0 \\ y &\ge 0 \end{aligned}$$

5. Maximize $P = 200x + 100y$ subject to
$$\begin{aligned} x + y &\ge 6 \\ 5x + y &\le 60 \\ x + 5y &\le 60 \\ x &\ge 0 \\ y &\ge 0 \end{aligned}$$

6. Maximize $P = 600x + 100y$ subject to
$$\begin{aligned} x + y &\ge 6 \\ 5x + y &\le 60 \\ x + 5y &\le 60 \\ x &\ge 0 \\ y &\ge 0 \end{aligned}$$

7. Minimize $C = 20x + 10y$ subject to
$$\begin{aligned} x + y &\ge 6 \\ 5x + y &\le 60 \\ x + 5y &\le 60 \\ x &\ge 0 \\ y &\ge 0 \end{aligned}$$

8. Minimize $C = 5x + 10y$ subject to
$$\begin{aligned} x + y &\ge 6 \\ 5x + y &\le 60 \\ x + 5y &\le 60 \\ x &\ge 0 \\ y &\ge 0 \end{aligned}$$

9. Maximize $P = x + y$
subject to
$$\begin{aligned} -x + 6y &\leq 30 \\ 3x + y &\leq 24 \\ x &\geq 0 \\ y &\geq 0 \end{aligned}$$

10. Minimize $C = x - y$
subject to
$$\begin{aligned} -x + 6y &\leq 30 \\ 3x + y &\leq 24 \\ x &\geq 0 \\ y &\geq 0 \end{aligned}$$

11. The Flite-Rite Company operates two factories, each producing three items for the airline industry, seatbelts, air valves, and pillows. The following table depicts each factory's daily output:

	Seat Belts	Air Valves	Pillows
Factory 1	100	200	400
Factory 2	100	500	200

The daily cost of operating factory 1 is \$10,000, and the daily cost of operating factory 2 is \$20,000. To meet demands, Flite-Rite must produce at least 5,000 seat belts, 15,000 air valves, and 16,000 pillows next quarter. How many days should each factory be in production in order to minimize cost if, in addition, the sum of the days both factories can be open per quarter cannot exceed 100?

12. The ElectroWave Company manufactures two types of hair dryers, economy and deluxe. Because of limited facilities, the company can produce no more than 40 economy dryers per week and no more than 60 deluxe dryers per week. The total number of units produced cannot exceed 80. If the profit for each economy dryer is \$10.00 and the profit for each deluxe dryer is \$15.00, find the number of units ElectroWave must produce to maximize profit. What is that profit?

8.6 The Simplex Algorithm (Optional)

The **simplex algorithm** gives us both a systematic approach to solving linear programming problems and a matrix-related method that can readily be adapted to computer use.

We begin by analyzing "maximum" problems in matrix form. For example, consider the problem

$$\text{Maximize} \quad P = 40x + 120y$$

$$\begin{aligned} \text{subject to} \quad -x + 4y &\le 0 \\ x + 2y &\le 150 \\ x &\ge 0 \\ y &\ge 0 \end{aligned}$$

If we consider the matrices

$$X = \begin{pmatrix} x \\ y \end{pmatrix} \qquad C = (40 \quad 120) \qquad A = \begin{pmatrix} -1 & 4 \\ 1 & 2 \end{pmatrix}$$

$$B = \begin{pmatrix} 0 \\ 150 \end{pmatrix} \tag{1}$$

then the problem can be put into the following matrix form:

$$\begin{aligned} \text{Maximize} \quad P &= C \cdot X \\ \text{subject to} \quad A \cdot X &\le B \\ X &\ge 0 \end{aligned}$$

where C is the row vector of coefficients in the optimization statement, X is a column vector of variables, A is the coefficient matrix, and B is a column vector of constants.

The first step in the simplex algorithm is to introduce new variables, called **slack variables**, that will enable us to transform inequalities into equations. That is, the inequality

$$-x + 4y \le 0$$

is equivalent to the equation

$$-x + 4y + s_1 = 0$$

where $s_1 \ge 0$ is a temporary variable used to take up the "slack" of the inequality. If two such variables are inserted into the system mentioned above, we have

$$\begin{aligned} \text{Maximize} \quad P &= 40x + 120y \\ \text{subject to} \quad -x + 4y + s_1 &= 0 \\ x + 2y + s_2 &= 150 \\ x &\ge 0 \\ y &\ge 0 \end{aligned}$$

So the first step in the simplex algorithm is to introduce slack variables.

Next, we represent the system 1 above in the **initial tableau**:

x	y	s_1	s_2	B
-1	4	1	0	0
1	2	0	1	150
-40	-120	0	0	0

Indicators

The elements in the last row, except the last number, are called **indicators** and that row can actually be interpreted as $P - 40x - 120y + 0s_1 + 0s_2 = 0$. (Notice that the coefficients of x and y are the *negatives* of their values in the original objective function.) Setting up this initial tableau is the second step in the simplex algorithm:

step 1. Introduce slack variables.
step 2. Establish the initial tableau.

Before we proceed with the rest of the algorithm, we present an example to reinforce the first two steps.

EXAMPLE 1

Construct the initial tableau for the following problem:

$$\begin{aligned} \text{Maximize} \quad & P = 16x + 10y + z \\ \text{subject to} \quad & 2x + y + z \leq 10 \\ & 3y + z \leq 8 \\ & y \leq x \\ & x, y, z \geq 0 \end{aligned}$$

Solution

First, notice that the third constraint is not in the required matrix form (1). We can rewrite $y \leq x$, however, as $-x + y \leq 0$. Now, we introduce the slack variables into the three constraints:

$$\begin{aligned} 2x + y + z + s_1 &= 10 \\ 3y + z + s_2 &= 8 \qquad \text{step 1} \\ -x + y + s_3 &= 0 \end{aligned}$$

The initial tableau is

step 2

x	y	z	s_1	s_2	s_3	B
2	1	1	1	0	0	10
0	3	1	0	1	0	8
−1	1	0	0	0	1	0
−16	−10	−1	0	0	0	0

The third step in the simplex algorithm is to locate the **pivot entry** in the tableau. The pivot entry is found in the following manner:

1. Look for the smallest (that is, the most negative) indicator. The column containing it is called the pivotal column.
2. Divide each *positive* entry of the pivotal column into the last column's entry. The row associated with the smallest resulting quotient is called the pivotal row.
3. The pivot entry (or **pivot**) is the number in the pivotal row and in the pivotal column.

To illustrate how to find the pivot, consider the following tableau:

x	y	s_1	s_2	B
2	3	1	0	12
1	1	0	1	60
−25	−40	0	0	0
	↑ Pivotal column			

Since −40 is the smallest indicator, the second column is the pivotal column. Next we examine the results of dividing each positive number in the pivotal column into the corresponding number in the last column:

x	y	s_1	s_2	B	
2	3	1	0	12	$12 \div 3 = 4$
1	1	0	1	60	$60 \div 1 = 60$
−25	−40	0	0	0	
	↑ Pivotal column				

Of 4 and 60, the smaller is 4, and thus the pivotal row is the first row. We highlight the pivot below:

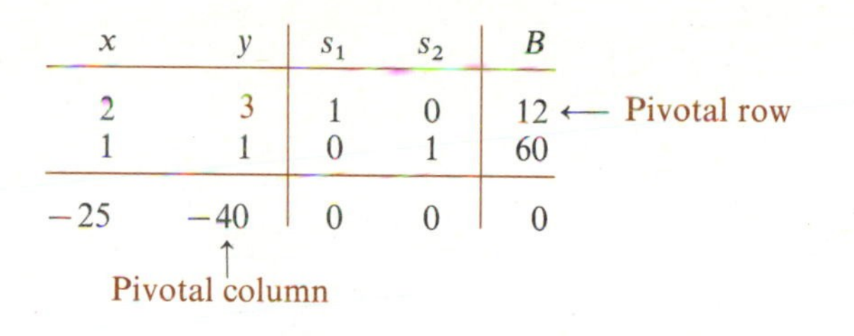

x	y	s_1	s_2	B	
2	3	1	0	12	← Pivotal row
1	1	0	1	60	
−25	−40	0	0	0	
	↑ Pivotal column				

Once the pivot is located, we perform the row transformations (from Chapter 7) to form a new tableau. Then this process of locating the pivot and performing row transformations is *repeated until all indicators are nonnegative.* The algorithm is summarized below:

THE SIMPLEX ALGORITHM

step 1. Introduce slack variables.

step 2. Establish the initial tableau.

step 3. Determine the pivot.

step 4. Perform row transformations. Convert the pivot to one by dividing by it and then convert all other numbers in the pivotal column to zero.

step 5. Repeat steps 3 and 4 until all indicators are nonnegative. When this occurs, the resulting tableau is called the **terminal tableau.** Properly interpreted, this solves our linear programming problem.

The next example takes us through a linear programming problem using the simplex method of solution.

EXAMPLE 2

Maximize $P = 3x + 5y$

subject to

$$2x + 4y \leq 60$$
$$x + y \leq 20$$
$$x \geq 0$$
$$y \geq 0$$

Solution

step 1. There are two slack variables, s_1 and s_2.

$$2x + 4y + s_1 = 60$$
$$x + y + s_2 = 20$$

step 2. The initial tableau is

x	y	s_1	s_2	B	
2	4	1	0	60	← $60 \div 4 = 15$
1	1	0	1	20	$20 \div 1 = 20$
−3	−5	0	0	0	
	↑				

step 3. The pivot is in the second column (because −5 is the smallest indicator) and in the first row (because $60 \div 4$ is smaller than $20 \div 1$).

step 4. Pivoting on 4 yields the tableau below. It is obtained by dividing row 1 by 4 and then adding suitable multiples of the first row to rows 2 and 3 so that the other entries in the pivotal column (1 and −5) become 0.

x	y	s_1	s_2	B	
0.5	1	0.25	0	15	$15 \div 0.5 = 30$
0.5	0	−0.25	1	5	← $5 \div 0.5 = 10$
−0.5	0	1.25	0	75	
↑					

step 5. We repeat the process; this time the pivot is in the second row, first column:

x	y	s_1	s_2	B
0	1	0.5	−1	10
1	0	−0.5	2	10
0	0	1	1	80

Finally, to solve the problem, we must interpret the terminal tableau. The last row represents the following equation:

$$P + s_1 + s_2 = 80 \qquad \text{or} \qquad P = 80 - s_1 - s_2$$

In order for P to be a maximum, we want s_1 and s_2 to be zero. So, $s_1 = 0$ and $s_2 = 0$ and the maximum profit is 80. From the first row, $y + 0.5s_1 - 1s_2 = 10$. But with the two slack variables equal to zero, we can deduce that $y = 10$. Similarly interpreting row 2, we find x to be 10. When $x = 10$ and $y = 10$ the maximum value of P is reached and found to be 80.

We conclude this section with another example of using the simplex algorithm to solve a linear programming problem.

EXAMPLE 3

Use the simplex algorithm to solve the problem of Example 2 of Section 8.5.

$$\text{Maximize} \quad P = 2a + b - c$$
$$\text{subject to} \quad 3a - b + 2c \le 24$$
$$a + b + c \le 22$$
$$a + 2b + 3c \le 36$$
$$a \ge 0$$
$$b \ge 0$$
$$c \ge 0$$

Solution

steps 1 and 2.

a	b	c	s_1	s_2	s_3	B
3	−1	2	1	0	0	24
1	1	1	0	1	0	22
1	2	3	0	0	1	36
−2	−1	1	0	0	0	0

step 3. The pivot is in the first column (because −2 is the smallest indicator) and in the first row (because 24 ÷ 3 is smaller than 22 ÷ 1 or 36 ÷ 1).

a	b	c	s_1	s_2	s_3	B	
3	−1	2	1	0	0	24	←
1	1	1	0	1	0	22	
1	2	3	0	0	1	36	
−2	−1	1	0	0	0	0	
↑							

step 4. First, multiply row 1 by $\frac{1}{3}$ to convert the pivot to 1:

a	b	c	s_1	s_2	s_3	B
1	$-\frac{1}{3}$	$\frac{2}{3}$	$\frac{1}{3}$	0	0	8
1	1	1	0	1	0	22
1	2	3	0	0	1	36
-2	-1	-1	0	0	0	0

Next, multiply row 1 by -1 and add that to rows 2 and 3; multiply row 1 by 2 and add that to row 4:

a	b	c	s_1	s_2	s_3	B	
1	$-\frac{1}{3}$	$\frac{2}{3}$	$\frac{1}{3}$	0	0	8	
0	$\frac{4}{3}$	$\frac{1}{3}$	$-\frac{1}{3}$	1	0	14	←
0	$\frac{7}{3}$	$\frac{7}{3}$	$-\frac{1}{3}$	0	1	28	
0	$-\frac{5}{3}$	$\frac{7}{3}$	$\frac{2}{3}$	0	0	16	
	↑						

step 5. Notice that this tableau still has a negative indicator, so we must repeat steps 3 and 4:

We convert the pivot $\frac{4}{3}$ to 1 by multiplying row 2 by $\frac{3}{4}$:

a	b	c	s_1	s_2	s_3	B
1	$-\frac{1}{3}$	$\frac{2}{3}$	$\frac{1}{3}$	0	0	8
0	1	$\frac{1}{4}$	$-\frac{1}{4}$	$\frac{3}{4}$	0	$\frac{21}{2}$
0	$\frac{7}{3}$	$\frac{7}{3}$	$-\frac{1}{3}$	0	1	28
0	$-\frac{5}{3}$	$\frac{7}{3}$	$\frac{2}{3}$	0	0	16

Now, multiply row 2 by $\frac{1}{3}$ and add that to row 1; multiply row 2 by $-\frac{7}{3}$ and add that to row 3 and multiply row 2 by $\frac{5}{3}$ and add to row 4. All this results in the terminal tableau:

a	b	c	s_1	s_2	s_3	B
1	0	$\frac{3}{4}$	$\frac{1}{4}$	$\frac{1}{4}$	0	$\frac{23}{2}$
0	1	$\frac{1}{4}$	$-\frac{1}{4}$	$\frac{3}{4}$	0	$\frac{21}{2}$
0	0	$\frac{7}{4}$	$\frac{1}{4}$	$-\frac{7}{4}$	1	$\frac{7}{2}$
0	0	$\frac{11}{4}$	$\frac{1}{4}$	$\frac{5}{4}$	0	$\frac{67}{2}$

We interpret the bottom row as follows:

$$P + \tfrac{11}{4}c + \tfrac{1}{4}s_1 + \tfrac{5}{4}s_2 = \tfrac{67}{2} \qquad \text{or}$$
$$P = \tfrac{67}{2} - \tfrac{11}{4}c - \tfrac{1}{4}s_1 - \tfrac{5}{4}s_2$$

So, P will be greatest when c, s_1, and s_2 are 0. Row 1 gives us

$$a + \tfrac{3}{4}c + \tfrac{1}{4}s_1 + \tfrac{1}{4}s_2 = \tfrac{23}{2}$$

Thus, $a = \tfrac{23}{2}$. Row 2 gives us

$$b + \tfrac{1}{4}c - \tfrac{1}{4}s_1 + \tfrac{3}{4}s_2 = \tfrac{21}{2}$$

Thus, $b = \tfrac{21}{2}$. Finally, the maximum value of P is $\tfrac{67}{2}$, and it is achieved when

$$a = \tfrac{23}{2}$$
$$b = \tfrac{21}{2}$$
$$c = 0$$

We conclude this section by warning the reader that we have merely scratched the surface of linear programming and the simplex algorithm. In fact, only limited types of problems can be solved with the material presented here. Only maximum problems, only tableaux where the entries in the B column are all positive, and only problems where the variables must be positive can be solved using the algorithm the way it is presented here. The curious reader should consult texts on the topic of linear programming for further details.

EXERCISES FOR SECTION 8.6

In Exercises 1 through 12, solve each linear programming problem (see Examples 1–3).

1. Maximize $P = 5x + 2y$
subject to
$$2x + 5y \le 6$$
$$3x + 7y \le 20$$
$$x \ge 0$$
$$y \ge 0$$

2. Maximize $P = 10x + 2y$
subject to
$$4x + 5y \le 6$$
$$6x + 7y \le 20$$
$$x \ge 0$$
$$y \ge 0$$

3. Maximize $P = 100x + 120y$
subject to
$$\begin{aligned} x + 2y &\le 100 \\ 3x + 2y &\le 140 \\ x &\le 40 \\ x &\ge 0 \\ y &\ge 0 \end{aligned}$$

4. Maximize $P = 100x + 2y$
subject to
$$\begin{aligned} x + 2y &\le 100 \\ 3x + 2y &\le 140 \\ x &\ge 0 \\ y &\ge 0 \end{aligned}$$

5. Maximize $P = 100x + 2y$
subject to
$$\begin{aligned} x + 2y &\le 100 \\ 3x + 2y &\le 140 \\ x &\le 40 \\ x &\ge 0 \\ y &\ge 0 \end{aligned}$$

6. Maximize $P = 10x + 5y$
subject to
$$\begin{aligned} x + 3y &\le 30 \\ 4x + 7y &\le 100 \\ x &\ge 0 \\ y &\ge 0 \end{aligned}$$

7. Maximize $P = 100x + 2y$
subject to
$$\begin{aligned} x + 2y &\le 100 \\ 3x + 2y &\le 140 \\ y &\le 20 \\ x &\le 40 \\ x &\ge 0 \\ y &\ge 0 \end{aligned}$$

8. Maximize $P = 10x + 5y$
subject to
$$\begin{aligned} x + 3y &\le 30 \\ 4x + 7y &\le 100 \\ x &\le 18 \\ x &\ge 0 \\ y &\ge 0 \end{aligned}$$

9. Maximize $P = x + 2y + 10z$
subject to
$$\begin{aligned} 3x + 2y &\le 14 \\ x + 10z &\le 102 \\ x + y + 5z &\le 56 \\ z &\le 10 \\ x &\ge 0 \\ y &\ge 0 \\ z &\ge 0 \end{aligned}$$

10. Maximize $P = x + 10y + 2z$
subject to
$$\begin{aligned} 3x + 2z &\le 14 \\ x + 10y &\le 102 \\ x + 5y + z &\le 56 \\ y &\le 10 \\ x &\ge 0 \\ y &\ge 0 \\ z &\ge 0 \end{aligned}$$

11. Maximize $P = 300x + 200y$
subject to
$$\begin{aligned} x + y &\le 10 \\ 2x + y &\le 12 \\ 4x + 3y &\le 30 \\ x &\ge 0 \\ y &\ge 0 \end{aligned}$$
(Compare with Exercise 1, Section 8.5.)

12. Maximize $P = 600x + 200y$
subject to
$$\begin{aligned} x + y &\le 10 \\ 2x + y &\le 12 \\ 4x + 3y &\le 30 \\ x &\ge 0 \\ y &\ge 0 \end{aligned}$$
(Compare with Exercise 2, Section 8.5.)

13. The ElectroWave Company manufactures two types of hair dryers, economy and deluxe. Because of limited facilities, the company cannot produce more than 60 deluxe dryers per week and no more than 40 economy dryers per week. The total number of dryers produced cannot exceed 80. If the profit for each economy dryer is $10.00 and the

profit for each deluxe dryer is \$15.00, find the number of units ElectroWave must produce to maximize profit. What is that profit?

14. A machine shop produces two types of firing pins for rifles, short and long. Because of its limited facilities, the company cannot produce more than 4000 short pins per week and cannot produce more than 6000 long pins per week. Also, the total number of pins produced weekly cannot exceed 8000. If it is known that the profit for each short pin is \$2.00 and for each long pin is \$3.00, how many of each pin should the company produce in order to maximize profit? What maximum profit is achieved? (Use the simplex method.)

15. The examples in this section were limited to two-variable or three-variable linear programming problems. The algorithm works, of course, for any number of variables or constraints. Solve the system below:

$$\begin{aligned} \text{Maximize} \quad & x + 2y + z + w \\ \text{subject to} \quad & x + y + w \leq 9 \\ & y + z \leq 10 \\ & x + w \leq 7 \\ & x + y + z - w \leq 16 \\ & x \geq 0 \\ & y \geq 0 \\ & z \geq 0 \\ & w \geq 0 \end{aligned}$$

In Exercises 16 through 20, solve the following linear programming problem. All that changes is the objective function, P. The constraints for each exercise are as follows:

$$\begin{aligned} \text{Maximize } P \text{ subject to} \quad & x + 4y \leq 200 \\ & 2x + y \leq 120 \\ & x \leq 50 \\ & x \geq 0 \\ & y \geq 0 \end{aligned}$$

16. $P = 10x + 20y$

17. $P = 30x + 10y$

18. $P = 10x + 50y$

19. $P = 20x + 10y$

20. Notice that in Exercise 19 above, the value of P that will be maximum is, indeed, unique. However, *unlike* each of Exercises 16 through 18, many different combinations of x and y will yield that same maximum value of P.

a. Show that each of the following pairs of values for x and y yields the same maximum value of P, 1200:

(1) $x = 40 \quad y = 40$

(2) $x = 50 \quad y = 20$

(3) $x = 45 \quad y = 30$

b. Solve the problem graphically.

8.7 Chapter Review

Chapter 8 presented various methods of solving systems of linear equations (equations whose graphs are straight lines) in two or more variables. Techniques for solving systems of two equations in two variables were presented first. They included solution by graphing, the method of substitution, Gaussian elimination, and the use of row transformations on the augmented matrix of the system.

Sections 8.3 and 8.4 dealt with solving systems of linear equations in more than two variables. Examples were presented for the three-equation, three-variable case, although the methods used can be applied to larger systems as well. The methods discussed were using row transformations of the augmented matrix, using the inverse of the coefficient matrix, and Cramer's rule.

The last two sections of the chapter dealt with a practical application of systems of linear equations to linear programming problems—problems dealing with maximizing or minimizing quantities like profits and costs. Some elementary applications of the simplex algorithm were presented as examples of the power of matrix methods in dealing with real-life problems.

VOCABULARY

linear equation
inconsistent equations
dependent equations
method of substitution
Gaussian elimination
equivalent system
augmented matrix
row transformations
Cramer's rule
coefficient matrix
linear programming
objective function
constraints
feasible solution
simplex algorithm
slack variables
initial tableau
indicators
pivot entry
terminal tableau

CHAPTER TEST

Solve each of the following systems of equations (a) by graphing and (b) by substitution or Gaussian elimination if a unique solution exists. If there is no unique solution, state whether the system is inconsistent or dependent.

1. $4x - 4y = 4$
$2x - 3y = -2$

2. $3x - 2y = 5$
$-6x + 4y = -10$

3. $3x + 2y = -2$
$3x + 4y = -3$

4. $x - y = 4$
$-x + y = -3$

Use row transformations to solve each of the following systems of equations. If there is no unique solution, state whether the system is inconsistent or dependent.

5. $x + y + z = 7$
$-2x + y - z = -4$
$3x - 5y + 2z = 1$

6. $2x + 3y - z = 1$
$3x + 5y + 2z = 8$
$-2x + 4y + 6z = 2$

7. $3x + 6y + 9z = 9$
$4x + 6y + 16z = 8$
$3x + 2y + 17z = 1$

8. Use Cramer's rule to solve the following system of equations, if possible.

$$-3x - 6y + 3z = -9$$
$$2x - 5y + 2z = 1$$
$$-3x + 4y + z = -2$$

9. Solve the following linear programming problem graphically.

Maximize $P = 20x + 30y$
subject to $x + y \le 80$
$x \le 40$
$y \le 60$
$x \ge 0$
$y \ge 0$

CHAPTER

9

We deal frequently with uncertainty in our daily lives. Whether we are trying to pick a horse for the next race, choose a daily number for the state lottery, or forecast tomorrow's weather, we are relying on an informed (more or less, depending on the circumstances) guess—a probability. Researchers also deal in probabilities. For example, scientists may try to simulate a particular set of environmental conditions in order to predict their long-term effects. In this chapter, we focus on the concepts of probability and simulations, along with some related topics from statistics.

Probability, Simulation, and Statistics

9.1 Choices, Permutations, and Combinations

In this section we begin to answer questions about "how many." That is, questions that investigate the number of different ways something can happen. "How many different passwords are there?" "How many different lottery tickets are there?" And so on.

Many college timesharing computer systems require that a student enter a password in order to access the system. At Mattatuck Community College a student's password consists of four *different* letters. How many different passwords are there? To address this "how many" question, or counting problem, we create a slot for each of the four letters:

____	____	____	____
first letter	second letter	third letter	fourth letter

There are 26 letters in the alphabet and, as we attempt to create four-letter "words," we have those 26 possibilities for the first letter, 25 for the second (since they must be different), 24 for the third, and 23 for the last. Thus, we have

26	25	24	23
first letter	second letter	third letter	fourth letter

There are $26 \cdot 25 \cdot 24 \cdot 23 = 358{,}800$ possible passwords.

An arrangement of this type, in which the order of the nonrepeated components (in this case, letters) is important, is called a **permutation**, or **ordered arrangement**. We say there are 358,800 permutations of 26 letters, using four at a time.

EXAMPLE 1

How many permutations are there of the letters C, A and B?

Solution

3	2	1
first letter	second letter	third letter

We are choosing any of the three (C or A or B) for the first letter; then there are two of the three left for the second position, and one for the third:

$$3 \cdot 2 \cdot 1 = 6$$

There are *six* permutations of the letters C, A and B.

An alternative solution to Example 1 is to list the possibilities in a **tree diagram**. This type of diagram is so named because it "branches out" all the possibilities. An appropriate tree diagram for Example 1 is shown in Figure 9.1. Notice that it shows six possible arrangements.

FIGURE 9.1
The tree diagram for permutations of the letters C, A and B.

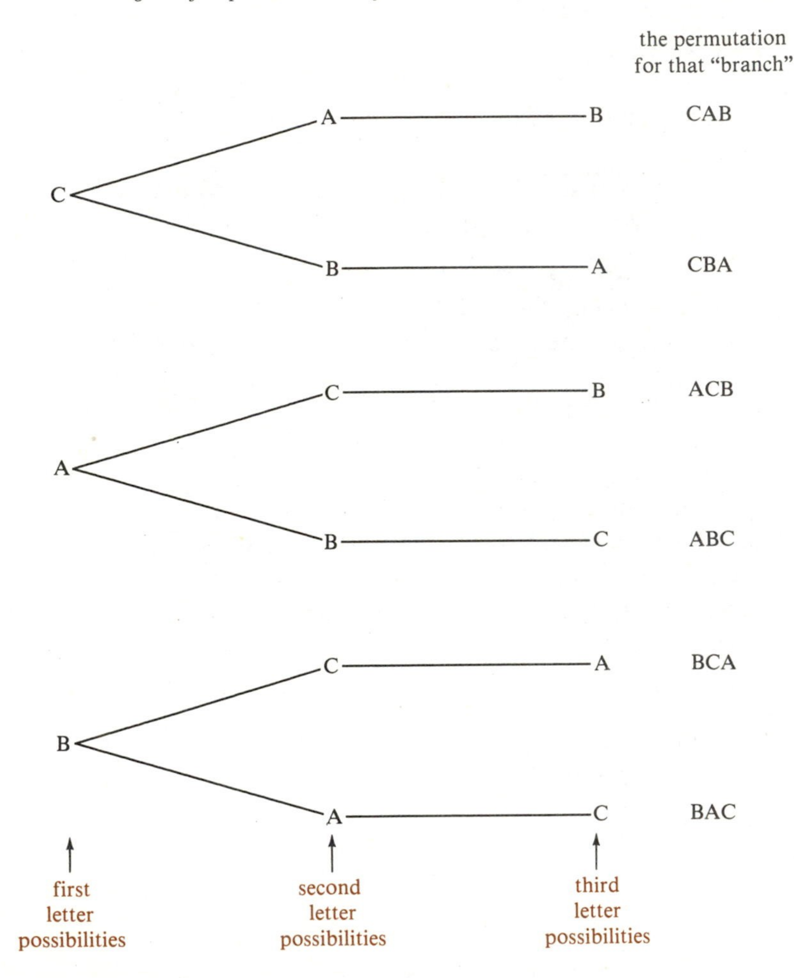

EXAMPLE 2
How many three-letter "words" can be formed using the letters A, B and C if repetition is allowed within any three-letter word?

Solution

Because repetition is allowed, there are three choices for the first letter, still three for the second (we can now repeat whatever the first letter was), and three for the third. In all, there are $3 \cdot 3 \cdot 3 = 27$ "words." A tree diagram for creating them appears in Figure 9.2.

FIGURE 9.2

The tree diagram for all three-letter words formed from A, B and C.

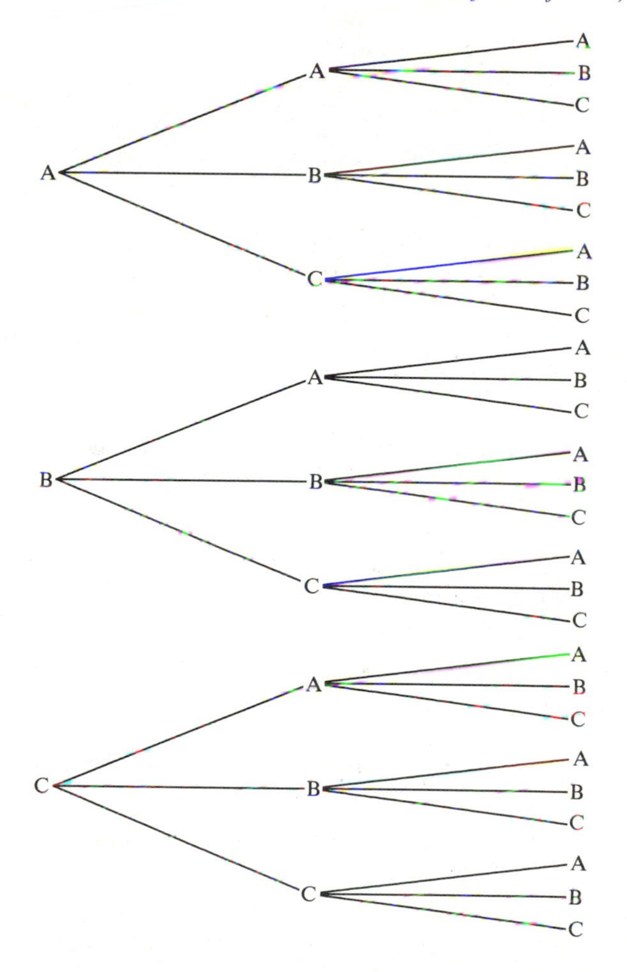

EXAMPLE 3

In 1980, the State of Connecticut began running out of license plates. At that time, a license plate consisted of two letters followed by four digits. Because the demand for plates exceeded the total possible number of combinations, the configuration was changed to three letters followed by three digits.

a. How many license plates are possible if each plate consists of two letters followed by four digits (the "old" plates)?

b. How many license plates are possible if each plate consists of three letters followed by three digits (the "new" plates)?

Solution

a.

26	26	10	10	10	10
letter	letter	digit	digit	digit	digit

There are 26 letters possible for each of the first two slots (we assume repetition is allowed) and 10 possible digits (0 through 9) for each of the last four slots:

$$26 \cdot 26 \cdot 10 \cdot 10 \cdot 10 \cdot 10$$
$$= 6{,}760{,}000 \text{ possible license plates}$$

b.

26	26	26	10	10	10
letter	letter	letter	digit	digit	digit

$$26 \cdot 26 \cdot 26 \cdot 10 \cdot 10 \cdot 10 = 17{,}576{,}000 \text{ possible license plates}$$

EXAMPLE 4

A studio is photographing a microcomputer company's new product line. A microcomputer, a disk drive, a monitor, and a printer are to be in the picture. How many different arrangements (and thus different pictures) can be made?

Solution

There are $4 \cdot 3 \cdot 2 \cdot 1$ different arrangements; that is, 24 different arrangements can be made.

Products such as $4 \cdot 3 \cdot 2 \cdot 1$ or $8 \cdot 7 \cdot 6 \cdot 5 \cdot 4 \cdot 3 \cdot 2 \cdot 1$ occur often enough in mathematics that they have a name. We call them **factorials** and use the symbol !:

$$3! = 3 \cdot 2 \cdot 1 = 6$$

FACTORIALS

$$0! = 1$$
$$1! = 1$$
$$2! = 2 \cdot 1 = 2$$
$$3! = 3 \cdot 2 \cdot 1 = 6$$
$$4! = 4 \cdot 3 \cdot 2 \cdot 1 = 24$$
$$5! = 5 \cdot 4 \cdot 3 \cdot 2 \cdot 1 = 120$$
$$\vdots$$
$$n! = n(n-1)(n-2) \cdot \ldots \cdot 3 \cdot 2 \cdot 1$$

Note: $n! = n(n-1)!$

EXAMPLE 5

Evaluate each of the following:

a. $0! + 3!$ **b.** $\frac{7!}{4!}$ **c.** $\frac{6!}{2!4!}$

Solution

a. $0! + 3! = 1 + 6 = 7$

b. $\frac{7!}{4!} = \frac{7 \cdot 6 \cdot 5 \cdot 4!}{4!} = 7 \cdot 6 \cdot 5 = 210$

c. $\frac{6!}{2!4!} = \frac{6 \cdot 5 \cdot 4!}{2 \cdot 1 \cdot 4!} = \frac{30}{2} = 15$

EXAMPLE 6

If a small computer uses a byte consisting of four bits, how many *bit patterns* (see Chapter 3) are possible?

Solution

There are two choices for each bit (either 0 or 1); thus there are $2 \cdot 2 \cdot 2 \cdot 2 = 16$ permutations:

0001	0101	1001	1101
0010	0110	1010	1110
0011	0111	1011	1111
0100	1000	1100	0000

0000 is a valid bit pattern.

Note that no factorials were used here because the same number of options (two) were available for each bit. We use factorials when the number of available choices is *reduced by one* each time a choice is made. Such a situation is illustrated by the next example.

EXAMPLE 7

If an account number consists of ten different digits, how many different account numbers are possible?

Solution

There is no reason that an account number could not begin with 0. Since account numbers are not used in arithmetic, they can be stored as labels or character strings (see Chapter 1). Thus, there are ten choices for the first number, nine for the second (since each digit must be different), eight for the third, and so on.

$$10 \cdot 9 \cdot 8 \cdot 7 \cdot 6 \cdot 5 \cdot 4 \cdot 3 \cdot 2 \cdot 1 = 10! = 3{,}628{,}800$$

There are $10! = 3{,}628{,}800$ permutations of ten different characters.

An essential property of a permutation of a group of characters is that *ordering* is important. Certainly the binary numbers 101 and 011 are not equal, even though they consist of the same characters (0, 1 and 1). There are instances, however, when order is *not* important, as the next example suggests.

EXAMPLE 8

Suppose four people apply for assignment to a committee consisting of just two people. The committee is distinguished only by who is a member, not by any particular hierarchy (such as president or vice-president). How many committees can be formed?

Solution

There are four choices for the first person chosen and three people left for the second, so it *appears* that the number of committees is $4 \cdot 3$. It does not matter, however, who is chosen first. Thus, we divide by the number of permutations of two things (2!). So the solution is

$$\frac{4 \cdot 3}{2!} = 6$$

This can be illustrated by letting A, B, C, D be the names of the applicants.

There are $4 \cdot 3 = 12$ permutations of four letters (names) taken two at a time:

AB	BA
AC	CA
AD	DA
BC	CB
BD	DB
CD	DC

We note that in this case AB = BA; they both represent the same committee. There are two permutations (2!) of each pair picked, so one must divide by 2! to eliminate the duplication.

If, in an arrangement of elements, order is not significant and only *which* elements are present is of importance, that arrangement is called a **combination** of the elements.

EXAMPLE 9

Find all two-letter permutations and all two-letter combinations of the letters A, B and C.

Solution

The permutations are

AB, AC, BA, BC, CA, CB

There are $3 \cdot 2 = 6$.

The number of combinations is

$$\frac{3!}{2!} = 3$$

The combinations are AB, AC and BC. They represent every possible group of two that can be chosen from the three letters A, B and C when *order is unimportant.*

EXAMPLE 10

Suppose in Example 8 that five people applied for the two-member committee. How many committees are possible?

Solution

$5 \cdot 4$ must be divided by the number of permutations of each pair to eliminate duplication.

$$\frac{5 \cdot 4}{2!} = 10$$

If A, B, C, D, and E represent the people, then BE is one permutation as is EB. Dividing by 2! eliminates that duplication.

EXAMPLE 11

If six people apply for a four-member committee and there are no specific committee assignments or officers, how many committees (combinations) are there?

Solution

There are $6 \cdot 5 \cdot 4 \cdot 3$ permutations (such as ABCD or BCDA). But ABCD = BCDA since position is not important. Dividing by 4! (the number of permutations of four things) eliminates duplication. Thus the number of committees, or combinations of six things taken four at a time, is

$$\frac{6 \cdot 5 \cdot 4 \cdot 3}{4!} = \frac{(6 \cdot 5 \cdot 4 \cdot 3)}{(4 \cdot 3 \cdot 2 \cdot 1)} \cdot \frac{(2 \cdot 1)}{(2 \cdot 1)}$$

Multiplying both numerator and denominator by $2 \cdot 1$ is, in effect, multiplying the fraction by 1. We thus have

$$\frac{6 \cdot 5 \cdot 4 \cdot 3 \cdot 2 \cdot 1}{(4 \cdot 3 \cdot 2 \cdot 1) \cdot (2 \cdot 1)} = \frac{6!}{4!2!} = \frac{6!}{4!(6-4)!}$$

This last expression is abbreviated in mathematics as $\binom{6}{4}$ and represents the number of combinations (or unordered arrangements) of six things taken four at a time.

COMBINATIONS

The number of combinations of n different objects, choosing r at a time, is given by

$$\binom{n}{r} = \frac{n!}{r!(n-r)!}$$

EXAMPLE 12

Eight people apply for a committee of five. How many possibilities are there if no specific assignments or positions are held?

Solution

Since the order of selection is not important, we must find the number of *combinations* of eight items taken five at a time (five committee members chosen from a group of eight people). The formula is

$$\binom{8}{5} = \frac{8!}{5!(8-5)!}$$

$$= \frac{8!}{5!3!}$$

$$= \frac{8 \cdot 7 \cdot \cancel{6} \cdot \cancel{5!}}{\cancel{5!} \cdot \cancel{3 \cdot 2 \cdot 1}}$$

$$= 56$$

EXAMPLE 13

Evaluate each of the following:

a. $\binom{5}{2}$ **b.** $\binom{6}{0}$ **c.** $\binom{8}{2} + \binom{8}{3}$

Solution

a. $\binom{5}{2} = \frac{5!}{2!3!} = \frac{5 \cdot 4 \cdot 3!}{2!\,3!} = 10$

b. $\binom{6}{0} = \frac{6!}{0!6!} = 1$

c. $\binom{8}{2} + \binom{8}{3} = \frac{8!}{2!6!} + \frac{8!}{3!5!}$

$$= \frac{8 \cdot 7 \cdot 6!}{2!\,6!} + \frac{8 \cdot 7 \cdot 6 \cdot 5!}{3!\,5!}$$

$$= 28 + 56$$

$$= 84$$

EXERCISES FOR SECTION 9.1

In Exercises 1 through 6, evaluate each expression, using a calculator where appropriate (see Examples 4 and 5).

1. 7! **2.** 4! + 3! **3.** 4! + 6! **4.** 10!

5. 15! (Calculate 15! by hand and then by calculator or computer. Are the answers exactly the same? Explain.)
6. 20!

In Exercises 7 through 11, simplify each expression by using factorials (see Examples 4 and 5).

7. $8 \cdot 7 \cdot 6 \cdot 5 \cdot 4 \cdot 3 \cdot 2 \cdot 1$
8. $4 \cdot 3 \cdot 2$
9. $10 \cdot 9 \cdot 8 \cdot 7 \cdot 6 \cdot 5!$
10. $N(N-1)(N-2)!$
11. $(N+2)(N-1)(N)(N+1)(N-2)!$

In Exercises 12 through 16, evaluate each expression, giving exact answers (see Examples 12 and 13).

12. $\binom{9}{4}$
13. $\frac{4!}{3!2!}$
14. $\frac{8!}{5!3!}$
15. $\binom{7}{0}$
16. $\binom{6}{3} + \binom{5}{2}$

For Exercises 17 through 28, see Examples 6 through 12.

17. How many arrangements of the five letters A, B, C, D, E are possible under the following conditions:
    a. Repetitions are allowed.
    b. Repetitions are *not* allowed.
    c. Repetitions are *not* allowed, and C must be the middle letter.
    d. Repetitions are not allowed, and C must be the first letter.
18. At a technical college, a student wishing to use the computer must enter a password consisting of two different letters followed by a four-digit code with no restrictions. How many different passwords are possible?
19. A student at the technical college of Exercise 18 has written a computer program that generates a trial password once every second. Use your answer to Exercise 18 to calculate the maximum number of minutes it would take for his program to match a correct password.
20. Suppose license plates consist of four letters followed by four digits, with repetitions allowed.
    a. How many different license plates are possible?
    b. If one license plate could be stamped every five minutes, how many hours would it take to complete the production of all possible license plates?
    c. Suppose the requirement that four letters be *followed* by four digits were dropped and *any* arrangement consisting of four letters and four

digits was permitted. How many license plates would now be possible?

21. An eight-bit computer "word" consists of eight "cells," each containing a binary digit (0 or 1). How many different bit patterns are possible for an eight-bit word?

22. Suppose nine people apply for a committee in which there are no specific officers or positions.
   a. If the committee is to have three members, how many different ways can the committee be filled?
   b. If the committee is to consist of six people, how many possible committees are there?
   c. What if the committee is to consist of nine people?
   d. If the nine applicants consist of five females and four males, how many committees of six can be formed with an equal number of males and females?

23. Six salespeople arrive at the same time in a city that has eight motels. In how many ways can they choose a motel if no two wish to spend the night in the same motel?

24. A girl has five close friends. In how many ways can she invite *one or more* of them to dinner?

25. How many rows are in a truth table (see Chapter 5) consisting of six simple statements (p, q, r, s, t and u)?

26. Seven people apply for jobs at a firm in which there are three vacant positions. In how many ways can these positions be filled?

27. Suppose a computer reserves 16 bits for each integer, with the leading bit being a sign bit. What is the largest factorial that can be stored as a pure integer type? Does the answer to this question explain why your evaluation of 15! (see Exercise 5) by computer or calculator was not exact?

28. Suppose that there are ten major brands of microcomputers for sale. If a school wanted to have a representative sample of four different brands, how many different samples of four could be considered?

9.2 Elementary Probability: An Introduction

Probability questions or problems occur often enough that most people have at least an intuitive feeling for the meaning of probability. We hear such expressions as "The probability of rain is greater than 1/2"; or "The probability of survival is 1/10"; or "The probability of winning the lottery is 1/100,000."

We shall now explore the concept of probability in greater detail and with greater precision than in this intuitive sense. We shall begin with a familiar situation:

When a coin is tossed, what is the probability of its coming up heads? If you did not know the answer, you could toss the coin many times under the same conditions and observe the results. A chart such as the following could then be constructed:

Possible Outcome	**Number of Occurrences**	**Relative Frequency**
heads	240	$\frac{240}{500} = 0.48$
tails	260	$\frac{260}{500} = 0.52$
	total 500	

It appears that the experimental or *empirical* results indicate a bias toward tails and one might say the coin is slightly biased, with the probability of heads 0.48. But what if you were to continue tossing the coin for another 1000 trials? The chart might then look like this:

Possible Outcome	**Number of Occurrences**	**Relative Frequency**
heads	739	$\frac{739}{1500}$
tails	761	$\frac{761}{1500}$
	total 1500	

The relative frequency of heads $739/1500 = 0.49266667$. So now you conclude that maybe the coin is fair, or nearly so, since the observed relative frequency of heads $= 0.49266667$ and the relative frequency of tails is approximately 0.5073333 ($= 761 \div 1500$).

Some questions should immediately come to mind:

1. How many trials are necessary to determine if the coin is fair?
2. What is the probability of getting heads?

Perhaps the best answer we can give to the first question is that the greater the number of trials, the more secure we can feel in the result. But an experiment conducted in this way cannot be used to determine definitive results. The results of repeated trials cannot be used to prove any conclusion *definitively*.

In answer to the second question, two possibilities are offered:

1. The relative frequency definition of probability states that the probability of an event can be approximated by evaluating

$$\frac{\text{number of successes (i.e. occurrences of event)}}{\text{number of trials}}$$

   This approximation is generally more accurate as the number of trials increases.

2. We can also examine the coin. We note its symmetry and balance and conclude that because it is balanced and symmetrical, the coin is fair. Then the probability of heads is $\frac{1}{2}$ (both heads and tails have the same chance of occurring).

EXAMPLE 1

What is the probability of getting two spots showing when you throw one die? Discuss two methods of justifying or determining your answer.

Solution

One method is to toss the die many times. This experimental method cannot be used to *prove* an answer, but it does help in *approximating* one. A chart such as the following could be used:

Face	Success
2	
not 2	
total	

Suppose that after 1000 trials (tosses of the die), your results were

Face	Success
2	150
not 2	850
total	1000

The table reveals that two spots came up 150 times out of a total of 1000 trials; so, using the relative frequency definition, we define the probability of two spots = 150/1000 = 0.15. How many times would you be willing to toss the die to obtain an approximation? Would you have any idea of the probability of your answer being correct? Certainly the more trials (tosses), the more secure you would feel about your answer. But that would still be a vague statement mathematically.

A second method would be to examine the physical characteristics of the die and note its symmetry and balanced weight. You might conclude that since there are six equally likely faces, the probability could be expressed as

$$\text{probability} = \frac{\text{number of equally likely ways of getting a success}}{\text{total number of equally likely possibilities}}$$

In this example, $\frac{1}{6}$ would be the answer.

There is also a third method of analysis, one that involves the concept of a **sample space**—the set of all possible outcomes (results) of an experiment. The list of possibilities is frequently enclosed in curly braces { }, since it is a set.

EXAMPLE 2

State sample spaces for the coin-tossing experiment and the die-tossing experiment.

Solution

For the coin, a possible sample space is

$$\{\text{heads, tails}\}$$

For the die, a possible sample space is

$$\{1, 2, 3, 4, 5, 6\}$$

EXAMPLE 3

Give a sample space for the experiment of tossing a fair die, and find the probability that the toss results in a number greater than 2.

Solution

$S = \{1, 2, 3, 4, 5, 6\}$ is a sample space. If the die is fair, then each face is equally likely because of the symmetry and balance of the die. There are four possibilities that are called successes, $\{3, 4, 5, 6\}$. The total number of equally likely possibilities is six, so

$$\text{probability(toss} > 2) = \frac{4}{6}$$

What if the die above were not fair? Then the sample space would still be $\{1, 2, 3, 4, 5, 6\}$, but it would not be accurate to assign equal weight to each possibility.

EXAMPLE 4

A die is weighted so that the face with six spots is twice as likely to come up as any of the other faces. Find the probability that a toss results in a six.

Solution

You could employ the experimental or empirical method of repeatedly tossing the die. However, you would be faced with a boring task and one that cannot result in a certain answer.

Let us use logical analysis instead. As we have seen, the sample space S is $S = \{1, 2, 3, 4, 5, 6\}$. But here each spot or point is *not* equally likely, since the six is twice as likely as the rest. We can "weight" each event or possibility as follows:

Event	1	2	3	4	5	6
Weight	x	x	x	x	x	$2x$

Here we have given the six twice the weight of the others. The probability P now becomes

$$P(6 \text{ spot}) = \frac{\text{weight of success}}{\text{total weight}} = \frac{2x}{7x} = \frac{2}{7}$$

Now that we have done a number of examples, it is appropriate to state some of the fundamental definitions and properties of probability. The *certain event* or "sure thing" is assigned a probability of 1. Any impossible event is assigned a probability of 0. Thus, all probabilities are between 0 and 1, inclusive.

PROPERTY 1:
$0 \leq P(E) \leq 1$, where E is any event.

An event that has a *single* outcome is called an **elementary** or **simple event**. The event that the toss of a single die results in a five is elementary since it involves a single outcome. The event that the toss will result in an even number is *not* elementary since there are *three* possible outcomes: $\{2, 4, 6\}$.

PROPERTY 2 (THE ADDITION PROPERTY):
If an event E consists of a finite number of distinct elementary events $E_1, E_2, E_3, \ldots, E_n$, then $P(E) = P(E_1) + P(E_2) + P(E_3) + \cdots + P(E_n)$.

That is, the probability of the event E is the sum of the probabilities of the simple events that make up E.

EXAMPLE 5

Suppose you toss a fair die. What is the probability that the uppermost face is even?

Solution

The sample space is $\{1, 2, 3, 4, 5, 6\}$. The event is $E = \{2, 4, 6\}$. (Note that this is *not* an elementary event.) Since the die is fair, we can assign equal weight to each elementary event:

Event	1	2	3	4	5	6
Weight	x	x	x	x	x	x

The event is $\{2, 4, 6\}$, whose probability, by the addition property, is the sum of the probabilities of the individual possibilities.

$$P(\text{even}) = \frac{x}{6x} + \frac{x}{6x} + \frac{x}{6x} = \frac{1}{6} + \frac{1}{6} + \frac{1}{6} = \frac{3}{6} = \frac{1}{2} = 0.5$$

EXAMPLE 6

Suppose a die is "loaded" so that even faces are three times as likely as odd faces. What is the probability that after a toss the uppermost face shows an even number of dots?

Solution

The sample space is still $S = \{1, 2, 3, 4, 5, 6\}$. The event "even" is $E = \{2, 4, 6\}$. Weights are assigned as follows:

Event	1	2	3	4	5	6
Weight	x	$3x$	x	$3x$	x	$3x$

The sum of all the weights is $x + 3x + x + 3x + x + 3x = 12x$. The probability of any single odd-numbered face is

$$\frac{x}{12x} = \frac{1}{12}$$

The probability of any single even-numbered face is

$$\frac{3x}{12x} = \frac{1}{4}$$

By the addition property the probability of the event $\{2, 4, 6\}$ is the sum of the probabilities of the simple events $\{2\}$, $\{4\}$, and $\{6\}$:

$$\frac{1}{4} + \frac{1}{4} + \frac{1}{4} = \frac{3}{4} \qquad \text{or} \qquad \frac{3x}{12x} + \frac{3x}{12x} + \frac{3x}{12x} = \frac{9x}{12x} = \frac{9}{12} = \frac{3}{4}$$

EXAMPLE 7

Five people are finalists in a contest in which two people will be selected for a free trip. How many winning pairs are possible, and what is the probability a particular pair will be chosen if random selection is used to pick the pair?

Solution

First find how many pairs are possible.

$$\binom{5}{2} = \frac{5!}{2!3!} = \frac{5 \cdot 4 \cdot 3 \cdot 2 \cdot 1}{(2 \cdot 1) \cdot (3 \cdot 2 \cdot 1)} = 10 \text{ possible pairs}$$

If the five people were labeled A, B, C, D, E, then the ten pairs are

AB	BC	CD	DE	AC
BD	CE	AD	BE	AE

A *sample space* would be {AB, AC, AD, AE, BC, BD, BE, CD, CE, DE}. Since the pair will be chosen at random, each pair has the same chance of winning; so the probability that any particular pair is chosen is $\frac{1}{10}$.

EXAMPLE 8

Two fair dice are thrown. What is the probability that the sum of the uppermost faces is (a) 6; (b) 7; (c) 13; and (d) 12?

Solution 1

One way to solve this problem is to estimate the probability of any sum by tossing a pair of dice many times and recording the results. Many tosses would be necessary before you might feel secure that the estimates are good.

An example might result in the following chart after 3600 throws of a pair:

Sum	2	3	4	5	6	7	8	9	10	11	12
Observed frequency	90	210	270	420	510	605	495	400	300	198	102

If you used this method, one would say you *simulated* the experiment of tossing a pair of dice.

a. Based on observed relative frequency, the probability of 6 is

$$P(6) = \frac{510}{3600}$$

b. $P(7) = \frac{605}{3600}$

c. $P(13) = \frac{0}{3600} = 0$

d. $P(12) = \frac{102}{3600}$

Solution 2

Since there are six possible faces for each die, there are $6 \times 6 = 36$ possible pairs:

(1, 6)	(2, 6)	(3, 6)	(4, 6)	(5, 6)	(6, 6)
(1, 5)	(2, 5)	(3, 5)	(4, 5)	(5, 5)	(6, 5)
(1, 4)	(2, 4)	(3, 4)	(4, 4)	(5, 4)	(6, 4)
(1, 3)	(2, 3)	(3, 3)	(4, 3)	(5, 3)	(6, 3)
(1, 2)	(2, 2)	(3, 2)	(4, 2)	(5, 2)	(6, 2)
(1, 1)	(2, 1)	(3, 1)	(4, 1)	(5, 1)	(6, 1)

The possible sums and their frequencies (number of occurrences in the list) are:

Sum	2	3	4	5	6	7	8	9	10	11	12
Frequency	1	2	3	4	5	6	5	4	3	2	1

If we observe also that the two dice are balanced and symmetrical, the above theoretical frequencies would be correct. Then the probabilities are as follows:

a. $P(6) = \frac{5}{36}$

b. $P(7) = \frac{6}{36} = \frac{1}{6}$

c. $P(13) = \frac{0}{36} = 0$

d. $P(12) = \frac{1}{36}$

Notice that the sum of all the probabilities is 1.

$$P(1) + P(2) + P(3) + \cdots + P(11) + P(12)$$
$$= \frac{1}{36} + \frac{2}{36} + \frac{3}{36} + \cdots + \frac{2}{36} + \frac{1}{36} = 1$$

It is often important to compare theoretical probabilities (solution 2) with observed probabilities (solution 1). This would be one way to decide if the tossed dice are fair. The differences between expected and observed results should not be too large. There are advanced methods (chi square statistical tests, for example) that make use of the discrepancies between theoretical (expected) and observed results. The computer can be especially helpful, however, when theoretical results are extremely difficult or impossible to obtain. The computer's use in such situations depends on a number of concepts. One of these is the random number, our next topic.

EXERCISES FOR SECTION 9.2

1. A purse contains two nickels, three dimes, one penny and one quarter.

- **a.** Find the probability that a person removing two coins from the purse will get two nickels.
- **b.** Find the probability that two coins chosen at random from the purse will add up to thirty cents.

For Exercises 2 and 3, refer to Examples 2 and 3.

2. State a sample space for tossing a fair die, and find the probability that the toss results in a number less than four.
3. Find the probability of tossing any one of the numbers one through six inclusive with an ordinary fair die.

For Exercises 4 through 6, refer to Examples 4 through 6.

4. Suppose that the die in Exercise 3 is *not* fair—a two is three times as likely to appear as any of the other faces. What is now the probability that a random toss will result in a number less than four?
5. A special die is loaded so that the probability of a face showing is proportional to the number of spots on the face. For example, a two is twice as likely as a one, a five is five times as likely as a one, and so on.
   a. State a sample space for this problem.
   b. Find the probability of tossing a three with this die.
6. A specially designed four-sided die is in the shape of a regular triangular pyramid. The face that lands down is the face that is counted. Suppose that the faces are numbered 1, 2, 3 and 4, and that the probability of each side facing down is proportional to the number on the face. Find the probability that the 3 face is down.

For Exercises 7 through 9, refer to Example 7.

7. Six people are finalists in a contest in which three people will be selected for a free trip.
   a. How many winning triplets are possible?
   b. What is the probability that a particular triplet will be chosen if random selection is used to choose the winners?
8. The access code to a computer system consists of five digits.
   a. What is the probability that the code is an even number (that is, the last digit is 0, 2, 4, 6 or 8)?
   b. What is the probability that the code is divisible by six (that is, it is an even number *and* it is divisible by three)?
9. A supervisor is in charge of eight people of equal ability and seniority. If she must lay off three, how many choices has she, and what is the probability that a particular person will be laid off?

For Exercises 10 and 11, refer to Example 8.

10. Find the probability of getting a sum of four with two fair dice.
11. Suppose you have *two* pyramid-shaped dice weighted like the one in Exercise 6. If you toss these dice, what is the probability that the sum of the down faces will be six?

9.3 Random Numbers and Simulations

If we wish to find the probability of rolling a sum of six with two fair dice, we can either calculate the probability using theoretical mathematics, or simulate the experiment and estimate the probability. The simulation would consist of tossing two fair dice many times or of using some means that *imitates* the throwing of the two dice. If you were to conduct a *real* simulation of the above experiment, you would not physically toss the two dice, but rather you would imitate the experiment by using some other means, possibly a computer. In order to conduct a simulation properly, random numbers are needed.

RANDOM AND PSEUDORANDOM NUMBERS

Random numbers are numbers that are selected without reference to any rule or pattern. Each number has exactly the same chance or probability of appearing in the list. Each number is chosen independently of any criteria.

Pseudorandom numbers are numbers generated by an algorithm that is free of any human bias but that is cyclic (at some time it repeats the previous cycle).

EXAMPLE 1

Illustrate a simple algorithm for generating pseudorandom numbers.

Solution

1. Begin with any digit (this initial number is called the **seed** of the pseudorandom number generator).
2. Multiply the number by 7 (not all primes will work well; for example, try 11.)
3. Discard any digits beyond the ten's position.
4. Record the result.
5. Repeat multiplying each new "product" by 7, until the seed appears again.

EXAMPLE 2

Use the algorithm of Example 1 to generate a sequence of "random" numbers.

Solution

Assume our seed is 3. The random numbers are circled:

$$3 \cdot 7 = 21 \quad 21 \cdot 7 = 147 \quad 47 \cdot 7 = 329 \quad 29 \cdot 7 = 203$$

the seed

We stop at 203 since the seed has appeared. Thus our sequence of "random" (actually pseudorandom) numbers is 21, 47, 29, 3

This algorithm has a very short cycle and was only used to illustrate a simple method for generating random numbers. If 13 had been used as the multiplier, the sequence would have had a longer cycle. This algorithm could be further modified to generate larger random numbers. More sophisticated methods exist to generate random numbers, but we will not discuss these here. The important points are that there are methods of obtaining random numbers and that random numbers are very useful in conducting simulations and in playing games. Some reasons for using simulations might be that the real-life simulation may be difficult or impossible to do because of environmental problems as in the case of simulating the path of a rocket or a nuclear reaction. Or it may be very inconvenient to conduct an actual experiment, such as observing the congestion of streets when some are designated one-way. Simulation might also be the best approach to a problem if the mathematical solution is unknown or very difficult to find.

Random numbers are useful in selecting **random samples**, samples free of bias. A random sample used in polls, for example, would mean the individuals selected were not chosen by favoritism. For example, if one wanted to know how people rated former President Carter's presidency, Democrats, Republicans, and Independents would have to have a fair representation in the sample—not just Democrats! Many erroneous results in polls and in experiments can be traced to the lack of a relevant, properly selected random sample. The study of selecting and using random samples, however, is beyond the scope of this text. We now offer some simple examples to illustrate the use of random numbers in simulations. The reader interested in further examples can find more advanced material under the topic of Monte Carlo methods.

Most computer languages have *intrinsic* or built-in pseudorandom number generators. These are usually named **RND**. Some RND functions take an argument, a value that determines the range of integers to be generated. For example, RND(50) would generate a simple random integral number in the range $1 \leq N \leq 50$. Each of the integers would have equal probability of appearing. In general, RND(N) would have an integral value between 1 and the value of its *argument*, N. (This function operates differently on different computers, however.

For example, the RND function works as described on a Radio Shack microcomputer, but not on an Apple.)

EXAMPLE 3

Use RND(N) to simulate the following:

a. Tossing of a single, fair die.
b. Tossing a fair coin.

Solution

a. D = RND(6). Since the six faces of a fair die each have an equal chance of appearing on top, we need to obtain integers in the range 1 to 6 inclusive. Because RND(6) generates a random integer between 1 and 6 inclusive, it could be used to imitate or simulate the tossing of the single die.

b. A fair coin has an equal chance of coming up heads or tails. There are two possibilities, each with equal probability. If we arbitrarily designate the integer 1 = heads, and 2 = tails, we can simulate the single coin toss by C = RND(2).

EXAMPLE 4

Use RND(N) to simulate the tossing of two dice and obtain the sum.

Solution

If

D1 = number appearing on one die and
D2 = number appearing on the other die;

then

$$\left.\begin{aligned} D1 &= RND(6) \\ D2 &= RND(6) \end{aligned}\right\} \quad \text{Simulate two single dice.}$$

$$SUM = D1 + D2$$

SUM has range $1 + 1 = 2$, to $6 + 6 = 12$ inclusive, or $2 \leq SUM \leq 12$.

EXAMPLE 5

Find an alternative method of using RND(N) to generate the integers 2 through 12 inclusive, as in the previous example.

Solution

$$SUM = RND(11) + 1$$

RND(11) would generate random integers between 1 and 11 inclusive. Adding 1 to the result would give the range 2 through 12.

EXAMPLE 6

Could the method of Example 5 be used to simulate the tossing of *two* fair dice?

Solution

Example 8 of Section 9.2 shows that the probabilities of the individual sums are not equal. The probability of SUM = 7, $P(7)$, equals $\frac{1}{6}$ while $P(2) = \frac{1}{36}$.

RND(11) produces all the random numbers 1 through 11 with *equal* probability of occurrence. Thus, RND(11) + 1 would produce sums 2 through 12 with equal probability. It could *not* properly be used to simulate the sum of two dice.

EXAMPLE 7

Use the method of Example 4 to estimate the probability of obtaining a sum of 6.

Solution

The simulation will consist of repeating the "tossing" of the two fair dice many times. A shortcoming of using simulations is that the answers are not precise or exact, only approximations. Generally, the larger the number of trials or repetitions, the more secure one can feel about the answer. Below is the pseudocode of a possible simulation for finding the probability of a sum of 6 using two fair dice.

We will need some accumulators and other variables:

S = sum of the two dice
N = number of trials in the simulation
F = number of favorable outcomes, i.e, sums of 6
T = counter of trials, or trial counter
EP = estimated probability of obtaining a sum of 6.

```
1. OBTAIN N  (* Number of trials to be used *)
2. F = 0     (* Initialize counter to count
              number of sums equal to 6 *
3. T = 1     (* Initialize trial counter *)
4. DOWHILE (T ≤ N)
5.   D1 = RND(6)  (* Simulate toss of fair die *)
6.   D2 = RND(6)
```

```
 7.     S = D1 + D2  (* Obtain sum of dice *)
 8.     IF S = 6 THEN F = F + 1  (* Increment counter
                                    if SUM = 6 *)
 9.     T = T + 1  (* Increment trial counter *)
10.  END DOWHILE
11.     EP = F/T  (* Relative frequency method is used
                     here: number of observed successes
                     divided by number of trials *)
12.  PRINT EP
13.  END.
```

The simulation of Example 7 could be repeated a number of times. The *average* (mean) of all the simulations performed gives a more reliable result than any *one* simulation.

EXAMPLE 8

Suppose a baseball player's statistics for a previous season were

Singles	Doubles	Triples	HR	AB	BA
50	5	0	10	200	.325

Ignoring walks, devise a simulation of a single at-bat using the computer function RND(N).

Solution

```
R = RND(200)                          (* There were 200 at-bats *)
IF R < 51 THEN PRINT "SINGLE"         (* He had 50 singles *)
ELSE
    IF R < 56 THEN PRINT "DOUBLE"     (* He had 5 doubles; 50 + 5 = 55 *)
    ELSE
        IF R < 66 THEN PRINT "HR"     (* He had 10 homeruns; 55 + 10 = 65 *)
        ELSE
            PRINT "OUT"
```

Since the previous history of the player indicates that 50/200 at-bats resulted in a single, any random integer less than 51 (≤ 50) is assigned a value of "single." Since the player's history indicates 5 doubles in 200 at-bats, we assign the random integers 51–55 to have a value of "double." Similarly 56 to 65 inclusive (ten integers) are assigned a value of "home run." No integers are assigned a value of "triple" since the player's history indicates no triples in 200 at-bats. A flowchart of this procedure follows:

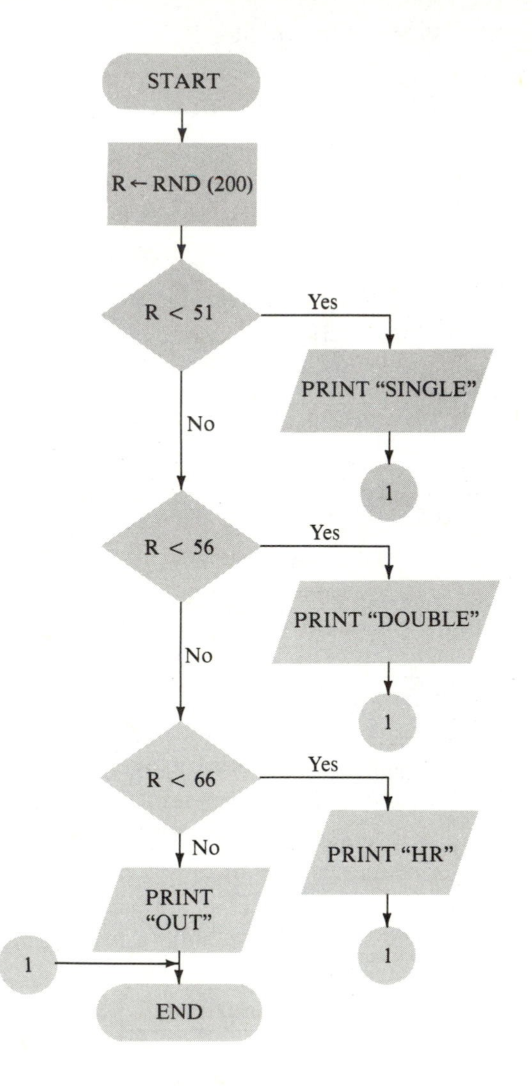

This simple simulation using previous player statistics could be the basis of a simulated baseball game. The programmer would have to obtain statistics on 18 players and modify and enhance the previous example. Many other applications and games are possible.

EXERCISES FOR SECTION 9.3

In Exercises 1 through 4, use the given seed and the given value of the multiplier to generate a sequence of pseudorandom numbers (see Examples 1 and 2).

1. seed = 2; multiplier = 7
2. seed = 3; multiplier = 13
3. seed = 3; multiplier = 11
4. seed = 2; multiplier = 5 (Does something unusual happen here?)
5. Some computers have random number generators that produce real numbers between 0 and 1 (excluding 0 and 1). Find a formula using the INT and RND functions that generates integers between 1 and 6 (*including* 1 and 6) from such a random number generator. (See Examples 3–5.)

For Exercises 6 and 7, refer to Example 7.

6. Using the RND function, design a simulation to imitate the tossing of three fair coins. Describe how your simulation could be used to estimate the probability of getting two heads and a tail.
7. In Connecticut, there is a state lottery game called Play 4. We describe below two different ways a person can win in this lottery. For each, try to find the exact probability, and then design a simulation to estimate the probability.
   a. *4-digit straight*: The four digits the player picks must match the winning number in *exact order*.
   b. *4-digit box*: The four digits chosen must be the same as the four winning digits, but they can be in *any order*.
8. Modify the flowchart of Example 8 to take into account *walks*.

9.4 Measures of Central Tendency

One of the data processing tasks that is often performed by computers is to present summary information. The main objective of this section is to explain ways of summarizing numerical data. Both business and government frequently need to extract meaningful information from large volumes of numeric data. The concept of *average* is very helpful in summarizing and condensing such data.

Average can signify **mean**, **median**, or **mode**. All three are **measures of central tendency**.

FIGURE 9.3

Finding the mean for a set of numbers.

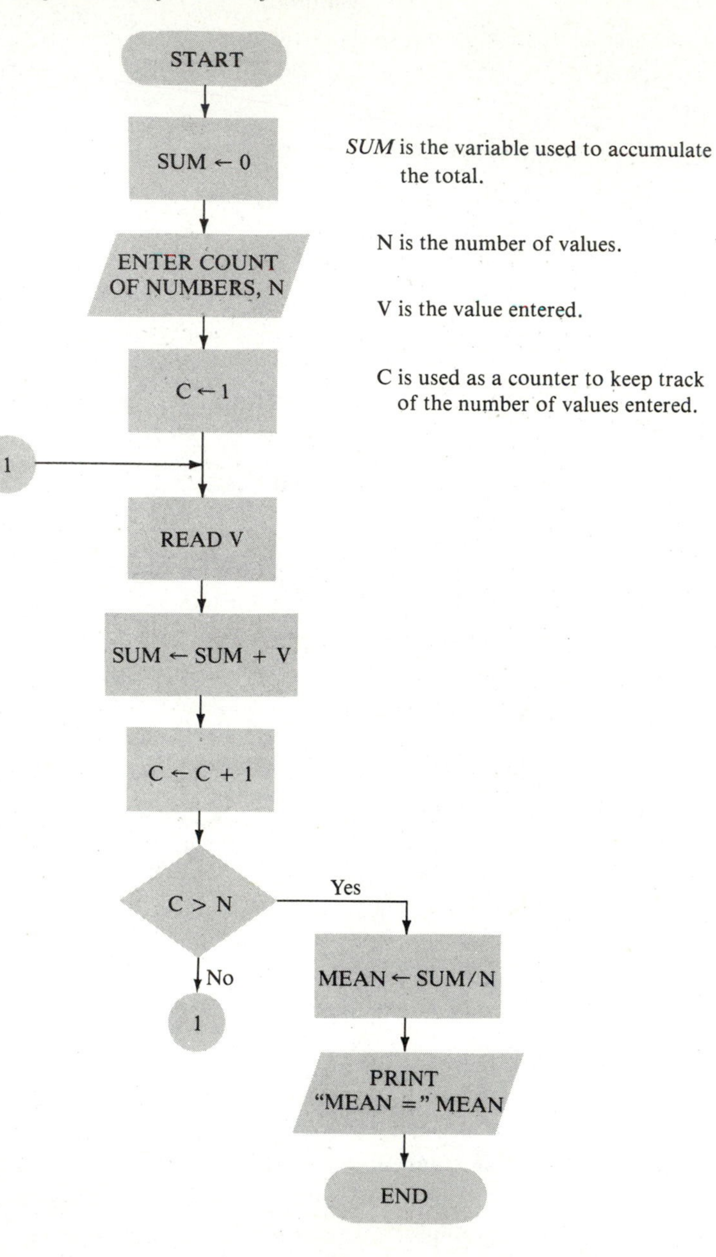

THE MEAN

The **mean** of a set of values is found by summing all of them and dividing by the number of values.

The procedure for finding the mean is shown in the flowchart of Figure 9.3.

A short trace through this flowchart for N = 4 follows:

C	N	Value	Sum	Average	
0	4	—	0	—	
1	4	20	20	—	
2	4	50	70	—	
3	4	30	100	—	
4	4	80	180	—	
5	4	80	180	180/4 = 45	The mean is 45.

EXAMPLE 1

Find the mean of 70, 80, 90, 30, and 60.

Solution

$$\text{SUM} = 70 + 80 + 90 + 30 + 60$$

$$\text{SUM} = 330$$

$$\text{MEAN} = \frac{330}{5} = 66$$

EXAMPLE 2

If a student earns grades of 70, 60, 80, and 40, what must the student earn on a fifth test to have a 75 "average"?

Solution

In order that the mean of five numbers be 75, their SUM must be $5 \cdot 75 = 375$. (Recall that MEAN = SUM/N; so N · (MEAN) = SUM. Here, N = 5 and MEAN = 75.) The sum of the given grades is 250 so the student must earn a grade of $375 - 250 = 125$ on the fifth test. In a real sense, such a test score is usually not possible, so the real answer is, it is **impossible** for this student to earn a 75 average.

EXAMPLE 3

Using the data from Example 2, find the score needed in a fifth test so that the student would have a 70 average.

Solution

MEAN = SUM/N or N · (MEAN) = SUM

5 · 70 =	350	Sum needed
	−250	Accumulated so far
	100	Needed on the fifth test

Notice that if the data are scattered or very variable, the mean does not convey too much information. Consider two sets of student grades both with a mean of 80.

set 1: 70, 75, 85, 90, 80

set 2: 50, 70, 80, 100, 100

The mean of 80 hides the fact that in set 2 there is a 50, but there are also two 100s.

A better example would be the following: Suppose the wages of a company's eleven employees are

$10,000 for ten hourly employees each earning 10,000

$100,000 for the president of the company

total wages = 10,000 · 10 + 100,000 = 200,000

average wage = $200,000 ÷ 11 = $18,181.82

Thus the mean wage is $18,181.82, which is highly misleading. The true *average* wage is not $18,181.82. When information is *shown* this way, the *median* may be a better measure of the "average" or central tendency.

THE MEDIAN

The **median** is found by first arranging the data in either ascending or descending order. Then the "middle" value is desi nated the median. Half the values are larger than the median, and half are smaller.

EXAMPLE 4

Find the median of 30, 60, 40, 20, and 90.

Solution

First arrange the numbers in order (sort) → 20, 30, 40, 60, 90

40 is the middle value, so 40 is the median. There are as many values below the median as above.

EXAMPLE 5

Find the median of 30, 60, 40, 20, 80, and 90.

Solution

First sort the values:

20, 30, 40, 60, 80, 90

When there are an even number of values, the *mean* of the two middle values is used.

$$\frac{40 + 60}{2} = 50 \qquad \text{The median}$$

Notice that the median 50 is not actually one of the given values and that there are as many values above 50 as below.

An IF-THEN statement customarily is used in a computer program that finds the median because two different formulas are used, depending on whether there are an even or odd number of values in the data set. You will be asked in the exercises to write an algorithm to find the median using an IF-THEN statement. Below we illustrate how it can be done without an IF statement; but first we must review the function INT(X) introduced in Chapter 2. Recall that INT(3) is 3; INT(3.9) is 3; and INT(3.1) is also 3. Thus INT *truncates* nonnegative real numbers.

EXAMPLE 6

Describe a method of finding the median of a data set without using any IF-THEN statements. Assume the data set has been sorted.

Solution

Let N be the number of numerical values in the data set and let

$$A = \text{INT}\left(\frac{1 + N}{2}\right)$$

$$B = \text{INT}\left(1 + \frac{N}{2}\right)$$

Then the median of a set of values will be the mean of the Ath and Bth terms in the list. Two illustrations:

a. Let the data set consist of 20, 30, 40, 60, 90

Here N is 5, so

$$A = \text{INT}\left(\frac{1+5}{2}\right) = \text{INT}\left(\frac{6}{2}\right) = \text{INT}(3) = 3$$

$$B = \text{INT}\left(1 + \frac{5}{2}\right) = \text{INT}(1 + 2.5) = \text{INT}(3.5) = 3$$

The mean of the Ath item (the third) and the Bth item (also the third) is

$$\frac{40 + 40}{2} = 40$$

So 40 is the median.

b. Let the data set be 20, 30, 55, 70, 80, 90
Here N is 6.

$$A = \text{INT}\left(\frac{1+6}{2}\right) = 3$$

$$B = \text{INT}\left(1 + \frac{6}{2}\right) = 4$$

We calculate the mean of items 3 and 4:

$$\frac{55 + 70}{2} = 62.5$$

The median is 62.5.

Those with access to Pascal might wish to modify the above method using the functions TRUNC(X) and ROUND(X). Finding the median offers one illustration of how functions can be useful in programming.

The median is often quoted in newspapers and business journals because it is less likely than the mean to be influenced by extreme values in the data. Recall the example earlier in the section on wages and how the mean was influenced by the president's $100,000 salary. The median of that example, $10,000, is more truly reflective of the "average" wage of an employee of the company.

The last measure of central tendency that we shall discuss is the **mode**.

THE MODE

The **mode** is the value that occurs with greatest frequency. The mode need not be unique; for example, a set of values may be **bimodal** (that is, it may have two modes). The mode is found by counting the number of occurrences of each value in the data set.

EXAMPLE 7

Find the mode in this data set: 7, 9, 2, 6, 7, 8, 7, 5

Solution

The mode is 7 since it is the value that occurs most often (three times).

EXAMPLE 8

Find the mode in this data set: 7, 9, 2, 5, 6, 7, 8, 7, 5, 8, 5

Solution

This set has two values that each occur three times: 7 and 5 are both modes since they both occur with highest frequency. Thus, the set is *bimodal* (two modes).

A set of data values may be **multimodal**, with more than two modes. The mode is used when a rough or fast estimate of central tendency is needed. Although it is not used as often as the mean or median, the mode does have application in business. For example, in the clothing or shoe industry, the modal sizes can be stressed by manufacturers.

EXAMPLE 9

Find the mean, median, and mode of the following data set:

5, 11, 3, 6, 2, 9, 8, 7, 6, 3, 2, 5, 4, 7, 6, 12, 3

Solution

The mean is found by summing the values and dividing by the number of values. The sum is 99. The number of values is 17. The mean is 99/17, or approximately 5.824. The median is found by first sorting the data:

2, 2, 3, 3, 3, 4, 5, 5, 6, 6, 6, 7, 7, 8, 9, 11, 12

N is 17, so the position of the middle item is INT($1 + N/2$), or ninth in the sorted list. So the median is 6.

The modes are 3 and 6; so the distribution is bimodal.

Let us consider a possible interpretation of the data of Example 9. The data could be the result of rolling a pair of dice 17 times and finding the sum of each roll. Which measure of central tendency might be most valuable in determining whether the dice rolled were fair? Would measures of central tendency be of *any* help in determining the fairness?

EXERCISES FOR SECTION 9.4

In Exercises 1 through 5, find the *mean*, *median* and *mode* of each data set [see Examples 1 (mean), 4 and 5 (median), 7 and 8 (mode), and 9 (all three)].

1. 2, 3, 2, 4, 1

2. 2, 4, 3, 8, 6, 7, 1, 5

3. 2, 1, 6, 2, 10, 4, 10

4. 7, 9, 15, 6, 11, 19, 9, 5, 9, 4

5. 7, 5, 2, -3, 6, 9, 11, 13, -9, 8, 5, 20

For Exercises 6, 7 and 8, refer to Examples 2 and 3.

6. Given the data set 7, 11, 5, 4, 6, 9, 2, 4, x, what must be the value of x to make the sum of the data set equal to 65?

7. If the sum of the following data set is to be 60, what must be the value of x?

$$5, -2, 3, 9, 11, 14, 16, x$$

8. A student earns grades of 60, 70, 80 and 80. What grade must the student earn on a fifth test to have a *mean* average of 75?

9. Given the sorted data 20, 30, 35, x, 40, 60, what must be the value of x if the *median* is 36?

10. A company has five hourly employees who earn yearly salaries of $10,000, $15,000, $13,000, $14,000, and $17,000. The annual salary of the company president is $121,000. Find the *mean* and *median* yearly salaries. Which is the more representative "average" wage?

11. Write an algorithm or set of instructions to find the median of a set of data, using an IF-THEN statement and the INT function. (See the discussion following Example 5.)

12. Design or write an algorithm that accepts as input ten real numbers (both positive and negative) and prints the mean of the positive numbers and the mean of the negative numbers. For example, if we use the five numbers -5, -3, 6, -7, and 11, the mean of the positives would be $(6 + 11)/2$, or 8.5. The mean of the negatives would be

$$\frac{-5 + (-3) + (-7)}{3} = -5$$

9.5 Frequency Distributions and Histograms

One way to determine whether the dice referred to in Example 9 of Section 9.4 are fair is to compare the expected frequencies determined by probability theory with the observed frequencies. In Section 9.2 (Example 8) the probability of each sum was calculated:

Sum	2	3	4	5	6	7	8	9	10	11	12
Frequency	1	2	3	4	5	6	5	4	3	2	1
Probability	$\frac{1}{36}$	$\frac{2}{36}$	$\frac{3}{36}$	$\frac{4}{36}$	$\frac{5}{36}$	$\frac{6}{36}$	$\frac{5}{36}$	$\frac{4}{36}$	$\frac{3}{36}$	$\frac{2}{36}$	$\frac{1}{36}$

The number of trials in our example is 17 so the expected frequencies can be calculated by multiplying the number of trials, 17, by the respective probabilities.

Sum	2	3	4	5	6	7	8	9	10	11	12
Expected Frequency	$\frac{17}{36}$	$\frac{34}{36}$	$\frac{51}{36}$	$\frac{68}{36}$	$\frac{85}{36}$	$\frac{102}{36}$	$\frac{85}{36}$	$\frac{68}{36}$	$\frac{51}{36}$	$\frac{34}{36}$	$\frac{17}{36}$

For example, the expected frequency for the sum 4 is found by multiplying

$$17 \cdot \frac{3}{36} = \frac{51}{36}$$

Number of trials · Probability of sum

If we now *round* our results to reflect the fact that frequencies must be whole numbers we get

Sum	2	3	4	5	6	7	8	9	10	11	12
Rounded Expected Frequency	0	1	1	2	2	3	2	2	1	1	0
Observed Frequency	2	3	1	2	3	2	1	1	0	1	1

The number of trials (17) is too small for a good determination of fairness; but the observed frequencies, called a **frequency distribution**, enable us to see relationships more easily than when the results were raw or unorganized. In fact, an advanced statistical test called the **chi square test** can help in determining whether the rolls are fair. The chi square test makes use of the observed frequency data and the expected frequency. A discussion of this test is beyond the scope of this text; the interested reader is referred to books on statistics.

Sometimes it is easier to draw conclusions from data if the data are not only organized by a numerical frequency distribution but also visually. A **histogram** is a graph of a frequency distribution in which equal intervals of values are marked on a horizontal axis and the frequency corresponding to each interval is indicated by the height of a rectangle having the interval as its base. Histograms can be produced by computers from a data set to help managers, scientists, and others see possible relationships and to organize data. VisiCalc, a popular spreadsheet program which does calculations, can produce histogramlike graphs using various symbols.

EXAMPLE 1

Draw a histogram of the data at the beginning of this section showing the result of 17 rolls of a pair of dice.

Solution

The frequency distribution of the results should be used for reference since the height of the histogram (Figure 9.4) is determined by the observed frequency.

FIGURE 9.4

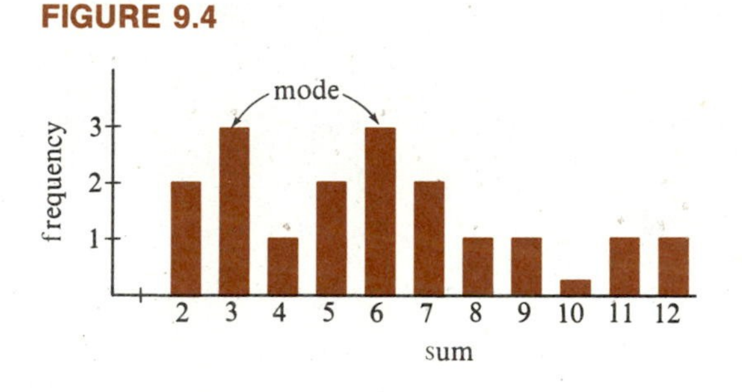

A computer program could superimpose the graph or histogram of the expected frequencies on the observed, and results could be compared visually. Notice that from the histogram the mode can also be determined by visual inspection.

9.6 Measures of Dispersion

If you are told that the mean annual temperature of a city is 80°, you may assume a nice yearly climate. But to determine whether you want to live there, you should also know how variable the temperature is. For example, some climates are fairly stable; others are highly variable. Thus, the mean or measure of central tendency may not convey enough information. As another example, the grades of two students on five tests are given below:

student 1: 80, 75, 85, 80, 70

student 2: 60, 100, 90, 90, 50

One student is a fairly consistent C+ student; the other student, also with a C+ average, shows potential for possible A work (since three test scores are 90 or above) but is very erratic.

In manufacturing, consistency is very important because tools or other items must be manufactured to precise specifications with little room for variance. Our last example used to illustrate the importance of variability is a coffee vending machine, which must reliably and consistently measure out a precise amount of coffee. This concern with variability is in fact an important consideration in selecting any machine used to fill packages to precise amounts. We will discuss two measures of variability in this section, the *range* and the *variance*.

The Range

RANGE

The **range** is the difference between the largest and smallest values in the data set.

EXAMPLE 1

Find the range of the following data: 7, 3, 9, 6, 5, 8, 11.

Solution

Find the maximum and the minimum values and subtract:

$$11 - 3 = 8$$

The range is 8.

EXAMPLE 2

Find the range of the following data: 7, 3, 9, 6, 5, 8, 11, 100.

Solution

The range is $100 - 3 = 97$.

Notice that the data of Examples 1 and 2 are exactly the same except for one value. This shows that the range is greatly influenced by even a single extreme score. We must thus consider an alternative measure of dispersion or variability that takes all the data into account and not just two values.

The Variance

The **variance** is a measure of *variability about the mean.* The variability is found by summing each value's distance or deviation from the mean, but for reasons that are beyond the scope of this book the variance must be improved. The following formula is therefore used:

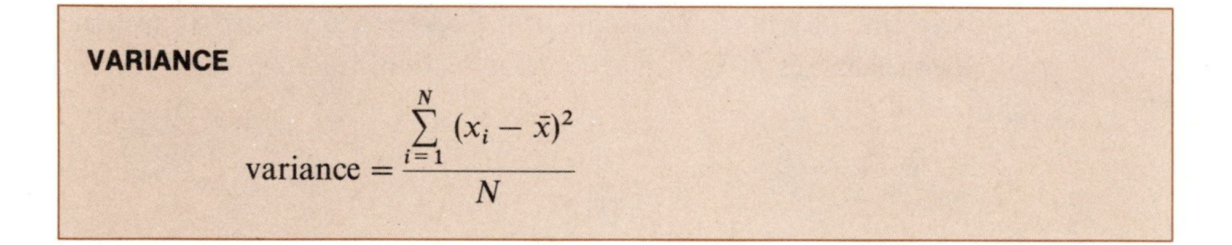

VARIANCE

$$\text{variance} = \frac{\sum_{i=1}^{N} (x_i - \bar{x})^2}{N}$$

The symbol $\sum$ stands for "sum." The formula indicates that we must subtract the mean $\bar{x}$ from each data item x_i, square that difference, and then add up all N of the squares. Use of this formula is demonstrated in the following example:

EXAMPLE 3

Find the variance of the following values: 9, 5, 4, 6.

Solution

The formula used is

$$\text{variance} = \frac{\sum_{i=1}^{N} (x_i - \bar{x})^2}{N}$$

The formula indicates that the deviation of each value from the mean, $\bar{x}$, is to be squared. The sum of all the squared deviations is then divided by the number of values. First we must calculate the mean:

$$\frac{9+5+4+6}{4} = \frac{24}{4} = 6 \qquad \text{Mean value}$$

Next, we construct a table.

i	x_i	$x_i - \bar{x}$	$(x_i - \bar{x})^2$
1	9	$9 - 6 = 3$	9
2	5	$5 - 6 = -1$	1
3	4	$4 - 6 = -2$	4
4	6	$6 - 6 = 0$	0
		sum =	14

$$\sum_{i=1}^{4} (x_i - \bar{x})^2 = 14 \text{ and } \frac{\sum_{i=1}^{4} (x_i - \bar{x})^2}{4} = \frac{14}{4} = \frac{7}{2} = 3.5 \qquad \text{The variance}$$

Used in this way, the variance is not very useful. It becomes more useful when variances of two data sets are compared.

EXAMPLE 4

Calculate the variances of the test scores of the two students:

student 1: 80, 75, 85, 80, 70

student 2: 60, 100, 90, 90, 50

Solution

The mean in both cases is 78 since the sum of each set of scores is 390 and $390/5 = 78$. For student 1:

i	x_i	$x_i - \bar{x}$	$(x_i - \bar{x})^2$
1	80	$80 - 78 = 2$	4
2	75	$75 - 78 = -3$	9
3	85	$85 - 78 = 7$	49
4	80	$80 - 78 = 2$	4
5	70	$70 - 78 = -8$	64
total	390	0	130

The variance is $130/5 = 26$.

Notice that the total in the third column is zero. This will always be the case, since the sum of N deviations from the mean is always zero. This can serve to check some of your work. If your sum does not equal zero, you made a mistake. However, the converse is not true—if your sum is zero, the deviations may not all be correct. For student 2:

i	x_i	$x_i - \bar{x}$	$(x_i - \bar{x})^2$
1	60	$60 - 78 = -18$	324
2	100	$100 - 78 = +22$	484
3	90	$90 - 78 = +12$	144
4	90	$90 - 78 = +12$	144
5	50	$50 - 78 = -28$	784
total	390	0	1880

$$\sum_{i=1}^{5} (x_i - \bar{x})^2 = 1880 \quad \text{and} \quad \text{variance} = \frac{1880}{5} = 376$$

Although the much higher variability of the second student's scores could be determined by visual inspection of the data, in most cases this is not possible. The variance is a good measure of variability that uses all the data.

Because all the deviations were squared in calculating the variance, another useful measure of variability is used to lessen the effect of the squarings. The **standard deviation** is the square root of the variance.

EXAMPLE 5

Find the standard deviations of the variances calculated in Examples 3 and 4, 3.5, 26 and 376.

Solution

The standard deviation is the square root of each variance.

$$\sqrt{3.5} = 1.87$$
$$\sqrt{26} = 5.099 \text{ or } 5.10$$
$$\sqrt{376} = 19.39$$

expressing each standard deviation to two decimal places.

Generally the variance or standard deviation alone is not very useful. Their usefulness is in comparing variability of data that can meaningfully be compared and in interpreting probability using a distribution such as the normal distribution (which we examine in the next section).

It is important to note that the formula for the variance and standard deviation for a *whole* set of data and the one used when sampling are not the same. If you see a formula with $N - 1$ as the divisor, the formula is for *sample data.* For further information, you should consult a statistics text.

The formula given for variance is the one based on the definition. A *calculation formula* is useful with relatively large sets of values. The calculation formula is given by

$$\text{variance} = \frac{\sum x_i^2 - \frac{1}{N}\left(\sum x_i\right)^2}{N}$$

EXAMPLE 6

Use the calculation formula to find the variance of 9, 5, 4, 6.

Solution

i	x_i	x_i^2
1	9	81
2	5	25
3	4	16
4	6	36
total	24	158

$$\sum x_i^2 = 158$$
$$\sum x_i = 24$$
$$\left(\sum x_i\right)^2 = 576$$
$$\text{var} = \frac{[158 - \frac{1}{4}(576)]}{4} = \frac{158 - 144}{4}$$
$$= \frac{14}{4} = \frac{7}{2} = 3.5$$

The difference between $\sum x_i^2$ and $(\sum x_i)^2$ is that in the first *each value* is squared and then added to the previous value; that is, it is the *sum of the squares.* The second expression, $(\sum x_i)^2$ is the *sum squared.* In the second formula, the squaring procedure is used only once.

The distance a value is from the *mean* in terms of standard deviations is very important in the study of statistics.

EXAMPLE 7

Find how many standard deviations above the mean 9 is in the data set 9, 5, 4, 6 used in Examples 3 and 5.

Solution

The standard deviation of 1.87 was calculated for this data in Example 5. The formula used to find distance from mean in terms of standard deviations is:

$$\frac{\text{value} - \text{mean}}{\text{standard deviation}} = \frac{9 - 6}{1.87} = \frac{3}{1.87} \approx 1.6$$

We can say that 9 is approximately 1.6 standard deviations above the mean.

EXAMPLE 8

Given the test scores, 60, 100, 90, 90, and 50 (see Example 4), find how many standard deviations below the mean the lowest score is.

Solution

These scores are of student 2 in Example 4. The standard deviation is 19.39 (see Example 5). The formula is

$$\frac{\text{value} - \text{mean}}{\text{S.D.}} = \frac{50 - 78}{19.39} = \frac{-28}{19.39} \approx -1.44$$

The minus sign indicates that the score is *below* the mean, and in fact is said to be *1.44 standard deviations below the mean.*

EXAMPLE 9

Scores on the SATs range from 200 to 800 with an original mean of 500 and a standard deviation of 100. How many standard deviations above the mean is a score of 750?

Solution

$$\frac{750 - 500}{100} = +2.5$$

The solution to Example 9 indicates that 750 is 2.5 standard deviations above the mean. This value can be converted to a percentile based on the *normal curve.* The distribution of SAT scores is approximately normal, and the mean and standard deviation of any normal distribution can be used to calculate percentiles, which can in turn be used to determine the percentage of scores above or below a given score. We shall see how this is done in Section 9.7.

EXERCISES FOR SECTION 9.6

1. Give a clear, specific example of why measures of variation are needed in addition to measures of central tendency.

In Exercises 2 and 3, find the mean, median, and range of the given data set (see Examples 1 and 2).

2. 8, 5, 6, 9, 2, 20, 13, 7

3. $-5, 3, -7, 9, 2, 11, 13, -9, -4, 6$

4. Use the formula given in the *definition* of variance to find the variance of the set of data, 9, 6, 2, 7, 1 (see Examples 3 and 4).
5. Use the *calculation* formula to find the variance of the data in Exercise 4 (see Example 6). Also, calculate the standard deviation (see Example 5).
6. For each of the sets of grades below, calculate the range, mean, variance, and standard deviation:

   student A: 70, 75, 90, 100, 75
   student B: 70, 80, 90, 95, 85

   Compare and contrast your results. Which is the more consistent student?
7. For each of student A's grades, calculate the number of standard deviations above or below the mean (see Examples 7–9).
8. If you received a score of 600 on the quantitative section of the SAT, how many standard deviations above or below the mean would your score be? Assume a mean of 500 and a standard deviation of 100.
9. Students in a physics class recently received the following test scores:

   90, 80, 70, 75, 60, 65, 60, 50, 50, 60, 65,
   75, 80, 80, 70, 65, 60, 50, 40, 30, 20, 90

   a. Find the *mean* score.
   b. Find the *median* score.
   c. Find the *variance.*
   d. Find the *standard deviation.*
   e. Find the number of students who scored *more than one* standard deviation *above* the mean.

f. Find the number of students who scored *within one* standard deviation of the mean.

g. Find the *mode.*

9.7 The Normal Curve

The distribution of scores on college entrance examinations is one of many distributions that approximate a normal curve. The **normal curve** is the graph of a **normal frequency distribution**, which approximates the measurements taken on many variables including height, weight, and I.Q.

The normal distribution curve is bell-shaped and symmetrical as shown in Figure 9.5. This curve is *symmetrical about the mean* with mean, mode, and median coinciding. The curve in the figure is called a **standard normal curve**. Each division along the horizontal axis represents one standard deviation. Most of the area under the curve is contained within three standard deviations of the mean, but the curve extends infinitely far in both horizontal directions.

Note the following major characteristics of standard normal curves:

1. The mean μ is at the origin on the horizontal axis.
2. Approximately 68% of the values lie within one standard deviation σ of the mean; similarly, 95% of the values lie within two standard deviations, and 99.7% within three standard deviations.

EXAMPLE 1

Find the percentile of a SAT score of 650 if the mean is assumed to be 500 and the standard deviation is 100.

FIGURE 9.5

The standard normal curve.

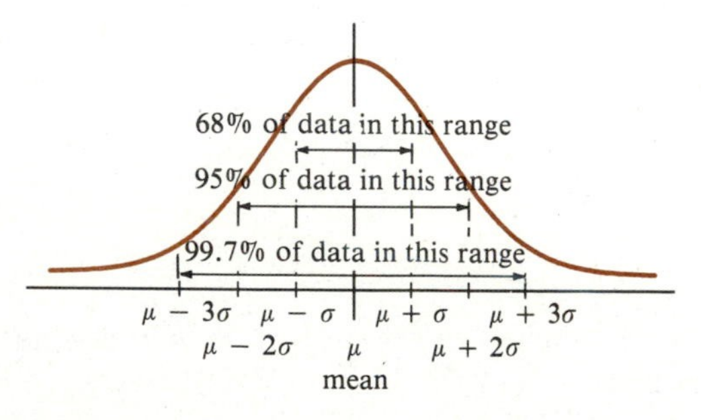

Solution

FIGURE 9.6

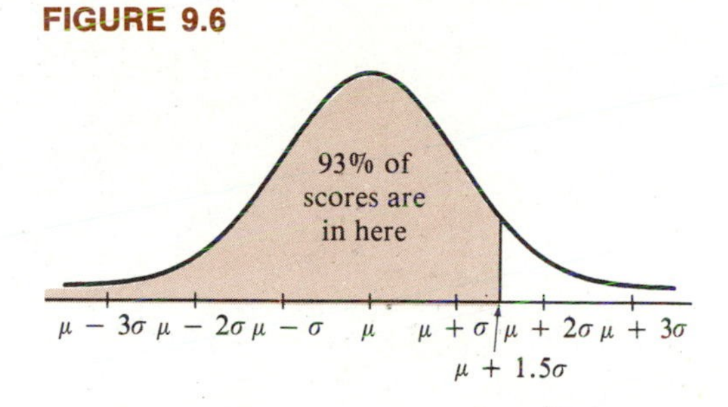

The shaded area of Figure 9.6 is the measure of percentile. We can only estimate it at this point. The first step is to convert the score of 650 to a standard score in terms of standard deviations. The formula is

$$\text{number of standard deviations} = \frac{\text{score} - \text{mean}}{\text{standard deviation}}$$

$$\frac{650 - 500}{100} = 1.5$$

So 650 is 1.5 standard deviations above the mean.

Figure 9.5 indicates that 95% of the scores in a normal distribution are within two standard deviations of the mean. A reasonable guess, then, would be that a score of 650 is in a percentile range of a little less than 95. A more systematic approach to problems of this type follows.

Example 1 uses the concept of a ***Z*-score**, or **standard score**. This is the number of standard deviations a piece of data is above or below the mean. In symbols, the Z-score is defined as follows

Z-SCORE

$$Z = \frac{x - \mu}{\sigma}$$

where x is a score obtained on a measure that has a normal distribution, μ is the mean of the distribution or set of measures, and σ is the standard deviation.

EXAMPLE 2

An aptitude test has a mean score of 60 and a standard deviation of 12. Assume that the scores are normally distributed. Find the corresponding Z-score for the following scores:

a. 78 **b.** 54

Solution

a. $Z = \frac{x - \mu}{S} = \frac{78 - 60}{12} = \frac{18}{12} = 1.5$

That is, a score of 78 is 1.5 standard deviations above the mean.

b. $Z = \frac{54 - 60}{12} = \frac{-6}{12} = -0.5$

A *negative* Z-score indicates a score *below* the mean. So, a score of 54 is 0.5 standard deviation below the mean.

EXAMPLE 3

A student scores 70 on a test that has a mean of 60 and standard deviation of 10. Another student scores 85 on a comparable test with mean of 75 and standard deviation of 8. Assuming normality for both tests, which student scored higher?

Solution

For the first student,

$$Z = \frac{70 - 60}{10} = \frac{10}{10} = +1$$

For the second student.

$$Z = \frac{85 - 75}{8} = \frac{10}{8} = +1.25$$

The second student scored higher, since a Z-score of 1.25 is greater than a Z-score of 1.00. This indicates that standard deviation is important in interpreting test scores.

We indicated in the discussion following Example 1 that Z-scores could be used to determine the percentile of a given score—that is, the percent of scores in the sample that are below that score. To use the Z-score in this way, we need a table giving the area under the normal distribution curve associated with specific Z-scores. Such data are given in Table 9.1.

TABLE 9.1

	Second Decimal Place in Z									
Z	.00	.01	.02	.03	.04	.05	.06	.07	.08	.09
0.0	.0000	.0040	.0080	.0120	.0160	.0199	.0239	.0279	.0319	.0359
0.1	.0398	.0438	.0478	.0517	.0557	.0596	.0636	.0675	.0714	.0753
0.2	.0793	.0832	.0871	.0910	.0948	.0987	.1026	.1064	.1103	.1141
0.3	.1179	.1217	.1255	.1293	.1331	.1368	.1406	.1443	.1480	.1517
0.4	.1554	.1591	.1628	.1664	.1700	.1736	.1772	.1808	.1844	.1879
0.5	.1915	.1950	.1985	.2019	.2054	.2088	.2123	.2157	.2190	.2224
0.6	.2257	.2291	.2324	.2357	.2389	.2422	.2454	.2486	.2517	.2549
0.7	.2580	.2611	.2642	.2673	.2704	.2734	.2764	.2794	.2823	.2852
0.8	.2881	.2910	.2939	.2967	.2995	.3023	.3051	.3078	.3106	.3133
0.9	.3159	.3186	.3212	.3238	.3264	.3289	.3315	.3340	.3365	.3389
1.0	.3413	.3438	.3461	.3485	.3508	.3531	.3554	.3577	.3599	.3621
1.1	.3643	.3665	.3686	.3708	.3729	.3749	.3770	.3790	.3810	.3830
1.2	.3849	.3869	.3888	.3907	.3925	.3944	.3962	.3980	.3997	.4015
1.3	.4032	.4049	.4066	.4082	.4099	.4115	.4131	.4147	.4162	.4177
1.4	.4192	.4207	.4222	.4236	.4251	.4265	.4279	.4292	.4306	.4319
1.5	.4332	.4345	.4357	.4370	.4382	.4394	.4406	.4418	.4429	.4441
1.6	.4452	.4463	.4474	.4484	.4495	.4505	.4515	.4525	.4535	.4545
1.7	.4554	.4564	.4573	.4582	.4591	.4599	.4608	.4616	.4625	.4633
1.8	.4641	.4649	.4656	.4664	.4671	.4678	.4686	.4693	.4699	.4706
1.9	.4713	.4719	.4726	.4732	.4738	.4744	.4750	.4756	.4761	.4767
2.0	.4772	.4778	.4783	.4788	.4793	.4798	.4803	.4808	.4812	.4817
2.1	.4821	.4826	.4830	.4834	.4838	.4842	.4846	.4850	.4854	.4857
2.2	.4861	.4864	.4868	.4871	.4875	.4878	.4881	.4884	.4887	.4890
2.3	.4893	.4896	.4898	.4901	.4904	.4906	.4909	.4911	.4913	.4916
2.4	.4918	.4920	.4922	.4925	.4927	.4929	.4931	.4932	.4934	.4936
2.5	.4938	.4940	.4941	.4943	.4945	.4946	.4948	.4949	.4951	.4952
2.6	.4953	.4955	.4956	.4957	.4959	.4960	.4961	.4962	.4963	.4964
2.7	.4965	.4966	.4967	.4968	.4969	.4970	.4971	.4972	.4973	.4974
2.8	.4974	.4975	.4976	.4977	.4977	.4978	.4979	.4979	.4980	.4981
2.9	.4981	.4982	.4982	.4983	.4984	.4984	.4985	.4985	.4986	.4986
3.0	.4987	.4987	.4987	.4988	.4988	.4989	.4989	.4989	.4990	.4990
3.1	.4990	.4991	.4991	.4991	.4992	.4992	.4992	.4992	.4993	.4993
3.2	.4993	.4993	.4994	.4994	.4994	.4994	.4994	.4995	.4995	.4995
3.3	.4995	.4995	.4995	.4996	.4996	.4996	.4996	.4996	.4996	.4997
3.4	.4997	.4997	.4997	.4997	.4997	.4997	.4997	.4997	.4997	.4998
3.5	.4998									
4.0	.49997									
4.5	.499997									
5.0	.4999997									

First, we make some observations about the graph of, and the area under, the normal curve shown below (Figure 9.7).

FIGURE 9.7

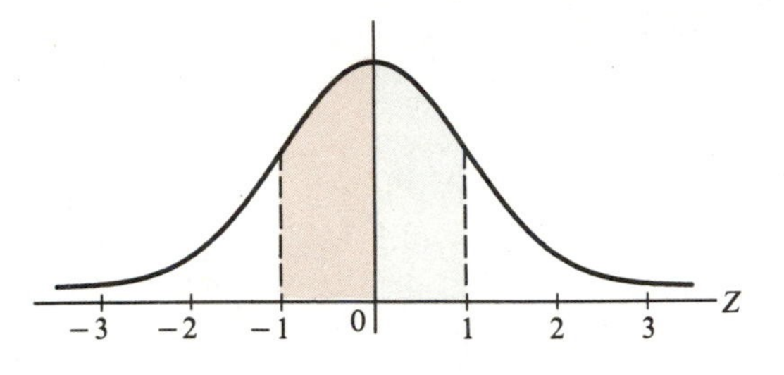

This shows the standard normal curve with mean at $Z = 0$ and standard deviation of 1. Since the curve is symmetrical about the mean (the vertical line drawn at $Z = 0$), the area from $Z = -1$ to $Z = 0$ is the same as the area from $Z = 0$ to $Z = +1$. The same observation holds for all corresponding positive and negative Z-scores. Negative Z-scores convey *only* that the score is below the mean. For this reason, a table of areas corresponding to Z-scores gives the areas measured *from the mean* only for positive values of Z. Suppose, for example, that we want to find the area under the normal curve between $Z = -0.5$ and $Z = 0$. We look in Table 9.1 in the extreme left-hand column for the value $Z = +0.5$. Now, look in the 0.5 row under the column labelled .00 (since we are looking for 0.500); the value there is .1915. Therefore, the area under the normal distribution curve from $Z = -0.5$ to $Z = 0$ is the same as the area from $Z = 0$ to $Z = 0.5$; and that area is 0.1915. This area is shown in the graph below (Figure 9.8).

FIGURE 9.8

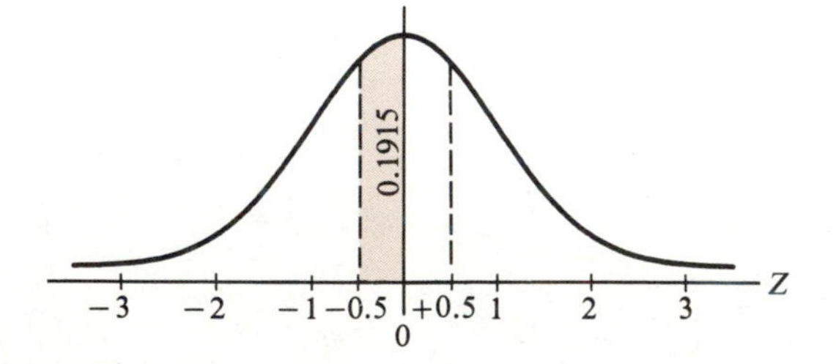

Now suppose we want to know what *percentile* a Z-score of -0.5 represents. First note that exactly *half* the scores lie to the left of $Z = 0$. If the area under the *whole* curve is 1, then the area to the left of $Z = 0$ is 0.5000 (or 0.5 or 1/2). We determined above that the area from $Z = -0.5$ to $Z = 0$ is 0.1915. The *percentile* represented by a Z-score of -0.5 is given by the area to the left of $Z = -0.5$, and this value is simply the area to the left of $Z = 0$ (0.5000) minus the area between $Z = -0.5$ and $Z = 0$. Thus the desired percentile, represented by the shaded area in the graph below (Figure 9.9) is

$$0.5000 - 0.1915 = 0.3085, \text{ or approximately } 31\%.$$

FIGURE 9.9

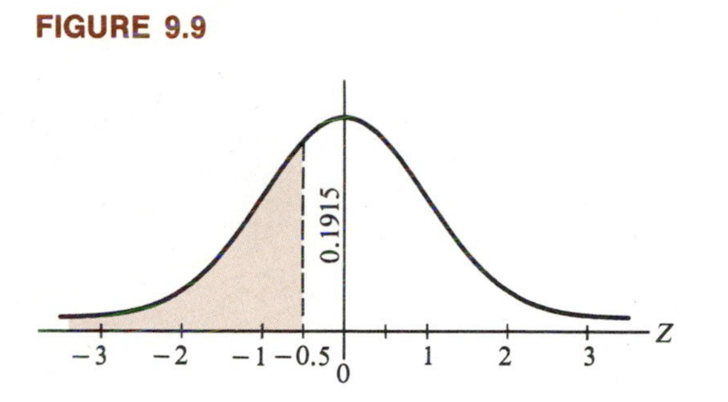

EXAMPLE 4

A student earns a score of 66 on a test that is based on the normal curve with a mean of 50 and a standard deviation of 10. Find the student's Z-score and percentile on the test.

Solution

$$Z\text{-score} = Z = \frac{\text{score} - \text{mean}}{\text{standard deviation}}$$

$$Z = \frac{66 - 50}{10} = \frac{16}{10} = +1.600$$

This means that the student scored 1.6 standard deviations above the mean. To find the percentile, look at Table 9.1 in the leftmost column. In the 1.6 row, look in the column labelled .00. The number there is .4452. A picture of this situation is given in Figure 9.10.

FIGURE 9.10

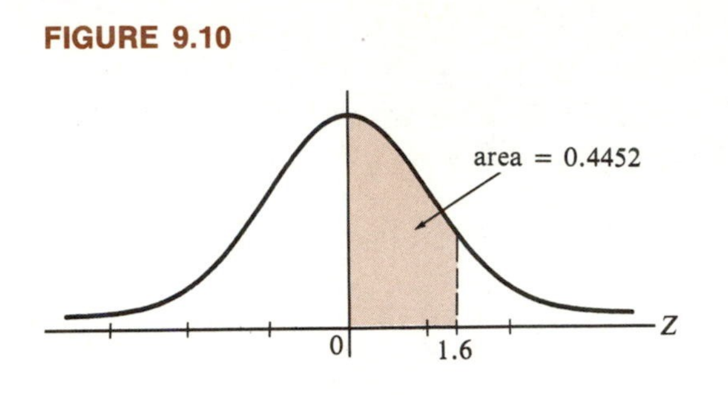

To find the required percentile, we must add 0.5000 (representing the area to the left of $Z = 0$) to the table value (Figure 9.11). Therefore, the student's percentile is

$$P = 0.5000 + 0.4452 = 0.9452 \approx 95\%$$

FIGURE 9.11

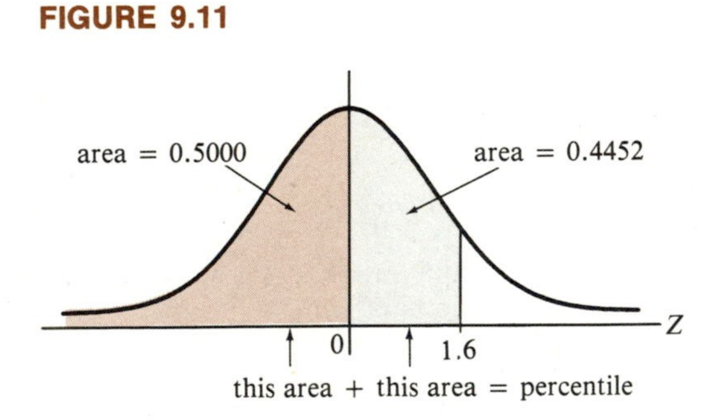

This means that the student obtained a score at or above the scores of 95% of the population tested.

EXAMPLE 5

A student obtains a score of 435 on a test that has a mean of 500 and a standard deviation of 100. Find the student's percentile.

Solution

Assuming that the test scores have a normal distribution, we first find the Z-score.

$$Z = \frac{435 - 500}{100} = \frac{-65}{100} = -0.65$$

The Z-score is *below* the standard mean, but we look up the value $+0.65$ in the table (since the area from $Z = 0$ to $Z = +0.65$ is the same as the area from $Z = -0.65$ to $Z = 0$). This time, we find 0.6 in the left-hand column (the Z column) and then look in the 0.6 row under the .05 column (note that $0.60 + 0.05 = 0.65$). The value in that location is .2422. This value gives the area under the normal curve from $Z = -0.65$ to $Z = 0$, as shown below (Figure 9.12).

FIGURE 9.12

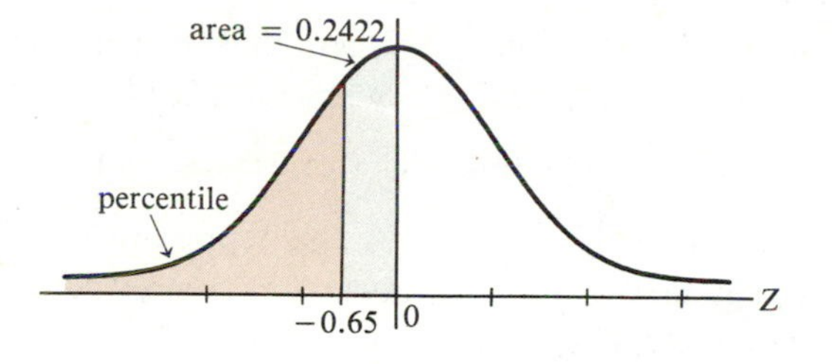

Since the percentile is represented by the area to the *left* of $Z = -0.65$, we must subtract 0.2422 from the area under the left half of the curve, 0.5000. Thus, the percentile is:

$$P = 0.5000 - 0.2422 = 0.2578 \approx 26\%$$

EXAMPLE 6

A set of measurements on the diameters of washers has been found to have a normal distribution, with mean 0.5 mm and standard deviation 0.001 mm. Find the percentage of washers with diameters between .499 mm and .5014 mm.

Solution

When a problem asks for the percent falling *between* two values, it is best to treat it as two separate parts. We begin by finding the percentage of washers having a diameter between 0.499 and the mean (the *percentile* associated with the value 0.499). This is represented by the graph of Figure 9.13.

FIGURE 9.13

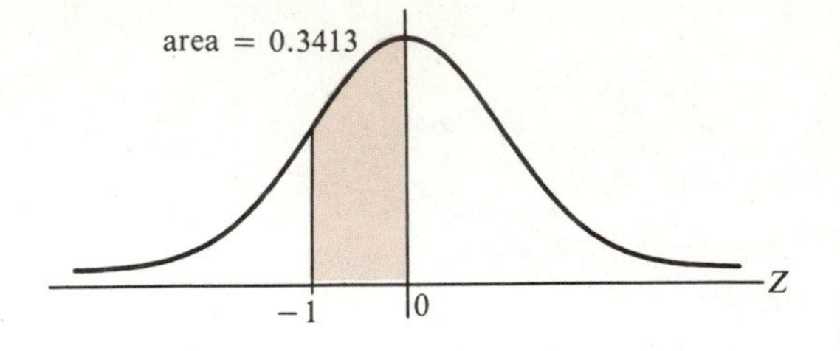

Begin by finding the Z-score.

$$Z = (0.499 - 0.5)/0.001 = -0.001/0.001 = -1$$

Looking in the table for $Z = -1$, we find that the area between $Z = -1$ and $Z = 0$ is 0.3413. Thus, approximately 34% of the washers have diameters between 0.499 and the mean.

Now we must find the percentage of washers that have diameters between the mean and 0.5014. First, we need the Z-score for 0.5014.

$$Z = (0.5014 - 0.5000)/0.001 = 0.0014/0.001 = +1.4$$

Table 9.1 gives a value of 0.4192 for $Z = 1.4$; this represents the area from $Z = 0$ to $Z = 1.4$, or the percentage of diameters between 0.5000 mm and 0.5014 mm— approximately 42%.

FIGURE 9.14

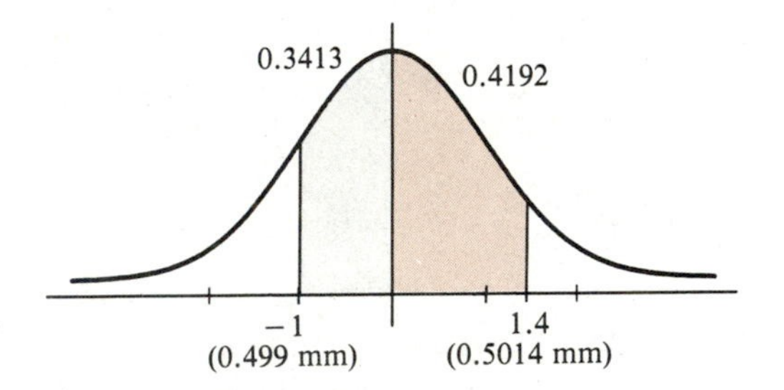

Finally, the graph of Figure 9.14 indicates that the percentage of washers with diameters between 0.499 mm and 0.5014 mm is found by adding the two percentiles found:

$$0.3413 + 0.4192 = 0.7605 \approx 76\%$$

We point out here that the concepts in this section were introduced primarily to illustrate one of the major uses of standard deviation and to acquaint the reader with the normal curve. A statistics text should be consulted for more complete information.

EXERCISES FOR SECTION 9.7

In Exercises 1 through 4, find the Z-score for each given SAT score, assuming a mean of 500 and a standard deviation of 100 (see Examples 1 and 2).

1. 400 **2.** 600 **3.** 350 **4.** 700

In Exercises 5 through 9, assume that an I.Q. test has a mean score of 100 and a standard deviation of 15. Find the Z-score of each of the given I.Q. scores (see Examples 1 and 2).

5. 85 **6.** 95 **7.** 100
8. 120 **9.** 130

In Exercises 10 through 18, find the corresponding percentile for each score in Exercises 1 through 9 (see Examples 3 and 4).

For Exercises 19 and 20, refer to Example 5.

19. The number of ounces of a particular cereal actually dispensed into 16-ounce boxes is normally distributed, with mean 16.2 ounces and standard deviation 0.2 ounce.
 a. What percentage of cereal boxes actually contains at least 16 ounces of cereal?
 b. What percentage of boxes contains less than 16 ounces of cereal?

20. Suppose that the height of men in the United States is approximately normally distributed, with a mean of 69 inches and a standard deviation of 3 inches. What percentage of U.S. males is taller than seven feet?

9.8 Chapter Review

This chapter dealt with three major topics: probability, simulations, and statistics. We first discussed probability, beginning with definitions and examples of permutations and combinations. The concepts of relative frequency and weighted probabilities, along with factorial notation (as in 7!) were important.

In Section 9.3, the concepts of random numbers and simulations were introduced. We learned how to generate a sequence of pseudorandom numbers from a seed and discussed some algorithms for using computers and the RND function to set up simulations.

Topics related to statistics were discussed in Sections 9.4 through 9.7. We began with measures of central tendency—the mean, median, and mode of a set of data. Measures of dispersion—including range, variance, and standard deviation—were introduced to supplement the measures of central tendency in summarizing data. Finally, we used the normal curve, along with the concept of a normal distribution, to illustrate an important application of the mean and standard deviation.

VOCABULARY

permutation
tree diagram
factorial
combination
probability
sample space
relative frequency
weight
random sample
simulation
random number generator
seed of a random number sequence
RND function
measure of central tendency
mean
median
mode
bimodal
multimodal
frequency distribution
histogram
range
variance
standard deviation
normal distribution
normal curve
percentile

CHAPTER TEST

1. How many permutations are there of the letters P, R, O, B, L, E, M?

2. Evaluate each of the following:

   a. $\frac{6! + 2!}{4!}$ b. $9! - 7!$

3. Ten talented dancers audition for a chorus.

a. If only four are to be chosen, how many "chorus lines" of four are possible?

b. Suppose the company decides to have a real chorus line and auditions 100 talented dancers. If the director chooses 40 dancers, how many possible lines are there? (Do *not* evaluate the answer to this question; just give the formula for finding it.)

4. A die is weighted so that the three-spot face is three times as likely as each of the other five faces. Find the probability of rolling each of the following:

   a. a three   b. a six

5. Explain what a simulation is and the role played in a simulation by random numbers.

6. Identify three measures of central tendency, and evaluate each for the following data:

   $$5, -3, 2, 9, 6, -3, 16, 4, 6, 8, -3, 4$$

7. Identify two measures of dispersion, and find each for the following data:

   $$8, 2, 9, 11, 10$$

   Show your work.

8. Using the data of Problem 7, find the Z-score for 9.

9. A student receives a score of 100 on a test in which the sum of *all* the fifteen student's scores is 1650. If the variance is 100, find the student's Z-score on this test, and his percentile.

10. Design a simulation to find the probability of getting a sum of 7 or 11 with two dice. Use a flowchart, pseudocode, or a well-written paragraph to explain your method.

CHAPTER

10

Simulations are used in business to predict or control inventory or to experiment with automobile and aircraft models. Such detailed simulations often involve elaborate equations, for which exact solutions are not always readily available. In this chapter we will examine some numerical methods for approximating solutions, to any degree of precision, of single equations and systems of linear equations.

Numerical Methods

Many real-life problems in science, engineering, business and other areas require methods of solution involving tedious calculations. In this chapter we investigate a number of mathematical methods that, although very useful, are difficult to implement without a calculator or computer. We shall, for example, find out how a table of square or cube roots is calculated and learn new methods for solving complicated equations that might arise in analyzing scientific experiments.

10.1 Introduction to Numerical Methods

Absolute and Relative Error

We begin our study of numerical methods by considering a problem we already know how to solve. In Chapter 2 we studied two methods of solving *quadratic equations*, equations of the form $ax^2 + bx + c = 0$: (1) by factoring and (2) by using the quadratic formula. Let's apply these methods to a specific problem.

EXAMPLE 1

Suppose an object is fired vertically upward with a speed of 160 ft/sec from ground zero. At the end of t seconds, it will reach a height given by the function

$$h(t) = 160t - 16t^2$$

How many seconds will it take to reach a height of 400 ft?

Solution

Since the height after t seconds is given by $160t - 16t^2$, we set $160t - 16t^2 = 400$ and solve for t. Recall that we must first write the equation in the form $ax^2 + bx + c = 0$.

$$160t - 16t^2 = 400$$

$$0 = 16t^2 - 160t + 400 \qquad \text{Add } -160t + 16t^2 \text{ to both sides.}$$

or

$$16t^2 - 160t + 400 = 0$$

$$16(t^2 - 10t + 25) = 0 \qquad \text{Factor out a common factor of 16.}$$

$$16(t - 5)(t - 5) = 0 \qquad \text{Factor } t^2 - 10t + 25.$$

$$t - 5 = 0 \qquad \text{Set factors equal to 0.}$$

$$t = 5 \qquad \text{Solution}$$

Therefore, the object will reach a height of 400 feet in 5 seconds.

This example illustrates the method of factoring to solve a quadratic equation. It gave us an *exact* solution, as we can see by substituting 5 for t in the original equation.

$$\begin{aligned} 160(5) - 16(5)^2 &= 400 \\ 800 - 16(25) &= 400 \\ 800 - 400 &= 400 \\ 400 &= 400 \quad \checkmark \end{aligned}$$

EXAMPLE 2

Find the time at which the object of Example 1 will reach 320 feet.

Solution

This time we set the formula for height after t seconds equal to 320, giving

$$160t - 16t^2 = 320$$

or

$$\begin{aligned} 16t^2 - 160t + 320 &= 0 \\ 16(t^2 - 10t + 20) &= 0 \qquad \text{Factor out the common factor of 16.} \\ t^2 - 10t + 20 &= 0 \qquad \text{Divide both sides by 16.} \end{aligned}$$

This equation cannot easily be solved by factoring, so we turn to the quadratic formula

$$t = \frac{-b \pm \sqrt{b^2 - 4ac}}{2a}$$

For our equation, $a = 1$; $b = -10$; and $c = 20$. So the formula gives

$$t = \frac{-(-10) \pm \sqrt{(-10)^2 - 4(1)(20)}}{2(1)}$$

$$t = \frac{10 \pm \sqrt{100 - 80}}{2}$$

$$t = \frac{10 \pm \sqrt{20}}{2} = \frac{10 \pm 2\sqrt{5}}{2} = \frac{\not{2}(5 \pm \sqrt{5})}{\not{2}}$$

$$t = 5 \pm \sqrt{5}$$

$t = 5 + \sqrt{5}$ gives $t \approx 7.24$ — This is read "t is approximately equal to 7.24."

$t = 5 - \sqrt{5}$ gives $t \approx 2.76$

Both answers are accurate to two decimal places. (There are two times at which the object is 320 feet above the ground because "what goes up must come down.")

The methods we used to solve the equations in Examples 1 and 2 are **explicit** methods. This means that there are formulas that *always* result in a solution, if one exists. Any equation of degree two can be solved by the quadratic formula; that is, by an explicit method. But what about equations of higher degree? The equation

$$x^5 = 2x^4 - 3x^2 + 1$$

cannot be solved by such a formula because no such formula exists! Later in this chapter, we shall illustrate some elementary numerical methods that can help in approximating roots of difficult equations or sets of equations.

Let us first make some observations about the examples we have just done.

1. The time when the object reached 400 feet in Example 1 is *unique* (the equation had only one solution) because 400 feet is the *maximum* height given by the formula $h(t) = 160t - 16t^2$.

2. In Example 2, 7.24 and 2.76 are *not exact* roots of $t^2 - 10t + 20 = 0$. We can see this by substituting in the equation

$$\begin{aligned}(7.24)^2 - 10(7.24) + 20 &\approx (52.42) - 72.4 + 20 \\ &\approx 0.02 \neq 0\end{aligned}$$

and

$$\begin{aligned}(2.76)^2 - 10(2.76) + 20 &\approx 7.62 - 27.6 + 20 \\ &\approx 0.02 \neq 0\end{aligned}$$

These "solutions" are really *approximations* given to two decimal places. An exact root makes an equation *exactly true.*

This second observation leads us to an important consideration: In order for much of the material on numerical methods to be understandable, we must agree on some fundamentals. First we need to decide what we shall mean by "accurate to d decimal places." We shall adopt the convention illustrated below by some examples dealing with accuracy to two decimal places ($d = 2$).

Assume 0.36 is the real answer to some problem.

Then 0.363 is accurate to two places.
0.367 is *not*, since $0.367 \approx 0.37$.
0.355 is *not* accurate to two places.
0.358 is accurate to two places.
0.354 is *not* accurate to two decimal places since $0.354 \approx 0.35$.

If we generalize these examples, then, we can say that if $|\text{error}| < 0.005$, then the answer is accurate to *two* decimal places. Absolute value is used so that we may ignore the sign of the error or discrepancy. If y is the real answer, the error is given by subtracting y from the approximation, x; that is, error $= x - y$. (We should note that some sources accept an error *equal to* 0.005—that is, $|\text{error}| \leq 0.005$—but we shall require that the error be strictly *less than* the limit.)

EXAMPLE 3

Use the definition of accuracy to two decimal places ($|\text{error}| < 0.005$) to justify the results above.

Solution

The real answer is 0.36, so $y = 0.36$.

For $x = 0.363$, $|x - y| = |0.363 - 0.36| = |0.003| = 0.003 < 0.005$; so 0.363 *is* accurate to two decimal places.

For $x = 0.367$, $|x - y| = |0.367 - 0.36| = |0.007| = 0.007 > 0.005$; so 0.367 *is not* accurate to two decimal places.

For $x = 0.355$, $|x - y| = |0.355 - 0.36| = |-0.005| = 0.005$; so 0.355 *is not* accurate to two decimal places.

For $x = 0.358$, $|x - y| = |0.358 - 0.36| = |-0.002| = 0.002 < 0.005$; so 0.358 *is* accurate to two decimal places.

For $x = 0.354$, $|x - y| = |0.354 - 0.36| = |-0.006| = 0.006 > 0.005$; so 0.354 *is not* accurate to two decimal places.

We shall further agree that three decimal place accuracy or *agreement to three decimal places* means

$$|x - y| < 0.0005$$

The absolute value of the error (approximate value − real value) is less than 0.0005.

Agreement to four places will mean that x and y differ in absolute value by less than 0.00005; in symbols,

$$|x - y| < 0.00005$$

For agreement to d places, the maximum error would have d zeros between the decimal point and the 5.)

EXAMPLE 4

Suppose the *real* solution to a problem is 6.437. Determine whether each of the following values is *accurate to three decimal places*:

a. 6.4376 **b.** 6.4367 **c.** 6.4372 **d.** 6.4365 **e.** 6.4363

Solution

The real answer is 6.437; so $y = 6.437$. Accuracy to three decimal places means $|x - y| < 0.0005$.

a. $x = 6.4376$; $|x - y| = |6.4376 - 6.437| = |0.0006| = 0.0006 > 0.0005$; so 6.4376 *is not* accurate to three decimal places.

b. $x = 6.4367$; $|x - y| = |6.4367 - 6.437| = |-0.0003| = 0.0003 < 0.0005$; so 6.4367 *is* accurate to three decimal places.

c. $x = 6.4372$; $|x - y| = |6.4372 - 6.437| = |0.0002| = 0.0002 < 0.0005$; so 6.4372 *is* accurate to three decimal places.

d. $x = 6.4365$; $|x - y| = |6.4365 - 6.437| = |-0.0005| = 0.0005$; so 6.4365 *is not* accurate to three decimal places.

e. $x = 6.4363$; $|x - y| = |6.4363 - 6.437| = |-0.0007| = 0.0007 > 0.0005$; so 6.4363 *is not* accurate to three decimal places.

We have been discussing *absolute* agreement or accuracy. In many circumstances, *relative accuracy* may be more important. Some examples will illustrate this point.

1. If the real root of an equation were 1000 and our methods estimated the root to be 1000.5, then the error would be relatively small compared to the size of the root.
2. If the real root of an equation were 0.5 and our methods estimated the root to be 0.501, then the error would be relatively large compared to the size (magnitude) of the root.

The **actual or absolute error** is defined to be |real root − calculated root|, if the sign is of no concern. Then the actual error for the first example is

$$|1000 - 1000.5| = 0.5$$

The actual error for the second example is

$$|0.5 - 0.501| = 0.001$$

We can see that the actual error in the first case is larger than the actual error in the second. However, an absolute error of 0.5, when compared to the size of the root (1000) is not as large as it appears.

For this reason, the concept of **relative error** is important. We can define it as follows:

$$\text{relative error} = \frac{\text{actual error}}{\text{real answer}}$$

Applying this definition to the first example, we find

$$\text{relative error} = \frac{0.5}{1000} = 0.0005$$

and in the second example we find

$$\text{relative error} = \frac{0.001}{0.5} = 0.002$$

The relative error in the first example is clearly less than the relative error in the second!

EXAMPLE 5

Calculate the relative error for each value in Examples 3 and 4.

Solution

For Example 3 the real answer is 0.36.

$x = 0.363$; actual error $= 0.003$

$$\text{relative error} = \frac{0.003}{0.36} \approx 0.0083$$

$x = 0.367$; actual error $= 0.007$

$$\text{relative error} = \frac{0.007}{0.36} \approx 0.0194$$

$x = 0.355$; actual error $= 0.005$

$$\text{relative error} = \frac{0.005}{0.36} \approx 0.0139$$

$x = 0.358$; actual error $= 0.002$

$$\text{relative error} = \frac{0.002}{0.36} \approx 0.0056$$

$x = 0.354$; actual error $= 0.006$

$$\text{relative error} = \frac{0.006}{0.36} \approx 0.0167$$

For Example 4 the real answer is 6.437.

a. $x = 6.4376$; actual error $= 0.0006$

$$\text{relative error} = \frac{0.0006}{6.437} \approx 0.0000932$$

b. $x = 6.4367$; actual error $= 0.0003$

$$\text{relative error} = \frac{0.0003}{6.437} \approx 0.0000466$$

c. $x = 6.4372$; actual error $= 0.0002$

$$\text{relative error} = \frac{0.0002}{6.437} \approx 0.0000311$$

d. $x = 6.4365$; actual error $= 0.0005$

$$\text{relative error} = \frac{0.0005}{6.437} \approx 0.0000777$$

e. $x = 6.4363$; actual error $= 0.0007$

$$\text{relative error} = \frac{0.0007}{6.437} \approx 0.0001087$$

The concept of error is very important in the study of numerical methods, since errors may be introduced by the finite precision of computer arithmetical operations and by the *iterative* techniques that are often used to solve equations.

EXERCISES FOR SECTION 10.1

In Exercises 1 through 4, solve the given quadratic equation (a) to two decimal places; and (b) to three decimal places (see Examples 1 and 2).

1. $x^2 - 3x - 1 = 0$

2. $x^2 + 5x + 2 = 0$

3. $7x^2 + 10x = 2$

4. $15x = 20 - 25x^2$

In Exercises 5 through 10, assume that the real answer is 2.43. Determine whether or not each approximation is accurate to two decimal places (see Example 3).

5. 2.437

6. 2.427

7. 2.434

8. 2.425

9. 2.423

10. 2.430

In Exercises 11 through 16, calculate the actual error and the relative error of each approximation (see Example 5).

11. true answer = 50
estimated answer = 50.1

12. true answer = 100
estimated answer = 100.7

13. true answer = 0.05
estimated answer = 0.051

14. true answer = 0.1
estimated answer = 0.17

15. true answer = 67527
estimated answer = 67528

16. true answer = 6752
estimated answer = 6753

10.2 Graphing and the Location Principle

Degree of Precision

Suppose the quadratic formula did not exist or was unknown to you. Could you obtain an approximation of the roots of $t^2 - 10t + 20 = 0$? (This is the equation from Example 2 of Section 10.1.) One way to obtain some information about the roots of an equation is to graph it. Since we are assuming here no knowledge of the quadratic formula, we shall graph the *function* $h(t)$ associated with this equation by plotting points.

$$h(t) = t^2 - 10t + 20$$

As in Chapter 2, we can graph the function by making a table of values.

t	$h(t)$
0	20
1	11
2	4
3	−1
4	−4
5	−5
6	−4
7	−1
8	4

Some questions arise immediately:

Why did we start at $t = 0$? Are we prejudiced against negative numbers? We started at $t = 0$ because, in the original problem, t represented time in seconds. Thus, the domain of the function (the set of possible values for t) is the set of nonnegative real numbers. In solving an equation, it is important to determine the domain of reasonable values at the start. This can save considerable work later, both in computer applications and in "pure" mathematics.

How do we decide where to stop? We stopped at $x = 8$ because the table of values indicated that we had found two roots! Since this is not immediately obvious, we'll explain. Notice that, during the time interval from $t = 2$ to $t = 3$, the value of $h(t)$ went from $+4$ to -1. If the graph of h is *continuous* (has no gaps), then it must pass through zero on the way from $+4$ to -1. That is, there must be some value of t between 2 and 3 for which

$$h(t) = t^2 - 10t + 20 = 0$$

Such a number is a root (solution) or *zero* of the equation $t^2 - 10t + 20 = 0$ (Figure 10.1).

FIGURE 10.1

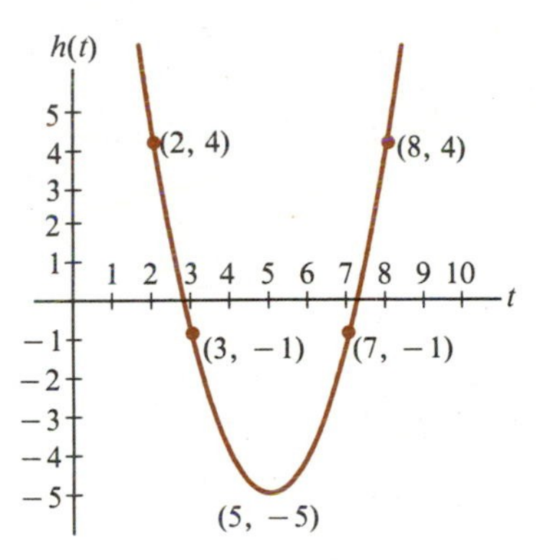

Again notice that during the time interval from $t = 7$ to $t = 8$, the graph of h passes from -1 to $+4$; thus a root of the equation must exist between $t = 7$ and $t = 8$.

If we are willing to cheat a little and recall our original values for t ($t = 2.76$ and $t = 7.24$), we see that we seem to be correct in our choices of intervals. However, our approximations through the table of values need refinement; they are not close enough to the actual roots. By constructing a careful graph, we may be able to get a better approximation. In any case, the graph will provide a better understanding of the process of finding roots (Figure 10.1).

The roots occur where the graph crosses the t-axis; that is, where $h(t) = 0$. This is why roots are sometimes called the *zeros* of an equation. We set the function $h(t) = t^2 - 10t + 20$ equal to zero to solve for such roots.

A very precise graph might give a better approximation than our initial estimate that:

> root 1 is between 2 and 3 (in symbols, $2 < \text{root } 1 < 3$); and
> root 2 is between 7 and 8 ($7 < \text{root } 2 < 8$).

One way of obtaining such a graph would be to refine the original table—choose values of t that are closer together, especially in the area where we expect to find a root.

EXAMPLE 1

In the interval from $t = 2$ to $t = 3$, find a closer approximation of root 1 of the equation $t^2 - 10t + 20 = 0$.

Solution

We begin by subdividing the interval between 2 and 3:

2 2.1 2.2 2.3 2.4 2.5 2.6 2.7 2.8 2.9 3

Make a table showing $h(t)$ for these values of t. A calculator or computer would be very useful for this job!

t	$h(t)$
2.1	3.41
2.2	2.84
2.3	2.29
2.4	1.76
2.5	1.25
2.6	0.76
2.7	+0.29
2.8	−0.16
2.9	−0.59

Examination of the table shows that $h(2.7) > 0$ and $h(2.8) < 0$. Since the function is continuous, it must, then, pass through zero between $t = 2.7$ and $t = 2.8$. So the equation has a root between 2.7 and 2.8.

This process can be continued by dividing the interval from 2.7 to 2.8 into hundredths and then further subdividing those intervals, and so on. However, this method is tedious at best. Other methods presented in this chapter will provide some relief.

The procedure we followed in Example 1 depends on the **location principle**.

THE LOCATION PRINCIPLE

If a function f is continuous (has no "holes") in the domain of interest and if $f(x_1)$ and $f(x_2)$ differ in sign (where $x_1 < x_2$), then a zero, or root, of the function is located between x_1 and x_2 ($x_1 <$ root $< x_2$).

Note that we did not state that a *unique root* exists in the interval. Consider the graph of Figure 10.2:

FIGURE 10.2

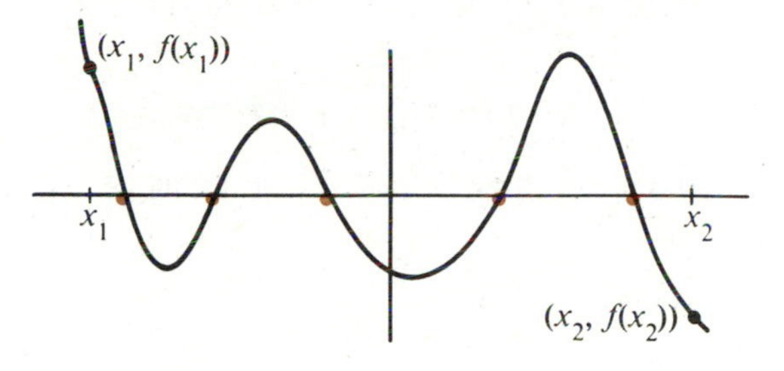

We have $f(x_1) > 0$ and $f(x_2) < 0$; so we are guaranteed a root. But here there are five!

Also be careful *not* to assume the *converse* of the above principle: If $f(r) = 0$ and r is between x_1 and x_2, then $f(x_1)$ and $f(x_2)$ necessarily differ in sign. For example, consider Figure 10.3.

FIGURE 10.3

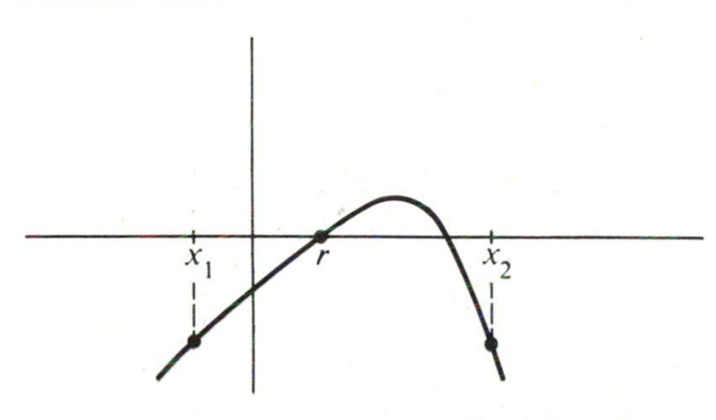

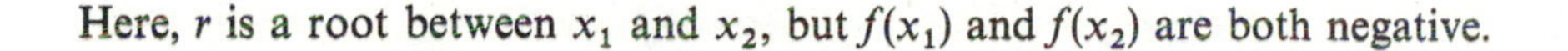

Here, r is a root between x_1 and x_2, but $f(x_1)$ and $f(x_2)$ are both negative.

EXAMPLE 2

Use the location principle to arrive at a closer approximation of the other root of the equation $t^2 - 10t + 20 = 0$.

Solution

If $h(t) = t^2 - 10t + 20$, we know that

$$h(7) = -1 < 0$$

and

$$h(8) = 4 > 0$$

Since all quadratic functions are continuous, the *location principle* indicates that there is a root between 7 and 8 since $h(7)$ and $h(8)$ differ in sign. As in Example 2 of Section 10.1, we divide this interval into smaller pieces:

7 7.1 7.2 7.3 7.4 7.5 7.6 7.7 7.8 7.9 8

Next, we construct a table for these values of t, stopping as soon as we find a change in the sign of $h(t)$:

t	$h(t)$
7.1	-0.59
7.2	-0.16
7.3	$+0.29$

Note that the sign of $h(t)$ changes between $t = 7.2$ and $t = 7.3$. Therefore, there is a root of the equation in the interval between 7.2 and 7.3.

We can get an even better approximation by subdividing *this* interval:

t	$h(t)$
7.21	-0.1159
7.22	-0.0716
7.23	-0.0271
7.24	$+0.0176$

Since the sign of $h(t)$ changes between $t = 7.23$ and $t = 7.24$, we know that there is a root between 7.23 and 7.24.

Before going on to discuss other methods of approximation, we make some observations about numerical methods in general and **iterative methods** (those involving repetition of some process) in particular.

In order to use numerical methods to their fullest potential, one must have a good knowledge of calculus. Since we do not assume such knowledge in this text, our treatment of numerical methods relies on less mathematically powerful tools. The techniques employed, however, form a good intuitive basis for further study of numerical methods.

Our approach to solving a problem is necessarily somewhat intuitive. We shall rely on the graph of a function to convince us that it is continuous in the region of interest. We shall also use graphing to show that a root exists in a particular interval and the location principle to help find the initial interval from which the search for the root begins.

Suppose we have found an interval in the domain of a function f such that $x_1 < x_2$ and $f(x_1)$ and $f(x_2)$ differ in sign. By making a rough sketch or by using a graphics package on the computer, we confirm that there is a single root in the interval. We then use an iterative process to arrive at an approximation of the desired root. Instead of finding a root with *one* application of a formula (such as the quadratic formula) we produce a set or *sequence* of successive values. We will want a **convergent sequence**, one in which each successive value produced comes closer to some fixed value. For example, the successive values 0.5372, 0.5343, 0.5339, 0.5333 are converging toward 0.533 and the values 0.3, 0.33, 0.333, 0.3333, . . . seem to be converging toward $\frac{1}{3}$. In contrast, the sequence 0.4, 0.45, 0.5, 0.6, 0.7, 0.9, 1.1, 1.9 appears at this stage to be *divergent*; there does not appear to be a unique number that the successive values approach.

An iterative method can produce a set of successive values that either converge or diverge depending on the initial values chosen for the process and on the nature of the equation or set of equations whose solution is sought. In general, an iterative method, unlike an explicit formula, does not *always* produce a root.

Two problems in using an iterative method are choosing the initial value (where to start the procedure) and deciding when to stop the process. To find an initial value, we shall use the location principle in conjunction with graphing. This will guarantee us an interval that contains a root (under our assumptions, unique). To decide when to terminate the process we shall at each stage compare the difference in absolute value (so we don't have to worry about sign) of two successive approximations. We'll stop when they are close enough together.

Two successive approximations to a root will be designated r_i and r_{i+1}, where r_{i+1} is the value following r_i. We shall agree that if the degree of precision required is to three decimal places, then $|r_i - r_{i+1}|$ must be less than 0.0005. If the degree of precision required is to four decimal places, then we must have $|r_i - r_{i+1}| < 0.00005$, and so on.

Let us pause to summarize what we know of iterative methods so far: We have seen that iterative methods do not always produce roots of an equation.

When an explicit formula for a root exists, it is certainly more reliable. There are, however, many equations (such as $x^5 - x^4 + 3x^2 + 1 = 0$) that cannot be solved by an explicit formula. For such an equation, we do the following:

1. decide on the domain of the function,
2. within that domain, find an interval in which a root exists, using the location principle and graphing,
3. terminate the iterative process when two successive approximations differ in absolute value by an amount decided on beforehand (the *degree of precision*, denoted by δ, the Greek letter *delta*). We express this condition symbolically by

$$|r_i - r_{i+1}| < \delta$$

where r_{i+1} is the approximation *after* r_i.

EXAMPLE 3

Find $\sqrt{7}$, correct to one decimal place.

Solution

We must set up an equation to express this problem. If $x = \sqrt{7}$, then

$$x^2 = 7$$

or

$$x^2 - 7 = 0$$

This second form is the equation whose roots we wish to find. Using the location principle, we create the following table:

x	$f(x)$
0	-7
1	-6
2	-3
3	$+2$

We see that there is a root between 2 and 3, sometimes expressed as the interval (2, 3), since $f(2) < 0$ and $f(3) > 0$. We say that 2 and 3 *bracket* the root or enclose the root. Since we must find the root to one decimal place, we need two successive approximations that differ by less than 0.05.

One iterative method used to solve this equation produced the following sequence of approximations:

Iteration	Estimate of Root	$\lvert r_i - r_{i+1} \rvert$
1	$r_1 = 2.5$	
2	$r_2 = 2.75$	$\lvert r_1 - r_2 \rvert = 0.25$
3	$r_3 = 2.625$	$\lvert r_2 - r_3 \rvert = 0.125$
4	$r_4 = 2.68$	$\lvert r_3 - r_4 \rvert = 0.055$
5	$r_5 = 2.65$	$\lvert r_4 - r_5 \rvert = 0.03$

$0.03 < 0.05$

We see from the table that the difference between the fifth root and the fourth root is $|2.68 - 2.65| = 0.03 < 0.05$. This last value, 2.65, is taken as $\sqrt{7}$, accurate to one decimal place. (The correct value is approximately 2.6457.)

The reader may well ask why we do not simply evaluate the equation at each approximation and stop when an approximation x_i results in $f(x_i) = 0$. We cannot generally employ this method in computer applications because of the finite storage of decimals and error accumulation mentioned in previous chapters. The value $f(x_i)$ will probably *never* exactly equal zero. Also observe that

$$2.68^2 - 7 = 0.1824$$

and

$$2.65^2 - 7 = 0.0225$$

That is, $f(2.68) \approx 0.18$ and $f(2.65) \approx 0.02$. Neither of these is exactly equal to zero, and neither functional value tells us that much about the precision of the associated root. By comparing two successive iterative values or approximations to the root, we can better control the degree of precision required.

However, we must point out again that there does not appear to be universal agreement on the condition for termination of an iterative process. Most texts do use the criterion of the absolute value of the difference between two successive approximations being less than one predetermined value, but others test the absolute value of the function evaluated at the potential root value. We may, then, choose either of the following two conditions for terminating an iterative procedure:

$$|r_i - r_{i+1}| < \delta$$ The difference between two approximations is close to zero.

or

$$|f(r_i)| < \delta$$ The functional value evaluated at a potential root is close to zero.

Here, δ represents the desired degree of precision.

Note that using the second condition does not allow the user to set a predetermined level of precision on the root value.

As we shall see, the problem of error analysis and precision is even more complicated than thus far presented. It leads to some final cautions.

If you are using a calculator or computer to do the exercises for this chapter, be careful when potential roots are negative. Suppose, for example, you wish to evaluate $(-3)^4$ or $(-2.57)^5$. If you enter in your calculator (-3), y^x, 4 or (-2.57), y^x, 5, you will get an error message.

The same is true using the BASIC language exponent operator, ↑ or ^. Under some conditions, statements like $y = (-3)$^4.5 or $y = (-2.57)$^5.2 will result in an error. This is because logarithmic functions are used to evaluate these expressions and the domain of a log function is the set of positive real numbers. Depending upon your choice of calculator or computer, you will have to make adjustments in such cases. Try PRINT (−3)^2.5 and PRINT (−3)^4 to see how your computer reacts.

EXERCISES FOR SECTION 10.2

In Exercises 1 through 4, use the location principle to find an approximation of a root of the given function in the given interval. Subdivide the interval into tenths to do so (see Examples 1 and 2).

1. $f(x) = -x^2 - 4x + 3$ in the interval $(0, 1)$
2. $f(x) = x^3 - 2x^2 + 10x - 3$ in the interval $(0, 1)$
3. $f(x) = x^5 - 2x^3 + x + 3$ in the interval $(-2, -1)$
4. $f(x) = x^3 - 3x^2 - x - 17$ in the interval $(4, 5)$

5. Use the location principle to arrive at an approximation of $\sqrt{5}$, that is, to find a root of the equation $x^2 - 5 = 0$ (see Examples 1–3).
6. Use the location principle to arrive at an approximation of $\sqrt[3]{6}$.

10.3 The Method of Bisection

The Bolzano Method

The methods of approximation used in Section 10.2, relying on successive refinements of the location principle, are both tedious and slow. In this section, we shall discuss a significantly faster iterative process, the method of bisection.

In order to apply the **method of bisection**, or **Bolzano method**, we must be dealing with a function f that is *continuous* (its graph has no breaks or gaps). We must also be able to find, by using a rough graph and the location principle, an interval in which a root lies. That is, we must bracket the root between two values, x_L and x_R, with the property that $f(x_L)$ and $f(x_R)$ differ in sign. If these two conditions are met—the function is continuous and we can find an interval containing a root—then the bisection method will give us values that converge toward the root.

If x_L (the *left* boundary for the interval) and x_R (the *right* boundary) bracket a root r, then the situation can be illustrated as shown in Figure 10.4.

FIGURE 10.4

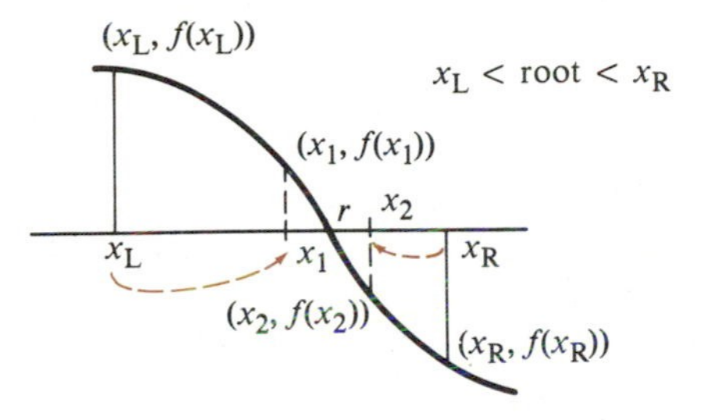

We begin by *bisecting* the interval (x_L, x_R) (dividing it in half) at the point x_1, with $x_1 = (x_L + x_R)/2$. We compare $f(x_1)$ with $f(x_L)$ and $f(x_R)$. Since it has the *same sign* as $f(x_L)$, we move the *left* endpoint of the interval to x_1 (that is, we replace x_L with x_1). This new interval, (x_1, x_R), contains a root of the function, since $f(x_1)$ and $f(x_R)$ *differ* in sign. Observe that the new interval is *half* the width of the old one since its left boundary is at the midpoint of the old interval.

Since the bisection method is an iterative process, we continue as we have begun. We find the midpoint, $x_2 = (x_1 + x_R)/2$, of the interval (x_1, x_R). We compare $f(x_2)$ with $f(x_1)$ and $f(x_R)$. This time $f(x_2)$ and $f(x_R)$ have the same sign, so we move the *right* endpoint of the interval to x_2 (we replace x_R with x_2). Again, the new interval (x_1, x_2) still contains the root, since $f(x_1)$ and $f(x_2)$ *differ* in sign. And again, the new interval (x_1, x_2) is half as wide as the old interval (x_1, x_R).

We can continue this process as long as necessary. Notice that the root is being bracketed by smaller and smaller intervals at each iteration (we are "closing in" on the root). Each midpoint $x_1, x_2, \ldots$ can be taken as a successive approximation of the root. We can stop the process when $|x_i - x_{i+1}|$ is less than some predetermined value δ (the degree of precision, as in Section 10.2). Or, since the method of bisection produces successively smaller intervals, we can stop when the width of the interval is less than δ, $|x_L - x_R| < \delta$.

In determining the required degree of accuracy, we use the criteria of Table 10.1, similar to the conventions used in Section 10.2.

TABLE 10.1

Accuracy to *d* Decimal Places	**Difference between Approximations of Root**	**Length of Interval**
$d = 1$	$\lvert r_i - r_{i+1}\rvert < 0.05$	$\lvert x_R - x_L\rvert < 0.1$
$d = 2$	$\lvert r_i - r_{i+1}\rvert < 0.005$	$\lvert x_R - x_L\rvert < 0.01$
$d = 3$	$\lvert r_i - r_{i+1}\rvert < 0.0005$	$\lvert x_R - x_L\rvert < 0.001$
$d = 4$	$\lvert r_i - r_{i+1}\rvert < 0.00005$	$\lvert x_R - x_L\rvert < 0.0001$
$\vdots$	$\vdots$	$\vdots$

Before illustrating the bisection method with an example, let us summarize what happens in a single iteration. Assume the degree of precision required is δ and the function f has a root in the interval (x_L, x_R).

1. Find the midpoint x_M of the interval. The formula is

$$x_M = (x_L + x_R)/2$$

   x_M will *replace one* of the endpoints.
2. Compare $f(x_M)$ with $f(x_L)$ and $f(x_R)$. If $f(x_M)$ and $f(x_L)$ have the *same sign*, replace x_L with x_M. That is, x_M becomes the new x_L, the new *left* endpoint. Otherwise replace x_R with x_M, making x_M the new *right* endpoint x_R.
3. Calculate the width of the new interval, |right endpoint − left endpoint|. If this width is less than δ, we're done, and x_M is the desired approximation of the root. If not, we return to step one.

EXAMPLE 1

Use the bisection method to solve the equation $t^2 - 10t + 20 = 0$ for t in the interval (2, 3).

Solution

Here, $f(t) = t^2 - 10t + 20$.

We begin by using the location principle (see Section 10.2) to find an interval that brackets the root. Recall that we found a unique root to be in the interval (2, 3). Let us agree that we want accuracy to two decimal places. Table 10.1 indicates that we want $|x_L - x_R| < 0.01$. We have

$$x_L = 2, \qquad f(x_L) = +4 > 0$$

and

$$x_R = 3, \qquad f(x_R) = -1 < 0$$

(See Figure 10.5.)

FIGURE 10.5

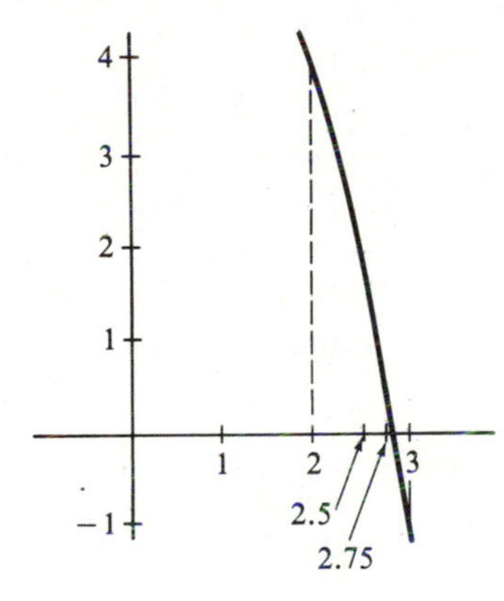

iteration 1:

1. $x_M = (x_L + x_R)/2 = (2 + 3)/2 = 5/2$
   $x_M = 2.5$
2. $f(x_M) = f(2.5) = 1.25 > 0$
   So $f(2.5)$ and $f(2)$ have the same sign. 2.5 becomes the new *left* endpoint, and the new interval is (2.5, 3).
3. $|x_R - x_L| = |3 - 2.5| = 0.5 > 0.01$, so we must continue.

iteration 2:

1. $x_M = (x_L + x_R)/2 = (2.5 + 3)/2 = 5.5/2$
   $x_M = 2.75$
2. $f(x_M) = 0.0625 > 0$
   So $f(2.75)$ and $f(2.5)$ have the same sign.
   2.75 becomes the new *left* endpoint, and the new interval is (2.75, 3) as in Figure 10.6.

FIGURE 10.6

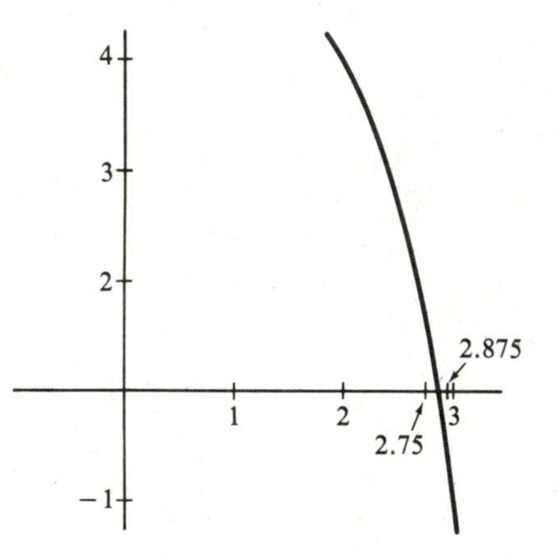

3. $|x_R - x_L| = |3 - 2.75| = 0.25 > 0.01$, so we must continue.

iteration 3:

1. $x_M = (x_L + x_R)/2 = (2.75 + 3)/2 = (5.75/2) = 2.875$
2. $f(2.875) \approx -0.484 < 0$.
   So $f(3)$ and $f(2.875)$ have the same sign.
   2.875 becomes the new *right* endpoint, and the new interval is (2.75, 2.875).
3. $|2.875 - 2.75| = 0.125 > 0.01$, so we must continue.

iteration 4:

1. $x_M = (2.75 + 2.875)/2 = 2.8125$
2. $f(2.8125) \approx -0.215 < 0$
   $f(2.8125)$ and $f(2.875)$ have the same sign, so 2.8125 becomes the new *right* endpoint. The new interval is (2.75, 2.8125).

3. $|2.8125 - 2.75| = 0.0625 > 0.01$, so we must continue.

iteration 5:

1. $x_M = (2.75 + 2.8125)/2 = 2.78125$
2. $f(2.78125) \approx -0.077 < 0$

   $f(2.78125)$ and $f(2.8125)$ have the same sign, so 2.78125 becomes the new *right* endpoint. The new interval is (2.75, 2.78125).
3. $|2.78125 - 2.75| = 0.03125 > 0.01$, so we must continue.

iteration 6:

1. $x_M = (2.75 + 2.78125)/2 = 2.765625$
2. $f(2.765625) \approx -0.0038 < 0$

   $f(2.765625)$ and $f(2.78125)$ have the same sign, so 2.765625 becomes the new *right* endpoint. The new interval is (2.75, 2.765625) as in Figure 10.7.

FIGURE 10.7

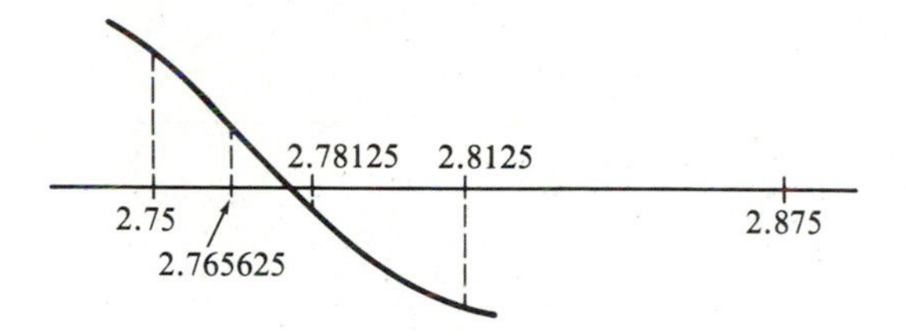

3. $|2.765625 - 2.75| \approx 0.016 > 0.01$, so we must continue.

iteration 7:

1. $x_M = (2.75 + 2.765625)/2 = 2.7578125$
2. $f(2.7578125) \approx 0.0274 > 0$

   $f(2.7578125)$ and $f(2.75)$ have the same sign, so 2.7578125 becomes the new *left* endpoint.

   The new interval is (2.7578125, 2.765625)
3. $|2.765625 - 2.7578125| \approx 0.008 < 0.01$, so we're done.

The approximate value of the root is the most recent value of x_M, correct to two decimal places. So the root is approximately 2.76. Whew!

A summary and a flowchart of the bisection method are given below:

1. Find the domain of values that is appropriate to the original problem from which the equation comes.
2. If the equation is not expressed in the form $f(x) = 0$, rewrite it in that form. For example if the equation is $5x^5 = 2x^2 - 3$, convert it to $5x^5 - 2x^2 + 3 = 0$.
3. If the function is continuous, use the location principle with the help of a calculator or computer to find an interval that contains the root.

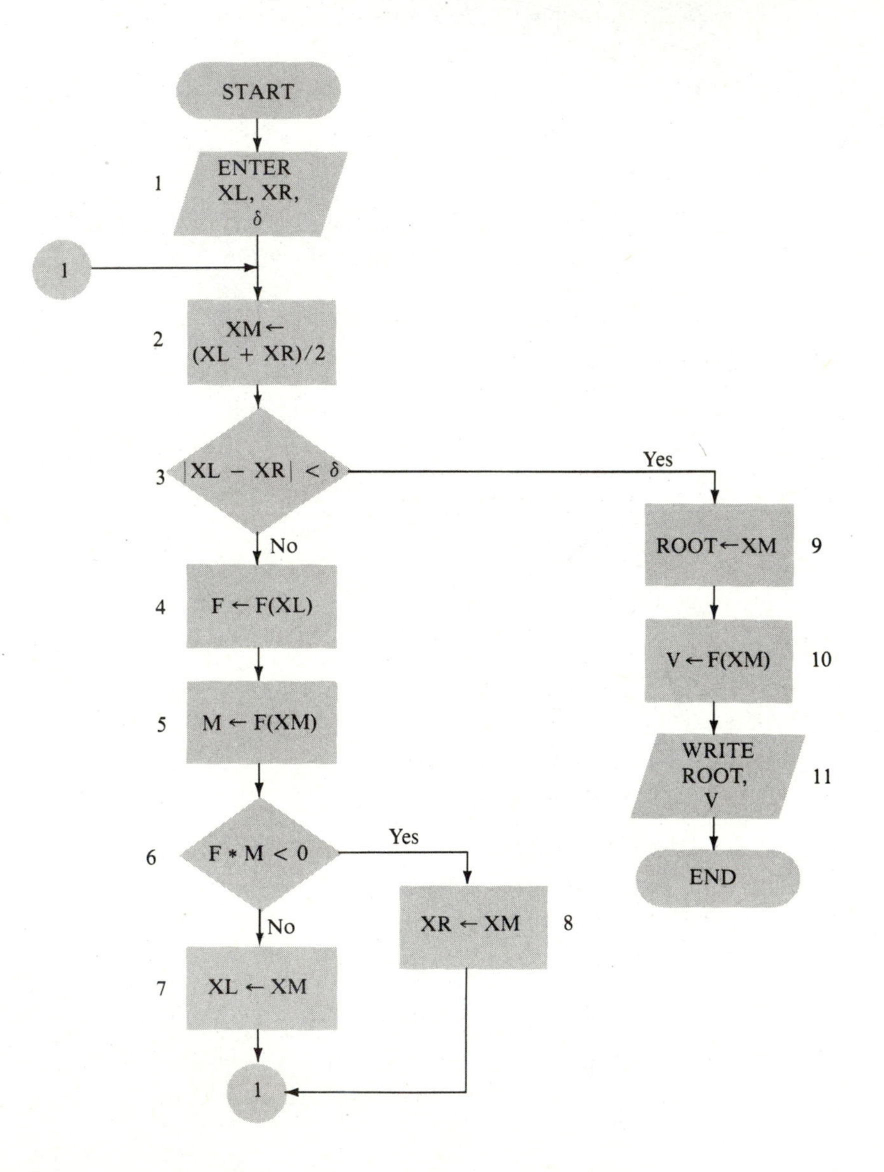

4. Graphing the function with a graphics package or by hand may provide additional information on the uniqueness of a root in an interval.
5. Determine the precision, δ, required. This may depend on the accuracy of the equipment that originally gave rise to the equation.
6. Use (x_1, x_2) or x_L = left boundary and x_R = right boundary for variables and make sure that $f(x_L)$ and $f(x_R)$ differ in sign. One way of expressing this is as follows:

$$f(x_L) \text{ and } f(x_R) \text{ have opposite signs if } f(x_L) \cdot f(x_R) < 0.$$

7. At the end of each iteration, the length of the interval is compared to δ. If $|x_R - x_L| < \delta$, then x_M, the midpoint of this last interval, is the root and the process stops. As an extra check, we may compare f(root) to zero and print the result.

The use of the width of the interval as a criterion for termination of the procedure in the bisection method is a classical convention, but the other criterion, successive differences in root approximations, might also be used. We have adopted here the more conventional approach.

The flowchart on page 492 applies the bisection method to this problem: Given the equation $f(x) = 0$, find a root in the interval (x_L, x_R).

EXAMPLE 2

Apply the flowchart and prior discussion to finding the roots of $x^3 - 3x + x^2 = 3$.

Solution

If the experiment or problem that gave rise to the equation allows for all real numbers, then the domain is all reals. If x represents some physical quantity that can only be positive then the domain would be $x > 0$. Let us assume all values are acceptable; thus the domain is all reals. Next we convert the given equation to $x^3 + x^2 - 3x - 3 = 0$ and apply the location principle. We shall consider only positive values and leave possible negative roots for the reader's consideration as an exercise.

$$f(x) = x^3 + x^2 - 3x - 3$$

The table of values is

x	$f(x)$
1	-4
2	$+3$
3	24
⋮	

Since $x^3 + x^2 - 3x - 3$ is continuous and $f(1) < 0$ and $f(2) > 0$, a real root exists between 1 and 2. We can thus choose XL = 1 and XR = 2. We shall assume that the root is unique in this interval. This can be checked by graphing.

Let $\delta = 0.1$ so that we will find the root to one decimal place.

1. $XL = 1, XR = 2, \delta = 0.1$
2. $XM = (1 + 2)/2 = 1.5$
3. $|XL - XR| = |1 - 2| \not< \delta$
4. $F = F(XL) = -4$ $\quad f(1) = -4$
5. $M = F(XM) = -1.875$ $\quad f(1.5) = -1.875$
6. $F * M \approx (-4)(-1.8) > 0$
7. $XL = 1.5$

* This is the end of one iteration *

2. $XM = 1.75$
3. $|XL - XR| = |1.5 - 2| = 0.5 \not< \delta$
4. $F = F(XL) = -1.875$ $\quad f(1.5) = -1.875$
5. $M = F(XM) = 0.17$ $\quad f(1.75) \approx 0.17$
6. $F * M = (-)(+) < 0$
7. $XR = 1.75$

* This is the end of the second iteration *

2. $XM = (1.5 + 1.75)/2 = 1.625$
3. $|XL - XR| = 0.25 \not< \delta$
4. $F = F(XL) = -1.875$ $\quad f(1.5) = -1.875$
5. $M = F(XM) = -0.094$ $\quad f(1.625) \approx -0.094$
6. $F * M = (-) * (-) > 0$
7. $XL = 1.625$

* End of the third iteration *

2. $XM = (1.625 + 1.75)/2 = 1.6875$
3. $|XL - XR| = |1.625 - 1.75| = 0.125 \not< \delta$
4. $F = F(XL) = -0.094$ $\quad f(1.625) \approx -0.094$
5. $M = F(XM) = -0.41$ $\quad f(1.6875) \approx -0.41$
6. $F * M = (-) * (-) > 0$
7. $XL = 1.6875$

* End of the fourth iteration *

2. $XM = (1.6875 + 1.75)/2 = 1.71875$
3. $|XL - XR| = |1.6875 - 1.75| = 0.0625 < 0.1$ $\quad$ Done!
9. $ROOT = XM = 1.71875$
10. $V = F(XM) \approx -0.125$ $\quad f(1.71875) \approx -0.125$
11. WRITE ROOT = 1.7 TO ONE DECIMAL PLACE
    $V \approx -0.125$

EXERCISES FOR SECTION 10.3

In these exercises use the method of bisection to solve each equation for a root in the indicated interval. Repeat iterations until the width of the interval is less than the value of δ specified.

1. $x^3 - 40x^2 - 196{,}875 = 0$ in the interval $(72, 78)$; $\delta = 0.0001$
2. $x^3 - 2x^2 + 10x - 3 = 0$ in the interval $(0, 1)$; $\delta = 0.00001$
3. $x^5 - 2x^3 + x + 3 = 0$ in the interval $(-2, -1)$; $\delta = 0.00001$
4. For the trigonometry fans, $-2 \sin x - x + 1 = 0$ in the interval $(0, 1)$; $\delta = 0.0001$
5. $x^3 - 3x^2 - x - 17 = 0$ in the interval $(4, 5)$; $\delta = 0.001$
6. Find a positive real fourth root of 17; use $\delta = 0.001$.
7. Find a positive square root of 12; use $\delta = 0.001$.
8. Solve the equation $x^3 = e^x$, where $e \approx 2.71828$ and $\delta = 0.0001$. Use the location principle to find a suitable interval.
9. A projectile thrown straight up from ground level with an initial velocity of 192 feet per second attains a height at time t seconds given by the formula

$$h(t) = -16t^2 + 192t$$

Find the time(s) when the height will be 100 feet (neglect air resistance).

10. An object is thrown vertically upward with an initial velocity of 200 feet per second. Its height after t seconds is given by the formula

$$h(t) = -16t^2 + 200t$$

Find the time(s) when it will be at a height of 200 feet.

In Exercises 11 through 13, (a) state the number of roots each equation has: (b) use the location principle to find an interval containing each root; and (c) use the method of bisection to approximate each root to three decimal places.

11. $2x^3 = 5x^2 + x - 6$ (Hint: all roots are between -2 and 4 inclusive.)
12. $2x^3 - 7x^2 - 5x - 1 = 0$ (Hint: all roots are between -2 and 5 inclusive.)
13. $x^4 - 6x^3 - 8x^2 + 23x - 6 = 0$ (Hint: all roots are between -3 and 8 inclusive.)

10.4 The Method of False Position

A numerical method related to the bisection method is the **method of false position**. It is an iterative method that is generally slightly faster than the method of bisection. Recall that the method of bisection always uses the *midpoint* of the interval bracketing the root (Figure 10.8).

FIGURE 10.8

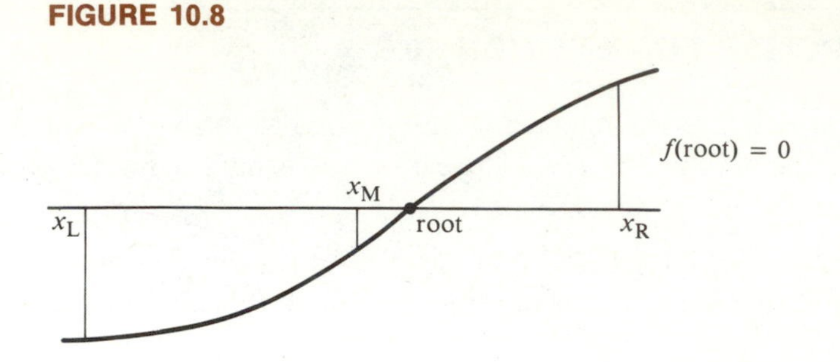

In using the method of false position, we draw the line segment joining the points $(x_L, f(x_L))$ and $(x_R, f(x_R))$. The point C (for *cross*) where this line crosses the x-axis replace x_M of the bisection method (Figure 10.9). Note that the y-coordinate of this point is 0.

FIGURE 10.9

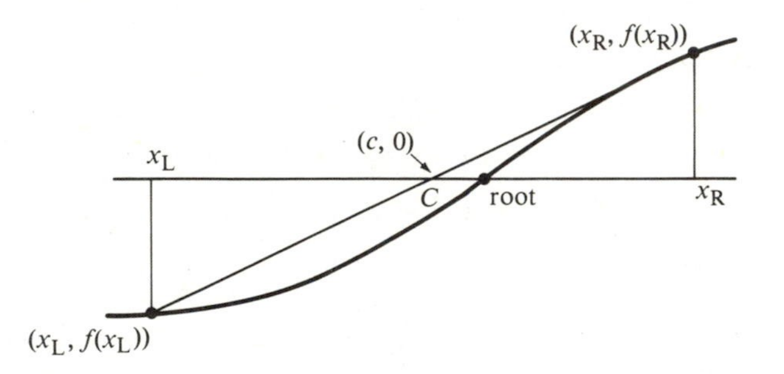

The formula for finding x_M was $x_M = (x_L + x_R)/2$. The formula for finding the x-coordinate of C is more complicated. The equation for the line joining $(x_L, f(x_L))$ and $(x_R, f(x_R))$ can be given by

$$y - f(x_L) = m(x - x_L) \qquad (1)$$

where m is the slope given by the formula

$$m = \frac{f(x_R) - f(x_L)}{x_R - x_L} \qquad (2)$$

The value of the slope m is found by dividing the difference of the y-coordinates of two points on the line, $f(x_R) - f(x_L)$, by the difference of the x-coordinates of the *same* points, $x_R - x_L$. Substituting the formula for m (2) in equation 1 above gives

$$y - f(x_L) = \frac{f(x_R) - f(x_L)}{x_R - x_L}(x - x_L)$$

or

$$y = f(x_L) + \frac{f(x_R) - f(x_L)}{x_R - x_L}(x - x_L) \qquad (3)$$

Since the y-coordinate of the point C is zero, we'll simply use c to represent the x-coordinate of that point. Since the point $(c, 0)$ is on the line, its coordinates must satisfy the equation of the line; so we substitute c for x and zero for y in equation 3. This results in the equation

$$0 = f(x_L) + \frac{f(x_R) - f(x_L)}{x_R - x_L}(c - x_L) \qquad (4)$$

Now we must solve equation 4 for c.

$$-f(x_L) = \frac{f(x_R) - f(x_L)}{x_R - x_L}(c - x_L)$$

$$-f(x_L) \cdot (x_R - x_L) = [f(x_R) - f(x_L)](c - x_L)$$

$$-f(x_L) \cdot (x_R - x_L) = [f(x_R) - f(x_L)] \cdot c - [f(x_R) - f(x_L)] \cdot x_L$$

$$x_L \cdot [f(x_R) - f(x_L)] - f(x_L) \cdot (x_R - x_L) = [f(x_R) - f(x_L)] \cdot c$$

$$\frac{x_L \cdot [f(x_R) - f(x_L)] - f(x_L) \cdot (x_R - x_L)}{f(x_R) - f(x_L)} = c$$

So

$$c = \frac{x_L \cdot f(x_R) - x_L \cdot f(x_L) - x_R \cdot f(x_L) + x_L \cdot f(x_L)}{f(x_R) - f(x_L)}$$

or

$$c = \frac{x_L \cdot f(x_R) - x_R \cdot f(x_L)}{f(x_R) - f(x_L)} \qquad (5)$$

Now that we have a formula for c, we can describe the method of false position. We begin, as in the method of bisection, by finding the domain of the function and using the location principle to find an interval which contains a root. We determine the degree of precision required, δ. Using x_L to represent the left boundary of the interval and x_R for the right boundary, we begin the iterations.

FIGURE 10.10

Flowchart for the method of false position.

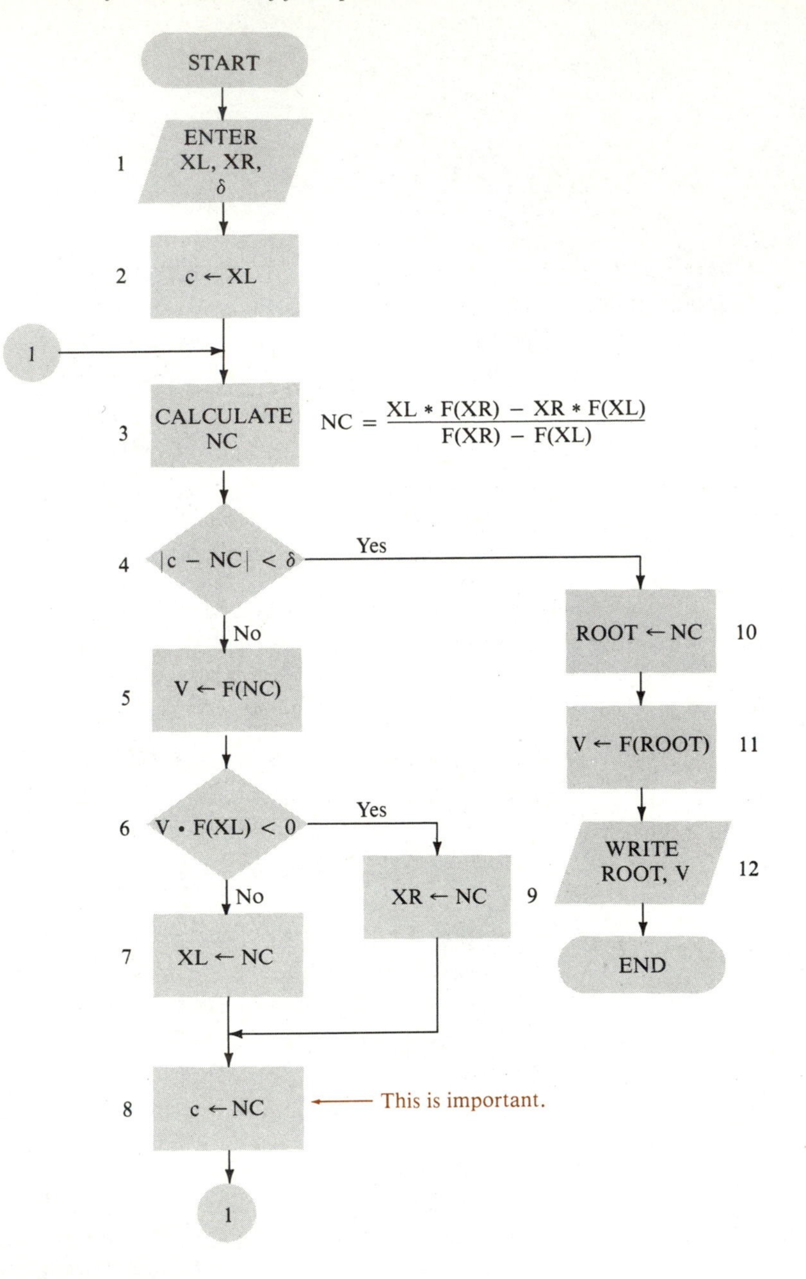

THE METHOD OF FALSE POSITION

1. Set the initial value of c to x_L. This is done simply because c has to start somewhere, so that we can compare a new value of c with the old one to determine whether or not we're done.
2. Find NC (for *new C*) by using equation 5:

$$NC = \frac{x_L \cdot f(x_R) - x_R \cdot f(x_L)}{f(x_R) - f(x_L)}$$

3. If the difference between c and NC is small enough ($|c - NC| < \delta$), then we're done. We can print out the approximation, which is the value of NC. If the difference is not small enough, we go on to step 4.
4. Compare the values $f(NC)$ and $f(x_L)$. If they have the same sign, NC becomes the new *left* endpoint of the interval. Otherwise, NC becomes the new *right* endpoint.
5. Replace the old value of c by NC and return to step 2.

Note that in this method, the successive approximations are c and NC, corresponding to r_i and r_{i+1} in the method of bisection. This algorithm is expressed by the flowchart in Figure 10.10.

Before we illustrate this method with an example, we review a fundamental point. The bisection method guarantees that the width of each successive interval is known (each interval is half the length of the preceding one). In determining when to terminate the process, we use the length of the interval as the *criterion* and use the last value of x_M as the *root*.

In the method of false position, c is calculated by a formula that does not always require c to be in an easily determined position on the interval. The length of each new interval that brackets the root is *not* in some fixed ratio to the length of the previous interval. We shall therefore end our procedure by using a more general criterion: When two successive approximations, c and NC, are "very close," the process is stopped and the last value for c is used as the root.

EXAMPLE 1

Given the equation $t^2 - 10t + 20 = 0$, use the method of false position to approximate the root between 2 and 3 to one decimal place.

Solution

Since we wish accuracy to one decimal place and we are comparing successive roots, the degree of precision is given by $\delta = 0.05$ according to

Table 10.1. We let

$$f(t) = t^2 - 10t + 20$$

and use the flowchart. Since the interval is (2, 3), we have

1. $x_L = 2$ and $x_R = 3$. Also, $\delta = 0.05$
   By substitution, $f(2) = 4$ and $f(3) = -1$.
   Initially, we let
2. $c = x_L = 2$
   Next, we calculate NC:

$$NC = \frac{x_L \cdot f(x_R) - x_R \cdot f(x_L)}{f(x_R) - f(x_L)}$$

3. $NC = \dfrac{2(-1) - 3(4)}{-1 - 4} = \dfrac{-2 - 12}{-5} = \dfrac{-14}{-5} = 2.8$
4. $|c - NC| = |2 - 2.8| = 0.8 > 0.05$; so we must continue.
5. $f(NC) = f(2.8) = -0.16$
   We let $V = -0.16$.
6. $V \cdot f(x_L) = -0.16 \cdot 4 = -0.64 < 0$; so step 9 is next.
9. x_R becomes NC, so $x_R = 2.8$.
   The interval is now (2, 2.8) as shown in Figure 10.11; so $x_L = 2$, $x_R = 2.8$.
8. c becomes 2.8, and we're ready to calculate NC again. Back to step 3.

(* End of the first iteration *)

FIGURE 10.11

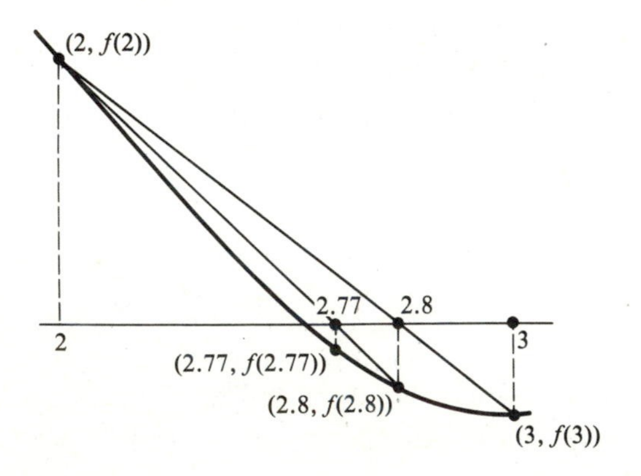

3. $NC = \dfrac{2(-0.16) - 2.8(4)}{-0.16 - 4} = \dfrac{-11.52}{-4.16} \approx 2.77$

4. $|c - NC| = |2.8 - 2.77| = 0.03 < 0.05$
   We're done. We proceed to step 10.

10. root = 2.77 (The current value of NC)
11. $V = f(\text{root}) = 2.77^2 - 10(2.77) + 20 = -0.0271$
12. PRINT root = 2.77, correct to one decimal place, and the function value is -0.0271.

If we continued this process, we would obtain a root of 2.76 after five iterations.

EXAMPLE 2

Given the function $f(x) = x^3 + x^2 - 3x - 3$, use the method of false position to find a root in the interval (1, 2), accurate to two decimal places.

Solution

Initially, let

$x_L = 1$ $\quad f(x_L) = -4$

$x_R = 2$ $\quad f(x_R) = 3$

$c = 1$

$\delta = 0.005$ $\quad$ For accuracy to two decimal places.

Using the flowchart,

3. $NC = \dfrac{1(3) - 2(-4)}{3 - (-4)} = \dfrac{11}{7} = 1.57$
4. $|c - NC| = |1 - 1.57| = 0.57 > \delta$; so we continue.
5. $V = f(NC) = -1.375$
6. $V \cdot f(x_L) = -1.375\ (-4) > 0$, so we go to step 7 next.
7. x_L becomes NC, which is 1.57; x_R is still 2; so the interval is (1.57, 2)
8. c becomes 1.57, and we go back to step 3.

(* End of iteration 1 *)

3. $NC = \dfrac{1.57(3) - 2(-1.375)}{3 - (-1.375)} = 1.705$
4. $|c - NC| = |1.57 - 1.705| = 0.135 > \delta$; so we continue.
5. $V = f(NC) = -0.251$

6. $V \cdot f(x_L) = -0.251(-1.375) > 0$; so we go to step 7.

7. x_L becomes NC, which is 1.705; the new interval is (1.705, 2).

9. c becomes 1.705, and we go back to step 3.

(* End of iteration 2 *)

If we continue this process, we will have the following values at the end of each iteration:

Iteration	x_L	x_R	NC	$\lvert c - NC \rvert$	
1	1.57	2	1.57	0.57	
2	1.705	2	1.705	0.135	
3	1.728	2	1.728	0.023	
4	1.731	2	1.731	0.003	This is $<\delta$, so we're done.

The root is the final value of *NC*; so the root is approximately 1.73 to two decimal places.

Although the flowchart does not indicate it, a computer program for the method of false position would use a loop with a maximum number of iterations, to prevent the possibility of an infinite loop. With an iterative technique such as the method of false position, there is no absolute guarantee that two successive approximations to the root will ever be within the precision or tolerance given. Thus the computer could continue calculating, possibly for a very long time. One way to handle this potential problem is to set a maximum number of iterations before the process begins. For those who program in BASIC a possibility might be

```
INPUT "ENTER MAX. NO. OF ITERATIONS"; M
FOR I = 1 TO M
   ⋮
NEXT I
```

In the loop would be the program corresponding to our flowchart.

One other point needs to be mentioned. Earlier, in Section 10.1, we mentioned relative error. In these examples we have been using an *absolute*, not a relative, error criterion to terminate the iteration process. However, you might wish to try the relative error criterion. Recall that the relative error is given by

$$\text{relative error} = \frac{\text{actual error}}{\text{real answer}} = \left| \frac{\text{real root} - \text{calculated root}}{\text{real (true) root}} \right|$$

Since we don't know the true root, we use

$$\left|\frac{r_i - r_{i+1}}{r_{i+1}}\right| < \delta$$

(Recall that r_{i+1} represents the next approximation *after* r_i.) This can also be written as

$$\left|\frac{r_i}{r_{i+1}} - 1\right| < \delta \qquad \text{or} \qquad \left|1 - \frac{r_i}{r_{i+1}}\right| < \delta$$

In the flowchart (box 4) we would have

$$\left|1 - \frac{c}{NC}\right| < \delta$$

EXERCISES FOR SECTION 10.4

In these exercises, solve each equation for a root in the indicated interval using the method of false position. Repeat iterations until the absolute value of the difference between two successive approximations is less than the required degree of precision, δ.

1. $x^3 - 40x^2 - 196875 = 0$ in the interval (72,78); $\delta = 0.00005$
2. $x^3 - 2x^2 + 10x - 3 = 0$ in the interval (0, 1); $\delta = 0.000005$
3. $x^5 - 2x^3 + x + 3 = 0$ in the interval $(-2, -1)$; $\delta = 0.000005$
4. For the trigonometry fans, $-2 \sin x - x + 1 = 0$ in the interval (0, 1); $\delta = 0.00005$
5. $x^3 - 3x^2 - x - 17 = 0$ in the interval (4, 5); $\delta = 0.0005$
6. Find a positive real fourth root of 17; $\delta = 0.0005$.
7. Find a positive square root of 12; $\delta = 0.0005$.
8. Solve the equation $x^3 = e^x$, where $e \approx 2.71828$ and $\delta = 0.00005$. Use the location principle to find a suitable interval.
9. A projectile thrown straight up from ground level with an initial velocity of 192 feet per second attains a height at time t seconds given by the formula

$$h(t) = -16t^2 + 192t$$

   Find the time(s) when the height will be 100 feet (neglect air resistance).

10. An object is thrown vertically upward with an initial velocity of 200 feet per second. Its height after t seconds is given by the formula

$$h(t) = -16t^2 + 200t$$

Find the time(s) when it will be at a height of 200 feet.

In Exercises 11 through 13, (a) state the number of roots each equation has; (b) use the location principle to find an interval containing each root; and (c) use the method of false position to approximate each root to three decimal places.

11. $2x^3 = 5x^2 + x - 6$ (Hint: All roots are between -2 and 4 inclusive.)
12. $2x^3 - 7x^2 - 5x - 1 = 0$ (Hint: All roots are between -2 and 5 inclusive.)
13. $x^4 - 6x^3 - 8x^2 + 23x - 6 = 0$ (Hint: All roots are between -3 and 8 inclusive.)

10.5 The Method of Successive Approximations

In this section we discuss one more iterative technique for finding roots of equations, the **method of successive approximations**. Recall that any iterative technique produces a sequence or succession of values. Our aim in using such a technique is to produce a sequence of approximations that "closes in on"—*converges to*—a solution or root of the equation.

The concept of **convergence** is somewhat difficult to pin down. Suppose x_1, $x_2, x_3, \ldots, x_i, x_{i+1}, \ldots, x_N, \ldots$ is a sequence of approximations produced by some iterative process. This sequence *converges* if successive approximations get closer and closer together; that is, if $|x_i - x_{i+1}|$ gets smaller as the value of i gets larger.

Consider, for example, the sequence of approximations of roots of the equation $x^3 - 3x + x^2 = 3$, produced in Example 2 of Section 10.3. For clarity, we'll list them in a table.

Iteration	Approximation	Difference
1	$x_1 = 1.5$	—
2	$x_2 = 1.75$	$\|x_1 - x_2\| = 0.25$
3	$x_3 = 1.625$	$\|x_2 - x_3\| = 0.125$
4	$x_4 = 1.6875$	$\|x_3 - x_4\| = 0.0625$
5	$x_5 = 1.71875$	$\|x_4 - x_5\| = 0.03125$
6	$x_6 = 1.734375$	$\|x_5 - x_6\| = 0.015625$
⋮	⋮	⋮

Since the values in the difference column become progressively smaller, the sequence of approximations $x_1, x_2, x_3, \ldots$ will eventually *converge* to a root of the equation. This means that the larger the number of iterations, the closer the approximation is to the *actual* root.

The method of successive approximations relies heavily on this concept of convergence. Unlike the other methods we have studied, it does not allow us to *assume* convergence. A rigorous approach to the task of *proving* convergence is beyond the scope of this text. Our approach here is essentially intuitive; if common sense tells us that a sequence is converging, we'll make that assumption.

As with the other iterative methods, we begin the method of successive approximations by making an educated guess at a root of the equation. This can usually be done by sketching the graph of the equation and/or by using the location principle to find an interval that contains a root. We choose some point in that interval, which we call x_0, as our first guess. We must also solve the original equation for x; that is, express it in the form $x = f(x)$. Our sequence of guesses (approximations) will then be

$$\begin{aligned} x_1 &= f(x_0) \\ x_2 &= f(x_1) \\ x_3 &= f(x_2) \\ x_4 &= f(x_3) \\ &\vdots \\ x_i &= f(x_{i-1}) \\ x_{i+1} &= f(x_i) \\ &\vdots \end{aligned}$$

Each new approximation is obtained by substituting the previous estimate for x in the right-hand side of the equation $x = f(x)$; thus $x_i = f(x_{i-1})$ for $i = 1, 2, \ldots$

This method is best explained by applying it to a specific example.

EXAMPLE 1

Use the method of successive approximations to find a root of the equation

$$x^2 - 10x + 20 = 0$$

in the interval (2, 3).

Solution

The location principle assures us that this equation has a root between 2 and 3 (see Example 1 in Section 10.2). We choose $x_0 = 2.5$, the midpoint of the interval, as a first guess.

Before proceeding any further, we must solve the equation for x.

$$x^2 - 10x + 20 = 0$$

$$x^2 + 20 = 10x \qquad \text{Add } 10x \text{ to both sides.}$$

$$(x^2 + 20)/10 = x \qquad \text{Divide both sides by 10.}$$

or

$$x = \tfrac{1}{10}(x^2 + 20) \qquad \text{The "working equation."}$$

We can now begin the iterative process:

$$x_0 = 2.5$$

To find x_1, we substitute x_0 (= 2.5) for x in the right-hand side of the working equation:

$$x_1 = \tfrac{1}{10}({x_0}^2 + 20) = \tfrac{1}{10}(2.5^2 + 20)$$
$$x_1 = 2.625$$

The next approximation, x_2, is found by substituting x_1 (= 2.625) for x in the same equation.

$$x_2 = \tfrac{1}{10}({x_1}^2 + 20) = \tfrac{1}{10}(2.625^2 + 20)$$
$$x_2 = 2.689$$

We continue as follows:

$$x_3 = \tfrac{1}{10}({x_2}^2 + 20) = \tfrac{1}{10}(2.689^2 + 20)$$
$$x_3 = 2.723$$

$$x_4 = \tfrac{1}{10}({x_3}^2 + 20) = \tfrac{1}{10}(2.723^2 + 20)$$
$$x_4 = 2.74$$

$$x_5 = \tfrac{1}{10}({x_4}^2 + 20) = \tfrac{1}{10}(2.74^2 + 20)$$
$$x_5 = 2.751$$

$$x_6 = \tfrac{1}{10}({x_5}^2 + 20) = \tfrac{1}{10}(2.75^2 + 20)$$
$$x_6 = 2.7568$$

$$x_7 = \tfrac{1}{10}({x_6}^2 + 20) = \tfrac{1}{10}(2.7568^2 + 20)$$
$$x_7 = 2.7599$$

$$x_8 = \tfrac{1}{10}({x_7}^2 + 20) = \tfrac{1}{10}(2.7599^2 + 20)$$
$$x_8 \approx 2.76$$

$$x_9 = \tfrac{1}{10}({x_8}^2 + 20) = \tfrac{1}{10}(2.76^2 + 20)$$
$$x_9 \approx 2.76$$

We have calculated x_8 and x_9 to two decimal places and found that they are approximately equal. So it seems that the root is 2.76. We can test it by substituting in the original equation.

$$x^2 - 10x + 20 = 0$$
$$(2.76)^2 - 10(2.76) + 20 = 0.0176 \approx 0$$

For our purposes, this is close enough. In fact, in Section 10.1, we used the quadratic formula to obtain 2.76 as a root of the given equation (see Example 2).

One possible problem with the method of successive approximations is that there may be more than one way to solve the given equation for x. Consider again

$$x^2 - 10x + 20 = 0$$

Suppose we had solved for x as follows:

$$x^2 = 10x - 20 \qquad \text{Add } 10x - 20 \text{ to both sides.}$$
$$x = \pm\sqrt{10x - 20}$$

Let us consider

$$x = +\sqrt{10x - 20}$$

Our approximations would be

$$x_1 = \sqrt{10x_0 - 20} = \sqrt{10(2.5) - 20} = \sqrt{5} \approx 2.24$$
$$x_2 = \sqrt{10(x_1) - 20} = \sqrt{10(2.24) - 20} = \sqrt{2.4} \approx 1.55$$
$$x_3 = \sqrt{10x_2 - 20} = \sqrt{10(1.55) - 20} = \sqrt{-4.5} \qquad \text{The square root of a negative number.}$$

So x_3 is not a real number. Clearly, this approach does *not* converge to a root.

It is possible that this problem arose because our initial guess was "bad." Suppose we try a different one, say $x_0 = 2.7$. We get the following sequence:

$$x_0 = 2.7$$
$$x_1 = 2.6457 \qquad \sqrt{10(2.7) - 20}$$
$$x_2 = 2.5411 \qquad \sqrt{10(2.6457) - 20}$$
$$x_3 = 2.326 \qquad \sqrt{10(2.5411) - 20}$$
$$x_4 = 1.806 \qquad \sqrt{10(2.326) - 20}$$
$$\vdots$$

This sequence is clearly moving **away from** the root, 2.76. (The next iteration will also get us in trouble—try it!) Notice too that the differences between successive approximations are getting larger instead of smaller.

These considerations lead to the conclusion that we may have to try more than one way of solving an equation for x before we hit on the "right" one. However, if we have a form of the equation that *does* lead to a convergent sequence, we don't have to be too concerned about our choice for x_0. In general, it can be shown that iterative methods are self-correcting; errors in choosing an initial value are not always serious. Iterative methods seem to learn from their past mistakes! An example will illustrate this.

EXAMPLE 2

Find a root of $x^2 - 10x + 20 = 0$, using $x_0 = 4$ as the initial value.

Solution

As in Example 1, we use

$$x = \tfrac{1}{10}(x^2 + 20)$$

We obtain the following sequence:

$$
\begin{aligned}
x_0 &= 4 \\
x_1 &= 3.6 && \tfrac{1}{10}(4^2 + 20) \\
x_2 &= 3.296 && \tfrac{1}{10}(3.6^2 + 20) \\
x_3 &= 3.086 && \tfrac{1}{10}(3.296^2 + 20) \\
x_4 &= 2.95 && \tfrac{1}{10}(3.086^2 + 20) \\
x_5 &= 2.87 && \tfrac{1}{10}(2.95^2 + 20) \\
x_6 &= 2.82 && \tfrac{1}{10}(2.87^2 + 20) \\
x_7 &= 2.79 && \tfrac{1}{10}(2.82^2 + 20) \\
x_8 &= 2.78 && \tfrac{1}{10}(2.79^2 + 20) \\
x_9 &= 2.77 && \tfrac{1}{10}(2.78^2 + 20) \\
x_{10} &= 2.77 && \tfrac{1}{10}(2.77^2 + 20)
\end{aligned}
$$

After *ten* iterations, we converged to a good approximation of the root, even starting with a bad guess. However, this example is *not* meant as a license to abandon reasonable analysis in finding a good initial guess. A good choice for x_0 usually speeds up the process of finding a root.

The flowchart of Figure 10.12 illustrates the successive approximation method.

FIGURE 10.12
The successive approximations method.

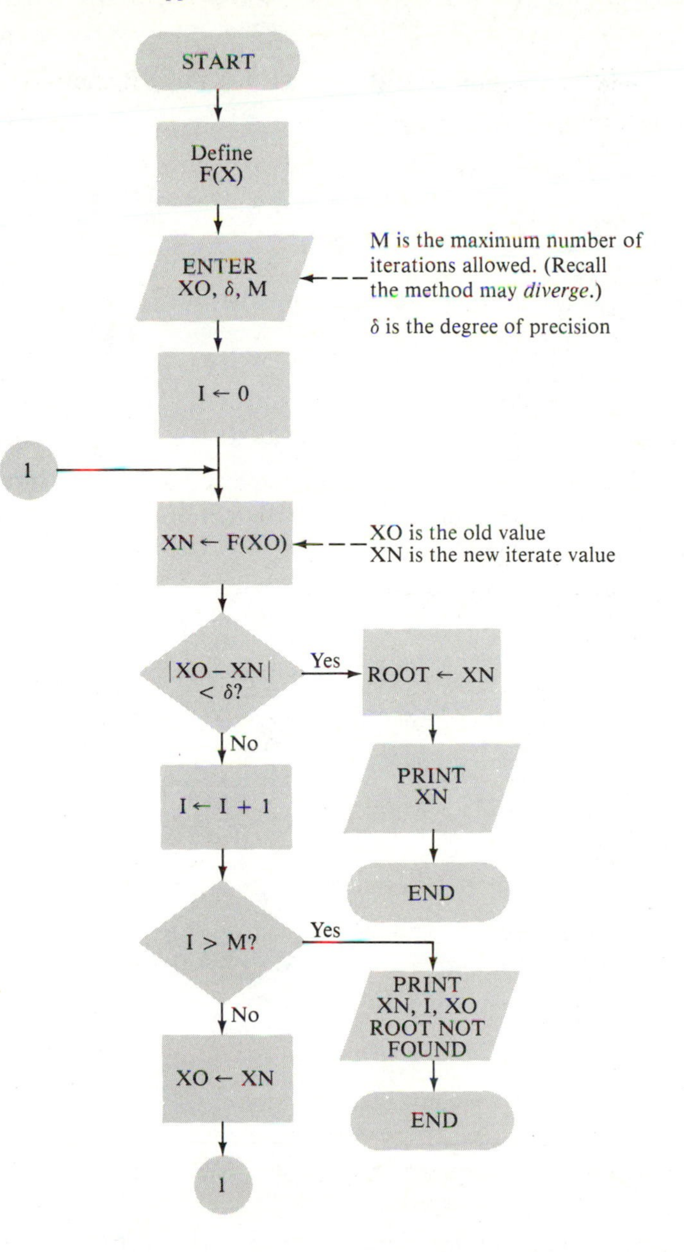

EXAMPLE 3

Solve $x^3 + x^2 - 3x - 3 = 0$ for a positive root between 1 and 2.

Solution

In Example 2 of Section 10.3 we verified that this equation has a root between 1 and 2. First, we solve for x.

$$x^3 + x^2 - 3x - 3 = 0$$
$$x^2 = 3x + 3 - x^3$$
$$x = \sqrt{3x + 3 - x^3}$$

We use $x_0 = 1.5$, the midpoint of the interval.

$$x_1 = 2.03 \qquad \sqrt{3(1.5) + 3 - (1.5)^3}$$
$$x_2 = 0.85 \qquad \sqrt{3(2.03) + 3 - (2.03)^3}$$
$$x_3 = 2.22 \qquad \sqrt{3(0.85) + 3 - (0.85)^3}$$

This sequence appears to be "jumping around," so we abandon this method. It does not seem to converge. (In fact, the next iteration produces the square root of a negative number.)

We try again, this time using a different form of the equation:

$$x^3 + x^2 - 3x - 3 = 0$$
$$x^3 = 3x + 3 - x^2$$
$$x = \sqrt[3]{3x + 3 - x^2}$$

$$x_0 = 1.5$$
$$x_1 = 1.738 \qquad \sqrt[3]{3(1.5) + 3 - (1.5)^2}$$
$$x_2 = 1.732 \qquad \sqrt[3]{3(1.738) + 3 - (1.738)^2}$$
$$x_3 = 1.732 \qquad \sqrt[3]{3(1.732) + 3 - (1.732)^2}$$

Since x_2 and x_3 are approximately equal, we stop here. The root is 1.732. If we substitute 1.732 into $x^3 + x^2 - 3x - 3 = 0$, we obtain approximately -0.00048, which is certainly close to zero.

EXERCISES FOR SECTION 10.5

In these exercises, use the method of successive approximations to solve each equation for a root in the indicated interval to the given *degree of precision* δ. (That is, repeat iterations until the absolute value of the difference between two successive approximations is less than δ.)

1. $x^3 - 40x^2 - 196875 = 0$ in the interval (72, 78); $\delta = 0.00005$
2. $x^3 - 2x^2 + 10x - 3 = 0$ in the interval (0, 1); $\delta = 0.000005$
3. $x^5 - 2x^3 + x + 3 = 0$ in the interval $(-2, -1)$; $\delta = 0.000005$
4. For the trigonometry fans, $-2 \sin x - x + 1 = 0$ in the interval (0, 1); $\delta = 0.00005$
5. $x^3 - 3x^2 - x - 17 = 0$ in the interval (4, 5); $\delta = 0.0005$
6. Find a positive real fourth root of 17; use $\delta = 0.0005$.
7. Find a positive square root of 12; use $\delta = 0.0005$.
8. Solve the equation $x^3 = e^x$, where $e \approx 2.71828$ and $\delta = 0.00005$. Use the location principle to find a suitable interval.
9. A projectile thrown straight up from ground level with an initial velocity of 192 feet per second attains a height at time t seconds given by the formula

$$h(t) = -16t^2 + 192t$$

Find the time(s) when the height will be 100 feet (neglect air resistance).

10. An object is thrown vertically upward with an initial velocity of 200 feet per second. Its height after t seconds is given by the formula

$$h(t) = -16t^2 + 200t$$

Find the time(s) when it will be at a height of 200 feet.

In Exercises 11 through 13, (a) state the number of roots each equation has; (b) use the location principle to find an interval containing each root; and (c) use the method of successive approximations to approximate each root to three decimal places.

11. $2x^3 = 5x^2 + x - 6$
(Hint: All roots are between -2 and 4 inclusive.)
12. $2x^3 - 7x^2 - 5x - 1 = 0$
(Hint: All roots are between -2 and 5 inclusive.)
13. $x^4 - 6x^3 - 8x^2 + 23x - 6 = 0$
(Hint: All roots are between -3 and 8 inclusive.)

10.6 The Gauss-Seidel Method for Solving Systems of Linear Equations

Thus far in this chapter, we have been describing methods of solving an equation in *one* unknown, such as

$$x^2 - 10x + 20 = 0$$

or

$$x^3 + x^2 - 3x - 3 = 0$$

We have discussed both iterative techniques and formula methods, noting that in most cases there is no general formula for equations above degree 3 or for equations involving trigonometric or other transcendental functions. When a formula for solving an equation exists, an exact solution can, at least in theory, be produced. Iterative methods are less reliable; they produce a sequence of approximations that may or may not converge to a solution.

Chapter 8 dealt with methods of solving systems of linear equations. An example is the system

$$\begin{aligned} 4x + y &= 11 \\ 2x + 5y &= 19 \end{aligned}$$

We used several approaches to solving such a system, including Gaussian elimination or substitution and Cramer's rule. These are formal methods that can produce an exact solution if one exists (recall that there are three possibilities: a unique solution, infinitely many solutions, or no solution). We shall now consider an iterative technique, called the **Gauss-Seidel method**, for solving a special class of systems of linear equations.

The Gauss-Seidel Method for Two Equations in Two Variables

We must first establish the conditions under which the Gauss-Seidel method can be applied. We begin by discussing a general system of two linear equations in two variables:

$$\begin{aligned} a_{11}x_1 + a_{12}x_2 &= c_1 \\ a_{21}x_1 + a_{22}x_2 &= c_2 \end{aligned}$$

This system has a solution that can be obtained by the Gauss-Seidel method if the associated matrix of coefficients (see Section 8.4) meets certain criteria

$$\begin{pmatrix} a_{11} & a_{12} \\ a_{21} & a_{22} \end{pmatrix}$$

Recall that the *major diagonal* of a matrix consists of the entries whose row number and column number are equal; they form a diagonal from upper left to lower right. If a system of equations has a unique solution then it can be solved by the Gauss-Seidel method when the matrix of coefficients has a *heavy major diagonal.* For the general matrix above, this means that *all* of the following conditions must be true:

1. $|a_{11}| \geq |a_{12}|$
2. $|a_{22}| \geq |a_{21}|$
3. *Strict inequality* ($>$, not $\geq$) must hold in *at least one* of the above inequalities.

EXAMPLE 1

Determine whether or not the system

$$4x + y = 11$$
$$2x + 5y = 19$$

can be solved by the Gauss-Seidel method.

Solution

The associated matrix is

$$\begin{pmatrix} 4 & 1 \\ 2 & 5 \end{pmatrix}$$

So

$$a_{11} = 4 \qquad a_{12} = 1$$
$$a_{21} = 2 \qquad a_{22} = 5$$

1. $|a_{11}| = 4$; $|a_{12}| = 1$; and $4 > 1$
2. $|a_{22}| = 5$; $|a_{21}| = 2$; and $5 > 2$
3. Strict inequality holds in both cases.

The given system fills all the necessary conditions, so the Gauss-Seidel method can be used.

This technique can be described most clearly by referring to a specific example. We'll use it to solve the system of Example 1. In doing so, we should see some similarities to the method of successive approximations.

$$4x + y = 11 \qquad (1)$$
$$2x + 5y = 19 \qquad (2)$$

step 1. Solve equation 1 for x and equation 2 for y:

$$x = \tfrac{1}{4}(11 - y) \qquad (1)$$
$$y = \tfrac{1}{5}(19 - 2x) \qquad (2)$$

step 2. Decide on the degree of precision. Suppose we choose two decimal places. As in Section 10.4, we need $\delta = 0.005$. We'll want two successive approximations of x *and* two successive approximations of y to differ by less than δ.

step 3. Choose initial values x_0 and y_0 for the variables. If the equations arose from some practical application, the context might suggest some reasonable guesses. Otherwise, it is common to choose $x_0 = 0$, $y_0 = 0$.

step 4. Begin the iterations. Each new value of x is obtained by substituting the *most recent value of* y (at the start, y_0) for the variable y in equation 1. Each new value of y is obtained by substituting the *most recent value of* x (at the start it will be x_1, if we find x_1 first) for the variable x in equation 2. We get the following sequence.

$$x_0 = 0$$
$$y_0 = 0$$
$$x_1 = \tfrac{1}{4}(11 - y_0) = \tfrac{1}{4}(11 - 0) = \tfrac{11}{4} = 2.75$$
$$y_1 = \tfrac{1}{5}(19 - 2x_1) = \tfrac{1}{5}[19 - 2(2.75)] = 2.7$$

We use x_1 here, since it was calculated above.

$$x_2 = \tfrac{1}{4}(11 - y_1) = \tfrac{1}{4}(11 - 2.7) = 2.075$$
$$y_2 = \tfrac{1}{5}(19 - 2x_2) = \tfrac{1}{5}[19 - 2(2.075)] = 2.97$$
$$x_3 = \tfrac{1}{4}(11 - y_2) = \tfrac{1}{4}(11 - 2.97) = 2.0075$$
$$y_3 = \tfrac{1}{5}(19 - 2x_3) = \tfrac{1}{5}[19 - 2(2.0075)] = 2.997$$
$$x_4 = \tfrac{1}{4}(11 - y_3) = \tfrac{1}{4}(11 - 2.997) = 2.00075$$
$$y_4 = \tfrac{1}{5}(19 - 2x_4) = \tfrac{1}{5}[19 - 2(2.00075)] = 2.9997$$
$$x_5 = \tfrac{1}{4}(11 - y_4) = \tfrac{1}{4}(11 - 2.9997) = 2.000075$$
$$y_5 = \tfrac{1}{5}(19 - 2x_5) = \tfrac{1}{5}[19 - 2(2.000075)] = 2.99997$$

At this point, we check to see whether we are done.

$$|x_4 - x_5| = |2.00075 - 2.000075| = 0.000675 < 0.005$$

and

$$|y_4 - y_5| = |2.997 - 2.99997| = 0.00027 < 0.005$$

Therefore, the process is complete. Our approximate solution,

$$x = 2.000075$$
$$y = 2.99997$$

suggests that the *actual* solution is

$$x = 2$$
$$y = 3$$

We can verify this by substituting these values for x and y in the original equations:

$4x + y = 11$	$2x + 5y = 19$
$4(2) + 3 = 11$	$2(2) + 5(3) = 19$
$8 + 3 = 11$	$4 + 15 = 19$
$11 = 11$ ✓	$19 = 19$ ✓

The Gauss-Seidel Method for Three Equations in Three Variables

Consider now a system of three linear equations in three variables.

$$a_{11}x_1 + a_{12}x_2 + a_{13}x_3 = c_1$$
$$a_{21}x_1 + a_{22}x_2 + a_{23}x_3 = c_2$$
$$a_{31}x_1 + a_{32}x_2 + a_{33}x_3 = c_3$$

We shall describe the Gauss-Seidel method for such a system; it is simply a generalization of the technique used for a system of two equations.

The conditions under which such a system can be solved by the Gauss-Seidel method are

1. $|a_{11}| \geq |a_{12}| + |a_{13}|$
2. $|a_{22}| \geq |a_{21}| + |a_{23}|$
3. $|a_{33}| \geq |a_{31}| + |a_{32}|$
4. *Strict inequality* must hold in *at least one* of the above cases.

All four conditions must hold in order for the Gauss-Seidel method to yield a solution, assuming that a unique solution exists.

EXAMPLE 2

If possible, use the Gauss-Seidel method to solve the following system:

$$10x + y + 2z = 11 \qquad (1)$$
$$2x + 5y - z = -13 \qquad (2)$$
$$x + y + 3z = 10.5 \qquad (3)$$

Obtain an approximation to two decimal places (that is, $\delta = 0.005$).

Solution

The matrix of coefficients is

$$\begin{pmatrix} 10 & 1 & 2 \\ 2 & 5 & -1 \\ 1 & 1 & 3 \end{pmatrix}$$

So

$$a_{11} = 10, \quad a_{12} = 1, \quad a_{13} = 2$$
$$a_{21} = 2, \quad a_{22} = 5, \quad a_{23} = -1$$
$$a_{31} = 1, \quad a_{32} = 1, \quad a_{33} = 3$$

1. $|a_{11}| = 10$; $|a_{12}| + |a_{13}| = 1 + 2 = 3$ and $10 > 3$
2. $|a_{22}| = 5$; $|a_{21}| + |a_{23}| = 2 + 1 = 3$ and $5 > 3$
3. $|a_{33}| = 3$; $|a_{31}| + |a_{32}| = 1 + 1 = 2$ and $3 > 2$
4. *Strict inequality* holds in *all three* cases.

Therefore, the Gauss-Seidel method can be used.

step 1. Solve equation 1 for x, equation 2 for y, and equation 3 for z.

$$x = \tfrac{1}{10}(11 - y - 2z) \qquad (1)$$
$$y = \tfrac{1}{5}(-13 - 2x + z) \qquad (2)$$
$$z = \tfrac{1}{3}(10.5 - x - y) \qquad (3)$$

step 2. $\delta = 0.005$

step 3. Choose the initial values

$$x_0 = 0$$
$$y_0 = 0$$
$$z_0 = 0$$

step 4. Begin the iterations. Note that to find the new value of a variable we use the *most recently calculated* values of the other two.

$$x_1 = \tfrac{1}{10}(11 - y_0 - 2z_0) = \tfrac{1}{10}(11 - 0 - 0) = \tfrac{11}{10} = 1.1$$
$$y_1 = \tfrac{1}{5}(-13 - 2x_1 + z_0) = \tfrac{1}{5}[-13 - 2(1.1) + 0] = -3.04$$

We use x_1 here since we just calculated it.

$$z_1 = \tfrac{1}{3}(10.5 - x_1 - y_1) = \tfrac{1}{3}[10.5 - 1.1 - (-3.04)] = 4.146 \approx 4.15$$

$$x_2 = \tfrac{1}{10}(11 - y_1 - 2z_1) = \tfrac{1}{10}[11 - (-3.04) - 2(4.15)] \approx 0.574$$
$$y_2 = \tfrac{1}{5}(-13 - 2x_2 + z_1) = \tfrac{1}{5}[-13 - 2(5.74) + 4.15] \approx -1.999$$
$$z_2 = \tfrac{1}{3}(10.5 - x_2 - y_2) = \tfrac{1}{3}[10.5 - 0.574 - (-1.999)] \approx 3.98$$

$$x_3 = \tfrac{1}{10}(11 - y_2 - 2z_2) = \tfrac{1}{10}[11 - (-1.999) - 2(3.98)] \approx 0.505$$
$$y_3 = \tfrac{1}{5}(-13 - 2x_3 + z_2) = \tfrac{1}{5}[-13 - 2(0.505) + 3.98] \approx -2$$
$$z_3 = \tfrac{1}{3}(10.5 - x_3 - y_3) = \tfrac{1}{3}[10.5 - 0.505 - (-2)] \approx 4$$

$$x_4 = \tfrac{1}{10}(11 - y_3 - 2z_3) = \tfrac{1}{10}[11 - (-2) - 2(4)] = 0.5$$
$$y_4 = \tfrac{1}{5}(-13 - 2x_4 + z_3) = \tfrac{1}{5}[-13 - 2(0.5) + 4] = -2$$
$$z_4 = \tfrac{1}{3}(10.5 - x_4 - y_4) = \tfrac{1}{3}[10.5 - 0.5 - (-2)] = 4$$

At this point, let's check to see whether we're done.

$$|x_3 - x_4| = |0.505 - 0.5| = 0.005 \nless 0.005$$
$$|y_3 - y_4| = |-2 - (-2)| = 0 < 0.005$$
$$|z_3 - z_4| = |4 - 4| = 0 < 0.005$$

Since $|x_3 - x_4| \nless \delta$, we must continue.

$$x_5 = \tfrac{1}{10}(11 - y_4 - 2z_4) = \tfrac{1}{10}[11 - (-2) - 2(4)] = 0.5$$
$$y_5 = \tfrac{1}{5}(-13 - 2x_5 + z_4) = \tfrac{1}{5}[-13 - 2(0.5) + 4] = -2$$
$$z_5 = \tfrac{1}{3}(10.5 - x_5 - y_5) = \tfrac{1}{3}[10.5 - 0.5 - (-2)] = 4$$

This time,

$$|x_4 - x_5| = |0.5 - 0.5| = 0 < 0.005$$
$$|y_4 - y_5| = |-2 - (-2)| = 0 < 0.005$$
$$|z_4 - z_5| = |4 - 4| = 0 < 0.005$$

We are done; the solution is the last set of x, y, z values:

$$x = 0.5$$
$$y = -2$$
$$z = 4$$

Check

If we substitute these values in the original equations, we find that they are correct.

$$10x + y + 2z = 11$$
$$10(0.5) + (-2) + 2(4) = 11$$
$$5 - 2 + 8 = 11$$
$$11 = 11 \checkmark$$

$$2x + 5y - z = -13$$
$$2(0.5) + 5(-2) - 4 = -13$$
$$1 - 10 - 4 = -13$$
$$-13 = -13 \checkmark$$

$$x + y + 3z = 10.5$$
$$0.5 + (-2) + 3(4) = 10.5$$
$$0.5 - 2 + 12 = 10.5$$
$$10.5 = 10.5 \checkmark$$

EXAMPLE 3

Solve the system of Example 2 using as initial values

$$x_0 = 1$$
$$y_0 = 1$$
$$z_0 = 1$$

Solution

The system is

$$\begin{aligned} 10x + y + 2z &= 11 && \text{or} && x = \tfrac{1}{10}(11 - y - 2z) && (1) \\ 2x + 5y - z &= -13 && \text{or} && y = \tfrac{1}{5}(-13 - 2x + z) && (2) \\ x + y + 3z &= 10.5 && \text{or} && z = \tfrac{1}{3}(10.5 - x - y) && (3) \end{aligned}$$

We know that the Gauss-Seidel method can be used, so we begin the iterations. We want to see what effect, if any, starting with different initial values will have.

$$x_0 = 1, \qquad y_0 = 1, \qquad z_0 = 1$$

$$x_1 = \tfrac{1}{10}[11 - 1 - 2(1)] = 0.8$$
$$y_1 = \tfrac{1}{5}[-13 - 2(0.8) + 1] = -2.72$$
$$z_1 = \tfrac{1}{3}[10.5 - 0.8 - (-2.72)] = 4.14$$

$$x_2 = \tfrac{1}{10}[11 - (-2.72) - 2(4.14)] = 0.544$$
$$y_2 = \tfrac{1}{5}[-13 - 2(0.544) + 4.14] \approx -1.99$$
$$z_2 = \tfrac{1}{3}[10.5 - 0.544 - (-1.99)] \approx 3.98$$

$$x_3 = \tfrac{1}{10}[11 - (-1.99) - 2(3.98)] \approx 0.502$$
$$y_3 = \tfrac{1}{5}[-13 - 2(0.502) + 3.98] \approx -2$$
$$z_3 = \tfrac{1}{3}[10.5 - 0.502 - (-2)] \approx 4$$

$$x_4 = \tfrac{1}{10}[11 - (-2) - 2(4)] = 0.5$$
$$y_4 = \tfrac{1}{5}[-13 - 2(0.5) + 4] = -2$$
$$z_4 = \tfrac{1}{3}[10.5 - 0.5 - (-2)] = 4$$

We're done since

$$|x_3 - x_4| = |0.502 - 0.5| = 0.002 < 0.005$$
$$|y_3 - y_4| = |-2 - (-2)| = 0 < 0.005$$
$$|z_3 - z_4| = |4 - 4| = 0 < 0.005$$

Notice that we *did* arrive at correct solutions even though we started with a different set of initial values. This is a further illustration of the observation in Section 10.5 that iterative techniques are generally self-correcting.

EXAMPLE 4

Use the Gauss-Seidel method, with $\delta = 0.005$, to solve the following system of equations:

$$2x + 5y + z = 5 \quad (1)$$
$$-x + 2z = 1.5 \quad (2)$$
$$3x + y + z = 0.7 \quad (3)$$

Solution

The matrix of coefficients is

$$\begin{pmatrix} 2 & 5 & 1 \\ -1 & 0 & 2 \\ 3 & 1 & 1 \end{pmatrix}$$

$$\begin{array}{lll} a_{11} = 2, & a_{12} = 5, & a_{13} = 1 \\ a_{21} = -1, & a_{22} = 0, & a_{23} = 2 \\ a_{31} = 3, & a_{32} = 1, & a_{33} = 1 \end{array}$$

1. $|a_{11}| = 2$; $|a_{12}| + |a_{13}| = 5 + 1 = 6$; so $|a_{11}| < |a_{12}| + |a_{13}|$.

This seems to indicate that the Gauss-Seidel method *cannot* be applied to this system since condition 1 is not met. However, all is not lost. Rearranging the equations will have no effect on the solution, so we'll try it. Keep in mind that we want a large coefficient for x in equation 1, for y in equation 2, and for z in equation 3 in order to get a "heavy" main diagonal. Let's try this ordering:

$$3x + y + z = 0.7 \quad (1)$$
$$2x + 5y + z = 5 \quad (2)$$
$$-x + 2z = 1.5 \quad (3)$$

Now the matrix of coefficients is

$$\begin{pmatrix} 3 & 1 & 1 \\ 2 & 5 & 1 \\ -1 & 0 & 2 \end{pmatrix}$$

$$\begin{array}{lll} a_{11} = 3, & a_{12} = 1, & a_{13} = 1 \\ a_{21} = 2, & a_{22} = 5, & a_{23} = 1 \\ a_{31} = -1, & a_{32} = 0, & a_{33} = 2 \end{array}$$

1. $|a_{11}| = 3$; $|a_{12}| + |a_{13}| = 1 + 1 = 2$ and $3 > 2$
2. $|a_{22}| = 5$; $|a_{21}| + |a_{23}| = 2 + 1 = 3$ and $5 > 3$
3. $|a_{33}| = 2$; $|a_{31}| + |a_{32}| = 1 + 0 = 1$ and $2 > 1$
4. *Strict inequality* holds in all three cases.

step 1. $x = \frac{1}{3}(0.7 - y - z)$ (1)

$y = \frac{1}{5}(5 - 2x - z)$ (2)

$z = \frac{1}{2}(1.5 + x)$ (3)

step 2. $\delta = 0.005$

step 3. Choose initial values:

$$x_0 = 0$$
$$y_0 = 0$$
$$z_0 = 0$$

step 4. To start off,

$$x_1 = \frac{1}{3}(0.7 - 0 - 0) = 0.233$$
$$y_1 = \frac{1}{5}[5.0 - 2(0.233) - 0] = 0.9066$$
$$z_1 = \frac{1}{2}(1.5 + 0.233) = 0.8666$$

Then

$$x_2 = \frac{1}{3}(0.7 - 0.9066 - 0.8666) = -0.3577$$

and so on.

We show the results of the iterations in table form, accurate to four decimal places.

i	x_i	y_i	z_i
0	0	0	0
1	0.2333	0.9066	0.8666
2	−0.3577	0.9697	0.5711
3	−0.2803	0.9979	0.6098
4	−0.3026	0.9991	0.5907
5	−0.2993	0.9999	0.6004
6	−0.3001	0.9999	0.5999
7	−0.2997	0.999	0.6000
8	−0.3000	0.999	0.5999
9	−0.2999	1	0.6000

The *solution* is

$$x = -0.3$$
$$y = 1$$
$$z = 0.6$$

We have shown in Example 3 more iterations than we really needed in order to provide a better illustration of the concept of convergence. Look, for example, at the x_i values:

$$
\begin{aligned}
x_1 &= 0.2333 \\
x_2 &= -0.3577 \qquad (< -0.3) \\
x_3 &= -0.2803 \qquad (> -0.3) \\
x_4 &= -0.3026 \qquad (< -0.3) \\
x_5 &= -0.2993 \qquad (> -0.3) \\
x_6 &= -0.3001 \qquad (< -0.3) \\
x_7 &= -0.2997 \qquad (> -0.3) \\
x_8 &= -0.3000 \qquad (= -0.3) \\
x_9 &= -0.2999 \qquad (> -0.3)
\end{aligned}
$$

The values from x_2 on alternate between being less than the true value (-0.3) and being greater than the true value. But they do get progressively *closer to* the true value—that is, they *converge* to -0.3.

All the systems of equations we have solved have involved two or three variables, but the Gauss-Seidel method can easily be generalized to solving a system of n equations in n variables, for any (reasonable) value of n.

Before we conclude this section on the Gauss-Seidel method, we should note that this section is about more than just a specialized technique for solving systems of linear equations. Its intent is also to clarify the concept of convergence to a root, even though we did not give a strictly theoretical treatment. A thorough treatment of the study of numerical analysis is beyond the scope of this book; but it is hoped that this intuitive approach will be valuable and instructive. The reader should be warned, however, that a deeper knowledge of theoretical mathematics is necessary in order to understand fully and apply correctly the techniques so far discussed in this section and chapter.

It should also be clear that a computer, or at least a calculator, is a very useful tool in applying numerical methods. Flowcharts were presented for some of the techniques. You may want to use them to write your own programs, both as a time-saving device and as a test of your understanding of the methods.

A Geometric Interpretation of the Gauss-Seidel Method

Consider the system of linear equations

$$
\begin{aligned}
3x + y &= 7 \\
x + 2y &= 9
\end{aligned}
$$

Note that

$$
\begin{aligned}
|a_{11}| &> |a_{12}| \\
|a_{22}| &> |a_{21}|
\end{aligned}
$$

So the Gauss-Seidel method should converge to a solution. Solving as before, we obtain

$$x = \tfrac{1}{3}(7 - y)$$
$$y = \tfrac{1}{2}(9 - x)$$

These equations will be used to implement the Gauss-Seidel method. Before beginning, we graph each line by plotting points:

$3x + y = 7$

x	y
2	1
3	−2

$x + 2y = 9$

x	y
3	3
1	4

The graph is shown in Figure 10.13.

FIGURE 10.13

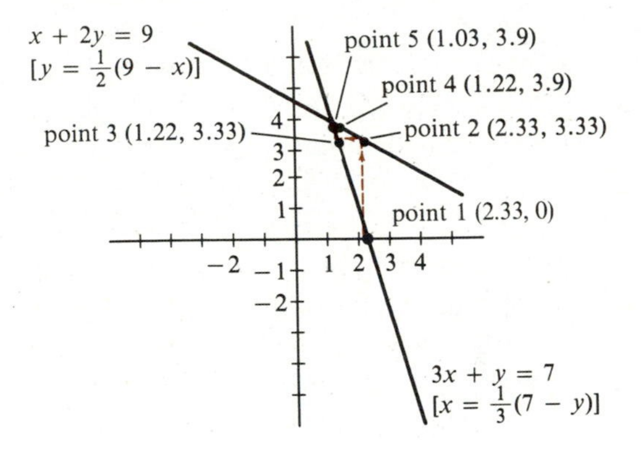

The following table is obtained by applying the Gauss-Seidel method:

I	x_i	y_i
0	0	0
1	2.33	3.33
2	1.22	3.89
3	1.03	3.985
4	1.03	3.985

The solution is

$$x = 1$$
$$y = 4$$

To show the convergence graphically, we begin by substituting $y = 0$ in the equation $x = \frac{1}{3}(7 - y)$ and obtaining $x_1 = 2.33$ (point 1). We take this x value and substitute it in $y = \frac{1}{2}(9 - x)$. This is done graphically by drawing a vertical line from the point $x = 2\frac{1}{3}$ on the x-axis to the line $y = \frac{1}{2}(9 - x)$. The lines meet at point 2. This yields $y_1 = 3.33$. We now substitute $y = 3.33$ in $x = \frac{1}{3}(7 - y)$ by drawing a horizontal line at $y = 3\frac{1}{3}$ across to $x = \frac{1}{3}(7 - y)$ (from point 2 to point 3). This has the effect of graphically substituting $y = 3\frac{1}{3}$ into $x = \frac{1}{3}(7 - y)$, yielding $x_2 = 1.222$ (point 3). Next we wish to substitute this latest value of x into $y = \frac{1}{2}(9 - x)$. We do this by drawing a vertical line at $x = 1.22$ (approximately!) to intersect $y = \frac{1}{2}(9 - x)$. Finding the intersection of this vertical line ($x = 1.22$) and $y = \frac{1}{2}(9 - x)$ has the effect of substituting $x = 1.22$ into $y = \frac{1}{2}(9 - x)$ obtaining $y_2 = 3.9$.

Continuing in this manner yields the sequence of points shown in the graph and in the table below.

x	y	**Point**
2.3	0	point 1
2.33	3.33	point 2
1.22	3.33	point 3
1.22	3.9	point 4
1.03	3.9	point 5

The graph of Figure 10.13 shows this sequence converging toward the point (1, 4).

It is also instructive to view the diverging case. Consider the linear system

$$x + y = 2 \qquad (1)$$
$$3x - y = 2 \qquad (2)$$

which is graphed in Figure 10.14.

FIGURE 10.14

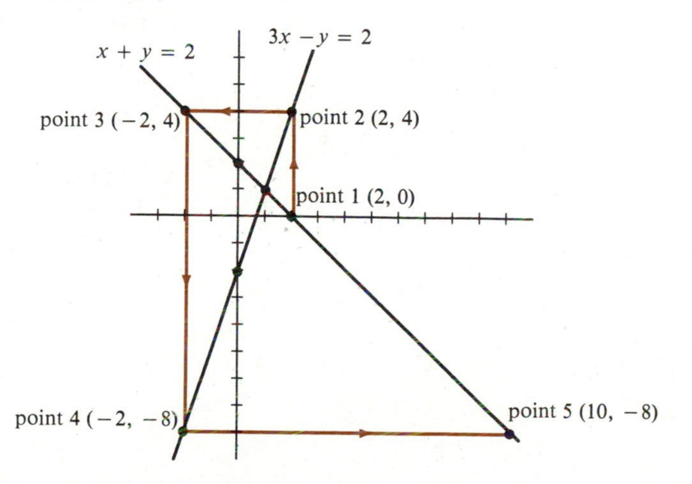

Solving for x and y yields:

$$x = 2 - y \qquad (1)$$
$$y = 3x - 2 \qquad (2)$$

We substitute $y_0 = 0$ in $x = 2 - y$, obtaining $x_1 = 2$. Next, a vertical line at $x = 2$ is drawn to intersect $y = 3x - 2$ at point 2. Here $y_1 = 4$. To substitute $y = 4$ into $x = 2 - y$ graphically we draw a horizontal line to intersect $x = 2 - y$ at point 3. We must now take this latest approximation for x, $x = -2$, and substitute into $y = 3x - 2$. This is done by drawing a vertical line at $x = -2$ to intersect $3x - 2 = y$ at point 4. This process is continued for one more point (point 5). It is clear, however, from the graph and the table below that the sequence of points is *not* converging to a single point.

x	y	**Point**
2	0	point 1
2	4	point 2
−2	4	point 3
−2	−8	point 4
10	−8	point 5

If we look back at the original system of equations, we will see that it did not satisfy the conditions for applying the Gauss-Seidel method.

EXERCISES FOR SECTION 10.6

In Exercises 1 through 7, determine whether the given system of linear equations satisfies the conditions for the Gauss-Seidel method. If possible, solve the system using the Gauss-Seidel method; if not, use a technique from Chapter 8. Solve each system to *two* decimal places. (Hint: In some cases, the order of the equations may have to be changed.)

1. $x - 3y = 7$
$2x + y = 0$

2. $3x - y = 7$
$2x + 5y = -1$

3. $x - 2y = 4$
$-2x + 4y = -8$

4. $x - 2y = 4$
$2x - 4y = 3$

5. $2y - z = 0.67$
$3x + y + 6z = 2.2$
$10x + 2y + 3z = 1$

6. $x + 3y + z = 3$
$x + y + 3z = -1$
$5x + y + 2z = 4$

7. $x - y + 3z = 6$
$2x + y + z = 5.5$
$x + 6y + z = 0.5$

8. Solve the following system (a) by graphing and (b) by the Gauss-Seidel method:

$$5x - y = -7$$
$$x - 3y = 7$$

9. Solve the following system (a) by Gaussian elimination (see Chapter 8) and (b) by the Gauss-Seidel method. Which method is preferable here?

$$2x + y - z = -2$$
$$x - 3y + z = 7$$
$$x + y + 2z = 6$$

10. A part-time craftsman makes three items labelled *A*, *B*, *C*. He made a total of 24 items, for which the total cost of raw materials was \$260. The raw material costs per unit were \$10 for item *A*, \$5 for item *B*, and \$15 for item *C*. The selling prices were \$18 for item *A*, \$9 for item *B*, and \$17 for item *C*, for a total profit of \$108.

    a. Set up a system of linear equations to solve this problem.

    b. Solve the system to find the number of each item made and sold.

11. Use the Gauss-Seidel method, if possible, to solve the following system of equations:

$$4x + y - z + w = 4$$
$$x + 6y + z - w = -6$$
$$2x - y + 5z + w = 16$$
$$x + y + 2z + 3w = 13$$

12. Given the following system of equations

$$4x - y - z = 8.5$$
$$x - 2y + 0.3z = -3.95$$
$$-0.5x + y + 2.5z = -2$$

test to see whether or not the Gauss-Seidel method can be applied here. If it can, find the roots, accurate to two decimal places by calculator or to three decimal places by computer.

10.7 Monte Carlo Techniques

In Chapter 9, we studied random numbers and simulations. In this section, we shall connect two seemingly unrelated topics—random numbers and numerical methods. We may have given the impression that numerical methods are used only to solve equations, but this is not the case. There are many other applications of numerical and iterative techniques. One of these is numerical integration, which deals with the problem of finding the area under a curve.

For example, consider the graph of $y = 1/x$ from $x = 1$ to $x = 2$, as shown below (Figure 10.15):

FIGURE 10.15

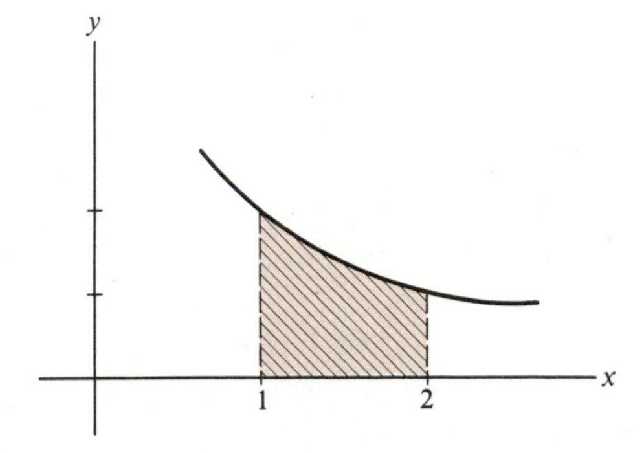

This area represents an important mathematical quantity, namely ln 2 (the natural logarithm of 2). Calculus offers formal, exact, techniques and various iterative techniques for finding areas under curves. In this section we shall describe a statistical technique using random numbers.

If we enclose the area under $y = 1/x$ from $x = 1$ to $x = 2$ in the rectangle whose four corners are (1, 0), (2, 0), (1, 1), and (2, 1), we may be able to approximate the area under the curve $y = 1/x$ (the shaded region).

FIGURE 10.16

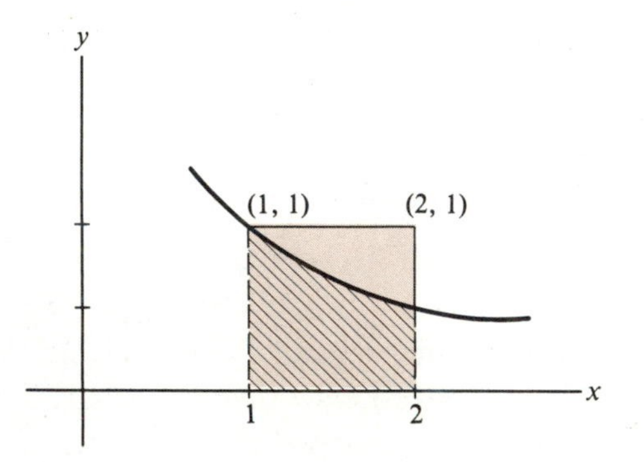

The exact area under the curve is proportional to the area of the rectangle. If a random point is chosen within the rectangle, the probability that it also lies under the curve is proportional to the area under the curve. We shall use this proportionality to obtain an estimate of the area. Such techniques are called **Monte Carlo techniques.**

We can explain this technique more concretely by the following example: Suppose the rectangular region shown in Figure 10.16 were used as a dartboard. If we threw darts *randomly* at the board, the proportion of the darts falling under the curve, in the shaded region, would be proportional to the area under the curve.

EXAMPLE 1

Suppose 100 darts are thrown at the board *randomly* and all land inside the rectangle. If we count 40 darts under the curve in the shaded region, what is the approximate area of the shaded region?

Solution

The 100 darts represent the area of the rectangle or $l \cdot w = 1 \cdot 1 = 1$. (Each side is one unit long.) If 40 darts fell under the curve, the proportion would be

$$\frac{40}{100} = \frac{A}{1} \Rightarrow A = 0.4$$

which means that the area is approximately 0.4.

There is a potential problem with this reasoning—the next time the 100 darts were thrown randomly *maybe* 30 or 60 darts would land under the curve. Simulations depend on *many trials*—many throws of 100 darts—after which an average is taken to get an estimate. No exact answer can be obtained with simulations or iterative techniques. It happens that the area under $y = 1/x$ can be found by formal calculus methods; but when no such techniques exist for a particular problem, simulations can give worthwhile approximations.

We now present an algorithm that could be used in conjunction with a calculator or computer to obtain the approximate area under the curve. In order to implement the algorithm we need to be able to generate random numbers. We shall use the computer's random number generator RND to generate real numbers between 0 and 1:

$$0 < \text{RND} < 1$$

We shall need to generate two random numbers in order to simulate (imitate) the throwing of a dart on our graph. One random number will represent the x-coordinate and the other will represent the y-coordinate. Thus (x, y) will be the coordinates of a randomly thrown dart on our graph. Since y must be between 0 and 1 in order for our dart to land within the confines of the bordering rectangle, we shall assign y to be RND, $y = \text{RND}$. However, x is between 1 and 2, so $x = \text{RND} + 1$.

In order to determine whether the "dart" or randomly chosen point is under the curve, we compare y to $1/x$. If $y < 1/x$, then we label this a "hit." After simulating the tossing of 100 darts, we compare the number of hits to 100 to obtain an estimate of the ratio of the area we wish to know to the known area of the rectangle.

$$\frac{\text{area to be found}}{\text{area of rectangle}} = \frac{\text{hits}}{100}$$

or, in general,

$$\frac{\text{area to be found}}{\text{area of known region}} = \frac{\text{number of hits}}{\text{number thrown}}$$

or

$$\text{area} = \frac{\text{hits} \cdot \text{area of known region}}{\text{total number of "darts" thrown}}$$

Applying this to Example 1, we find

$$A \approx \frac{40 \cdot 1}{100} = 0.4$$

A flowchart for the simulation to obtain an estimate for the area under the curve $y = 1/x$ from $x = 1$ to 2 is given in Figure 10.17.

To obtain a reliable estimate, this procedure would have to be repeated a number of times and an average (mean) taken. The number of trials needed is a topic for operations research and will not be taken up here. However, this intuitive approach should give you some idea of Monte Carlo and simulation techniques. For example, you should be aware that, in general, the more "darts" thrown at the area, the better the approximation will be. In a very large number of trials, the effect of an occasional chance aberration is cancelled out by the sheer number of trials. It is also better to have ten experiments of 100 trials each and take the average than to have one experiment of 1000 trials. The reason for this is studied in statistics. (The interested reader can find a more thorough treatment of these techniques in books on Monte Carlo technique or simulations.) We tried 1000 trials, not 100, and obtained 0.686 ($\ln 2 = 0.693$). We also tried 10 trials of 100 darts each and got: 0.60, 0.70, 0.66, 0.62, 0.61, 0.76, 0.77, 0.67, 0.76. The mean estimate here is 0.688, but note the fluctuations. The reader may wish to calculate the standard deviation of these approximations and compare them for each experiment or a different computer.

FIGURE 10.17

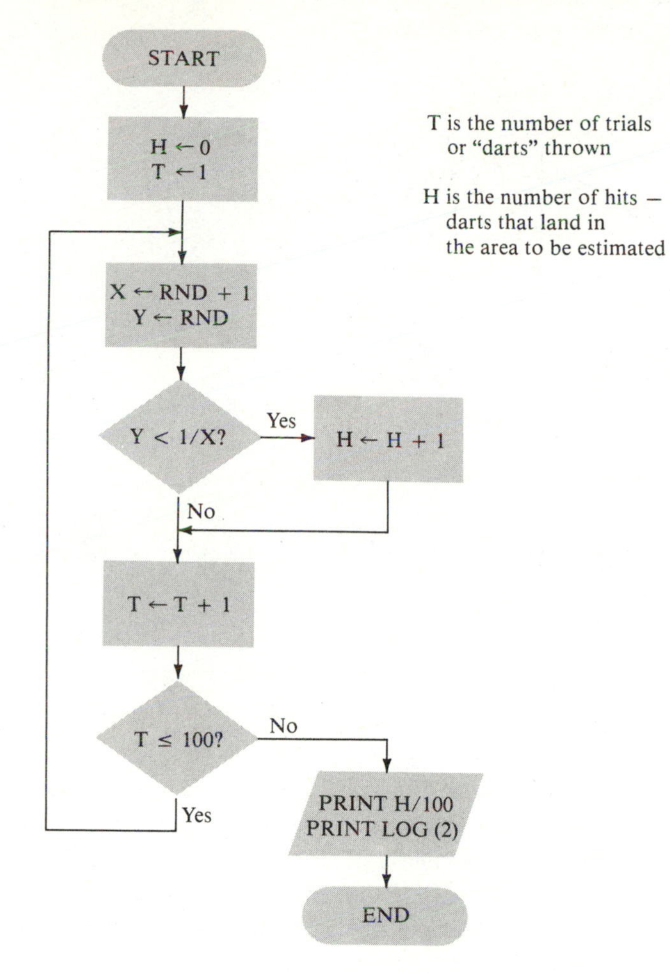

EXERCISES FOR SECTION 10.7

For these exercises use the random number generator of a computer or calculator.

1. Estimate the natural log of 3 ($\ln 3 = 1.0986$). Recall that the area under the curve of $y = 1/x$ from 1 to N is $\ln N$.
2. Estimate the natural log of 0.5. (The real value is $\ln 0.5 = -0.693$.) How can area be negative? Can you devise a solution?

3. Find an estimate for the area under the curve $x^2 + y^2 = 1$, for $x = 0$ to 1 and above the x-axis as shown.

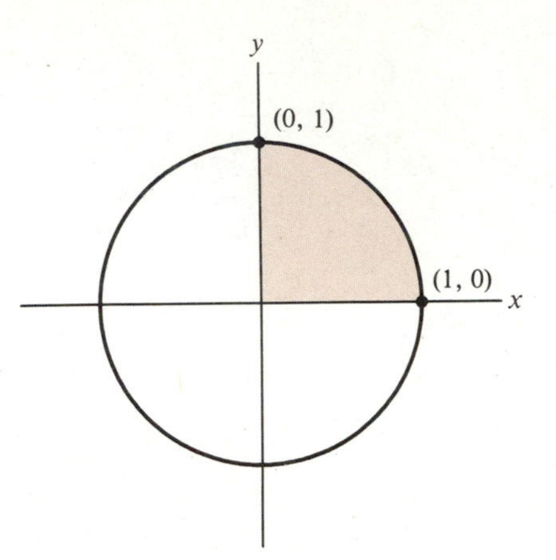

Can your estimate of the area of the quarter circle be used to obtain an approximation of π? How?

10.8 Chapter Review

After a brief review of formulas for solving equations, this chapter presented several numerical methods, involving iterative techniques, for estimating the roots of an equation. The iterative methods discussed included the location principle (Section 10.2); the method of bisection (Section 10.3); the method of false position (Section 10.4); the method of successive approximations (Section 10.5); and the Gauss-Seidel method (Section 10.6).

The concept of convergence was discussed in detail and illustrated in the context of each of the numerical methods presented. The chapter concluded with a brief introduction to Monte Carlo methods as an illustration of a simple application of random numbers to a numerical problem.

VOCABULARY

accuracy to d decimal places
error
actual or absolute error
relative error
root (zero) of an equation
location principle

iterative method
iteration
convergence
degree of precision
method of bisection
continuous function
method of false position
method of successive approximations
Gauss-Seidel method
Monte Carlo technique
random number generator

CHAPTER TEST

1. Explain iterative methods.

2. Use the location principle to bracket the negative root of $x^3 = 2x^2 + 5x - 9$ between two consecutive integers.

3. Use the method of bisection to solve for the negative root of $x^3 = 2x^2 + 5x - 9$ to one decimal place if using a calculator or to two decimal places if using a computer.

4. Use the method of successive approximations to solve the equation of Problem 3.

5. Use Gaussian elimination and then the Gauss-Seidel method to solve the system

$$\begin{aligned} 2x + y + 5z &= 20 \\ -x + 3y + z &= -10 \\ 8x + y - z &= 2 \end{aligned}$$

6. The perimeter of a rectangle is 33 and its area is 67.5.

   a. Set up an equation in *one* variable to find the lengths of the sides.
   b. If possible, solve the equation analytically, and obtain or express an exact answer.
   c. Use some iterative method to solve the equation to two decimal places. Compare these results to the answer obtained in part b.

Appendix: Answers to Odd-Numbered Exercises

CHAPTER 1

Section 1.1

1. **a.** integer, real, complex **b.** yes **3.** **a.** integer, real, complex **b.** yes
5. **a.** integer, real, complex **b.** yes **7.** **a.** integer, real, complex **b.** yes
9. **a.** real, complex **b.** yes **11.** **a.** integer, real, complex **b.** yes **13.** **a.** undefined **b.** no
15. **a.** integer, real, complex **b.** yes **17.** **a.** integer, real, complex **b.** yes
19. **a.** real, complex **b.** no; decimal expansion is infinite **21.** **a.** real, complex **b.** no
23. **a.** real, complex **b.** yes **25.** **a.** real, complex **b.** yes **27.** true **29.** true
31. true **33.** true **35.** false **37.** -3, integer, real, complex
39. complex, imaginary, results in a computer error
41. complex, imaginary, results in a computer error **43.** 5, integer, real, complex
45. real, complex, the value can only be approximated **47.** 13, integer, real, complex
49. undefined, results in a computer error **51.** complex, imaginary, results in a computer error
53. complex, imaginary, results in a computer error

Section 1.2

1. valid **3.** valid **5.** valid **7.** not valid **9.** not valid **11.** valid
13. not valid **15.** valid **17.** 2.6 **19.** 2.2 **21.** 26
23. cannot be done; SUM1 is declared to be of type INTEGER, and X/2 = 2.5, which is not an integer
25. SUM2 = 2.5

Section 1.3

1. 13 **3.** 9 **5.** 1 **7.** 1 **9.** 14 **11.** 4 **13.** 2 **15.** 39 **17.** 54
19. 9 **21.** 20 **23.** 5 **25.** -5172 **27.** 9 **29.** -2.25 **31.** -2
33. -0.5 **35.** 36 **37.** 8 **39.** -12 **41.** $(a + b) \cdot (c + d)$ **43.** $(2 \cdot x + y)/3$
45. $x + 3 \cdot (x + 1)/(2 \cdot y)$

Section 1.4

1. **a.** $1.562 \cdot 10^3$ **b.** 1.562E+03 **3.** **a.** $1.75 \cdot 10^2$ **b.** 1.75E+02
5. **a.** $5.73 \cdot 10^8$ **b.** 5.73E+08 **7.** **a.** $-2.3 \cdot 10^6$ **b.** -2.3E+06
9. **a.** $-1.2 \cdot 10^3$ **b.** -1.2E+03 **11.** **a.** $1.45 \cdot 10^{-6}$ **b.** 1.45E-06
13. **a.** $-5.76 \cdot 10^{-2}$ **b.** -5.76E-02 **15.** **a.** $-5.002 \cdot 10^{-3}$ **b.** -5.002E-03
17. **a.** $8 \cdot 10^4$ **b.** 8E+04 **19.** **a.** $9.73 \cdot 10^{-7}$ **b.** 9.73E-07
21. **a.** $-4.002 \cdot 10^{-8}$ **b.** -4.002E-08 **23.** **a.** $3.005 \cdot 10^3$ **b.** 3.005E+03

25. a. $-6.002 \cdot 10^{-6}$ b. $-6.002E-06$ 27. a. $0.18 \cdot 10^3$ b. $0.18E+03$
29. a. $0.573 \cdot 10^9$ b. $0.573E+09$ 31. a. $-0.578 \cdot 10^6$ b. $0.578E+06$
33. a. $0.350 \cdot 10^{-5}$ b. $0.350E-05$ 35. a. $-0.3546 \cdot 10^2$ b. $-0.3546E+02$
37. a. $-0.8 \cdot 10^5$ b. $-0.8E+05$ 39. a. $-0.5002 \cdot 10^5$ b. $-0.5002E+05$
41. $1 \cdot 10^{-3} = 1E-03$ 43. $1 \cdot 10^{-9} = 1E-09$ 45. 3 47. 5 49. 2 51. 3
53. 3 55. 5 57. a. 14.99 b. 0.000333 59. a. 3.997 b. 0.00075

Section 1.5

1. a. 100 b. 100 3. a. 141 b. 142 5. a. 96.1 b. 96.1 7. a. 68.6 b. 68.7
9. a. 4.88 b. 4.88 11. a. 39.6 b. 39.6 13. a. 34 b. 33.9
15. a. 240 b. 240 17. no
19. a. -0.49 b. The value $\sqrt{896}$ was truncated in Exercise 18. In evaluating $R^2 - 30R + 1$, truncation also occurred.

Chapter Test

1. a. integer, rational, real, complex b. complex (imaginary) c. rational, real, complex d. integer, rational, real, complex e. rational, real, complex f. rational, real, complex g. irrational, real, complex h. complex 2. a. 32 b. 0 c. 5.2 d. 4 e. 14
3. a. (a) $4.32 \cdot 10^2$ (b) $4.32E+02$ b. (a) $4.32 \cdot 10^7$ (b) $4.32E+07$ c. (a) $-4.32 \cdot 10^3$ (b) $-4.32E+03$ d. (a) $4.32 \cdot 10^{-4}$ (b) $4.32E-04$ e. (a) $4.32 \cdot 10^{-1}$ (b) $4.32E-01$ 4. a. 3 b. 4 c. 2
5. result = 21; relative error = .0047393 ($=\frac{1}{210}$) 6. a. yes b. no c. no 7. $x = 4$
8. A zip code is a number being used as a label. The value 6759 will be stored in the computer's memory.
9. a. 22 b. undefined

CHAPTER 2

Section 2.1

1. 1 3. -1 5. 10000 7. 1/10000 or (0.0001) 9. $\frac{9}{16}$ 11. 0.25 13. $4^3 = 64$
15. $10^5 = 100{,}000$ 17. $2a^2b^5$ 19. b^6 21. $-3a^3$ 23. 2^{4x} 25. 3
27. 5^{2a-1} 29. $\dfrac{7x^3}{6y^2}$ 31. $-\frac{9}{2}x$ 33. $-3ab$ 35. $\frac{2}{3}x^4y^3$ 37. x^8 39. $9y^6$
41. $4a^4b^2$ 43. $\dfrac{x^6y^2}{4}$ 45. $-\dfrac{27a^6b^3}{64z^9}$ 47. $\dfrac{a^{6b}}{8c^{12x}}$ 49. 1 51. $-\frac{1}{8}$ 53. 10
55. 1 57. $\frac{9}{4}$ 59. 4 61. $1/2^3$ ($=\frac{1}{8}$) 63. y^2/x^4 65. b^4/a^4 67. 1
69. $\dfrac{1}{x^2y^2}$ 71. $\dfrac{2}{3a^6}$ 73. 10 75. 5 77. 100 79. 8 81. -8 83. $\frac{1}{27}$
85. $x^2/2$ 87. $2/x^2$ 89. $x^{1/4}$ ($=\sqrt[4]{x}$) 91. $2x^2$ 93. $-2a^2$ 95. $\dfrac{xy^2z^3}{a^2b^5}$
97. Method 1

Section 2.2

1. $x = 6$ 3. $x = -3$ 5. $y = 4$ 7. $y = -4$ 9. $x = 2$ 11. $x = 3$
13. $x = -4.5$ 15. $x = 2$ 17. $x = \frac{1}{10}$ 19. $x = -1$ 21. $x = \frac{4}{3}$

23. infinitely many solutions **25.** no solution **27.** infinitely many solutions
29. $x = -\frac{7}{13}$ **31.** 25 **33.** 0 **35.** 144 **37.** 576 **39.** 0 **41.** 4
43. $\frac{1}{2}, -2$ **45.** $\frac{5}{4}$ **47.** $\frac{5}{3}, \frac{1}{3}$ **49.** $\frac{4}{3}, -\frac{4}{3}$ **51.** $-\frac{2}{3}$ **53.** $-2, -3$ **55.** 5
57. $-3, \frac{1}{4}$ **59.** $-5, -\frac{7}{3}$ **61.** no real solution

63.
```
10  REM  BASIC PROGRAM TO SOLVE A QUADRATIC EQUATION
50  INPUT "ENTER THE VALUE OF A "; A
60  INPUT "ENTER THE VALUE OF B "; B
70  INPUT "ENTER THE VALUE OF C "; C
80  IF A = 0 THEN GOTO 160
90  LET D = B^2 - 4 * A * C
100 IF D < 0 THEN GOTO 180
110 LET R1 = (-B + D^(1/2))/(2 * A)
120 LET R2 = (-B - D^(1/2))/(2 * A)
130 PRINT "ROOT #1 IS "; R1
140 PRINT "ROOT #2 IS "; R2
150 GOTO 999
160 PRINT "THE EQUATION IS NOT QUADRATIC."
170 GOTO 999
180 PRINT "THE EQUATION HAS NO REAL SOLUTION."
999 END
```

Section 2.3

1. 2 **3.** 12 **5.** 7 **7.** 32 **9.** 1 **11.** $10(4x^2 - 25)$ **13.** no common factor
15. $15y^2(x^3 + 2z)$ **17.** $a(25ab + 4c^2)$ **19.** $4y(x^2 + 12x - 28)$ **21.** $3y^2z(5x^2z - 10x + 1)$
23. $2xy^2(25x^2 + 70x + 49)$ **25.** $x^2 + 4x - 5$ **27.** $6x^2 - x - 2$ **29.** $49x^2 - 1$
31. $4x^2 + 12x + 9$ **33.** $-6x^2 + 37x - 35$ **35.** $9x^2 - 4$ **37.** $(2x + 5)(2x - 5)$
39. $(x + 5)(x + 1)$ **41.** $(3y + 5)(y - 4)$ **43.** $5(3x - 2)(2x + 1)$ **45.** $4(x - 2)(x + 14)$
47. $4(5x + 2)(5x - 2)$ **49.** $y^2(3x + 5)(3x + 5)$ or $y^2(3x + 5)^2$ **51.** $\frac{2}{3}, -\frac{1}{2}$ **53.** $\frac{5}{6}, -\frac{5}{6}$
55. $1, -5$ **57.** $4, -4$ **59.** $2, -4$ **61.** $\frac{4}{3}, -\frac{3}{2}$ **63.** $7, -\frac{1}{2}$
65. **a.** $101 \times 99 = (100 + 1)(100 - 1) = 9999$ **b.** $52 \times 48 = (50 + 2)(50 - 2) = 2496$

Section 2.4

1. $\frac{2y}{3x}$ **3.** $\frac{3x^2y}{2}$ **5.** $\frac{3x}{x + 1}$ **7.** $\frac{x - y}{3xy}$ **9.** $-(x + 3)$ or $-x - 3$ **11.** $\frac{5(x - 2)}{2x}$

13. $\frac{x + 3}{x - 3}$ **15.** $\frac{2x - 1}{x - 3}$ **17.** $\frac{x^2 + 6x + 7}{x^2(x + 3)}$ **19.** $\frac{3x + 4}{x - 1}$ **21.** xy^2 **23.** $\frac{12}{x^2}$

25. $\frac{3(x + 2)}{2(x - 2)}$ **27.** 1 **29.** $\frac{(3x + 2)(x + 1)}{2(2x + 3)}$ **31.** $\frac{1}{x^3}$ **33.** $\frac{5z}{8x^2y^2}$ **35.** $\frac{x - 3}{3x}$

37. $\frac{a - 1}{6a(b - 2)}$ **39.** $\frac{1}{y + 1}$ **41.** $\frac{2x + 5}{3y}$ **43.** 2 **45.** $\frac{15}{4x}$ **47.** $\frac{x^2}{(x + 2)^2}$

49. $\frac{2x^2 - 74}{(x - 7)(x - 5)}$ **51.** $\frac{12 - x^2}{2(x + 2)^2}$ **53.** $\frac{x^2 + 10x - 25}{(x + 5)^2(x - 5)}$ **55.** $\frac{5}{(x + 3)(x - 2)}$

57. $\frac{x(x^2 + 4x + 12)}{(x - 2)(x^2 + 2x + 4)}$ **59.** $\frac{x^3 + x^2 + 1}{x}$

Section 2.5

1. $\frac{5}{2}(=2.5)$ **3.** 16 **5.** 2, −2 **7.** 5 **9.** $0, -\frac{1}{5}$ **11.** 1 **13.** $\frac{2}{3}$

15. $\frac{12}{5}(=2.4)$ **17.** 2 **19.** −10 **21.** $\frac{3}{4}, -1$ **23.** no solution **25.** 3, −3

27. $\frac{7}{3}$ **29.** $-\frac{7}{2}(=-3.5), 1$ **31.** $m = \dfrac{Fd^2}{kM}$ **33.** $C = \dfrac{5(F-32)}{9}$ or $C = \dfrac{5}{9}F - \dfrac{160}{9}$

35. $y = \dfrac{xz}{z-x}$ **37.** $x = \dfrac{yz}{y+z}$ **39.** $x = \dfrac{Ftg + wy}{w}$ or $x = \dfrac{Ftg}{w} + y$

Section 2.6

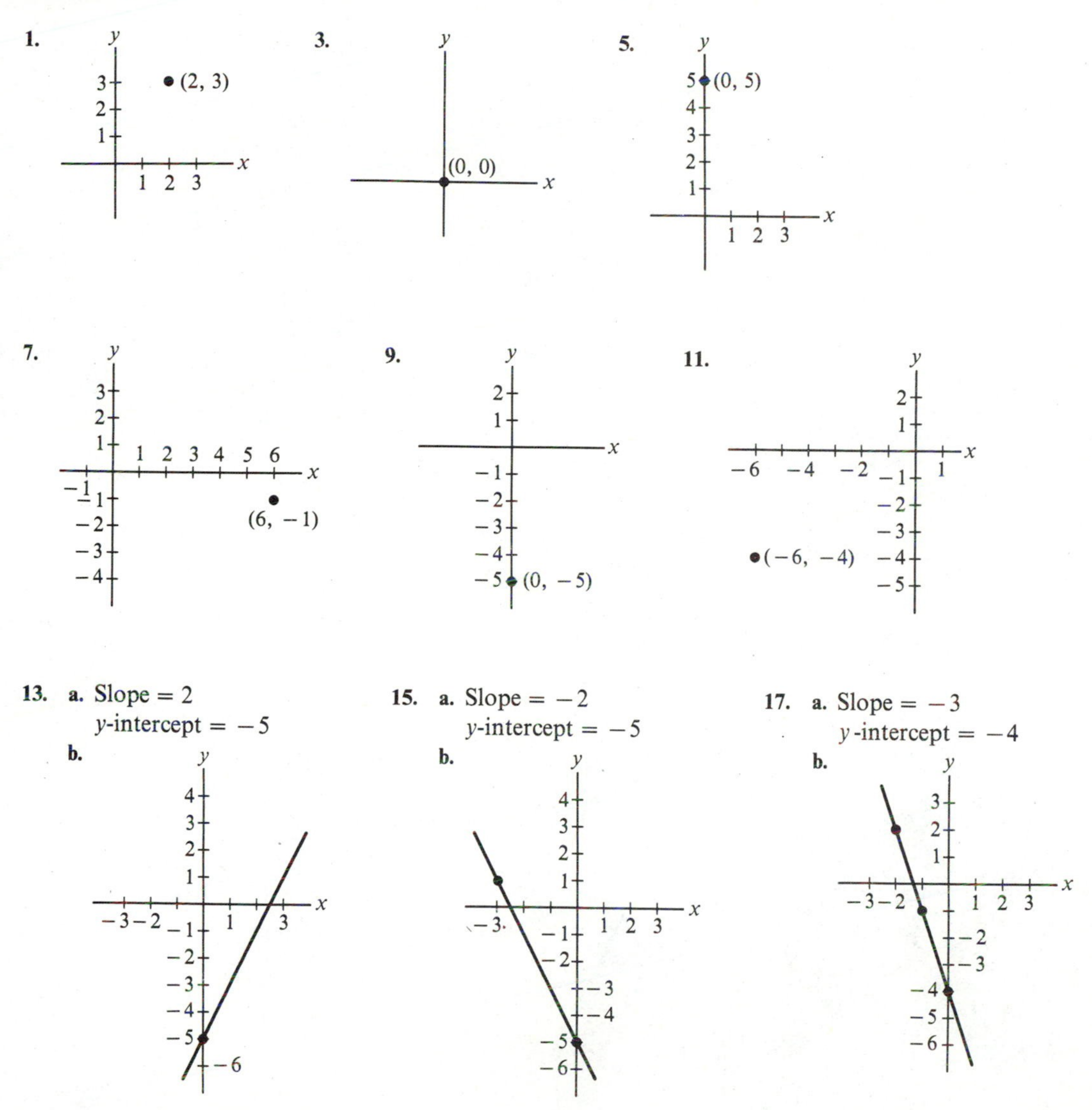

19. **a.** Slope $= -3$
y-intercept $= 4$

b.

21. **a.** Slope $= -\frac{3}{4}$
y-intercept $= 2$

b.

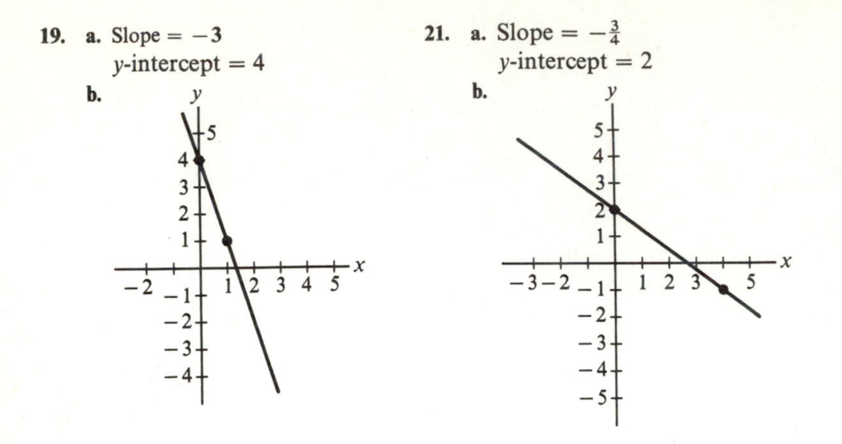

23. **a.** Slope $= \frac{1}{2}$
y-intercept $= 0$

b.

25. **a.** Slope $= -\frac{2}{3}$
y-intercept $= 0$

b.

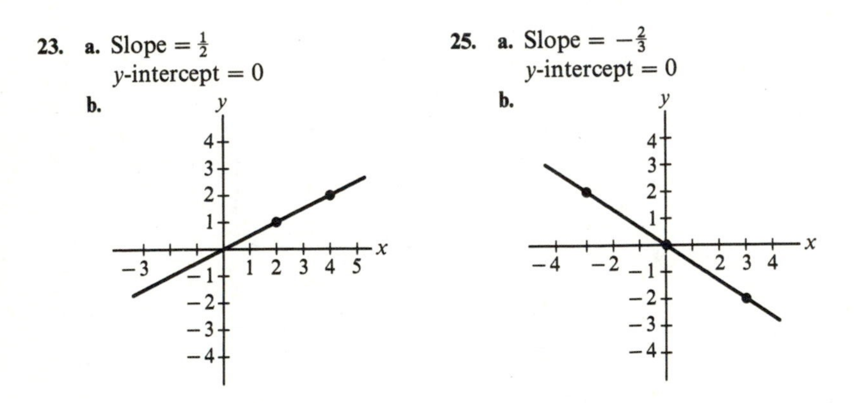

27.

29.

31.

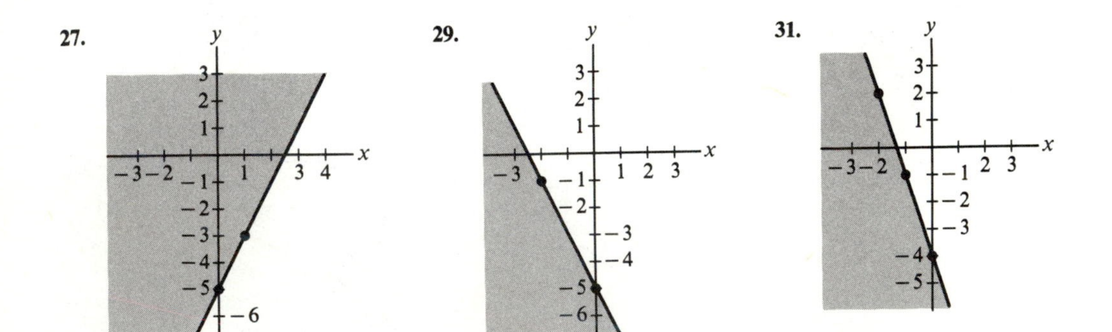

33.

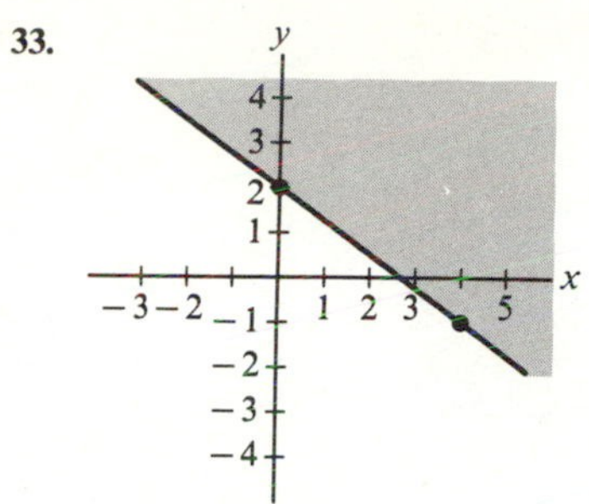

35.

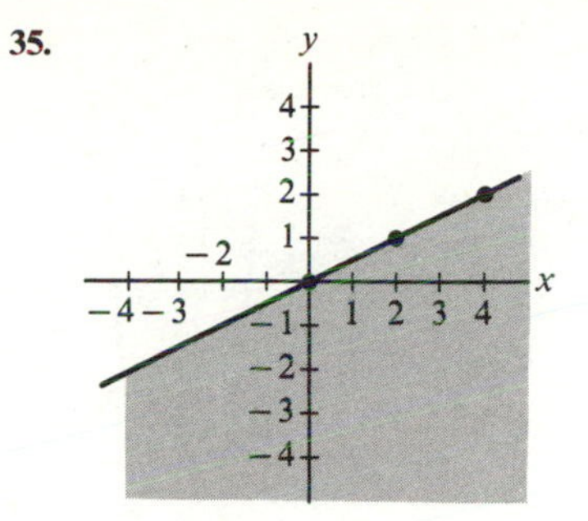

37.

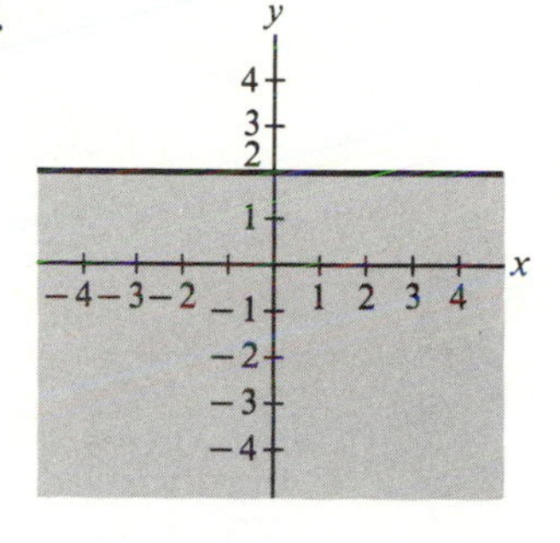

39. **a.** $F = 212°$
b. $F = -40°$
c.

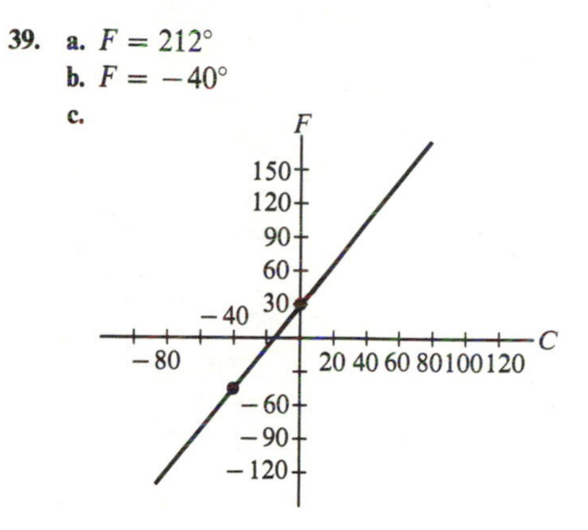

Section 2.7

1. vertex = (3, 0)
concave up

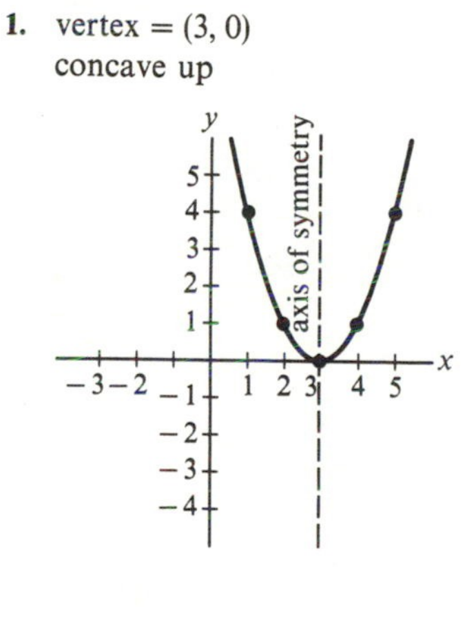

3. 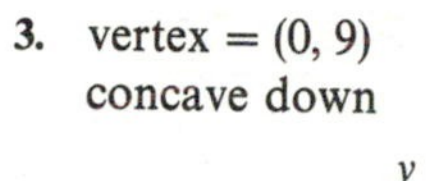
vertex = (0, 9)
concave down

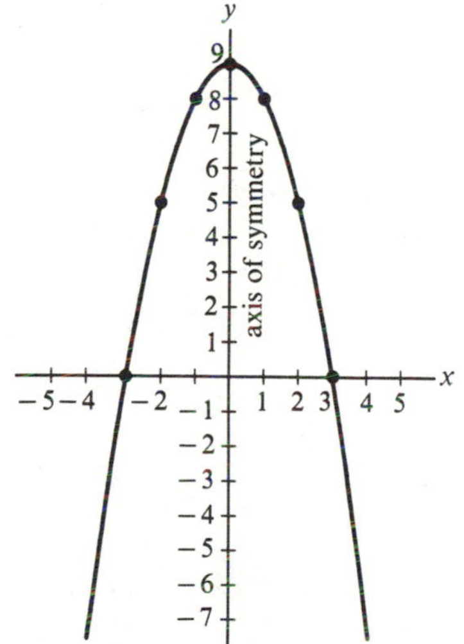

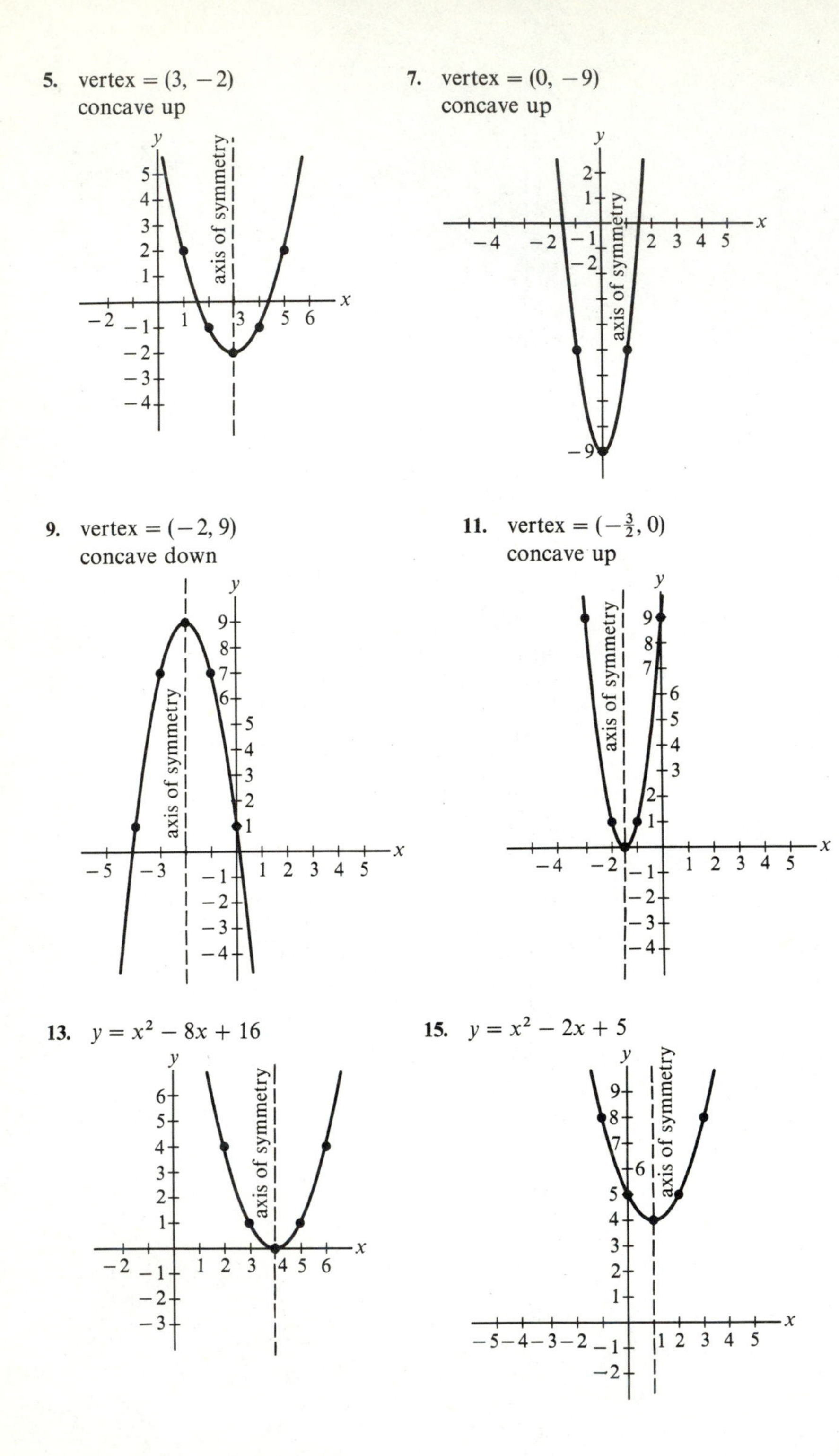

5. vertex = (3, −2)
concave up
axis of symmetry
7. vertex = (0, −9)
concave up
axis of symmetry
9. vertex = (−2, 9)
concave down
axis of symmetry
11. vertex = $(-\frac{3}{2}, 0)$
concave up
axis of symmetry
13. $y = x^2 - 8x + 16$
axis of symmetry
15. $y = x^2 - 2x + 5$
axis of symmetry

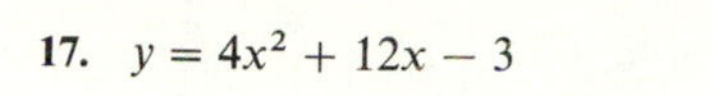

17. $y = 4x^2 + 12x - 3$

19. $y = -x^2 - 4x + 5$

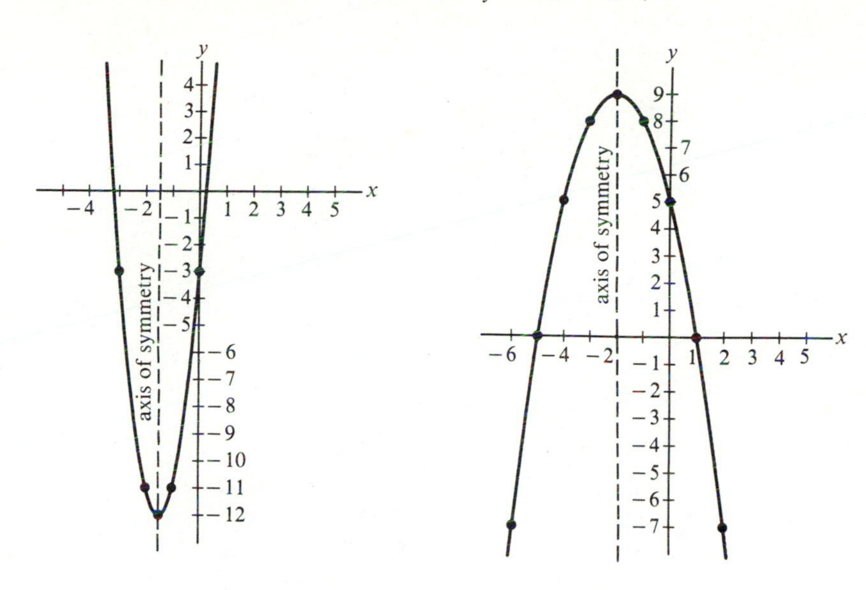

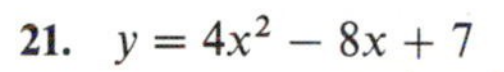

21. $y = 4x^2 - 8x + 7$

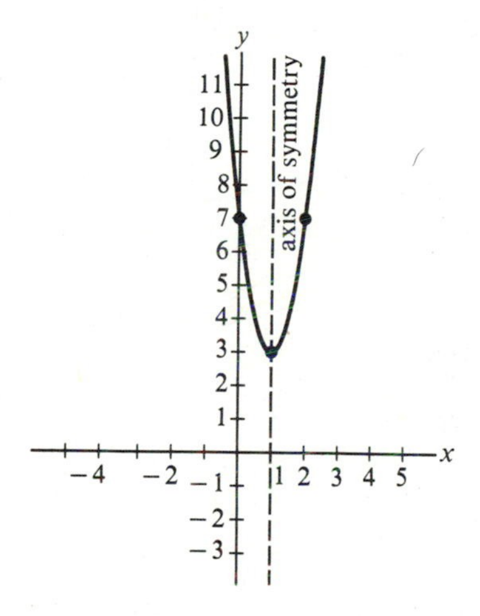

Section 2.8

1. function **3.** not a function **5.** function

7. domain = all real numbers
range = real numbers less than or equal to 8

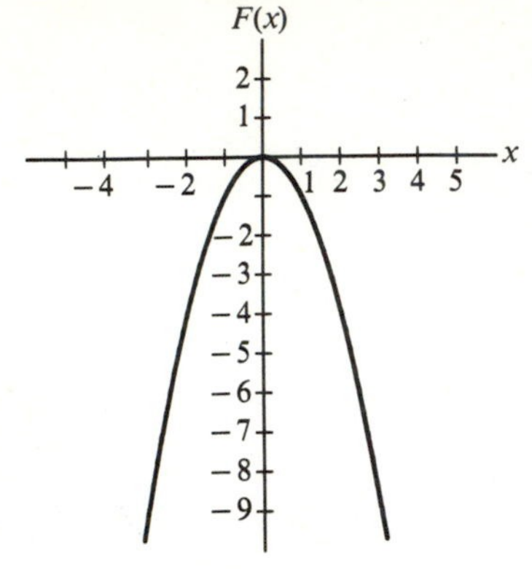

9. domain = all real numbers
range = all real numbers

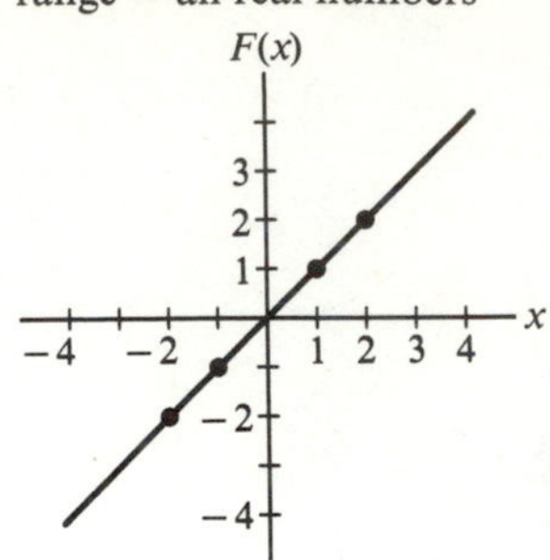

11. domain = all real numbers
range = all real numbers less than or equal to 8

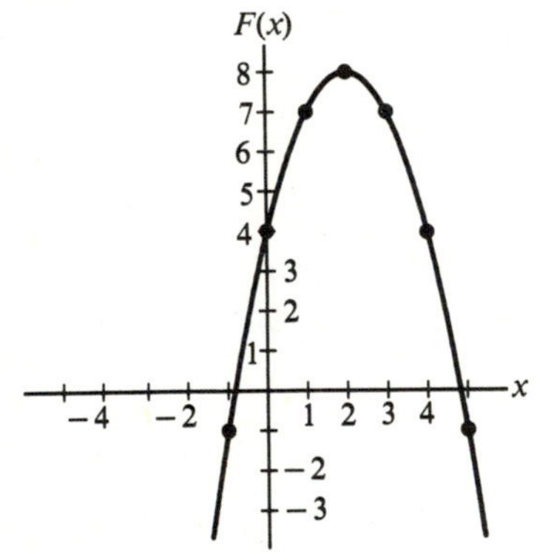

13. domain = all real numbers
range = the number 4

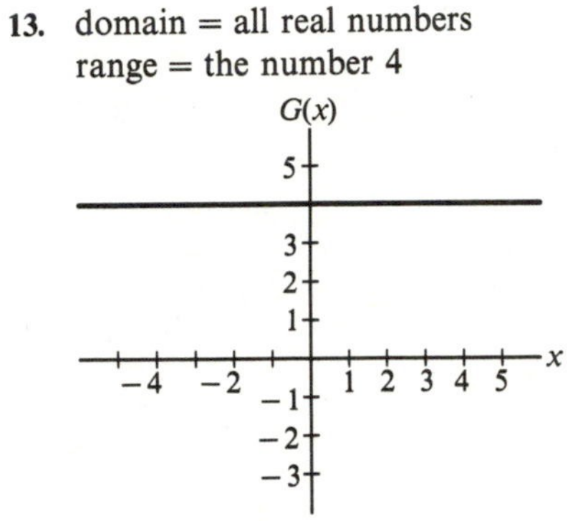

15. DEF FNF(X) = −(X^2) 17. DEF FNF(X) = X
19. DEF FNF(X) = −(X^2) + 4 * X + 4 21. DEF FNG(X) = 4
23. **a.** $M(0) = 0$; $M(2) = 2$; $M(3) = 0$; $M(8) = 2$; $M(9) = 0$; $M(10) = 1$
b. yes; the range is the set of integers 0, 1 and 2, {0, 1, 2}

Chapter Test

1. $2^6 = 64$ 2. 3 3. 4 4. $\dfrac{1}{9a^4b^6}$ 5. $\dfrac{x^6y^3}{27z^3}$ 6. $\dfrac{b^3}{a^6c^{15}}$ 7. $\dfrac{a^6c^{15}}{b^3}$
8. $\dfrac{x^{8/3}y^2}{z^4}$ 9. $x = \frac{7}{3}$ 10. $x = -5$ 11. no solution 12. $x = -\frac{3}{2}$
13. infinitely many solutions 14. $1, \frac{1}{4}$ 15. no real solution 16. 1 17. $\frac{2}{3}, -3$
18. $2, -2$ 19. $(x + 5)(x - 1)$ 20. $4y^2(3x - 1)(3x + 2)$ 21. $3(2x + 3)(2x - 3)$
22. $(5x + 2)(5x + 2)$ or $(5x + 2)^2$ 23. $xy^2z(6x^2 + 3x - 2)$ 24. $\frac{1}{4}, 1$ 25. $\frac{2}{3}, -3$

26. $\frac{7}{6}, -\frac{7}{6}$ **27.** $-\frac{1}{3}$ **28.** $\frac{1}{ac}$ **29.** $x - 2$ **30.** $\frac{a^3b^2}{c^3d^2}$ **31.** $\frac{x+3}{(x-3)(x+1)}$

32. 4 **33.** $\frac{5y+3x}{x^2y^3}$ **34.** $\frac{3x - xy + y^2}{(x+y)(x-y)}$ **35.** $\frac{3x^2}{(x+3)(x-3)(x-6)}$ **36.** $\frac{7}{8}$

37. no real solution

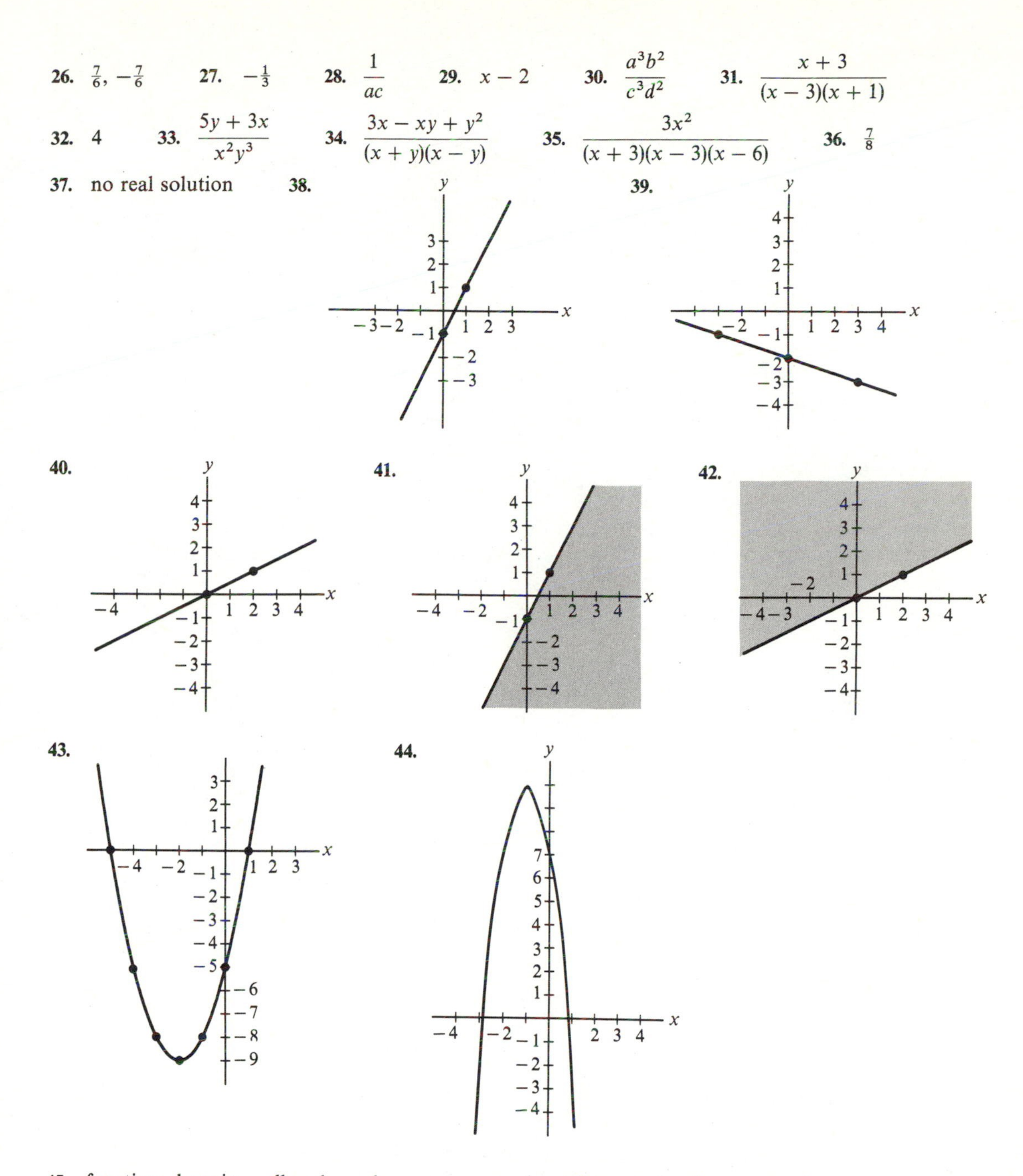

45. function; domain = all real numbers; range = real numbers greater than or equal to −3(Reals > = −3)

46. function; domain = all real numbers; range = nonnegative real numbers (reals > = 0)

47. not a function **48.** function; domain = all real numbers; range = {2}

CHAPTER 3

Section 3.1

1. $(2 \cdot 10^2) + (8 \cdot 10^1) + (9 \cdot 10^0)$ 3. $(8 \cdot 10^2) + (1 \cdot 10^1) + (2 \cdot 10^0)$
5. $(5 \cdot 10^4) + (0 \cdot 10^3) + (2 \cdot 10^2) + (3 \cdot 10^1) + (8 \cdot 10^0)$
7. $(9 \cdot 10^2) + (0 \cdot 10^1) + (0 \cdot 10^0) + (6 \cdot 10^{-1}) + (5 \cdot 10^{-2})$
9. $(1 \cdot 10^3) + (0 \cdot 10^2) + (0 \cdot 10^1) + (2 \cdot 10^0) + (0 \cdot 10^{-1}) + (1 \cdot 10^{-2})$
11. $(0 \cdot 10^{-1}) + (0 \cdot 10^{-2}) + (2 \cdot 10^{-3}) + (4 \cdot 10^{-4})$ 13. 13 15. 9 17. 23 19. 81
21. 61 23. 2.75 25. 1.4375 27. 0.84375 29. 2.09375

Section 3.2

1. 11100_2 3. 1101100111_2 5. 10011_2 7. 101000001_2 9. 11111111_2
11. 100000001_2 13. $.1001_2$ 15. $.111_2$ 17. $.10001_2$ 19. $.000011_2$
21. $.0101100110_2$ 23. $.000110011_2$ 25. 11100.0101100110_2 27. 1001000.000110011_2
29. 1110.1_2
31. **b.** When Z = 1, Q is assigned the highest power of 2 that is less than or equal to the value of X. **c.** It must be less than 2^{31}. **d.** method 1
33. On the Apple II computer, **a.** result = .321000099 **b.** result = .432100296
The decimal parts of these numbers do not convert to terminating binary "decimals." Also, note that not all computers give exactly the same results.

Section 3.3

1. 111_2 3. 1010_2 5. 10010_2 7. 11111_2 9. 1010101_2 11. 7417
13. 8997 15. 87607 17. 148 19. 2817 21. 0100_2 23. 0110010_2
25. 01011100110_2 27. 101_2 29. 100101110_2 31. 11001_2 33. 11110_2
35. 101011111_2 37. 100100010010_2 39. 110_2 41. 1001_2 43. 100_2 R111_2
45. **a.** The nines complement of y, y', is $99999 - y$. **b.** $x + y' = x + 99999 - y$
c. $x + y' - 100000 + 1 = x + 99999 - y - 100000 + 1$ **d.** $\underbrace{x + y' - 100000 + 1}_{\text{end-around-carry}} = x - y$

Section 3.4

1. 133 3. 15 5. 134 7. 1051 9. 2353 11. 13.75 13. 5.703125
15. 67.130859375 17. 443_8 19. 377_8 21. 401_8 23. 1760_8 25. 4315_8
27. 10263_8 29. 6572_8 31. 3341_8 33. 2667_8 35. 2473_8 37. 6130_8
39. 12067_8 41. 27664_8 43. 201_8 R 17_8 45. 314_8 R 24_8

Section 3.5

1. 419 3. 2745 5. 41259 7. 40961 9. 48879 11. 43981 13. $3B_H$
15. 11_H 17. AA_H 19. $1CO_H$ 21. $1CB_H$ 23. $1082A_H$ 25. $BBBB_H$
27. $116E_H$ 29. $8CCE_H$ 31. $3E0A5_H$ 33. $560C4A2_H$ 35. $1A5_H$ 37. 47_8
39. 133_8 41. 35.7_8 43. 74.04_8 45. 101111001_2 47. 11101.010101_2
49. 1010000101.000010111_2 51. CB_H 53. $5B_H$ 55. $43.E8_H$ 57. $C.2A_H$
59. 100000010_2 61. 101100011100_2 63. 10101010.000010111100_2
65. $3FF_H$ = 1023 bytes

Section 3.6

1. a. 10000111_2 **b.** 207_8 **c.** 87_H **3. a.** 1011011000_2 **b.** 1330_8 **c.** $2D8_H$
5. a. 10000000000_2 **b.** 2000_8 **c.** 400_H **7.** 0111 0011 **9.** 0100 0101 0001
11. 0001 0000 0000 0000 **13. a.** 0101 0111 0101 0011 **b.** 1111 0111 1111 0011
15. a. 0101 0100 0101 0101 0101 0001 **b.** 1111 0100 1111 0101 1111 0001
17. a. 0101 0111 0101 0010 0101 1000 **b.** 1111 0111 1111 0010 1111 1000 **19.** TONND
21. TON ND **23.** D067
25. a. 1011 0011 1010 0001 1010 1110 1010 0100 1011 1001
b. 1110 0010 1100 0001 1101 0101 1100 0100 1110 1000
27. a. 1010 1010 1011 0101 1010 1101 1011 0000 **b.** 1101 0001 1110 0100 1101 0100 1101 0111
29. 100 0001 100 0100 100 0100 010 0000 011 0001 011 0010 **31.** 011 0110 011 1110 011 0011

Chapter Test

1. a. 1101_2 **b.** 15_8 **c.** D_H **2. a.** 1111100_2 **b.** 174_8 **c.** $7C_H$ **3.** 22 **4.** 134
5. 539 **6.** 1.375 **7.** 101000_2 **8.** 1110_2 **9.** 1000001_2 **10.** 1011_2
11. 1303_8 **12.** 3667_8 **13.** 20316_8 **14.** $163C8_H$ **15.** $B1D84_H$ **16.** $143C0_H$
17. a. 313_8 **b.** CB_H **18. a.** 473_8 **b.** $13B_H$ **19. a.** 33.22_8 **b.** $1B.48_H$
20. 0111 0011 0010 0101

CHAPTER 4

Section 4.1

1. no **3.** no **5.** yes **7.** no **9.** yes **11.** compound **13.** compound
15. compound **17.** compound **19.** compound **21.** true **23.** false **25.** true

Section 4.2

1. $y <> 5$ ($y \neq 5$) **3.** $A < 1$ **5.** I did write that program. (or, I wrote that program.)
7. conjunction: $p \wedge q$ - Today is tomorrow's yesterday and today was yesterday's tomorrow.
disjunction: $p \vee q$ - Today is tomorrow's yesterday or today was yesterday's tomorrow.
9. conjunction: $p \wedge q$ - $0 < x$ and $x < 5$ (or $0 < x < 5$)
disjunction: $p \vee q$ - $0 < x$ or $x < 5$
11. false **13.** true **15.** false **17.** true **19.** true
21. Mama did not play bass. **23.** Mama played bass and Daddy sang tenor.
25. Mama played bass and Daddy did not sing tenor.
27. IF ($A <> 1$) AND ($A <> -1$) THEN PRINT "HALLELUJAH!"

Section 4.3

1. If Mama played bass, then Daddy sang tenor.
3. Mama played bass if and only if Daddy did not sing tenor.
5. If Mama played bass, then Daddy sang tenor; and if Daddy sang tenor, then Mama played bass.
7. Either Mama played bass or Daddy did not sing tenor, but not both.
9. statement 1 is false **11.** statement 3 is true **13.** statement 5 is false
15. statement 7 is false **17.** $z \rightarrow q$ **19.** $p \rightarrow (\sim q)$ **21.** $z \leftrightarrow (\sim p)$

23.

p	q	$p \vee q$	$(p \downarrow q)$ $\sim(p \vee q)$	$(\sim p) \wedge (\sim q)$
T	T	T	F	F
T	F	T	F	F
F	T	T	F	F
F	F	F	T	T

the same

Section 4.4

1.

p	q	$\sim p \to q$
T	T	T
T	F	T
F	T	T
F	F	F

neither

3.

p	q	$p \vee (p \to q)$
T	T	T
T	F	T
F	T	T
F	F	T

tautology

5.

p	$p \veebar p$
T	F
F	F

self-contradiction

7.

p	q	r	$(p \wedge q \vee r) \wedge (\sim p)$
T	T	T	F
T	T	F	F
T	F	T	F
T	F	F	F
F	T	T	T
F	T	F	F
F	F	T	T
F	F	F	F

neither

9.

p	q	r	$\sim[(\sim q \vee r) \to (p \wedge r)]$
T	T	T	F
T	T	F	F
T	F	T	T
T	F	F	F
F	T	T	F
F	T	F	F
F	F	T	F
F	F	F	F

neither

11. logically equivalent **13.** logically equivalent **15.** not logically equivalent

Section 4.5

1. $x < 7$ **3.** $x > 2$ **5.** $x < -2$ **7.** $x > -2$ and $x < 3$ $(-2 < x < 3)$

9. $x > 0$ and $x < 1$ $(0 < x < 1)$

11. A = age; condition: $A > 16$ and $A < 25$
IF ($A > 16$) AND ($A < 25$) THEN PRINT "FREE ADMISSION"

13. A = age; I = income; condition: $A > 65$ or $I < 12000$
IF ($A > 65$) OR ($I < 12000$) THEN PRINT "REDUCED TUITION"

15. N = number; condition: ($N > 18$ and $N < 35$) and $N < > 25$
IF ($N > 18$ AND $N < 35$) AND $N < > 25$ THEN PRINT "NUMBER REJECTED"

17. 70 IF ($X = 7$) AND ($Y = 4$) THEN PRINT "ACCEPT"

Section 4.6

1. false **3.** true **5.** false **7.** $\{\ \}, \{a\}, \{b\}, \{a, b\}$ **9.** $\{\ \}$ **11.** $\{2, 3, 4, 5, 6, 7\}$

13. $\{\ \}$ **15.** $\{7\} = D$ **17.** $\{1, 8, 9\}$ **19.** $U = \{1, 2, 3, 4, 5, 6, 7, 8, 9\}$

21. **a.** $\{2, 5\}$ **b.** $\{4, 7\}$ **c.** $\{1, 4, 7, 8, 9\} = \tilde{A}$ **d.** $\{3, 4, 6, 7\} = B$

Chapter Test

1. $p \wedge \sim q$ 2. $q \vee p$ 3. $p \rightarrow (\sim q)$ 4. $q \leftrightarrow p$ 5. $\sim(\sim q)$ [or q]
6. It is cold and someone will build a fire.
7. It is cold and we will not have a picnic, or someone will build a fire.
8. If it is cold and we will have a picnic, then someone will build a fire.
9. It is not cold if and only if we will not have a picnic.
10. It is not true that no one will build a fire. (Someone will build a fire.)

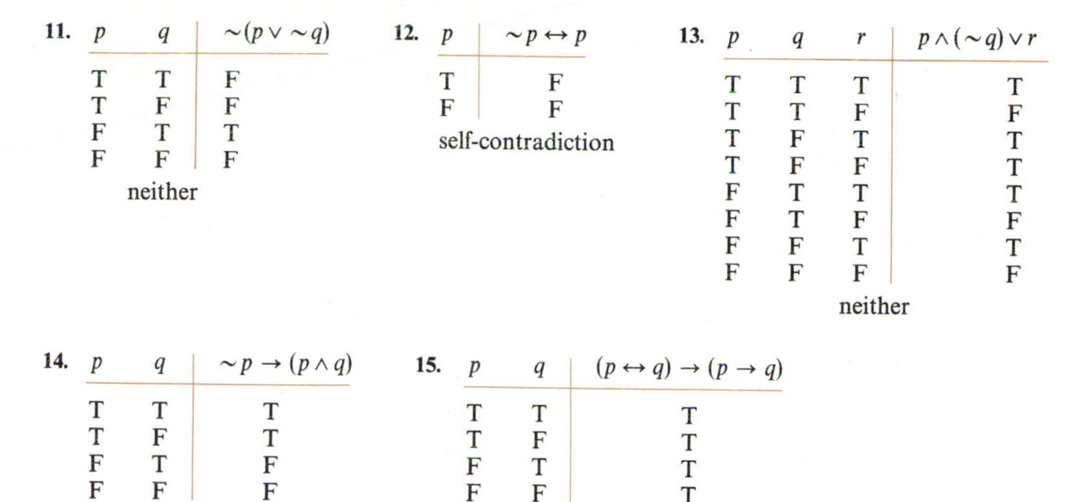

11.

p	q	$\sim(p \vee \sim q)$
T	T	F
T	F	F
F	T	T
F	F	F

neither

12.

p	$\sim p \leftrightarrow p$
T	F
F	F

self-contradiction

13.

p	q	r	$p \wedge (\sim q) \vee r$
T	T	T	T
T	T	F	F
T	F	T	T
T	F	F	T
F	T	T	T
F	T	F	F
F	F	T	T
F	F	F	F

neither

14.

p	q	$\sim p \rightarrow (p \wedge q)$
T	T	T
T	F	T
F	T	F
F	F	F

neither

15.

p	q	$(p \leftrightarrow q) \rightarrow (p \rightarrow q)$
T	T	T
T	F	T
F	T	T
F	F	T

tautology

16. logically equivalent 17. not logically equivalent
18. $\{\ \}, B, \{q\}, \{c\}, \{z\}, \{q, c\}, \{q, z\}, \{c, z\}$
19. **a.** $\{a, b, c, q, s, z\}$ **b.** $\{a\}$ **c.** $\{b, c, q, s, z\}$ **d.** $\{a, b, c, e, i, o, q, s, u, z\}$ **e.** $\{\ \}$
20. **a.** true **b.** false **c.** true

CHAPTER 5

Section 5.1

1. 1 3. 1 5. 1 7. 0 9. 1 11. B 13. $(C \cdot A) \cdot B$ 15. $A \cdot B$
17. A 19. A

21. 1. "+" is commutative
 2. associative property of "+"; distributive property
 3. $A + A' = 1$, since A' is the complement of A
 4. 1 is the identity for "+"
 5. commutative and associative properties of "+"
 6. $B + B' = 1$ by the definition of complement; $1 + A = 1$ by Table 5.1

23. **a.** 111 **b.** 111 **c.** 111 **d.** 000 **e.** 111 25. **a.** 1 **b.** 12 **c.** 2

Section 5.2

1. **a.** $p \vee (q \wedge r)$

b.

p	q	r	$p \vee (q \wedge r)$
T	T	T	T
T	T	F	T
T	F	T	T
T	F	F	T
F	T	T	T
F	T	F	F
F	F	T	F
F	F	F	F

3. **a.** $p \wedge [(q \wedge r) \vee q \vee (\sim p)]$

b.

p	q	r	$p \wedge [(q \wedge r) \vee q \vee (\sim p)]$
T	T	T	T
T	T	F	T
T	F	T	F
T	F	F	F
F	T	T	F
F	T	F	F
F	F	T	F
F	F	F	F

5. **a.** $[p \vee (q \wedge r) \vee (\sim r)] \wedge q$

b.

p	q	r	$[p \vee (q \wedge r) \vee (\sim r)] \wedge q$
T	T	T	T
T	T	F	T
T	F	T	F
T	F	F	F
F	T	T	T
F	T	F	T
F	F	T	F
F	F	F	F

7. **a.** $(p \vee r) \wedge (q \vee r)$

b.

p	q	r	$(p \vee r) \wedge (q \vee r)$
T	T	T	T
T	T	F	T
T	F	T	T
T	F	F	F
F	T	T	T
F	T	F	F
F	F	T	T
F	F	F	F

9. **a.** $[(\sim a) \vee (b \wedge c)] \wedge (\sim b) \wedge [(a \wedge c) \vee (\sim b \vee \sim c)]$

b.

a	b	c	$[(\sim a) \vee (b \wedge c)] \wedge (\sim b) \wedge [(a \wedge c) \vee (\sim b \vee \sim c)]$
T	T	T	F
T	T	F	F
T	F	T	F
T	F	F	F
F	T	T	F
F	T	F	F
F	F	T	T
F	F	F	T

11.

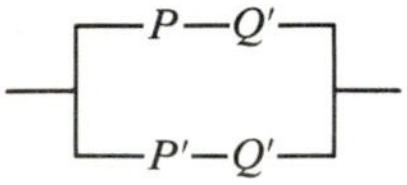

P—Q′

P′—Q′

13.

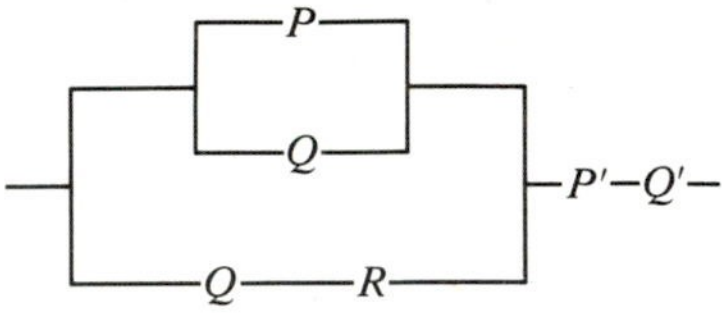

P

Q

—P′—Q′—

Q——R

15.

P

Q

—R—

P

Q

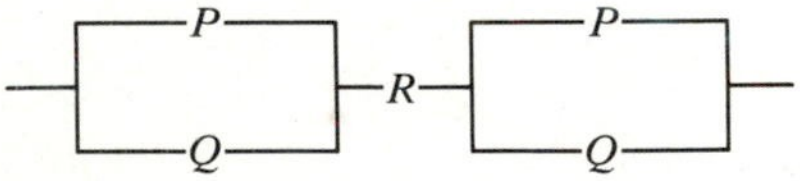

17. a.

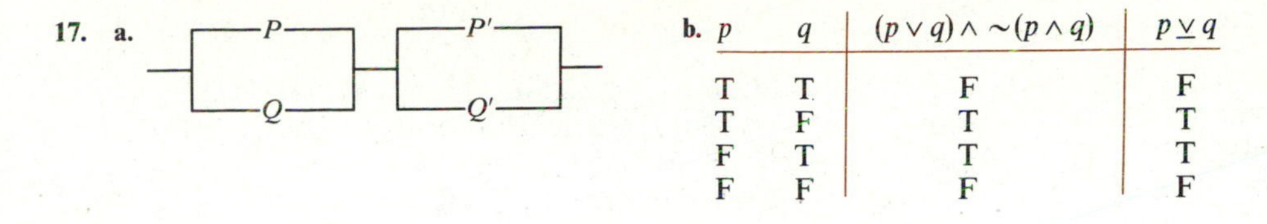

b.

p	q	$(p \vee q) \wedge \sim(p \wedge q)$	$p \underline{\vee} q$
T	T	F	F
T	F	T	T
F	T	T	T
F	F	F	F

c. The truth tables indicate that the networks are equivalent.

19. $P \cdot (Q + R')$ 21. $(P + Q) \cdot (Q \cdot Q' + P')$ 23. $(P \cdot Q' + R) \cdot (P' \cdot R' + Q)$

25. An equivalent logical expression is $(p \wedge \sim p) \vee (q \wedge \sim q)$.

p	q	$(p \wedge \sim p) \vee (q \wedge \sim q)$
T	T	F
T	F	F
F	T	F
F	F	F

The truth table has all F's; therefore, the network represents a self-contradiction.

27. —P'—Q'— 29. Note: C represents the chairman's switch.

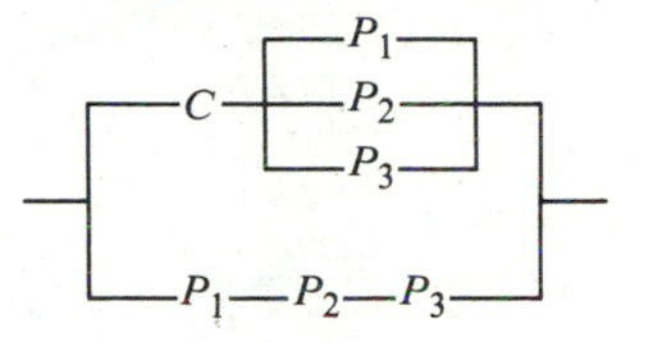

Section 5.3

1. —P— 3. 5. —P'—Q— 7. 9. 11. —P— 13. 15. —P'—Q— 17. 19.

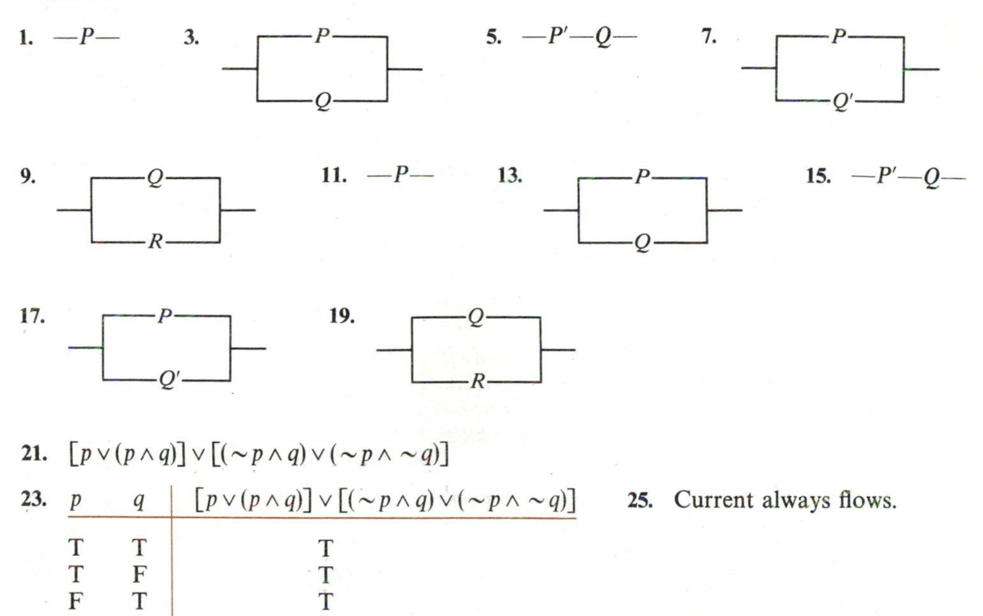

21. $[p \vee (p \wedge q)] \vee [(\sim p \wedge q) \vee (\sim p \wedge \sim q)]$

23.

p	q	$[p \vee (p \wedge q)] \vee [(\sim p \wedge q) \vee (\sim p \wedge \sim q)]$
T	T	T
T	F	T
F	T	T
F	F	T

25. Current always flows.

Section 5.4

1. 11010011 **3.** 11011111 **5.** 10101011 **7.** 11101111 **9.** 11111111
11. 01010000 **13.** 11010001 **15.** 00000000 **17.** 00000000 **19.** 11111111
21. 01101100 **23.** 11011101 **25.** 00011000

27. **a.** $(P \cdot Q') + (Q' \cdot R)$

b. A: 00001111
B: 00110011
C: 01010101
output: 00000100

c.

29. **a.** $P \cdot Q \cdot R + P \cdot Q' \cdot R + P' \cdot Q$

b. P: 00001111
Q: 00110011
R: 01010101
output: 00110101

c.

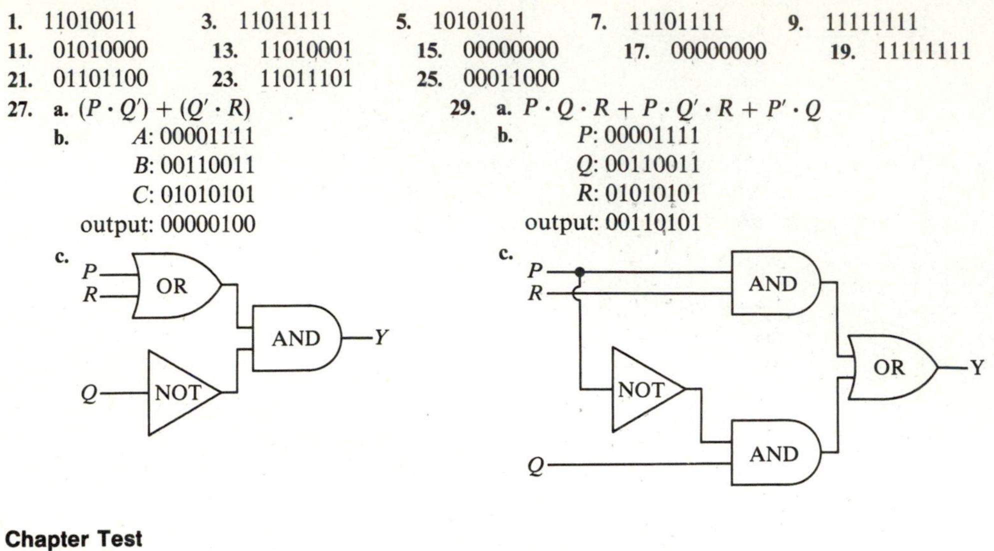

Chapter Test

1. A' **2.** $A + B$

3. **a.** $[p \vee (q \wedge r)] \wedge (\sim p)$ **b.** $(P + Q \cdot R) \cdot P'$ **c.**

p	q	r	$[p \vee (q \wedge r)] \wedge (\sim p)$
T	T	T	F
T	T	F	F
T	F	T	F
T	F	F	F
F	T	T	T
F	T	F	F
F	F	T	F
F	F	F	F

Current flows only when switch P is open and switches Q and R are closed.

4. **a.** $(p \vee q) \wedge r \wedge (p \vee q)$ **b.** $(P + Q) \cdot R \cdot (P + Q)$ **c.**

p	q	r	$(p \vee q) \wedge r \wedge (p \vee q)$
T	T	T	T
T	T	F	F
T	F	T	T
T	F	F	F
F	T	T	T
F	T	F	F
F	F	T	F
F	F	F	F

Current flows when (1) all three switches are closed; (2) switches P and R are closed; or (3) switches R and Q are closed.

5. 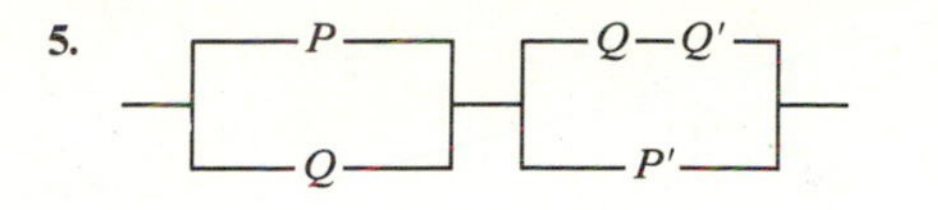

6. $(P + Q) \cdot (Q \cdot Q' + P')$

7. —— P' —— Q ——

8. 11011111 **9.** 10010001 **10.** 00101000

11. **a.** $[A \cdot (B + C)] + [(B + C) \cdot A']$

b.
A: 00001111
B: 00110011
C: 01010101
output: 01110111

c. 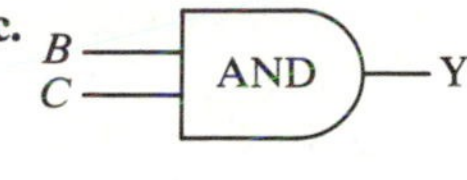

CHAPTER 6

NOTE: The algorithms given below are *possible* solutions to the problems presented in the exercises for this chapter. A given algorithm will *not*, however, represent the only possible solution.

Section 6.1

1.
1. Obtain PRICE (∗ Price of calculator ∗)
2. TAX = 0.07 ∗ PRICE
3. GROSS AMT = PRICE + TAX
4. Record GROSS AMT
5. END

3.
1. Obtain scores for GAME1, GAME2, GAME3
2. SUM = GAME1 + GAME2 + GAME3
3. AVERAGE = SUM/3
4. Record AVERAGE
5. END

5.
1. Obtain A and B
2. NUM = A + 3 ∗ B (∗ Numerator of the fraction ∗)
3. Y = NUM/2
4. Record Y
5. END

7.
1. Obtain A, B, and C
2. If C < > 0 then do
   a. NUM = A + 3 ∗ B
   b. Y = NUM/C
   c. Record Y
   End
3. ELSE Print "Can't divide by 0"
4. END

9.
1. Set SUBTOT = 0; set TOTTAXABLE = 0 (∗ Initialize ∗)
2. Set COUNT = 1 (∗ Initialize item counter ∗)
3. DOWHILE (COUNT < = 10)
4. Obtain ITEM, COST (∗ description & cost of item ∗)
5. IF ITEM is non-food THEN TOTTAXABLE = TOTTAXABLE + COST
6. SUBTOT = SUBTOT + COST (∗ done for *all* items ∗)
7. COUNT = COUNT + 1
8. END DOWHILE
9. TAX = 0.06 ∗ TOTTAXABLE
10. TOTAL = SUBTOT + TAX
11. Record TOTAL
12. END

11. **a.**
1. Set COUNT = 1 (∗ Initialize counter ∗)
2. DOWHILE (COUNT < = 10)
3. Obtain A, B, C

```
    4.  IF C < > 0 THEN DO
          a. NUM = A + 3 * B
          b. Y = NUM/C
          c. Record Y
          END DO
    5.  ELSE PRINT "Can't divide by 0"
    6.  COUNT = COUNT + 1
    7. END DOWHILE
    8. END
 b. 1. Obtain N      (* Here N is now known in advance *)
    2. Set COUNT = 1
    3. DOWHILE (COUNT < = N)
    Continue as in steps 3 through 8 in part a above.
```

```
13.  1. Obtain N      (* Number of employees *)
     2. Set E = 1     (* E counts employees *)
     3. DOWHILE (E < = N)
     4.   Obtain HOURS, RATE, TR       (* TR is the tax rate *)
     5.   IF HOURS > 40 THEN
            DO
            a. EXTRA = HOURS - 40
            b. OTPAY = 1.5 * RATE * EXTRA
            c. GROSS = 40 * RATE + OTPAY
            END DO
     6.   ELSE GROSS = HOURS * RATE
     7.   TAX = TR * GROSS
     8.   NET = GROSS - TAX
     9.   Record NET
    10.   E = E + 1
    11. END DOWHILE
    12. END
```

Section 6.2

```
1.  1. Obtain N      (* Number of employees *)
    2. E = 1         (* Initialize counter *)
    3. GSUM = 0; NETSUM = 0      (* Initialize accumulators *)
    4. DOWHILE (E < = N)
    5.     Obtain HOURS, RATE, TR
    6.     GROSS = HOURS * RATE
    7.     GSUM = GSUM + GROSS      (* GSUM is sum of gross pays)
    8.     TAX = GROSS * TR
    9.     NET = GROSS - TAX
   10.     NETSUM = NETSUM + NET     (* NETSUM is sum of net pays)
   11.     Record GROSS, NET
   12.     E = E + 1
   13. END DOWHILE
   14. Record GSUM, NETSUM
   15. END
```

3.

Pass	E	N	HOURS	Value of RATE	GROSS	SUM
1	1	4	40	7	280	280
2	2	4	25	8	200	480
3	3	4	30	6.5	195	675
4	4	4	20	12	240	915

MEAN GROSS PAY = $228.75

5. a.

C	N	I	NI	PRICE	BILL	TOTAL
1	3	1	2	10	10	0
1	3	2	2	12	22	22
2	3	1	4	5	5	22
2	3	2	4	6	11	22
2	3	3	4	17	28	22
2	3	4	4	12.5	40.5	62.5
3	3	1	3	5.5	5.5	62.5
3	3	2	3	7	12.5	62.5
3	3	3	3	8	20.5	83

b. $83.00

7.

```
 1. ZCNT = 0       (* Counts occurrences of D = 0 *)
 2. PCNT = 0       (* Counts occurrences of D > 0 *)
 3. NCNT = 0       (* Counts occurrences of D < 0 *)
 4. READ A, B, C
 5. DOWHILE (A < > 0)
 6.        D = B^2 - 4 * A * C
 7.        PRINT D
 8.        IF D = 0 THEN ZCNT = ZCNT + 1
 9.        ELSE IF D > 0 THEN PCNT = PCNT + 1
10.             ELSE NCNT = NCNT + 1
11.        READ A, B, C
12. END DOWHILE
13. Record ZNCT, PCNT, NCNT
14. END
```

9.

```
 10  INPUT N
 20  LET TTL = 0
 30  FOR C = 1 TO N
 40      LET BILL = 0
 50      INPUT NI
 60      FOR I = 1 TO NI
 70          INPUT PRICE
 80          LET BILL = BILL + PRICE
 90      NEXT I
100      PRINT BILL
110      TTL = TTL + BILL
120  NEXT C
130  PRINT TTL
140  END
```

Section 6.3

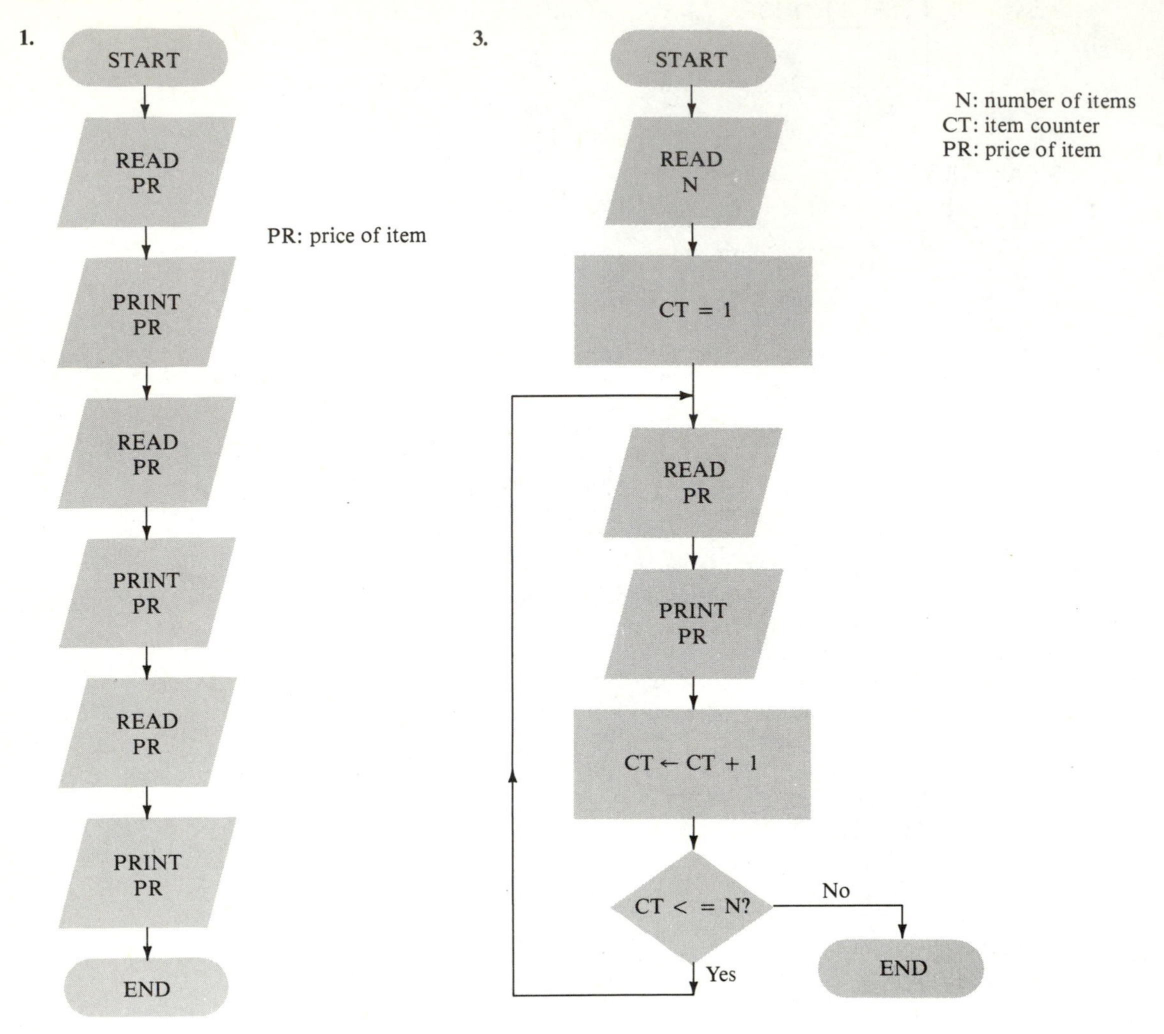

1.
START
READ
PR
PR: price of item
PRINT
PR
READ
PR
PRINT
PR
READ
PR
PRINT
PR
END
3.
START
N: number of items
CT: item counter
PR: price of item
READ
N
CT = 1
READ
PR
PRINT
PR
CT ← CT + 1
CT < = N?
No
Yes
END

5.

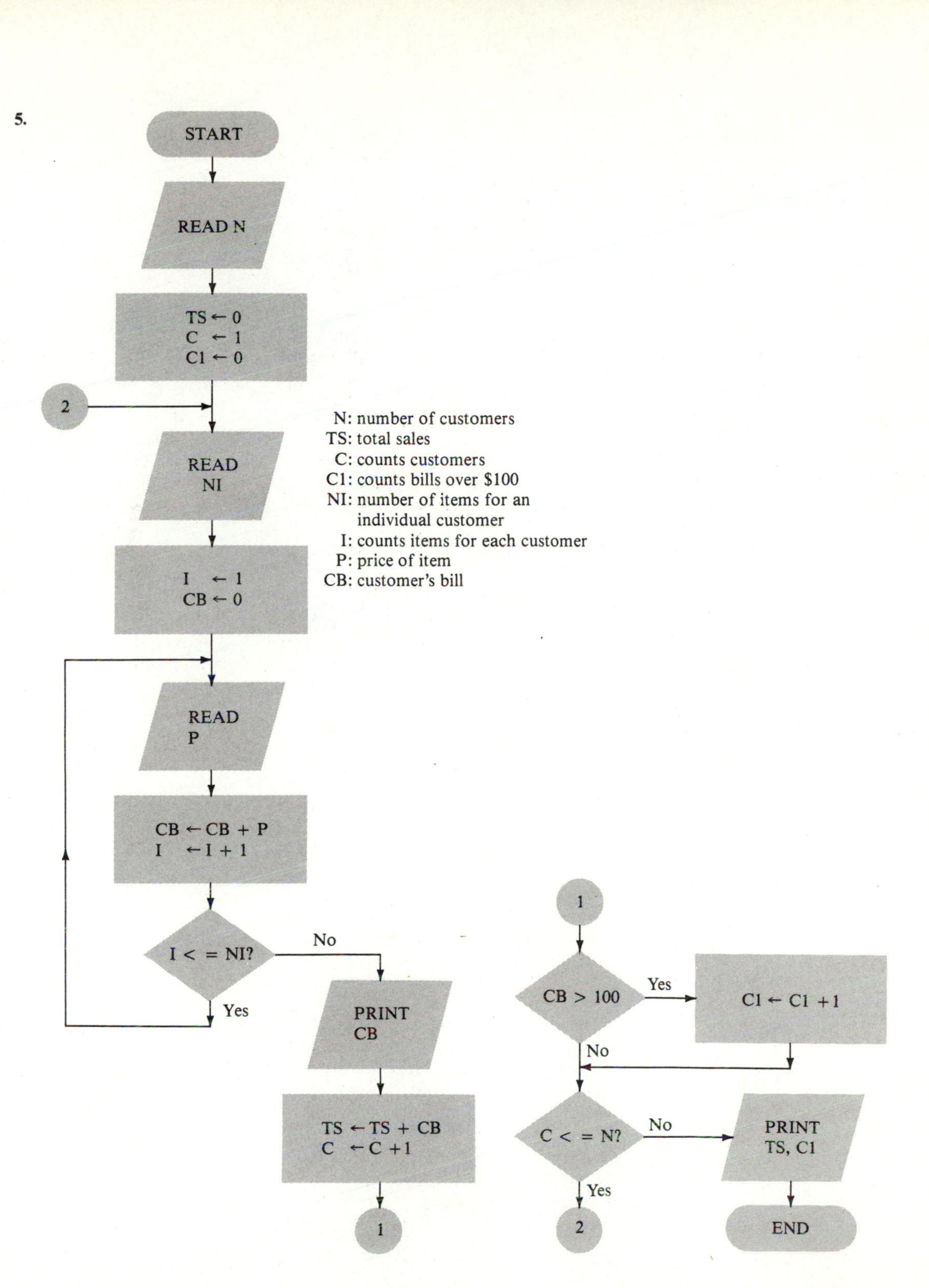

START
READ N
TS ← 0
C ← 1
C1 ← 0
2
N: number of customers
TS: total sales
C: counts customers
C1: counts bills over $100
NI: number of items for an individual customer
I: counts items for each customer
P: price of item
CB: customer's bill
READ NI
I ← 1
CB ← 0
READ P
CB ← CB + P
I ← I + 1
I < = NI?
No
Yes
PRINT CB
TS ← TS + CB
C ← C +1
1
1
CB > 100
Yes
C1 ← C1 +1
No
C < = N?
No
PRINT TS, C1
Yes
2
END

7\. a.

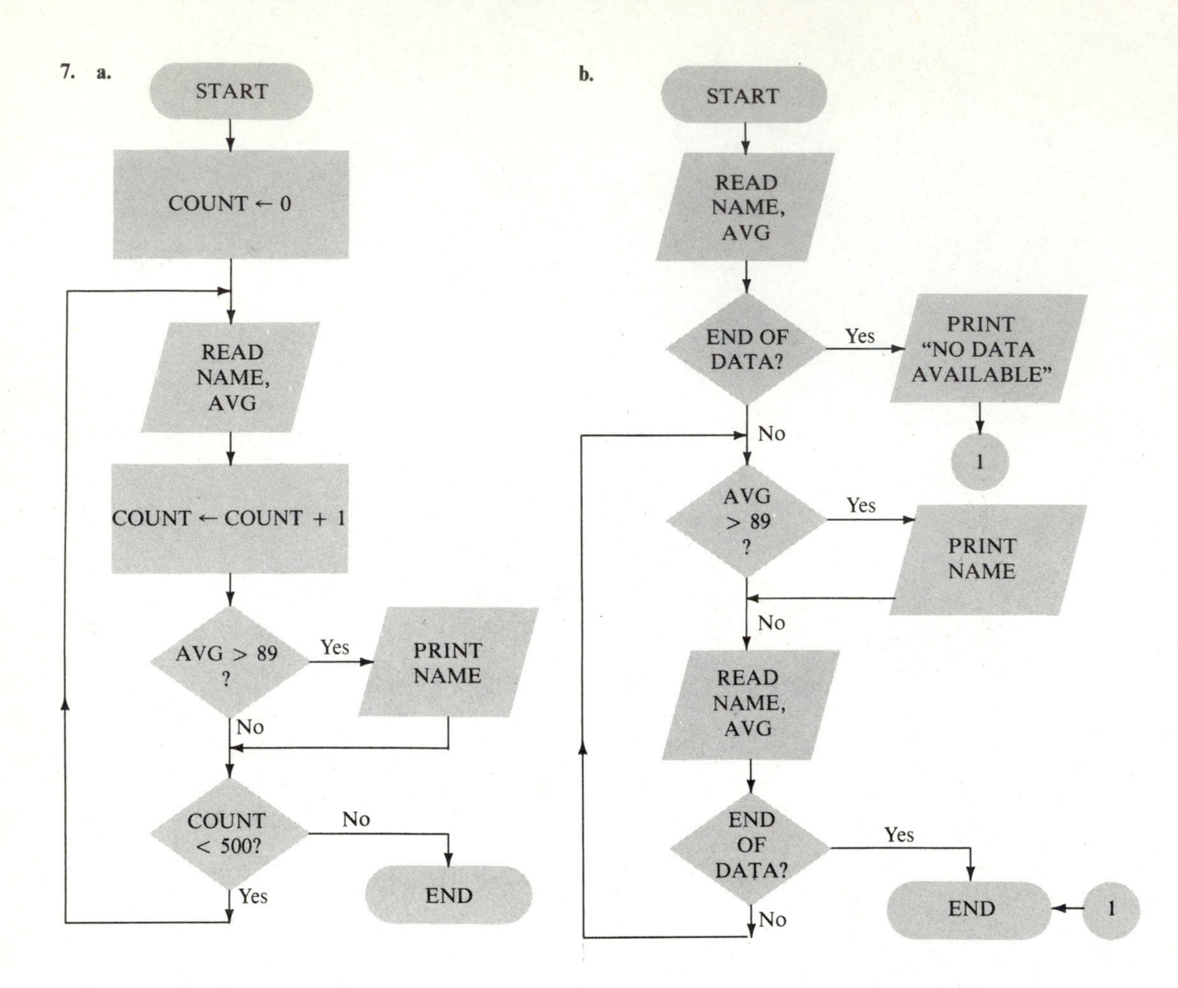

7. a.
START
COUNT ← 0
READ NAME, AVG
COUNT ← COUNT + 1
AVG > 89 ?
Yes
PRINT NAME
No
COUNT < 500?
No
END
Yes
b.
START
READ NAME, AVG
END OF DATA?
Yes
PRINT "NO DATA AVAILABLE"
1
No
AVG > 89 ?
Yes
PRINT NAME
No
READ NAME, AVG
END OF DATA?
Yes
END
1
No

9.

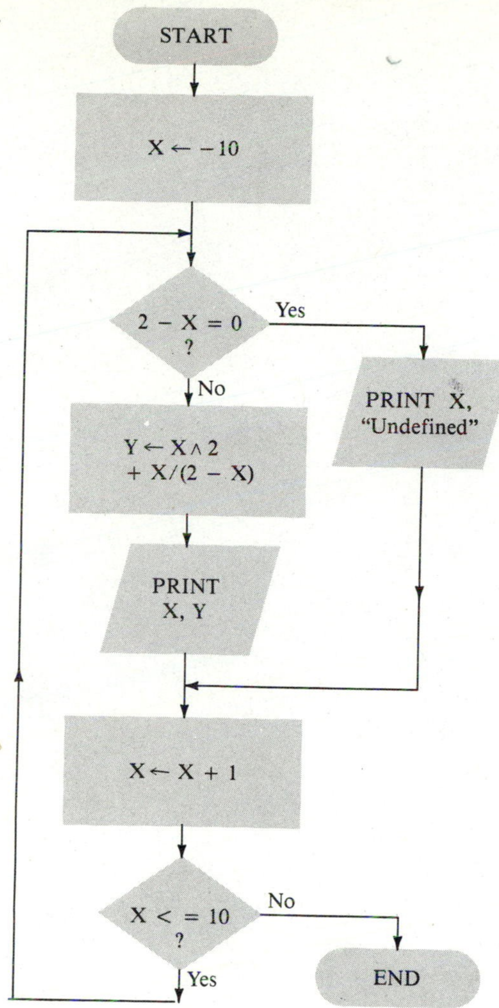

START
X ← −10
2 − X = 0
?
Yes
No
PRINT X,
"Undefined"
Y ← X ∧ 2
+ X/(2 − X)
PRINT
X, Y
X ← X + 1
X < = 10
?
No
Yes
END

11.

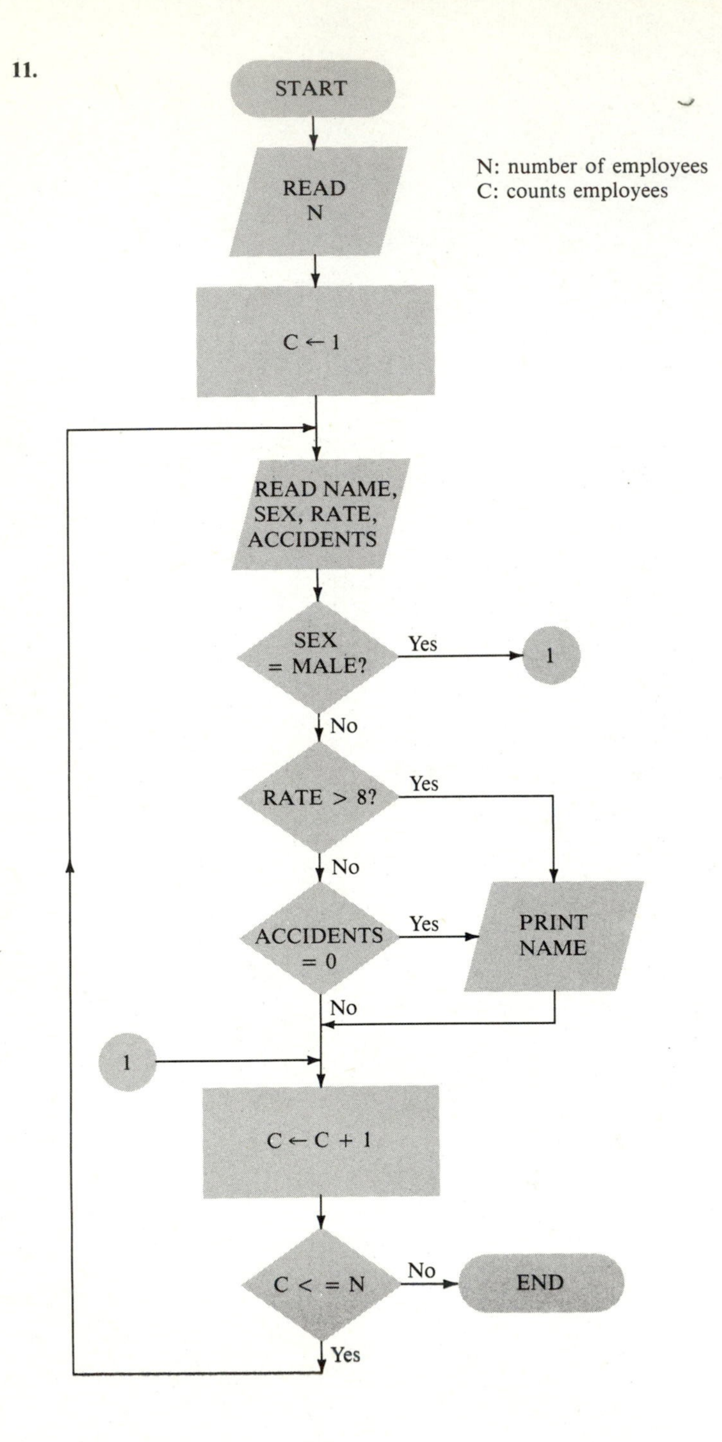
START
READ
N
N: number of employees
C: counts employees
C ← 1
READ NAME,
SEX, RATE,
ACCIDENTS
SEX
= MALE?
Yes
1
No
RATE > 8?
Yes
No
ACCIDENTS
= 0
Yes
PRINT
NAME
No
1
C ← C + 1
C < = N
No
END
Yes

Chapter Test

1. a. (1) initialize the accumulator T to 0
   (2) read in a value and assign it to the variable X
   (3) decision: determine whether X is negative
   If not, continue with
      (4) add the value of X to the accumulator T
      (5) add 1 to N, a counter of all the nonnegative values read
      (A) go back to step (2)
   If X is negative, do
      (6) assign the result of dividing T by N to the variable Y
      (7) Print the value of Y

   b. Add to box 1: N ← 0

   c. It acts as an end-of-data sentinel or "trailer" value.

   d.

Pass	X	T	N	Y
1	8	8	1	—
2	2	10	2	—
3	4	14	3	—
4	6	20	4	—
5	−2	20	4	5

   The number 5 would be printed.

2. a. (1) read in a value and assign it to the variable N—the number of problems to be done
   (2) will be used to initialize the counter I
   (3) read in values for X and Z, in that order
   (4) assign the result of calculating $2X + X/3$ to the variable Y
   (5) print the value of Y
   (6) add 1 to the counter I
   (7) determine whether I is less than or equal to N; if it is, return to step (3); if not, end.

   b. box 2 [I ← 1]

   box 4 [Y ← 2 * X + X^3/Z]

   c.
```
1. Obtain N, the number of problems
2. Set the value of I to 1     (* I counts problems *)
3. DOWHILE (I <= N)
4.     Obtain X, Z
5.     Y = 2 * X + X^3/Z
6.     Print Y
7.     I = I + 1
8. END DOWHILE
9. END
```

   d. after box 3

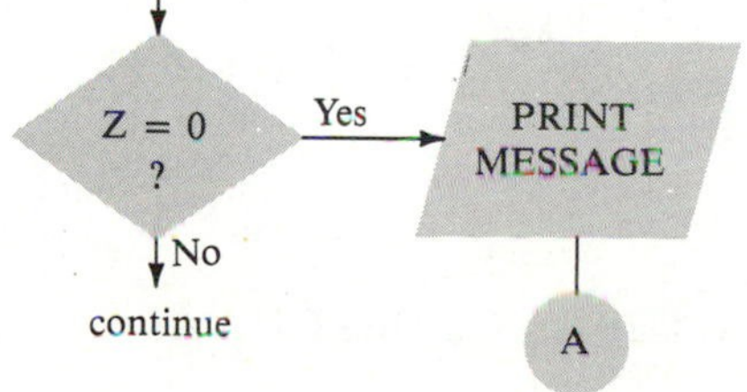

3.

```
 1. COUNT = 1      (* Counts number of salespeople *)
 2. DOWHILE (COUNT < = 5)
 3.     Obtain NAME, SALES, BSAL     (* BSAL is base salary *)
 4.     IF SALES < 10000 THEN RCOM = 0     (* RCOM is commission rate *)
 5.     ELSE IF SALES < 20000 THEN RCOM = 0.05
 6.          ELSE IF SALES < 40000 THEN RCOM = 0.10
 7.               ELSE RCOM = 0.15
 8.     COM = RCOM * SALES (* COM is commission amount *)
 9.     SAL = SALES + COM (* SAL is salary *)
10.     Print NAME, SALES, SAL, COM
11.     COUNT = COUNT + 1
12. END DOWHILE
13. END
```

4. Add to 1: COUNT = 1; TOTAL = 0 (* TOTAL = commission total *)
Add between 8 and 9: TOTAL = TOTAL + COM
Add between 12 and 13: Print TOTAL

CHAPTER 7

Section 7.1

1. 3 **3.** 5 **5.** 5 **7.** false **9.** true **11.** true

13. $(1 \quad 7 \quad 13 \quad 16 \quad 19)$ **15.** $(-1 \quad 3 \quad 7 \quad 8 \quad 9)$ **17.** $\begin{pmatrix} 1 \\ 2 \\ 3 \\ 7 \end{pmatrix}$

19. $(1 \quad \frac{9}{2} \quad 8 \quad 10 \quad 12)$ **21.** $\begin{pmatrix} 6 \\ 12 \\ 18 \\ 42 \end{pmatrix}$ **23.** 3 **25.** -3 **27.** 9 **29.** 48

31. **a.** $X = (1.2 \quad 1.3 \quad 2.6 \quad 1.6 \quad 2.1 \quad 1.9)$ **b.** $Y = \begin{pmatrix} 28 \\ 26 \\ 30 \\ 30 \\ 42 \\ 28 \end{pmatrix}$ **c.** $X \cdot Y = 334.8$ **33.** 13

35. **a.** $535.00 **b.** 572.45 **c.** 572.45

37.

```
10 DIM A(5), C(5)
20 REM     STATEMENTS 30 - 50 ACCEPT 5 VALUES OF A AS
```

```
25   REM        INPUT FROM THE KEYBOARD
30   FOR I = 1 TO 5
35       PRINT "ENTER QUANTITY OF ITEM #"; I
40       INPUT A(I)
50   NEXT I
60   REM     STATEMENTS 70 – 90 ACCEPT 5 VALUES OF C AS
65   REM        INPUT FROM THE KEYBOARD
70   FOR I = 1 TO 5
75       PRINT "ENTER COST OF ITEM #"; I
80       INPUT C(I)
90   NEXT I
95   LET SUM = 0
96   REM     MULTIPLY VECTORS A AND C
100  FOR I = 1 TO 5
110      LET M = A(I) * C(I)
120      LET SUM = SUM + M
130  NEXT I
200  PRINT "TOTAL INVOICE = "; SUM
1000 END
```

Section 7.2

1. 2×3 **3.** 1×3 **5.** 5×1 **7.** 5 **9.** 3 **11.** 4 **13.** 4
15. 0 **17.** 8 **19.** 5 **21.** 6 **23.** 3 **25.** 7

Section 7.3

1. $\begin{pmatrix} 13 & 9 & 4 \\ 2 & 13 & 30 \end{pmatrix}$ **3.** $\begin{pmatrix} 20 & 24 & -8 \\ 0 & 12 & 40 \end{pmatrix}$ **5.** cannot be done **7.** $\begin{pmatrix} 17 & 20 \\ 7 & 49 \\ 24 & 75 \end{pmatrix}$

9. cannot be done **11.** $\begin{pmatrix} 5.1 & 1 \\ 1.2 & 5 \\ 6.3 & 8 \end{pmatrix}$ **13.** $\begin{pmatrix} 19 & 8 \\ 63 & 59 \end{pmatrix}$ **15.** cannot be done

17. cannot be done **19.** cannot be done **21.** $\begin{pmatrix} 4 & 5 & -2 \\ 3 & 1 & 6 \\ 0 & 0 & 8 \end{pmatrix} = E$ **23.** $\begin{pmatrix} 28 & 103 & 206 \\ 56 & 206 & 412 \\ 84 & 309 & 618 \end{pmatrix}$

25. $\begin{pmatrix} 13 & 5 \\ 52 & 33 \\ 48 & 40 \end{pmatrix}$ **27.** $\begin{pmatrix} 17 & 41 & 53 \\ 12 & 32 & 43 \\ 8 & 16 & 24 \end{pmatrix}$ **29.** $\begin{pmatrix} 29 & 290 \\ 34 & 340 \\ 14 & 140 \end{pmatrix}$ **31.** $\begin{pmatrix} 11 & 8 \\ 20 & 19 \end{pmatrix}$

33. $\begin{pmatrix} 31 & 25 & 6 \\ 15 & 16 & 48 \\ 0 & 0 & 64 \end{pmatrix}$ **35.** $\begin{pmatrix} \text{tot. prod. cost} - \text{Belmont} & \text{tot. shipping cost} - \text{Belmont} \\ \text{tot. prod. cost} - \text{Wtby.} & \text{tot. shipping cost} - \text{Wtby.} \end{pmatrix}$

37. \$38.125 or \$38.13–the combined production costs of all three books.

39. $\begin{pmatrix} 1 & 26 \\ 0 & 27 \end{pmatrix}$ 41. $\begin{pmatrix} 0.112 & 0.888 \\ 0.111 & 0.889 \end{pmatrix}$ 43. $\begin{pmatrix} 0.94 & 8.06 \\ 0.93 & 8.07 \end{pmatrix}$

45. $\begin{pmatrix} 1 & 3 & 5 \\ 2 & 4 & 6 \end{pmatrix}$ 47. $\begin{pmatrix} 1 & 0 \\ 5 & -7 \end{pmatrix}$

Section 7.4

1. not square 3. order = 2 5. order = 4 7. yes 9. yes 11. yes

13. no 15. yes 17. $\begin{pmatrix} -3 & 2 \\ 2 & -1 \end{pmatrix}$ 19. $\begin{pmatrix} -5 & 7 \\ -2 & 3 \end{pmatrix}$ 21. singular

23. $\begin{pmatrix} 0 & 0 & \frac{1}{4} \\ 0 & \frac{1}{4} & 0 \\ \frac{1}{4} & 0 & 0 \end{pmatrix}$ 25. singular 27. $\begin{pmatrix} 70/33 & 20/33 \\ 10/33 & 50/33 \end{pmatrix}$

29. $\begin{pmatrix} 192/121 & 76/121 & 46/121 \\ 106/121 & 168/121 & 38/121 \\ 100/121 & 90/121 & 150/121 \end{pmatrix}$

Section 7.5

1. 9 3. -10 5. -8 7. -15 9. 21 11. $\begin{pmatrix} 4 & -1 \\ -11 & 3 \end{pmatrix}$

13. $\begin{pmatrix} 2 & -1 \\ -7 & 4 \end{pmatrix}$ 15. $\begin{pmatrix} 2 & -1 \\ -3 & 2 \end{pmatrix}$ 17. $\begin{pmatrix} -\frac{1}{5} & \frac{2}{5} \\ \frac{3}{5} & -\frac{1}{5} \end{pmatrix}$ 19. $\begin{pmatrix} 6/26 & -4/26 \\ 2/26 & 3/26 \end{pmatrix}$

21. **a.** -5 **b.** 5 **c.** 37 23. **a.** -80 **b.** -80 **c.** 20

25. **a.** -24 **b.** -24 **c.** -39 27. $\begin{pmatrix} 3/37 & -5/37 & -5/37 \\ 5/37 & 4/37 & 4/37 \\ 21/37 & -35/37 & 2/37 \end{pmatrix}$

29. $\begin{pmatrix} 17/39 & -2/39 & -7/39 \\ -9/39 & 24/39 & 6/39 \\ -8/39 & 17/39 & 1/39 \end{pmatrix}$

Section 7.6

1. Change: 990 DATA 1, 5, 0, 7, −3, 4, 1, 6, 7, 9

3. Changes:
```
10  DIM A(4, 3), B(3, 2)
20  REM  A IS A 4-BY-3 MATRIX AND B IS 3-BY-2
 :
100 FOR R = 1 TO 4
 :
```

```
         200  FOR R = 1 TO 3
              ⋮
         990  DATA 1, 2, 1, 5, 0, 0, 7, -4, 3, 3, 6, 9
         995  DATA 8, 3, 4, 9, 6, -5
```

5. Changes:

```
          10  DIM A(3, 2), B(2)
          20  REM  A IS A 3-BY-2 MATRIX AND B IS 1-BY-2
              ⋮
         100  FOR R = 1 TO 3
         110      FOR C = 1 TO 2
              ⋮
         150  REM  MATRIX B IS READ IN—STATEMENTS 200-220
         200  FOR C = 1 TO 2
         210      READ B(C)
         220  NEXT C
        (omit 230, 240)
         990  DATA 1, 0, 0, 1, 5, 7, 8, 9
```

7. Changes:

```
          10  DIM A(3, 4), B(2, 3)
         100  FOR R = 1 TO 3
         110      FOR C = 1 TO 4
              ⋮
         210  FOR C = 1 TO 3
              ⋮
```

9. Changes:

```
          10  DIM A(3, 4), B(3)
              ⋮
        (omit 200)
         210  FOR C = 1 TO 3
         220      PRINT "COMPONENT #"; C; "OF B IS"
         225      INPUT B(C)
         230  NEXT C
        (omit 240)
              ⋮
```

11.

```
1. FOR R = 1 TO N
2.     FOR C = 1 TO N
3.         A(R, C) = K * A (R, C)
4.     NEXT C
5.   NEXT R
6. END
```

13.

```
  ⋮
700  FOR R = 1 TO NR
710      FOR C = 1 TO NC
720          LET X(R, C) = K1 * A(R, C) + K2 * B(R, C)
730      NEXT C
740  NEXT R
  ⋮
```

15.

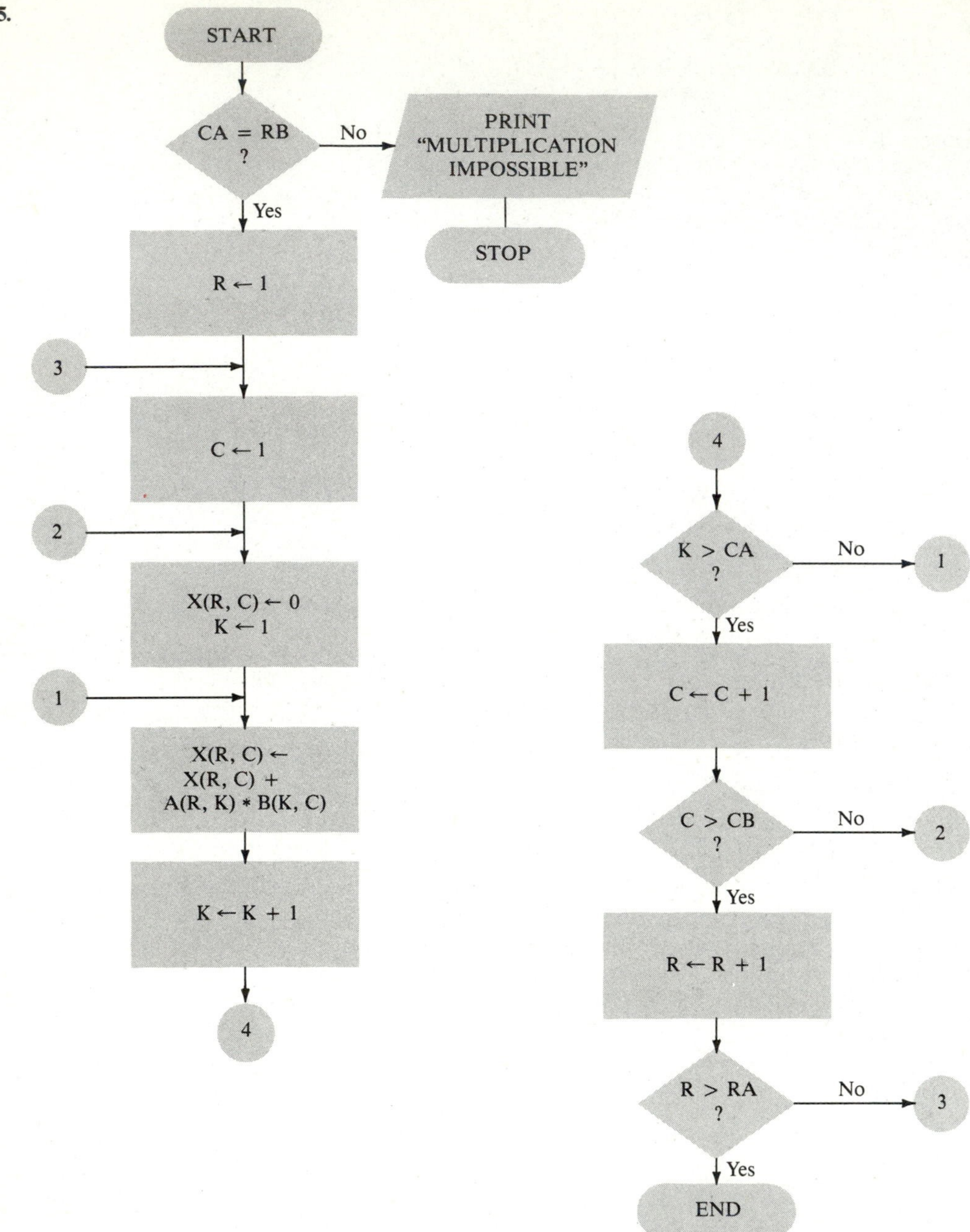

START
CA = RB ?
No
PRINT "MULTIPLICATION IMPOSSIBLE"
STOP
Yes
R ← 1
3
C ← 1
2
X(R, C) ← 0
K ← 1
1
X(R, C) ←
X(R, C) +
A(R, K) * B(K, C)
K ← K + 1
4
4
K > CA ?
No
1
Yes
C ← C + 1
C > CB ?
No
2
Yes
R ← R + 1
R > RA ?
No
3
Yes
END

17. a. $A \cdot B = \begin{pmatrix} 63 & 75 & 11 \\ 14 & 8 & -2 \end{pmatrix}$ b. multiplication is impossible c. $A \cdot B = 35$

19. Replace 810–860 with 810 MAT PRINT X

Chapter Test

1. $(1 \quad -1 \quad 12)$ 2. 20 3. $(-2 \quad 8 \quad -6)$ 4. $(-1 \quad 4 \quad -3)$

5. dimensions are 4×3; M is not a square matrix 6. a. 5 b. -6 c. 4 d. -4

7. $\begin{pmatrix} -9 & 6 & -12 \\ -3 & 0 & -15 \\ -18 & 18 & 9 \\ 12 & -6 & 21 \end{pmatrix}$ 8. $(-1 \quad 0 \quad -3)$ 9. $\begin{pmatrix} 1 & 0 & 0 \\ 0 & 1 & 0 \\ 0 & 0 & 1 \end{pmatrix}$

10. $\begin{pmatrix} 2 & 2 \\ 3 & 5 \end{pmatrix}$ 11. $\begin{pmatrix} 1 & 2 \\ 3 & 4 \end{pmatrix}$ 12. $|A| = -2$; $|C| = 0$

13. $A^{-1} = \begin{pmatrix} -2 & 1 \\ \frac{3}{2} & -\frac{1}{2} \end{pmatrix}$; C has no inverse.

14. a. 6 b. $\begin{pmatrix} 2 & -\frac{3}{2} & 2 \\ -\frac{1}{3} & \frac{1}{2} & -\frac{2}{3} \\ -1 & 1 & -1 \end{pmatrix}$ c. $\begin{pmatrix} 2 & -\frac{3}{2} & 2 \\ -\frac{1}{3} & \frac{1}{2} & -\frac{2}{3} \\ -1 & 1 & -1 \end{pmatrix}$

CHAPTER 8

Section 8.1

1. $x = 2$
   $y = 4$

3. $x = 1$
   $y = 2$

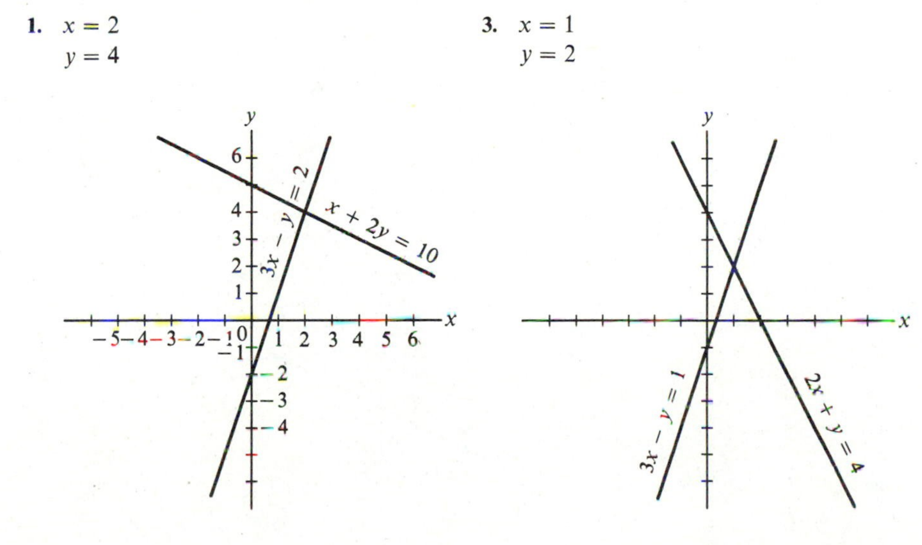

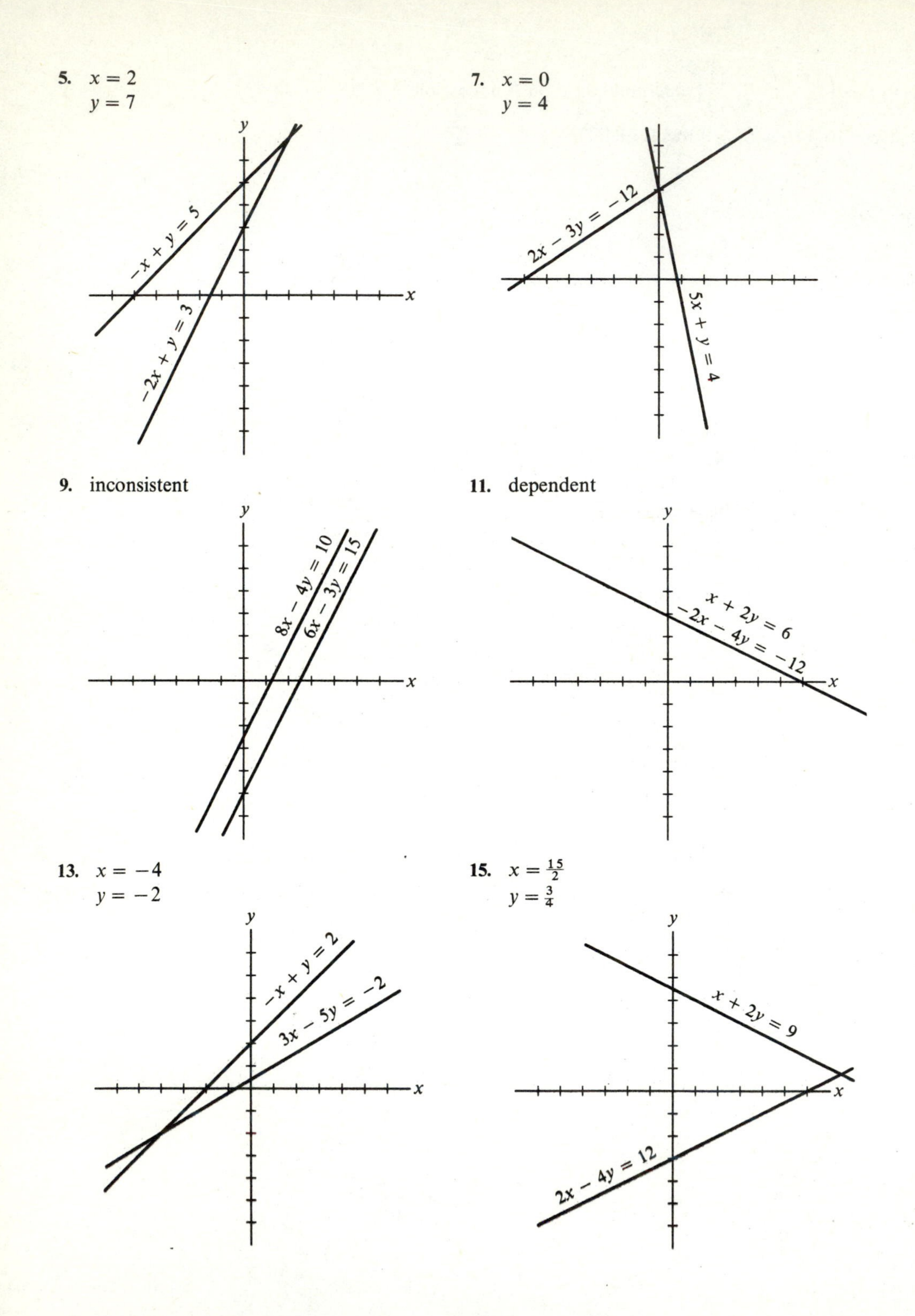

5. x = 2
y = 7
−x + y = 5
−2x + y = 3
7. x = 0
y = 4
2x − 3y = −12
5x + y = 4
9. inconsistent
8x − 4y = 10
6x − 3y = 15
11. dependent
x + 2y = 6
−2x − 4y = −12
13. x = −4
y = −2
−x + y = 2
3x − 5y = −2
15. x = 15/2
y = 3/4
x + 2y = 9
2x − 4y = 12
x
y

17. $x_1 = 0$
$x_2 = 4$

19. inconsistent

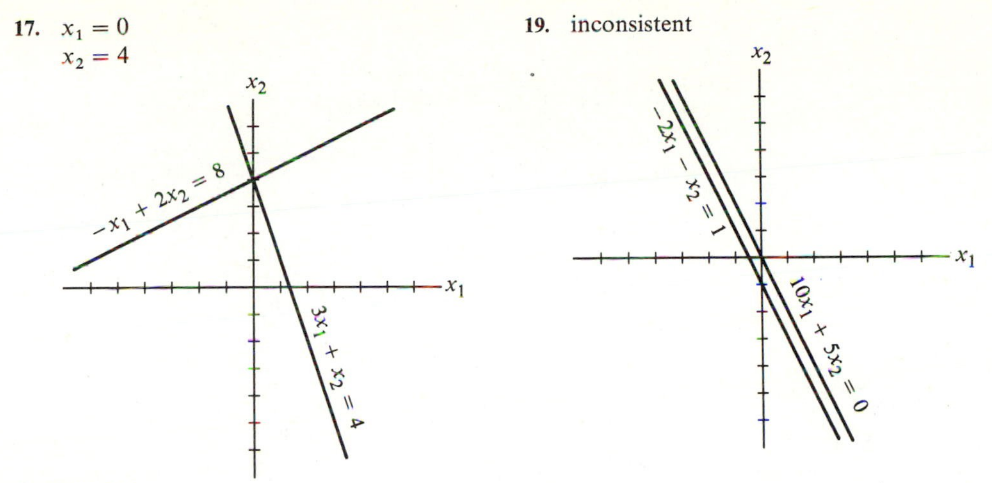

Section 8.2

1. $x = 5;\ y = 1$ **3.** $x = 0;\ y = 0$ **5.** $x = -4;\ y = 10$ **7.** $x = 5;\ y = 1$
9. $x = 0;\ y = 0$ **11.** $x = -4;\ y = 10$ **13.** $x = 1;\ y = 5$ **15.** $x = -2;\ y = 3$
17. $x = 10;\ y = 30$ **19.** $x = 0;\ y = -4$ **21.** 12; 7 **23.** adult = \$5.00; child = \$3.00

Section 8.3

1. $x = \frac{22}{5}$, $y = \frac{13}{5}$, $z = -\frac{38}{5}$

3. $x_1 = 1$, $x_2 = 2$, $x_3 = -1$

5. $x = -4$, $y = 1$, $z = 1$

7. dependent

9. $x_1 = \frac{26}{25}$, $x_2 = \frac{44}{25}$, $x_3 = -\frac{64}{25}$

11. $x_1 = -\frac{1}{3}$, $x_2 = \frac{2}{3}$, $x_3 = -\frac{8}{9}$

13. $x_1 = -\frac{89}{17}$, $x_2 = \frac{60}{17}$, $x_3 = \frac{260}{17}$

15. $x_1 = 5$, $x_2 = 5$, $x_3 = 5$

17. No. The system of equations associated with this problem is *dependent*; there is not a unique solution.

Section 8.4

1. $x = -2$, $y = 3$

3. $x = \frac{1}{2}$, $y = 4$

5. $x = \frac{4}{29}$, $y = \frac{39}{29}$

7. $x = 1$, $y = 4$, $z = 3$

9. $x = 1$, $y = 1$, $z = -2$

11. $x = 10$, $y = 37$, $z = 4$

13. $x_1 = 1$, $x_2 = 2$, $x_3 = 3$

15. $x = 6$, $y = 2$, $z = 0$

17. no solution ($D = 0$)

19. no solution ($D = 0$)

21. Start with

$$a_1x + b_1y = c_1 \qquad (1)$$
$$a_2x + b_2y = c_2 \qquad (2)$$

$$-a_1a_2x - a_2b_1y = -a_2c_1 \qquad \text{Multiply 1 by } -a_2.$$
$$a_1a_2x + a_1b_2y = a_1c_2 \qquad \text{Multiply 2 by } a_1.$$
$$a_1b_2y - a_2b_1y = a_1c_2 - a_2c_1 \qquad \text{Add the resulting equations.}$$
$$(a_1b_2 - a_2b_1)y = a_1c_2 - a_2c_1 \qquad \text{Solve for } y.$$
$$y = \frac{a_1c_2 - a_2c_1}{a_1b_2 - a_2b_1}$$

Section 8.5

1. maximum $P = 2100$ when $x = 3$, $y = 6$

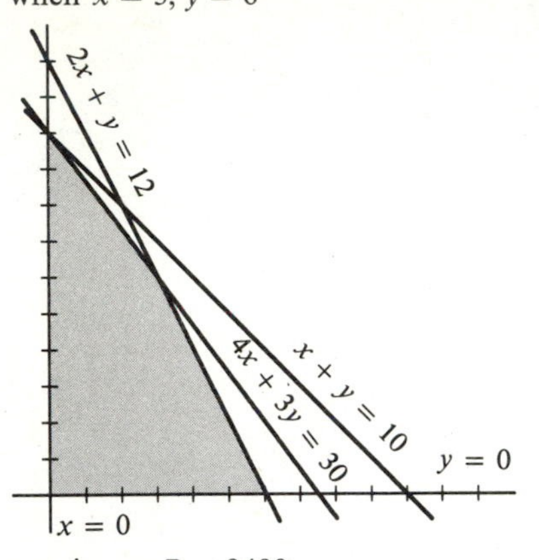

3. Graphs are the same as in 1. maximum $P = 2500$ when $x = 0$, $y = 10$

5. maximum $P = 2400$ when $x = 12$, $y = 0$

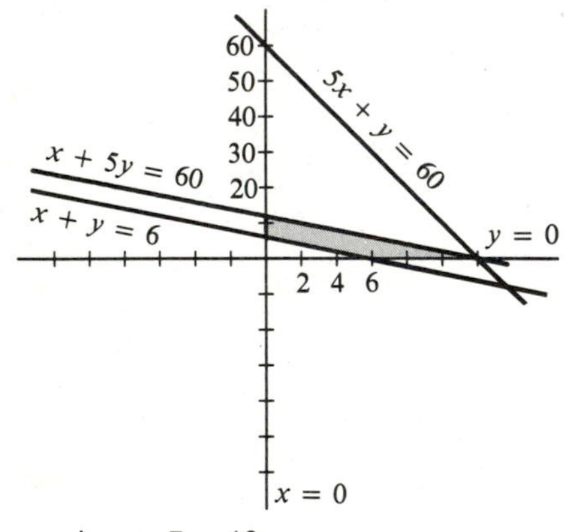

7. Graphs are the same as in 5. minimum $C = 60$ when $x = 0$, $y = 6$

9. maximum $P = 12$ when $x = 6$, $y = 6$

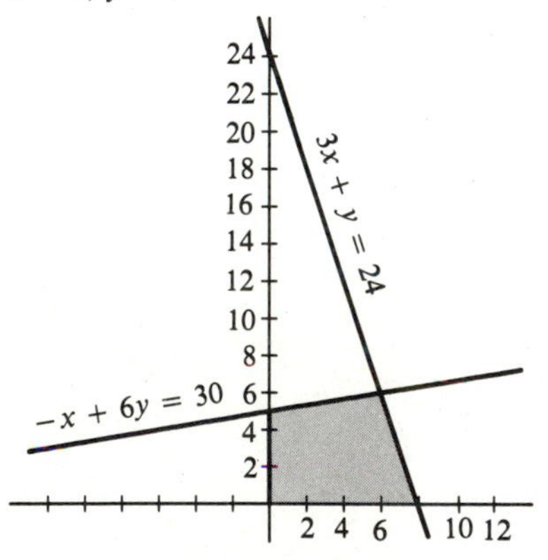

11. minimum $C = 680{,}000$
when factory 1 is open 34 days and factory 2 is open 17 days

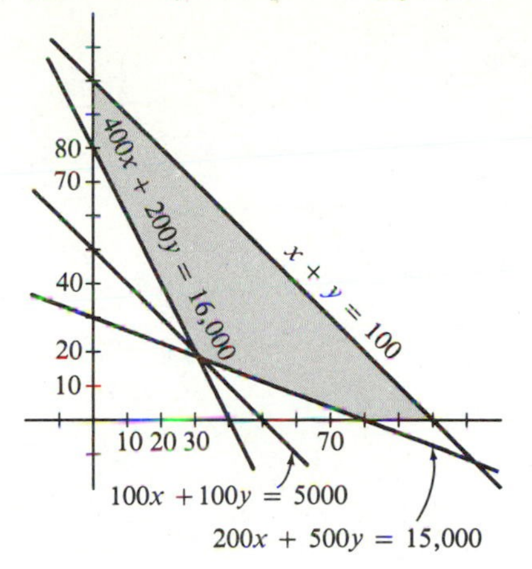

Section 8.6

1. Maximum value of P is 15; it occurs when $x = 3$, $y = 0$.
3. Maximum value of P is 6800; it occurs when $x = 20$, $y = 40$.
5. Maximum value of P is 4020; it occurs when $x = 40$, $y = 10$.
7. Maximum value of P is 4020; it occurs when $x = 40$, $y = 10$.
9. Maximum value of P is 112; it occurs when $x = 0$, $y = 6$, $z = 10$.
11. Maximum value of P is 2100; it occurs when $x = 3$, $y = 6$.
13. Maximum profit is 1200; it occurs when the company produces 80 deluxe dryers and no economy dryers.
15. Maximum value is 19; it occurs when $x = 6.5$, $y = 2$, $z = 8$, and $w = 0.5$.
17. Maximum value of P is 1700; it occurs when $x = 50$, $y = 20$.
19. Maximum value of P is 1200; it occurs when $x = 50$, $y = 20$.

Chapter Test

1. $x = 5$
   $y = 4$

2. dependent

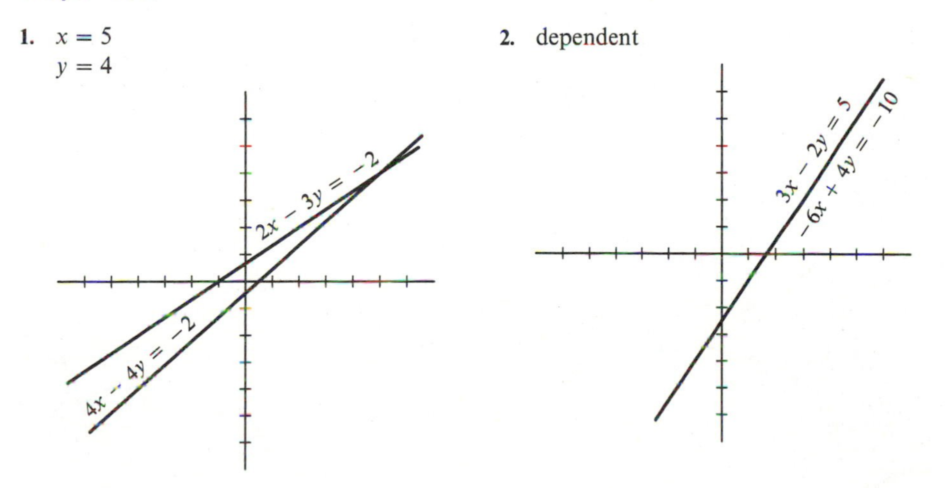

3. $x = -\frac{1}{3}$
$y = -\frac{1}{2}$

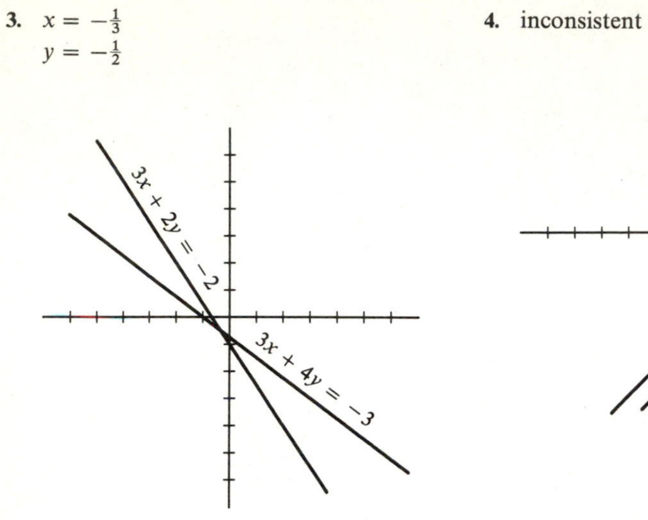

4. inconsistent

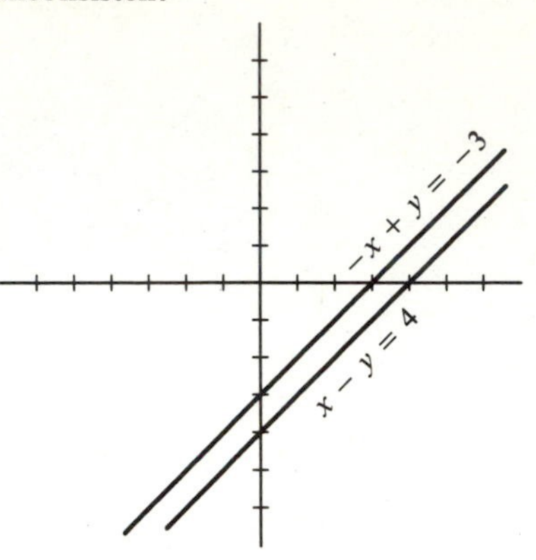

5. $x = 1$
$y = 2$
$z = 4$

6. $x = 3$
$y = -1$
$z = 2$

7. dependent

8. $x = 43/22$, $y = 9/11$, $z = 13/22$

9. maximum value of $P = 2200$;
when $x = 20$, $y = 60$

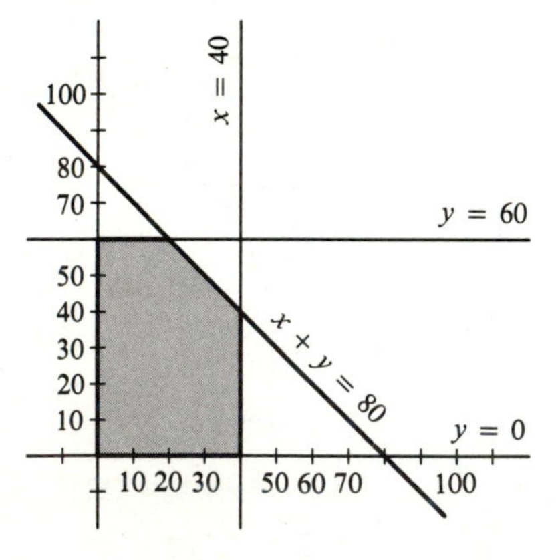

CHAPTER 9

Section 9.1

1. 5040 **3.** 744

5. by hand: 1,307,674,368,000
by calculator: 1.3076744E + 12 = 1,307,674,400,000

The answers are not identical because of the limited capacity of the calculator to store the digits of a single number.

7. 8! **9.** 10! **11.** $(N + 2)!$ **13.** 2 **15.** 1 **17. a.** 3125 **b.** 120 **c.** 24 **d.** 24
19. 108,334 minutes ($108{,}333\frac{1}{3}$) **21.** 256 patterns **23.** 20,160 **25.** 64 **27.** 8!

Section 9.2

1. a. 1/21 **b.** 2/21 **3.** 1/6

5 a.

Event	1	2	3	4	5	6
Weight	$1x$	$2x$	$3x$	$4x$	$5x$	$6x$

b. 1/7 **7. a.** 20 **b.** 1/20

9. 56 choices; probability = 7/56 (= 1/8) **11.** 25/100 (= 1/4)

Section 9.3

1. 14, 98, 86, 2 **3.** 33, 63, 93, 23, 53, 83, 13, 43, 73, 3 **5.** INT(6 * RND) + 1
7. a. 1/10000 **b.** 24/10000 (= 3/1250)

Section 9.4

1. mean = 2.4, median = 2, mode = 2 **3.** mean = 5; median = 4; modes: 2, 10
5. mean ≈ 6.17; median = 6.5; mode = 5 **7.** $x = 4$ **9.** $x = 37$

11. Let N = number of items.
If N/2 = INT(N/2) then let A = N/2, B = N/2 + 1. The median is the mean of item #A and item #B.
If N/2 < > INT(N/2) then let A = INT((N+1)/2). The median is item #A.

13.

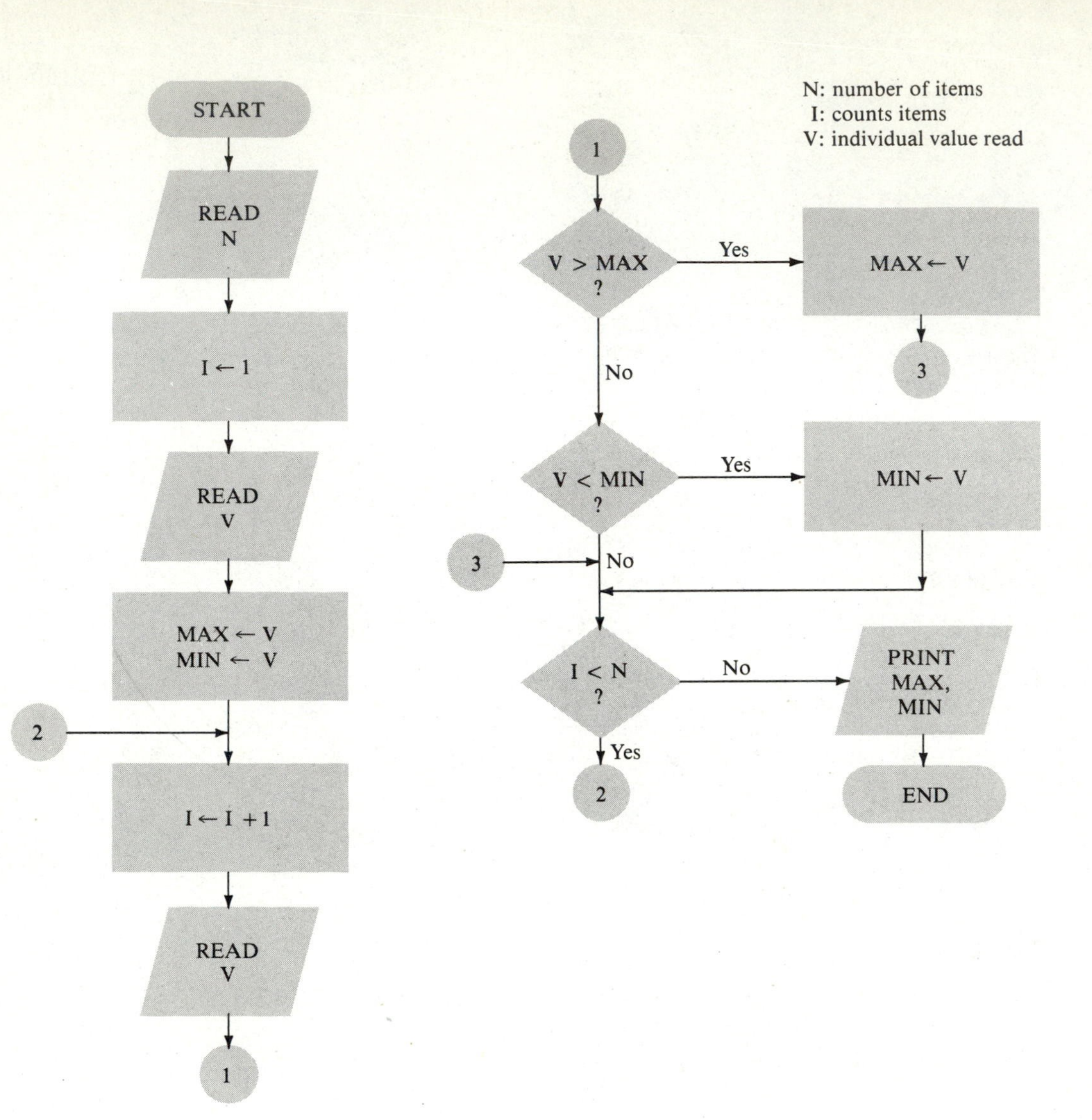

N: number of items
I: counts items
V: individual value read
START
READ
N
I ← 1
READ
V
MAX ← V
MIN ← V
2
I ← I + 1
READ
V
1
1
V > MAX
?
Yes
MAX ← V
3
No
V < MIN
?
Yes
MIN ← V
3
No
I < N
?
No
PRINT
MAX,
MIN
END
Yes
2

Section 9.6

3. mean = 1.9; median = 2.5; range = 22 **5.** variance = 9.2; standard deviation ≈ 3.033

7.

Grade	Number of Standard Deviations
70	−1.07
75	−0.624
90	0.71
100	1.604
75	−0.624

9. **a.** mean ≈ 62.95 ≈ 63 **b.** median = 65 **c.** variance ≈ 306 **d.** stand. dev. ≈ 17.5 **e.** 2 **f.** 17 **g.** 60

Section 9.7

1. −1 **3.** −1.5 **5.** −1 **7.** 0 **9.** 2 **11.** 84.1%
13. 97.7% **15.** 37.45% **17.** 90.8% **19.** **a.** 84.1% **b.** 15.9%

Chapter Test

1. 5040 **2.** **a.** $30\frac{1}{12} \approx 30.083$ **b.** 357,840 **3.** **a.** 210 **b.** $\binom{100}{40} = \frac{100!}{40!60!}$
4. **a.** 3/8 = 0.375 **b.** 1/8 = 0.125 **6.** mean = 4.25; median = 4.5; mode = −3
7. range = 9; variance = 10; standard deviation ≈ 3.16 **8.** 0.32
9. Z-score = −1; percentile = 15.9%

10.

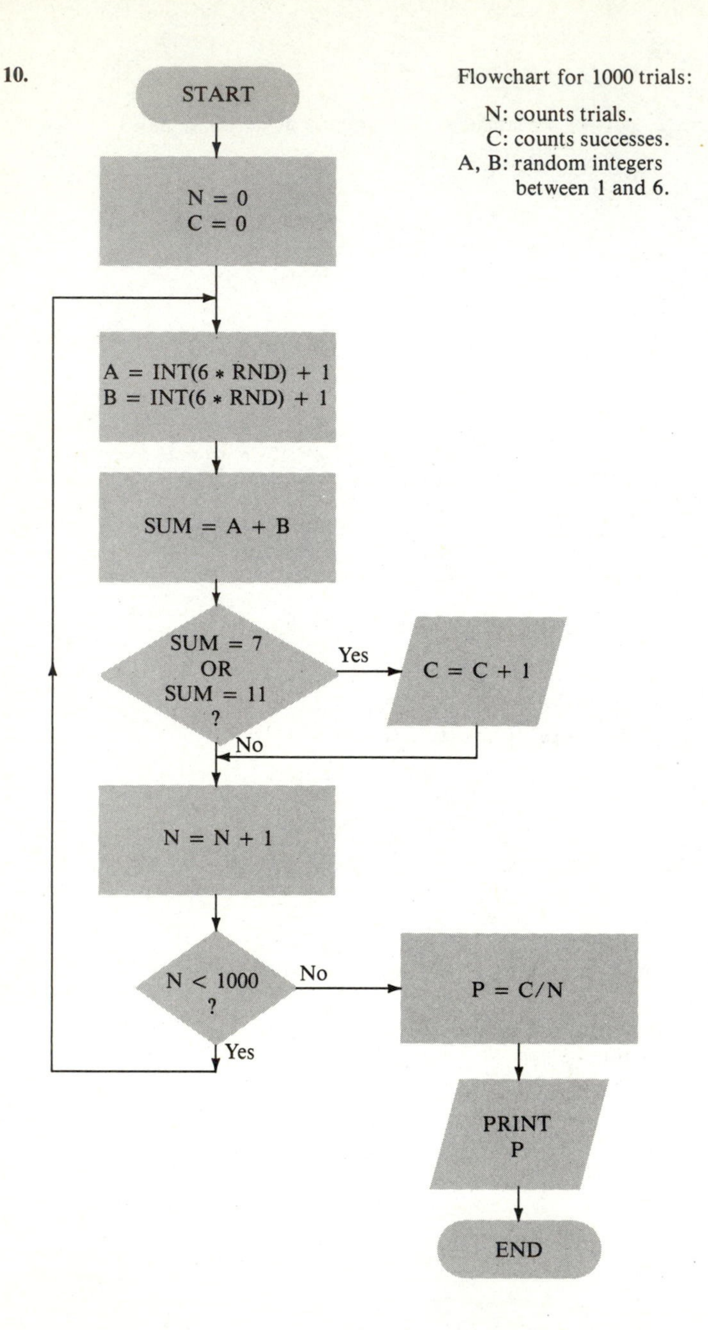

Flowchart for 1000 trials:

N: counts trials.
C: counts successes.
A, B: random integers between 1 and 6.

CHAPTER 10

All answers for this chapter are *approximate.* Your answer may differ slightly from the given answer because of a difference in the accuracy of the computer or calculator being used.

Section 10.1

1. a. 3.30, −0.30 b. 3.303, −0.303 3. a. 0.18, −1.61 b. 0.178, −1.606 5. no
7. yes 9. no 11. actual error = 0.1; relative error = 0.002
13. actual error = 0.001; relative error = 0.02 15. actual error = 1; relative error ≈ 0.00001481

Section 10.2

1. There is a root between 0.6 and 0.7. 3. There is a root between −1.6 and −1.5.
5. to one decimal place, $\sqrt{5} \approx 2.2$; to two decimal places, $\sqrt{5} \approx 2.23$; to three decimal places, $\sqrt{5} \approx 2.236$

Section 10.3

1. 74 3. −1.5468 5. 4.201 7. 3.4641016
9. 0.546, 11.454, accurate to three decimal places
11. a. three roots b. intervals: (−2, 0), (1, 1.75), (1.75, 2.5) c. root approximations: −1, 1.5, 2
13. a. four roots b. intervals: (−3, −2), (0, 1), (1, 2), (6, 7)
c. root approximations: −2.30, 0.299, 1.30, 6.70

Section 10.4

1. 74.99 3. −1.54681 5. 4.201 7. 3.464 9. 0.546, 11.454
11. a. three roots b. intervals: (−2, 0), (1, 1.75), (1.75, 2.5) c. root approximations: −1, 1.5, 2
13. a. four roots b. intervals: (−3, −2), (0, 1), (1, 2), (6, 7)
c. root approximations: −2.30, 0.299, 1.30, 6.70

Section 10.5

1. 74.99 3. −1.54681 5. 4.201 7. 3.464 9. 0.546, 11.454
11. a. three roots b. intervals: (−2, 0), (1, 1.75), (1.75, 2.5) c. root approximations: −1, 1.5, 2
13. a. four roots b. intervals: (−3, −2), (0, 1), (1, 2), (6, 7)
c. root approximations: −2.30, 0.299, 1.30, 6.70

Section 10.6

1. $x = 1, y = -2$ 3. method cannot be used; dependent 5. $x = -0.1, y = 0.5, z = 0.33$
7. $x = 2.49, y = -0.50, z = 1$
9. a. $x = 1, y = -1, z = 3$ b. $x = 1.02, y = -0.987, z = 2.98$ Method a is preferable.
11. Gaussian elimination yields $x = 1, y = -1, z = 2, w = 3$. The Gauss-Seidel method cannot be used.

Chapter Test

2. There is a root between −2 and −3. 3. −2.20 4. −2.20
5. Gaussian elimination: $x = \frac{4}{3}, y = -\frac{13}{3}, z = \frac{13}{3}$
Gauss-Seidel: $x = 1.33, y = -4.33, z = 4.33$
6. a. $x^2 - 16.5x + 67.5 = 0$ b. length = 9; width = 7.5 c. 7.50 and 9.00

Index

D

E

F

N

O

P

Q

R